Digital Signal Analysis

PRENTICE HALL SIGNAL PROCESSING SERIES

Alan V. Oppenheim, Editor

Second Edition

Digital Signal Analysis

SAMUEL D. STEARNS
Sandia National Laboratories

DON R. HUSH
The University of New Mexico

PRENTICE HALL, Englewood Cliffs, New Jersey 07632

Library of Congress Cataloging-in-Publication Data

Stearns, Samuel D.
   Digital signal analysis / Samuel D. Stearns, Don R. Hush. — 2nd ed.
     p. cm.
   Includes bibliographical references.
   ISBN 0-13-213117-X
    1. Signal processing — Digital techniques. I. Hush, Don R.
. II. Title.
TK5.102.5S698   1990
621.382′23 — dc20

Editorial/production supervision
  and interior design: Maria McColligan
Manufacturing buyer: Denise Duggan

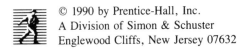 © 1990 by Prentice-Hall, Inc.
A Division of Simon & Schuster
Englewood Cliffs, New Jersey 07632

Printed in the United States of America
10 9 8 7 6 5 4 3 2 1

ISBN   0-13-213117-X

Prentice-Hall International (UK) Limited, *London*
Prentice-Hall of Australia Pty. Limited, *Sydney*
Prentice-Hall Canada Inc., *Toronto*
Prentice-Hall Hispanoamericana, S.A., *Mexico*
Prentice-Hall of India Private Limited, *New Delhi*
Prentice-Hall of Japan, Inc., *Tokyo*
Simon & Schuster Asia Pte. Ltd., *Singapore*
Editora Prentice-Hall do Brasil, Ltda., *Rio de Janeiro*

We have seen but a small part of His works, and there remain many mysteries greater still.

—Ecclesiasticus, 190 B.C.

This second edition is further dedicated to the cause of peace for all humankind. We hope that its contents will be used for this purpose.

Contents

Contents

Contents

Contents

Foreword

The information age in which we are living has underlined the importance of signal processing, while the development of solid-state integrated circuits, especially in the form of general-purpose minicomputers, has made practical the many theoretical advantages of digital signal processing. It is for these reasons that a good book on digital signal processing is welcomed by a wide range of people including engineers, scientists, computer experts, and applied mathematicians.

Although most signal processing is now done digitally, much of the data originates as continuous, analog signals, and often the result of the digital signal processing is converted back to analog form before it is finally used. Thus the complex and often difficult to understand relationships between the digital and analog forms of signals need to be examined carefully. These relationships form a recurring theme throughout this book.

The theory of digital signal processing involves a considerable amount of mathematics. The nonmathematically inclined reader should not be put off by the number of formulas and equations in the book, since the author has been careful to motivate and explain the physical basis of what is going on and at the same time avoid unnecessarily fancy mathematics and artificial abstractions.

It is a pleasure, therefore, to recommend this book to the serious student of digital signal processing. It is carefully written and illustrated by many useful examples and exercises, and the material is selected to cover the relevant topics in this rapidly developing field of knowledge.

R. W. Hamming
Bell Laboratories
Murray Hill, N.J.

Preface

to the Second Edition

Our primary goal for this second edition is the same as for the first: to provide a text on the fundamentals of signal analysis for the engineering and physical science community. We still believe in the validity of the original approach to the subject, which was to relate continuous and digital signals through the process of sampling in the first part of the text, and then to emphasize the techniques of digital signal analysis and processing in the rest of the text.

Readers of the first edition will notice several major additions and revisions in this edition which, we hope, will bring the text up to date and make it more useful. For example, the portable transform and filter design software modules included with the first edition were widely used, so this time we have included a larger library of portable modules in FORTRAN 77, in the text and also on a floppy disk. These modules cover more applications and are meant to be used to perform useful operations in practice as well as to demonstrate the signal processing operations described in the text.

We have also made major changes in the chapters on discrete transforms, FIR and IIR filtering, and spectral estimation, and we have added new chapters on the z-transform and on least-squares system design. We have added discussions of lattice structures and parametric spectral estimation — subjects that were not in the first edition. With these changes, we hope that the reader will find here a self-contained text that covers the major areas essential to a vocation in digital signal analysis and processing.

We acknowledge with thanks our indebtedness to many students and friends who helped produce this second edition, including Dale R. Breding, Bock-Sim Chia, Kurt Conover, Douglas F. Elliott, Glenn R. Elliott, Terry L. Hardin, Claude S. Lindquist, Caryl V. Peterson, John M. Salas, and Charles C. Stearns. We also thank Betty J. Hawley and Lori Jackson for their help in typing and editing the text.

Finally, we wish to express special gratitude to Professor Wolfgang Hilberg for the German translation of this book.

Samuel D. Stearns
Don R. Hush

Preface
to the First Edition

This text presents the fundamentals of signal analysis for the engineer, mathematician, and computing scientist. The emphasis is on digital signal analysis, with analog or continuous signal analysis occupying a supporting role.

The similarities between digital and continuous signals and signal processing systems are emphasized in most of the chapters. It is believed that by studying the analysis of both digital and continuous linear systems in a single course, the student will discover the fundamental similarities between the two types of systems and come to the realization that the same mathematical disciplines are applicable to both.

The text is divided roughly into four subject areas, not all equal in length. Chapters 1, 2, and 3 are devoted mainly to a review of the classical areas of linear signal analysis, including basic formulas and operations, least squares, Fourier series, continuous transforms, transfer functions, etc. The classical concepts are covered with an emphasis on aspects particularly applicable to digital signal analysis. Chapters 4 through 7 primarily deal with sampling, analog-to-digital conversion, and digital spectral analysis, with Chapter 6 devoted to the fast Fourier transform. Relationships between continuous and sampled waveforms and spectra are emphasized in these chapters.

Chapters 8 through 12 are devoted to topics on the synthesis and analysis of digital signal processing systems. The z-transform is introduced and used to discuss nonrecursive and recursive systems, digital transfer functions, relationships to and simulations of continuous transfer functions, and digital filter design. Finally, Chapters 13 and 14 deal with sampled random time series and the computation of power spectra.

The text has been developed from a one-year course in digital signal analysis taught at the University of New Mexico beginning in 1969, and at Sandia Laboratories in Albuquerque, New Mexico, starting in 1970. Most of the students in these courses have been electrical engineers with good backgrounds in the "classical areas" mentioned above, but some have been mathematics or computing science majors who are less familiar with operational methods, spectral analysis, and so on. The latter students, particularly those strong in mathematics, have fared very well in the course.

For the engineer, mathematician, or computing scientist, the course is meant to be essentially self-contained. The traditional engineering courses in transform theory and linear system analysis are useful but not essential prerequisites. The only prerequisites are a reasonable sophistication in mathematics, such as might be obtained from a course in advanced calculus, and a modest amount of problem-solving experience with any computer, large or small. It is the intent, or at least the hope, of the author that this may serve as a text for courses similar to those mentioned above and that engineers in the field may find it a useful reference to formulas and techniques of digital signal analysis.

No significant portion of the theory developed herein is due to the author, who has only attempted to bring together the work of many others into a single treatment. In developing a text such as this, one cannot help but appreciate the many great minds, from Fourier through Hamming, whose ideas constitute the modern theories of linear analysis.

The author is also very much indebted to many friends, without whose help and encouragement this text would never have been written. He is especially grateful for the counsel of Professors Lambert H. Koopmans, Donald R. Morrison, and Bruce R. Peterson at the University of New Mexico, as well as to Professor Richard W. Hamming of Bell Telephone Laboratories. Special thanks are also due to Dr. Alan B. Campbell of Sandia Laboratories and to Tim S. McDonald of the Dikewood Corporation, Albuquerque, New Mexico for their patient reviews of the entire manuscript. In addition, the author wishes to thank Sandia Laboratories for the use of their facilities and his many coworkers there who have reviewed and commented on portions of the text, including Dr. Winser E. Alexander, Dr. Ronald D. Andreas, Nolan A. Bourgeois, Robert H. Croll, Jr., Rondall E. Jones, James R. Kelsey, Thomas F. Marker, Dr. Donald H. Schroeder, Dr. Michael A. Soderstrand, and Dr. George P. Steck. Finally, the author owes a debt of gratitude to Noelle, Jeni, Laurie, and Chip for their love and encouragement, to the embarrassingly large number of patient ladies who have typed and retyped the manuscript, and to the many students whose questions and ideas are woven into the fabric of the text.

Samuel D. Stearns
Albuquerque, N.M.

Digital Signal
Analysis

1

Introduction

1.1 MODERN SIGNAL ANALYSIS

The term *signal analysis* generally refers to the science of analyzing or interpreting the signals produced by time-varying physical processes. The signals themselves may be transient in nature, occurring only over a short time duration, they may be periodic (repetitious), or they may be random and unpredictable. Methods of signal analysis apply to all of these types of signals, examples of which are shown in Fig. 1.1.

The applications of signal analysis reach into many different areas and disciplines: communications, control systems, biology and medicine, physics and astronomy, chemistry, seismology, mechanical vibration and shock studies, fluid dynamics, and radar design. All are fields in which signal analysis is presently helping to advance the state of the art. Nor is this list complete; signal analysis is constantly being applied in new ways and in new areas.

Many modern applications of signal analysis have been due to the tremendous growth and diversification of digital processing equipment. Today, digital computing systems come in all sizes and prices; what was once recognizable as a computer

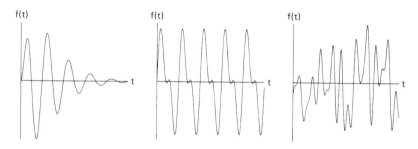

Figure 1.1 Transient, periodic, and random signals.

per se has now become a signal processing component, often dedicated to a specific task in a complex system and neither recognized nor used as a general-purpose computer. Digital signal processing systems have become reliable, rugged, and portable; they are used in locations and environments inaccessible to humans.

Some typical operations and systems within the scope of digital signal analysis are illustrated in Figs. 1.2 through 1.4. Figure 1.2 suggests some of the most common operations used in the analysis of signals. A physical quantity — such as a displacement, velocity, acceleration, intensity, force, pressure, temperature, color, charge, or repetition rate — is converted to an electrical potential by a transducer. Then, usually at regular intervals of time, the transducer output is digitized, that is, converted to *numbers* by the analog-to-digital converter, sometimes called an A-D converter or ADC. (The ADC and its operation are discussed in Chapter 4.) The

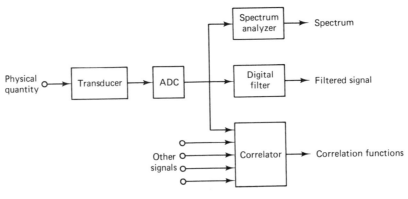

Figure 1.2 Some of the operations in digital signal analysis.

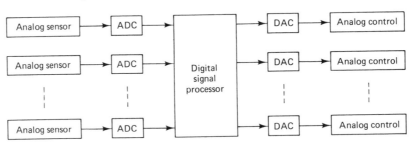

Figure 1.3 Digital control or communication system.

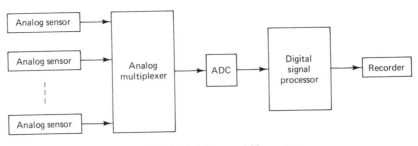

Figure 1.4 Digital data acquisition system.

sequence of numbers or *sample* sequence produced by the ADC might be recorded for further processing or, as suggested in Fig. 1.2, processed in "real time" without recording.

One common operation on the sample set is the computation of the *spectrum,* which generally gives the distribution of amplitude, phase, power, or energy over frequency. Some of the properties of the spectrum and its computation are discussed in Chapters 2, 3, 5, 6, 13, and 15.

Another common operation in digital signal analysis is *digital filtering,* which is used generally to produce an output sample set from the original or input sample set and in so doing to accomplish some purpose, such as eliminating high frequencies, reducing noise, and so on. Digital filtering is discussed mainly in Chapters 8, 9, and 12.

The third operation shown in Fig. 1.2 is *correlation,* which is a special process of comparing signals with themselves or with other signals. Correlation is discussed mainly in Chapter 13. All of these operations can be accomplished in a general-purpose computer, or special-purpose processors may be designed.

A typical digital control or communication system is shown in Fig. 1.3. In this system analog signals are converted to digital form and then filtered or processed in some manner, either together or separately, by the digital signal processor. The processor then produces digital outputs, each of which is converted to an analog control signal by a digital-to-analog converter (DAC).

Figure 1.4 shows a typical data acquisition system. Here an alternative to the input in Fig. 1.3 is shown, in which the incoming analog signals are multiplexed, (i.e., sampled in succession) and then converted to sample sequences using a single ADC. The signal processor then performs filtering and other required operations on the data before storing it in a digital recorder.

In the chapters that follow, the reader will find many subjects basic to signal analysis. Much of the mathematics and many of the approaches will be found to be the same regardless of whether *digital* or *continuous* signals are being analyzed. There are, of course, basic differences between these two types of signals and the systems involved, as illustrated in Fig. 1.5. The analog or continuous system is generally one that processes a continuous time-varying physical quantity, $f(t)$, and produces a similar quantity, $g(t)$. These quantities are typically voltages, currents displacements, angles, and so on. On the other hand, as described here, the *digital* system processes a sequence of numbers $f_0 f_1 f_2 \cdots$ to produce an output number sequence $g_0 g_1 g_2 \cdots$.

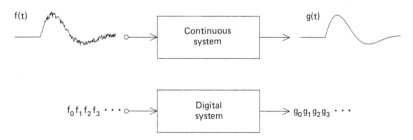

Figure 1.5 Continuous and digital systems.

A great deal of the discussion in the following chapters is based on these elementary concepts and on one additional basic concept: A common theoretical ground between digital and continuous systems is established by treating the number sequences processed by the digital system as if they were *samples of continuous signals,* which in turn are (or could be) processed by an analog or continuous system. This idea is illustrated in Fig. 1.6. The sample sequence $[f_m] = [f_0, f_1, f_2, \ldots]$ is derived from the continuous function $f(t)$ by assembling the values of $f(t)$ at $t = 0, T, 2T, \ldots$ into an ordered sequence. Using this concept, one can then draw a comparison between a continuous signal processing system and a digital signal processing system whose outputs are identical at the sampling points.

Figure 1.6, in fact, suggests the connection between the three types of signals in Fig. 1.1 and the number sequence in Fig. 1.5: It is the *sample sequences* derived from these continuous signals that are being processed in the digital system. For example, a complete signal processing system might consist of an ADC connected to a digital processor, which in turn is connected to a DAC to produce a continuous output as in Fig. 1.3.

On the other hand, the output of the digital system may be desired in digital form, or an ADC may be used at the output of an analog system, and so on. The main point here is that the mathematical approaches for analyzing these systems and signals are very much the same — one can study continuous and digital signal analysis most profitably by considering both at the same time.

As a final example in this introductory discussion, Fig. 1.7 provides a comparison between simple continuous and digital systems with similar performance characteristics. In the upper half of the figure a continuous transient waveform is the input to a simple *RC* integrating circuit (*R* for resistance and *C* for capacitance), a circuit familiar to most engineers. For a time constant (*RC*) of 1.67 ms, the output waveform is shown at the upper right. The lower half of the figure shows the result of processing the sample set of the input waveform (using an interval of 0.2 ms from one sample to the next) in a simple digital system. The symbol z^{-1} stands for a delay of one sampling interval, so in this case the digital processor, at each sampling instant, simply multiplies the previous output sample by 7.85, adds the result to the present input sample, and multiplies the sum by 0.113 to produce the present output sample. The analysis of systems such as these is covered in later chapters. This

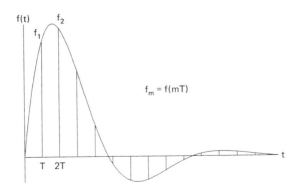

Figure 1.6 Sampled continuous signal.

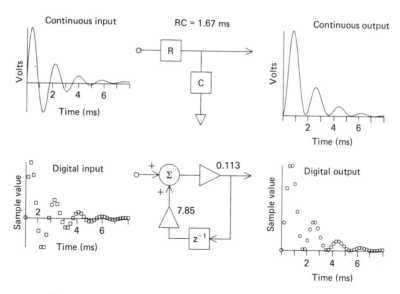

Figure 1.7 Continuous and digital systems with similar performance.

simple example serves only to illustrate the similarity between appropriately chosen digital and continuous systems, a phenomenon emphasized throughout this book.

1.2 SOME USEFUL FORMULAS

The chapters ahead proceed from a review of some fundamental mathematical concepts such as the principle of least squares, orthogonality, and so on, to the major subjects of signal analysis. To begin the review of fundamental mathematics, this section contains some basic formulas of algebra, trigonometry, and calculus used extensively throughout later chapters. These and other useful formulas and tables can be found in standard handbooks such as those listed in the references at the end of this chapter.

Trigonometric identities are often required in the manipulation of Fourier series and transforms, and throughout harmonic analysis in general. Some of the most common identities are listed in Table 1.1[†]

The *hyperbolic functions* are used in the design of some of the digital filters discussed in Chapter 12. Some useful hyperbolic relationships are summarized in Table 1.2.

The *geometric series* is used repeatedly in signal analysis and elsewhere to evaluate functions in closed form. Its basic form is

$$1 + x + x^2 + x^3 + \cdots + x^{N-1} = \sum_{n=0}^{N-1} x^n$$
$$= \frac{1 - x^N}{1 - x} \tag{1.17}$$

[†]The symbol "j" denotes $\sqrt{-1}$, a common usage in engineering, where "i" is used for instantaneous current.

TABLE 1.1   Trigonometric Identities

$$\left.\begin{array}{l} \sin(-\alpha) = -\sin\alpha \\ \cos(-\alpha) = \cos\alpha \end{array}\right\} \qquad (1.1)$$

$$\left.\begin{array}{l} \sin(\alpha + \beta) = \sin\alpha\cos\beta + \cos\alpha\sin\beta \\ \cos(\alpha + \beta) = \cos\alpha\cos\beta - \sin\alpha\sin\beta \end{array}\right\} \qquad (1.2)$$

$$\left.\begin{array}{l} 2\sin\alpha\sin\beta = \cos(\alpha - \beta) - \cos(\alpha + \beta) \\ 2\cos\alpha\cos\beta = \cos(\alpha + \beta) + \cos(\alpha - \beta) \\ 2\sin\alpha\cos\beta = \sin(\alpha + \beta) + \sin(\alpha - \beta) \end{array}\right\} \qquad (1.3)$$

$$\left.\begin{array}{l} \sin\alpha + \sin\beta = 2\sin\dfrac{1}{2}(\alpha + \beta)\cos\dfrac{1}{2}(\alpha - \beta) \\[2mm] \cos\alpha + \cos\beta = 2\cos\dfrac{1}{2}(\alpha + \beta)\cos\dfrac{1}{2}(\alpha - \beta) \end{array}\right\} \qquad (1.4)$$

$$\left.\begin{array}{l} \sin 2\alpha = 2\sin\alpha\cos\alpha \\ \cos 2\alpha = 2\cos^2\alpha - 1 \end{array}\right\} \qquad (1.5)$$

$$\left.\begin{array}{l} \sin\dfrac{\alpha}{2} = \pm\sqrt{\dfrac{1}{2}(1 - \cos\alpha)} \\[3mm] \cos\dfrac{\alpha}{2} = \pm\sqrt{\dfrac{1}{2}(1 + \cos\alpha)} \end{array}\right\} \qquad (1.6)$$

$$\left.\begin{array}{l} \cos^2\alpha = \dfrac{1}{2}(\cos 2\alpha + 1) \\[2mm] \sin^2\alpha = 1 - \cos^2\alpha \end{array}\right\} \qquad (1.7)$$

$$\left.\begin{array}{l} \sin\alpha = \dfrac{1}{2j}(e^{j\alpha} - e^{-j\alpha}) \\[2mm] \cos\alpha = \dfrac{1}{2}(e^{j\alpha} + e^{-j\alpha}) \end{array}\right\} \qquad (1.8)$$

$$e^{j\alpha} = \cos\alpha + j\sin\alpha \qquad (1.9)$$

TABLE 1.2   Hyperbolic Functions

$$\left.\begin{array}{l} \sinh(-\alpha) = -\sinh\alpha \\ \cosh(-\alpha) = +\cosh\alpha \end{array}\right\} \qquad (1.10)$$

$$\left.\begin{array}{l} \sinh(\alpha + \beta) = \sinh\alpha\cosh\beta + \cosh\alpha\sinh\beta \\ \cosh(\alpha + \beta) = \cosh\alpha\cosh\beta + \sinh\alpha\sinh\beta \end{array}\right\} \qquad (1.11)$$

$$\left.\begin{array}{l} 2\sinh\alpha\cosh\beta = \sinh(\alpha + \beta) + \sinh(\alpha - \beta) \\ 2\cosh\alpha\cosh\beta = \cosh(\alpha + \beta) + \cosh(\alpha - \beta) \\ 2\sinh\alpha\sinh\beta = \cosh(\alpha + \beta) - \cosh(\alpha - \beta) \end{array}\right\} \qquad (1.12)$$

$$\cosh^2\alpha - \sinh^2\alpha = 1 \qquad (1.13)$$

$$\left.\begin{array}{l} \sinh\alpha = \dfrac{1}{2}(e^{\alpha} - e^{-\alpha}) \\[2mm] \cosh\alpha = \dfrac{1}{2}(e^{\alpha} + e^{-\alpha}) \end{array}\right\} \qquad (1.14)$$

$$e^{\alpha} = \sinh\alpha + \cosh\alpha \qquad (1.15)$$

$$\left.\begin{array}{l} \sinh^{-1}\alpha = \log_e(\alpha + \sqrt{\alpha^2 + 1}) \\ \cosh^{-1}\alpha = \log_e(\alpha + \sqrt{\alpha^2 - 1}) \end{array}\right\} \qquad (1.16)$$

There are N terms in the sum, and N is the exponent in the numerator of the closed form. If the magnitude of x, $|x|$, is less than 1, the infinite geometric series converges to

$$\sum_{n=0}^{\infty} x^n = \frac{1}{1-x}; \qquad |x| < 1 \tag{1.18}$$

More complex forms of Eqs. 1.17 and 1.18 are sometimes difficult to recognize. For example,

$$\sum_{n=0}^{N-1} e^{-jan} = \sum_{n=0}^{N-1} (e^{-ja})^n$$

$$= \frac{1 - e^{-jaN}}{1 - e^{-ja}} \tag{1.19}$$

is easily seen as a geometric series, but

$$\sum_{n=0}^{\infty} nx^n = x\frac{d}{dx}\left(\sum_{n=0}^{\infty} x^n\right)$$

$$= x\frac{d}{dx}\left(\frac{1}{1-x}\right); \qquad |x| < 1 \tag{1.20}$$

$$= \frac{x}{(1-x)^2}$$

is more difficult to recognize as a geometric series in its original form.

Vectors and matrices are often used in signal analysis to represent the state of a system at a particular time, a set of signal values, a set of linear equations, and so on. A vector is a linear array of real or complex numbers, for example,

$$\mathbf{X} = \begin{bmatrix} x_1 \\ x_2 \\ \vdots \\ x_N \end{bmatrix} \tag{1.21}$$

and a matrix is a rectangular array, for example,

$$\mathbf{A} = \begin{bmatrix} a_{11} & \cdots & a_{1N} \\ \vdots & & \vdots \\ a_{M1} & \cdots & a_{MN} \end{bmatrix} \tag{1.22}$$

The *sum* of two such arrays can be taken provided both have the same number of rows and columns. Each element of the sum is the sum of corresponding elements:

$$\text{if } \mathbf{C} = \mathbf{A} + \mathbf{B}$$

$$\text{then } c_{mn} = a_{mn} + b_{mn} \tag{1.23}$$

Example:

$$\begin{bmatrix} 1 & 0 \\ -1 & -2 \end{bmatrix} + \begin{bmatrix} 1 & 3 \\ 2 & 2 \end{bmatrix} = \begin{bmatrix} 2 & 3 \\ 1 & 0 \end{bmatrix} \qquad (1.24)$$

The *product* of two arrays can be taken provided that the number of columns in the first array equals the number of rows in the second. Each element of the product array is found by proceeding across a row of the first array and down a column of the second, multiplying corresponding terms and summing products:

$$\text{if } \mathbf{C} = \mathbf{A} \cdot \mathbf{B}$$

$$\text{then } c_{mn} = \sum_{k=1}^{N} a_{mk} b_{kn} \qquad (1.25)$$

Example:

$$\begin{bmatrix} -1 & 1 \\ 0 & 1 \\ 1 & 1 \end{bmatrix} \cdot \begin{bmatrix} 1 \\ 2 \end{bmatrix} = \begin{bmatrix} 1 \\ 2 \\ 3 \end{bmatrix} \qquad (1.26)$$

The *identity matrix* (I) is a square array having ones on the diagonal and zeros elsewhere. With $N = 3$, for example,

$$\mathbf{I} = \begin{bmatrix} 1 & 0 & 0 \\ 0 & 1 & 0 \\ 0 & 0 & 1 \end{bmatrix} \qquad (1.27)$$

From the definition of the product, the identity matrix is seen to have the following property when used with any square matrix $\mathbf{A}$:

$$\mathbf{A} \cdot \mathbf{I} = \mathbf{I} \cdot \mathbf{A} = \mathbf{A} \qquad (1.28)$$

The main use of arrays in signal analysis is in the representation of *linear equations*. For example,

$$\begin{bmatrix} c_{11} & \cdots & c_{1N} \\ \vdots & & \vdots \\ c_{N1} & \cdots & c_{NN} \end{bmatrix} \cdot \begin{bmatrix} x_1 \\ \vdots \\ x_N \end{bmatrix} = \begin{bmatrix} y_1 \\ \vdots \\ y_N \end{bmatrix} \qquad (1.29)$$

represents a set of N equations in which $\mathbf{C}$ is the coefficient matrix, $\mathbf{X}$ is an unknown vector, and $\mathbf{Y}$ is a known vector. In compact form the equations may be written $\mathbf{CX} = \mathbf{Y}$. Several procedures are used to solve Eq. 1.29 for the unknown vector $\mathbf{X}$ [see Gerald (1970) as well as Chapter 14 of this book].

For $N = 2$, the *inverse* coefficient matrix, which must be multiplied by $\mathbf{Y}$ to obtain $\mathbf{X}$, is

$$\begin{bmatrix} c_{11} & c_{12} \\ c_{21} & c_{22} \end{bmatrix}^{-1} = \frac{1}{c_{11}c_{22} - c_{12}c_{21}} \begin{bmatrix} c_{22} & -c_{12} \\ -c_{21} & c_{11} \end{bmatrix} \qquad (1.30)$$

We could verify this result by showing that $\mathbf{CC}^{-1} = \mathbf{I}$.

1.3 THE CHAPTERS AHEAD

The next three chapters cover subjects fundamental to the analysis of all types of signals and linear systems, whether digital or continuous. The reader should be familiar with at least Chapters 2 and 3 before proceeding to Chapter 5 and beyond, because the Fourier series, transforms, transfer functions, and so on, are used throughout the remaining chapters.

After Chapter 3, the emphasis is on subjects fundamental to continuous or digital signal analysis and on comparisons between continuous and digital signal processing. Chapter 4 is on sampling and analog-to-digital conversion. Chapters 5 through 7 are generally on the subject of spectral analysis. Chapters 8 through 12 are all on "digital filtering" in the broad sense of the term.

Finally, Chapters 13 through 15 cover some of the concepts used most often in least-squares design and in the analysis of random signals.

Since the original publication of this book, a number of texts on signal analysis and digital signal processing have become available. A number of these that are considered to be good references to supplement and extend the material in this book are listed below in the references.

REFERENCES

AHMED, N., and NATARAJAN, T., *Discrete-Time Signals and Systems*. Reston, Va.: Reston, 1983.

AUTONIOU, A., *Digital Filters: Analysis and Design*. New York: McGraw-Hill, 1979.

BIRKHOFF, G., and MACLANE, S., *A Survey of Modern Algebra*, Chap. 10. New York: Macmillan, 1950.

BURINGTON, R. S., *Handbook of Mathematical Tables and Formulas*. New York: McGraw-Hill, 1965.

CADZOW, J. A., *Foundations of Digital Signal Processing and Data Analysis*. New York: Macmillan, 1987.

CROCHIERE, R. E., and RABINER, L. R., *Multirate Digital Signal Processing*. Englewood Cliffs, N.J.: Prentice-Hall, 1983.

DWIGHT, H. B., *Tables of Integrals and Other Mathematical Data*. New York: Macmillan, 1961.

GERALD, C. F., *Applied Numerical Analysis*. Reading, Mass.: Addison-Wesley, 1970.

HAMMING, R. W., *Digital Filters*. Englewood Cliffs, N.J.: Prentice-Hall, 1975.

KAPLAN, W. K., *Advanced Calculus*, Chap. 6. Reading, Mass.: Addison-Wesley, 1952.

KELLY, L. G., *Handbook of Numerical Methods and Applications*, Chaps. 7 and 8. Reading, Mass.: Addison-Wesley, 1967.

OPPENHEIM, A. V. (ed.), *Applications of Digital Signal Processing*. Englewood Cliffs, N.J.: Prentice-Hall, 1978.

OPPENHEIM, A. V., and SCHAFER, R. W., *Digital Signal Processing*. Englewood Cliffs, N.J.: Prentice-Hall, 1975.

OPPENHEIM, A. V., and WILLSKY, A. S., *Signals and Systems*. Englewood Cliffs, N.J.: Prentice-Hall, 1983.

PARKS, T. W., and BURRUS, C. S., *Digital Filter Design*. New York: Wiley, 1987.

PERLIS, S., *Theory of Matrices*. Reading, Mass.: Addison-Wesley, 1956.

RABINER, L. R., and GOLD, B., *Theory and Applications of Digital Signal Processing*. Englewood Cliffs, N.J.: Prentice-Hall, 1975.

STEARNS, S. D., and DAVID, R. A., *Signal Processing Algorithms*. Englewood Cliffs, N.J.: Prentice-Hall, 1988.

TRETTER, S. A., *Introduction to Discrete-Time Signal Processing*. New York: Wiley, 1976.

2

Review of Least Squares, Orthogonality, and the Fourier Series

2.1 INTRODUCTION

The three subjects reviewed in this chapter are of fundamental interest in signal analysis. They are closely related to each other, as shown in the discussion that follows. The introduction of orthogonal functions will be seen to alter and simplify the process of finding the "least-squares fit" of a linear function to another function or to a set of data. Further, the Fourier series is a series of orthogonal functions, as well as an important application of the principle of least squares. The list of references contains some excellent texts on these subjects.

2.2 THE PRINCIPLE OF LEAST SQUARES

The least-squares principle is used extensively throughout engineering and statistics. It is widely applicable (but not always appropriate) whenever a "goodness-of-fit" measure is needed.

Figure 2.1 illustrates the use of the least-squares principle in the continuous case. A "desired" function, $f(t)$, is to be approximated as closely as possible by an "actual" function $\hat{f}(c, t)$ in which c is a parameter that can be adjusted to obtain the best approximation. (In general there are many parameters to adjust instead of just one, but only one is assumed for this simple illustration.) The *total squared error* of $\hat{f}(c, t)$ is then a function of c and is given by

$$E^2(c) = \int_{-\infty}^{\infty} [f(t) - \hat{f}(c, t)]^2 \, dt \qquad (2.1)$$

Presumably, a minimum value of $E^2(c)$ can be obtained by adjusting c appropriately.

Figure 2.2 illustrates an analogous case for discrete data. Here, in effect, $f(t)$ is given only at a finite set of values of t. The set of values of $f(t)$ is called the *sample sequence* and is designated $[f_0, f_1, f_2, \ldots, f_{N-1}]$ so there are N samples in all. Each sample f_n is taken at a specific time t_n as shown in the figure. Again there is a

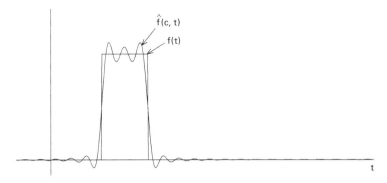

Figure 2.1 Actual and desired functions, $\hat{f}(c,t)$ and $f(t)$.

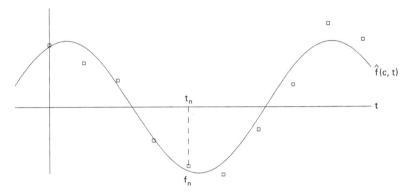

Figure 2.2 Sample set $[f_n]$ and fitted function $\hat{f}(c,t)$.

family of functions $\hat{f}(c,t)$ having (in this illustration) the single parameter c and there is an equation analogous to Eq. 2.1 for the total squared error:

$$E^2(c) = \sum_{n=0}^{N-1} [f_n - \hat{f}(c,t_n)]^2 \tag{2.2}$$

Again, as in the continuous case, the least-squares principle implies an adjustment of c to minimize $E^2(c)$. [Sometimes each term in the sum in Eq. 2.2 is multiplied by a weight $w_n \geq 0$ to account for the relative accuracy or credibility of the measurement f_n (see Hildebrand, 1956, or Kelly, 1967, for example). Here, all of these weights are assumed to be equal.]

Where digital computers are concerned, the case of discrete samples rather than the continuous case is usually applicable. Also, as mentioned, an adjustment of not just one but a set of parameters usually is available to minimize E^2. A general linear form for the "adjustable" function $\hat{f}(t)$ is

$$\hat{f}(c,t) \equiv \hat{f}(t) = \sum_{m=0}^{M-1} c_m \phi_m(t) \tag{2.3}$$

Here, $\hat{f}(t)$ is a linear combination of a set of functions $[\phi_0, \phi_1, \ldots, \phi_{M-1}]$, and there are M parameters $c_0, c_1, \ldots, c_{M-1}$ to be adjusted to minimize E^2.

Note that *only* the set $[c_m]$ is to be adjusted, the set of functions $[\phi_m]$ having been assigned. For example, a familiar case is

$$\phi_m = t^m; \qquad m = 0, 1, \ldots, M - 1 \qquad (2.4)$$

and the coefficients (c_m) are adjusted so that the polynomial

$$\hat{f}(t) = c_0 + c_1 t + c_2 t^2 + \cdots + c_{M-1} t^{M-1} \qquad (2.5)$$

is the *least-squares polynomial* of degree $M - 1$.

In the general case, with $\hat{f}(t)$ as in Eq. 2.3, the total squared error as expressed in Eq. 2.2 becomes

$$E^2([c_m]) = \sum_{n=0}^{N-1} \left[f_n - \sum_{m=0}^{M-1} c_m \phi_{mn} \right]^2 \qquad (2.6)$$

so that E^2 depends on the set of coefficients $[c_m]$. Here, to simplify the notation, ϕ_{mn} is used to represent $\phi_m(t_n)$. Obtaining a minimum value of E^2 now involves setting the partial derivatives of E^2 with respect to c_k equal to zero for all k:

$$\frac{\partial E^2}{\partial c_k} = -2 \sum_{n=0}^{N-1} \phi_{kn} \left[f_n - \sum_{m=0}^{M-1} c_m \phi_{mn} \right] = 0 \qquad (2.7)$$

or, rewriting Eq. 2.7,

$$\sum_{n=0}^{N-1} \sum_{m=0}^{M-1} c_m \phi_{mn} \phi_{kn} = \sum_{n=0}^{N-1} f_n \phi_{kn}; \qquad k = 0, 1, \ldots, M - 1 \qquad (2.8)$$

Equation 2.8 evidently represents a set of M linear equations in the M unknown coefficients $c_0, \ldots, c_{M-1}$. The matrix form equivalent to Eq. 2.8 is

$$\begin{bmatrix} \sum \phi_{0n} \phi_{0n} & \cdots & \sum \phi_{0n} \phi_{M-1,n} \\ \vdots & & \\ \sum \phi_{M-1,n} \phi_{0n} & \cdots & \sum \phi_{M-1,n} \phi_{M-1,n} \end{bmatrix} \cdot \begin{bmatrix} c_0 \\ \vdots \\ c_{M-1} \end{bmatrix} = \begin{bmatrix} \sum f_n \phi_{0n} \\ \vdots \\ \sum f_n \phi_{M-1,n} \end{bmatrix} \qquad (2.9)$$

in which each sum goes from $n = 0$ to $n = N - 1$. As with any set of linear equations, Eq. 2.9 may or may not yield a unique solution for the set $[c_m]$ that minimizes E^2. The number of solutions depends on the nature of the function set $[\phi_n]$ as well as the sample set $[f_n]$.

Example 2.1

Let $\hat{f}(t)$ be the least-squares polynomial with $\phi_m = t^m$ as in Eqs. 2.4 and 2.5 and suppose that the sample set consists of only three samples as follows:

n	0	1	2
t_n	-2	0	1
f_n	2	1	0

For this case, then, $N = 3$ and Eq. 2.9 becomes the following for $M = 1, 2,$ and 3: For $M = 1$:

$$3c_0 = 3 \qquad (2.10)$$

For $M = 2$:

$$\begin{bmatrix} 3 & -1 \\ -1 & 5 \end{bmatrix} \cdot \begin{bmatrix} c_0 \\ c_1 \end{bmatrix} = \begin{bmatrix} 3 \\ -4 \end{bmatrix} \tag{2.11}$$

For $M = 3$:

$$\begin{bmatrix} 3 & -1 & 5 \\ -1 & 5 & -7 \\ 5 & -7 & 17 \end{bmatrix} \cdot \begin{bmatrix} c_0 \\ c_1 \\ c_2 \end{bmatrix} = \begin{bmatrix} 3 \\ -4 \\ 8 \end{bmatrix} \tag{2.12}$$

When solved for the coefficients $[c_m]$, these equations give $\hat{f}_1(t)$, $\hat{f}_2(t)$, and $\hat{f}_3(t)$ for $M = 1$, 2, and 3 as follows:

$$\hat{f}_1(t) = 1 \tag{2.13}$$

$$\hat{f}_2(t) = \frac{11}{14} - \frac{9}{14} t \tag{2.14}$$

$$\hat{f}_3(t) = 1 - \frac{5}{6} t - \frac{1}{6} t^2 \tag{2.15}$$

These least-square solutions are shown in Fig. 2.3 and illustrate a fairly general rule:

If the number of samples N is greater than the number of functions M (as for $\hat{f}_1$ and $\hat{f}_2$ in the example), the total squared error E^2 is in general nonzero. If $N = M$, there usually is one solution ($\hat{f}_3$ in the example) for which $E^2 = 0$, and if $N < M$, there is a family of such solutions.

The main case of interest in applying the least-squares principle and Eq. 2.9 is of course the former of the three, that is, where $N > M$ and E^2 is nonzero when minimized. (*Note:* For a general form of Eq. 2.9 with polynomial functions, see the answer to Exercise 23.)

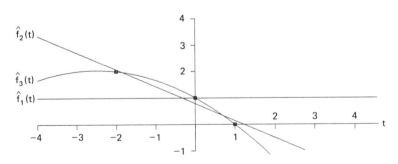

Figure 2.3 Least-squares solutions for Example 2.1.

A general formula for the minimum value of E^2 is obtainable from Eq. 2.6. A more realistic measure of goodness of fit, however, is the *minimum mean squared error,* which is derived by first writing E^2/N from Eq. 2.6:

$$
\begin{aligned}
\frac{E^2}{N} &= \frac{1}{N} \sum_{n=0}^{N-1} \left(f_n - \sum_{m=0}^{M-1} c_m \phi_{mn} \right)^2 \\
&= \frac{1}{N} \sum_{n=0}^{N-1} \left(f_n^2 - 2f_n \sum_{m=0}^{M-1} c_m \phi_{mn} + \sum_{k=0}^{M-1} c_k \sum_{m=0}^{M-1} c_m \phi_{mn} \phi_{kn} \right)
\end{aligned}
\tag{2.16}
$$

If the last term in Eq. 2.16 is summed first over m, second over n, and lastly over k, and if Eq. 2.8 is used to substitute for this last term, after summing over n, an equation for the minimum value of E^2/N can be obtained:

$$
\begin{aligned}
\frac{E_{\min}^2}{N} &= \frac{1}{N} \sum_{n=0}^{N-1} \left(f_n^2 - 2f_n \sum_{m=0}^{M-1} c_m \phi_{mn} + \sum_{k=0}^{M-1} c_k f_n \phi_{kn} \right) \\
&= \frac{1}{N} \sum_{n=0}^{N-1} f_n \left(f_n - \sum_{m=0}^{M-1} c_m \phi_{mn} \right) \equiv \frac{1}{N} \sum_{n=0}^{N-1} f_n (f_n - \hat{f}_n)
\end{aligned}
\tag{2.17}
$$

Equation 2.17 is thus a general formula for the minimum mean-squared error.

When the number (M) of least-squares coefficients is large, the linear equations 2.9 generally require a computer solution. Most scientific computer libraries contain subroutines to solve linear equations, and the reader is urged to utilize one of those standard routines for this purpose. However, if such a routine is not readily available, a very simple routine called SPSOLE is available in Appendix B. SPSOLE is discussed further in Chapter 14 and is meant primarily for the least-squares design equations in Chapter 14, but its simplicity makes it useful for other applications as well, even though it may indicate a "singularity error" where other routines do not.

To use SPSOLE, one must place the information in Eq. 2.9 in an *augmented matrix* (**DD**) that is declared *double precision* in the main program. The general form of this augmented matrix is

$$
\mathbf{DD} = \begin{bmatrix} \sum \phi_{0n} \phi_{0n} & \cdots & \sum \phi_{0n} \phi_{M-1,n} & \sum f_n \phi_{0n} \\ \vdots & & \vdots & \vdots \\ \sum \phi_{M-1,n} \phi_{0n} & \cdots & \sum \phi_{M-1,n} \phi_{M-1,n} & \sum f_n \phi_{M-1,n} \end{bmatrix}
$$

As in Eq. 2.9, all sums go from $n = 0$ through $n = N - 1$. The augmented matrix has M rows and $M + 1$ columns, and SPSOLE returns the solution, that is, the least-squares coefficient vector, in the last column. Exercises 24–26 contain examples of the use of SPSOLE.

Regarding the form of the function set $[\phi_m]$, it is especially useful when Eq. 2.9 can be solved explicitly for the coefficients $[c_m]$ without having to solve the $M \times M$ system of equations. This solution can be obtained trivially when the function set is *orthogonal,* and therefore the subject of orthogonal functions is of special interest here.

2.3 ORTHOGONAL FUNCTIONS

Orthogonality is an important and fundamental mathematical property. It is of interest here because of its influence on the solution to Eq. 2.9 for the least-squares coefficients.

The word "orthogonal" stems originally from a Greek word meaning "right-angled." Two intersecting lines, for example, are said to be orthogonal if they are perpendicular to each other. More generally, two N-dimensional vectors

$$\mathbf{U} = (u_1, u_2, \ldots, u_N)$$

$$\mathbf{V} = (v_1, v_2, \ldots, v_N)$$

are orthogonal (or perpendicular) if their *inner product* defined by

$$\mathbf{U} \cdot \mathbf{V} = u_1 v_1 + u_2 v_2 + \cdots + u_N v_N$$

is equal to zero. As illustrated in Fig. 2.4, when the inner product vanishes, the line segments representing the vectors in three dimensions are mutually perpendicular. This follows immediately upon applying the Pythagorean theorem to the right triangle with sides $\mathbf{U}, \mathbf{V}$, and $\mathbf{U} + \mathbf{V}$:

$$|\mathbf{U} + \mathbf{V}|^2 = |\mathbf{U}|^2 + |\mathbf{V}|^2$$

Therefore,

$$(u_1 + v_1)^2 + (u_2 + v_2)^2 + (u_3 + v_3)^2 = u_1^2 + u_2^2 + u_3^2 + v_1^2 + v_2^2 + v_3^2$$

and

$$u_1 v_1 + u_2 v_2 + u_3 v_3 = 0$$

The concept of orthogonal *functions* follows directly from that of orthogonal vectors. Let $[f_n]$ and $[g_n]$ be sample sets of two functions $f(t)$ and $g(t)$ as in the pre-

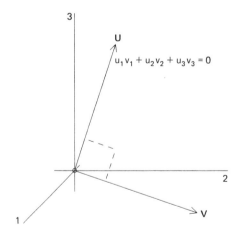

Figure 2.4   Orthogonal vectors $\mathbf{U}$ and $\mathbf{V}$ with $\mathbf{U} \cdot \mathbf{V} = 0$.

ceding section. The two functions are then said to be orthogonal with respect to the sampling points $[t_n]$ if, and only if,

$$\sum_{n=0}^{N-1} f_n g_n = 0 \qquad (2.18)$$

Thus, to preserve the original notion of vectors forming a right angle, the sample sets $[f_n]$ and $[g_n]$ must be viewed as vectors in N-dimensional space, where N is the number of samples. In statistics, because of this view, the number of samples N is called the number of *degrees of freedom* of the set $[f_n]$, and the N-dimensional space is called the *signal space*.

For two continuous functions, the definition of orthogonality is analogous to Eq. 2.18. The functions $f(t)$ and $g(t)$ are orthogonal over the interval (a, b) if, and only if,

$$\int_a^b f(t)g(t)\, dt = 0 \qquad (2.19)$$

Here, to retain the notion of perpendicular vectors, one must assume a space with an infinite number of dimensions. Visualization of perpendicular vectors therefore becomes obscure, and Eq. 2.19 is often taken per se as the basic definition of orthogonality.

The property of orthogonality leads to a simple solution for the least-squares coefficients in Eq. 2.9. First, following Eq. 2.18, the M functions $[\phi_m; m = 0, 1, \ldots, M - 1]$ are *mutually orthogonal* over the set of N points $[t_n; n = 0, 1, \ldots, N - 1]$ if they are pairwise orthogonal, that is, if

$$\sum_{n=0}^{N-1} \phi_{mn} \phi_{kn} = 0; \qquad m \neq k \qquad (2.20)$$

The effect on Eq. 2.9 is immediately evident, since the condition $m \neq k$ is met for all off-diagonal elements in the $M \times M$ matrix, and thus only the diagonal elements are nonzero. The solution for least-squares coefficients follows from either Eq. 2.8 or 2.9:

$$c_k = \frac{\displaystyle\sum_{n=0}^{N-1} f_n \phi_{kn}}{\displaystyle\sum_{n=0}^{N-1} \phi_{kn}^2}; \qquad k = 0, 1, \ldots, M - 1 \qquad (2.21)$$

(Note that the denominator in Eq. 2.21 is nonzero as long as at least one ϕ_{kn} is nonzero.) Thus, if the functions $[\phi_m]$ are orthogonal, the solution for least-squares coefficients is greatly simplified, and in fact the least-squares fit can be accomplished one step at a time, proceeding through successive values of k.

When the number of sample points increases without bound and $f(t)$ is given continuously in the interval from $t = a$ to $t = b$, and when the functions $[\phi_m]$ are orthogonal as in Eq. 2.19, the least-squares coefficients are found essentially by replacing the sums in Eq. 2.21 with integrals:

$$c_k = \frac{\displaystyle\int_a^b f(t)\phi_k(t)\,dt}{\displaystyle\int_a^b \phi_k^2(t)\,dt}; \qquad k = 0, 1, \ldots, M - 1 \qquad (2.22)$$

Equation 2.22 simply reexpresses Eq. 2.21 for the continuous case, with the continuous variable t replacing the discrete subscript n as an argument.

In summary, the solution for least-squares coefficients is greatly simplified when the approximating function $\hat{f}(t)$ is a linear combination of orthogonal functions. Also, it is important to note that the set of points on which two functions are defined is just as important as the functions themselves in determining orthogonality—two functions that are orthogonal with respect to one interval or set of points are, in general, *not* orthogonal with respect to another set of points.

2.4 THE FOURIER SERIES

In engineering, the well-known Fourier series is an important application of the foregoing concepts of least-squares and orthogonality. An understanding of how these concepts apply with the Fourier series is especially important in the discrete case, where sample sequences and digital signal analysis are involved.

The Fourier series itself is widely applicable in engineering. It provides a unique way to express any periodic function in terms of its components at discrete frequencies, giving explicitly the frequency composition of the function.

Figure 2.5 is an illustration of a periodic function, $f_p(t)$, composed of components at three frequencies (one of which is zero). The components of $f_p(t)$, each with its own amplitude and phase, are sinusoidal and occur only at multiples (*harmonics*) of the *fundamental frequency*, which is the frequency at which $f_p(t)$ recurs.

The formulas for the Fourier series and the Fourier coefficients can be developed logically by finding a least-squares approximation to an arbitrary periodic function,

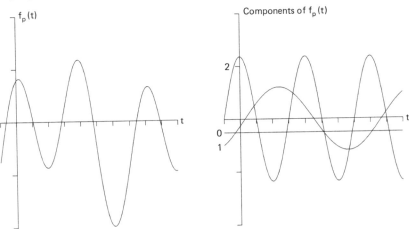

Figure 2.5 Illustration of $f_p(t)$ composed of three components.

$f_p(t)$, in the form of Eq. 2.3. In particular, let the least-squares approximation to $f_p(t), \hat{f}_p(t)$, be expressed in any one of the following three forms:

$$\hat{f}_p(t) = \frac{a_0}{2} + \sum_{k=1}^{K} (a_k \cos k\omega_0 t + b_k \sin k\omega_0 t) \tag{2.23}$$

$$= \frac{A_0}{2} + \sum_{k=1}^{K} A_k \cos(k\omega_0 t + \alpha_k) \tag{2.24}$$

$$= \sum_{k=-K}^{K} c_k e^{jk\omega_0 t} \tag{2.25}$$

in which ω_0 is the fundamental frequency, determined so that the period of the approximated function $f_p(t)$ is $2\pi/\omega_0$, and in which k is the harmonic number and the a's, b's, A's, α's, and c's are least-squares coefficients of $\hat{f}_p(t)$. The complete equivalence of the expressions above, which are three of the most common forms of the Fourier series, is demonstrated in Table 2.1, where the coefficients are given in terms of each other. Since the three expressions are equivalent, Eq. 2.23 only is used in the following development.

Since both $f_p(t)$ and $\hat{f}_p(t)$ are periodic, a least-squares fit can be obtained over any fundamental period (or set of periods), and in particular over the interval $(-\pi/\omega_0, \pi/\omega_0)$. The approximated function can be known either at a set of sample points in this interval or completely throughout the interval.

In either case the orthogonality of the functions composing $\hat{f}(t)$ is of interest. In Eq. 2.23 these functions are $\cos k\omega_0 t$ and $\sin k\omega_0 t$ with k going from zero to K. The orthogonality of these functions with respect to N *regular (equally spaced) points in the interval* $(-\pi/\omega_0, \pi/\omega_0)$ is expressed as follows:

$$\sum_{n=0}^{N-1} \cos k\omega_0 t_n \cos m\omega_0 t_n = 0; \quad k \neq m \tag{2.26}$$

TABLE 2.1. Relationships between Fourier Coefficients[a]

Coefficient $(k \geq 0)$	In Terms of:		
	a_k, b_k	A_k, α_k	c_k, c_{-k}
a_k	a_k	$A_k \cos \alpha_k$	$c_k + c_{-k}$ (except $a_0 = 2c_0$)
b_k	b_k	$-A_k \sin \alpha_k$	$j(c_k - c_{-k})$
A_k	$(a_k^2 + b_k^2)^{1/2}$	A_k	$2(c_k c_{-k})^{1/2}$
α_k	$-\tan^{-1}\left(\dfrac{b_k}{a_k}\right)$	α_k	$\tan^{-1}\left[\dfrac{\text{Im}(c_k)}{\text{Re}(c_k)}\right]$
c_k	$\dfrac{1}{2}(a_k - jb_k)$	$\left(\dfrac{A_k}{2}\right)e^{j\alpha_k}$	c_k
c_{-k}	$\dfrac{1}{2}(a_k + jb_k)$	$\left(\dfrac{A_k}{2}\right)e^{-j\alpha_k}$	c_{-k}

[a]$b_0 = \alpha_0 = 0$; Im, imaginary part; Re, real part.

$$\sum_{n=0}^{N-1} \sin k\omega_0 t_n \sin m\omega_0 t_n = 0; \qquad k \neq m \tag{2.27}$$

$$\sum_{n=0}^{N-1} \sin k\omega_0 t_n \cos m\omega_0 t_n = 0 \tag{2.28}$$

Here a member of the set of N regular points $[t_n]$ can, without loss of generality, be assumed to be

$$t_n = -\frac{\pi}{\omega_0} + \frac{n}{N} \cdot \frac{2\pi}{\omega_0}$$

$$= \left(\frac{n}{N} - \frac{1}{2}\right) \frac{2\pi}{\omega_0}; \qquad n = 0, 1, \ldots, N - 1 \tag{2.29}$$

so that the samples of $f(t)$ are taken as illustrated in Fig. 2.6. Note carefully that the Nth sampling point is *just before* π/ω_0, rather than at π/ω_0.

The validity of Eqs. 2.26 through 2.28 can be demonstrated by completing the following four steps:

Step 1: Substitute Eq. 2.29 for t_n into all three equations.

Step 2: Reexpress the three equations by making substitutions of the form

$$\sin x = \frac{1}{2j} \left(e^{jx} - e^{-jx}\right); \tag{2.30}$$

$$\cos x = \frac{1}{2} \left(e^{jx} + e^{-jx}\right) \tag{2.31}$$

Step 3: Show that, to prove the three equations for $k \neq m$, it is sufficient to prove that

$$\sum_{n=0}^{N-1} e^{\pm j(2\pi n i/N)} = 0; \qquad i = 1, 2, \ldots, 2K - 1 \tag{2.32}$$

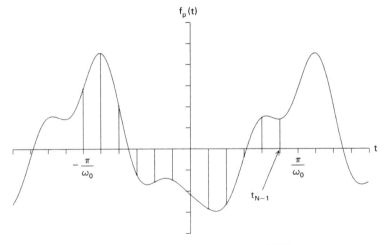

Figure 2.6 Samples over one cycle of $f_p(t)$.

Review of Least Squares, Orthogonality, and the Fourier Series   Chap. 2

Step 4: By summing Eq. 2.32, show that it is true with the condition that

$$N \geq 2K \qquad (2.33)$$

The completion of these steps, including the case $k = m$ in Eq. 2.28, is left for the exercises. The essential result here is that

---

The sine and cosine functions of the Fourier series are orthogonal over a single fundamental period of $f_p(t)$ provided that the number of equally spaced samples N is at least twice the highest harmonic number of the series.

There is thus a lower bound on N but no upper bound, and the components of $\hat{f}_p(t)$ can also be said in the limit to be orthogonal over the continuous interval from $-\pi/\omega_0$ to π/ω_0. Therefore, Eqs. 2.26 through 2.28 can be expressed as integrals corresponding to the sums.

The statement following Eq. 2.15 places a restriction on N slightly beyond that of Eq. 2.33. Since the number of adjustable coefficients in Eqs. 2.23 through 2.25 is $M = 2K + 1$, the number of samples N must be at least $2K + 1$. If $N = 2K + 1$, the error is zero and the Fourier series passes through the sample points; if $N > 2K + 1$, the Fourier series provides a minimum total squared error.

The lower bound $N \geq 2K + 1$ *will be assumed from here on,* although, since $M = 2K + 1$ is odd, the singular case where $N = M$ and N is even is excluded by the form of Eqs. 2.23 through 2.25. (That is, these equations as given contain an odd number of coefficients.) This singular case can be accommodated by removing one of the final components from Eq. 2.23.

With the above restrictions on the number of samples N, the Fourier coefficients a_k and b_k in Eq. 2.23 can be formulated in the simplified form of Eq. 2.21; first, the denominator of Eq. 2.21 is either

$$\sum_{n=0}^{N-1} \phi_{kn}^2 = \sum_{n=0}^{N-1} \cos^2 k\omega_0 t_n; \qquad k = 0, 1, \ldots, K \qquad (2.34)$$

or

$$\sum_{n=0}^{N-1} \phi_{kn}^2 = \sum_{n=0}^{N-1} \sin^2 k\omega_0 t_n; \qquad k = 1, 2, \ldots, K \qquad (2.35)$$

In either case, after substituting Eq. 2.29 for t_n, the denominator of Eq. 2.21 for $k = 0$ to $N/2$ becomes

$$\sum_{n=0}^{N-1} \phi_{kn}^2 = \begin{cases} N; & k = 0 \\ \dfrac{N}{2}; & 0 < k < \dfrac{N}{2} \end{cases} \qquad (2.36)$$

Hence, for the least-squares coefficients in Eq. 2.23, Eq. 2.21 gives

$$a_k = \frac{2}{N} \sum_{n=0}^{N-1} f_{pn} \cos k\omega_0 t_n; \qquad k = 0, 1, \ldots, K \tag{2.37}$$

$$b_k = \frac{2}{N} \sum_{n=0}^{N-1} f_{pn} \sin k\omega_0 t_n; \qquad k = 1, 2, \ldots, K \tag{2.38}$$

with f_{pn} being a sample of the periodic function $f_p(t)$ taken at time t_n as defined in Eq. 2.29. Formulas for the other coefficients in Table 2.1 can be derived by applying the formulas in the table. For example,

$$c_k = \frac{1}{2} (a_k - jb_k)$$

$$= \frac{1}{N} \sum_{n=0}^{N-1} f_{pn} (\cos k\omega_0 t_n - j \sin k\omega_0 t_n)$$

$$= \frac{1}{N} \sum_{n=0}^{N-1} f_{pn} e^{-jk\omega_0 t_n} \tag{2.39}$$

The more familiar formulas for the case where $f_p(t)$ is known continuously in the interval $-\pi/\omega_0 \leq t < \pi/\omega_0$ are derived in the same manner and are completely analogous to the discrete formulas; for example,

$$a_k = \frac{\omega_0}{\pi} \int_{-\pi/\omega_0}^{\pi/\omega_0} f_p(t) \cos k\omega_0 t \, dt \tag{2.40}$$

$$b_k = \frac{\omega_0}{\pi} \int_{-\pi/\omega_0}^{\pi/\omega_0} f_p(t) \sin k\omega_0 t \, dt \tag{2.41}$$

$$c_k = \frac{\omega_0}{2\pi} \int_{-\pi/\omega_0}^{\pi/\omega_0} f_p(t) e^{-jk\omega_0 t} \, dt \tag{2.42}$$

These equations are clearly related to Eqs. 2.37 through 2.39. If dt is allowed to become finite in such a way that there are N subintervals in the total interval from $-\pi/\omega_0$ to π/ω_0, then, in relating an integral to a sum, one would substitute

$$dt = \frac{2\pi}{N\omega_0} \tag{2.43}$$

Thus Eq. 2.40 follows from Eq. 2.37, and so on.

EXERCISES

1. Adjust the parameter p so that the line $y = px + 3$ is, in the least-squares sense, closest to the function

$$f(x) = x^2 - x + 3$$

in the interval $0 \leq x \leq 4$.

2. Adjust p so that $y = px + 3$ is closest to the following set of points:

x	0	1	3	4
y	3	3	9	15

Hint: There is only one parameter to adjust: $M = 1$ in Eqs. 2.3 and 2.9.

3. Write Eq. 2.9 for the case where:
    (a) $\hat{f}(x) = c_0 + c_1 x$, a line
    (b) $\hat{f}(x) = c_0 + c_1 x + c_2 x^2$, a parabola
    (c) $\hat{f}(x) = c_0 + c_1 x + \cdots + c_k x^k$
    (d) $\hat{f}(x) = c_0 + c_1 e^x + c_2 e^{2x}$
    (e) $\hat{f}(x) = c_0 + c_1 \sin x + c_2 \sin 2x$

4. Find and plot the least-squares straight line for the following points:

t	0	1	2	4
$f(t)$	8	6	3	0

5. Find and plot the least-squares parabola for the following points:

x	0	1	2	4
y	0	1	5	9

6. Show how the minimum mean-squared-error formula, Eq. 2.17, is derived from Eq. 2.6.

7. Using Eq. 2.17 along with the given answer to Exercise 4, find the minimum mean-squared error for Exercise 4.

8. Similarly, find the minimum mean-squared error for Exercise 5.

9. Find the least-squares coefficients in $y = c_0 + c_1 e^x + c_2 e^{-x}$ for the following data:

x	-1	0	1	2
y	1	0	1	2

10. Specify the family of intervals along the t-axis over which the functions $\sin 2\pi t$ and $\cos 2\pi t$ are orthogonal.

11. Prove that the functions x and x^2 are orthogonal in the interval $-a \leq x \leq a$.

12. Given the smoothing formula $\hat{f}(t) = c_1 t + c_2 t^2$ along with the following data:

t	-1	$-1/2$	0	1/2	1
$f(t)$	2	1	0	1	2

    (a) Write Eq. 2.9 for this case.
    (b) Find the least-squares values of c_1 and c_2 *using the result of Exercise 11* rather than Eq. 2.9.

13. Compute the least-squares coefficients for $\hat{f}(t) = c_1 t + c_2 t^2$ assuming $\hat{f}(t)$ is used to approximate $f(t) = \cos 2\pi t$ in the interval $-1 \leq t \leq 1$.

14. Derive the relationships in Table 2.1:
    (a) For a_k and b_k in terms of c_k.
    (b) For c_k in terms of a_k and b_k.
    (c) For a_k and b_k in terms of A_k and α_k.

15. For Eqs. 2.26 through 2.28:

    (a) Perform steps 1 and 2 in the text.

    (b) Give the proof required in step 3.

    (c) Perform step 4.

16. A Fourier series is to be fitted to the following points:

t	0	1	2
$f(t)$	1	0	1

    (a) What is the fundamental frequency ω_0?

    (b) What is the Fourier series, Eq. 2.23 with $K = 0$?

    (c) What is the Fourier series, Eq. 2.23 with $K = 1$?

17. Prove that the continuous or discrete Fourier series for an *even* periodic function, for which $f_p(t) = f_p(-t)$, is a cosine series.

18. Prove that the continuous or discrete Fourier series for an *odd* periodic function, for which $f_p(t) = -f_p(-t)$ is a sine series.

19. Derive the continuous Fourier series for the square wave shown.

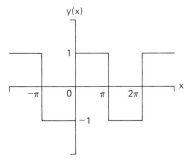

20. Derive the continuous Fourier series for the triangular function shown. Using the derivative of this series, obtain the answer to Exercise 19.

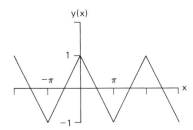

21. Derive the continuous Fourier series for the rectangular pulse train illustrated.

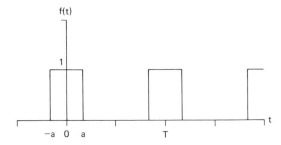

22. Assume that the triangular wave in Exercise 20 is sampled at $x = -\pi, -\pi/2, 0$, and $\pi/2$ and, given the samples, derive the Fourier series with the maximum number of terms. Compare this with the series for Exercise 20.

23. Write Eq. 2.9 explicitly for a least-squares polynomial of the form

$$\hat{f}(t) = c_0 + c_1 t + c_2 t^2 + \cdots + c_{M-1} t^{M-1}$$

fitted to the sample sequence

$$[f_n] = [f_0, f_1, f_2, \ldots, f_{N-1}]$$

obtained at times

$$[t_n] = [0, T, 2T, \ldots, (N-1)T]$$

24. Use the routine SPSOLE in Appendix B to solve the following system of equations:

$$\begin{bmatrix} 1 & 2 & 3 \\ 2 & 3 & 4 \\ 3 & 4 & 6 \end{bmatrix} \begin{bmatrix} c_0 \\ c_1 \\ c_2 \end{bmatrix} = \begin{bmatrix} 2 \\ 2 \\ 3 \end{bmatrix}$$

To see how the routine works, insert a statement into SPSOLE to print the augmented matrix after each row operation.

25. Use the routine SPSOLE in Appendix B to fit a least-squares cubic polynomial to the following data, which have been sampled regularly at $t = 0, 2, 4, \ldots, 14$ s:

$$[f_n] = [-1, 0, 1, 2, 1, 0, 0, 1]$$

(a) Print the augmented matrix.
(b) Find the least-squares coefficients.
(c) Plot $\hat{f}(t)$ versus t for $-5 \le t \le 20$. Put the sample points $[f_n]$ on the same plot.

26. The intensity, I, of a certain extraterrestrial source is known to vary sinusoidally around a constant value, C_0, with a period of 1000 years. We have the following history:

Date	Intensity	Date	Intensity	Date	Intensity
1/1/1900	167	1/1/1930	121	1/1/1960	74
1/1/1910	152	1/1/1940	105	1/1/1970	58
1/1/1920	136	1/1/1950	89	1/1/1980	42
				1/1/1990	27

(a) Use SPSOLE to fit the function

$$\hat{I}(t) = C_0 + C_1 \sin \omega_0 t + C_2 \cos \omega_0 t$$

to these data, let t = years from 1/1/1900, and choose ω_0 appropriately. Find C_0, C_1, and C_2.
(b) Plot $\hat{I}(t)$ from year 1900 through year 2999. Show the 10 history points on the plot.
(c) Predict the intensity on January 1, 2050. Discuss the accuracy of this prediction.

SOME ANSWERS

1. $p = 2$ **2.** $p = 2.54$ **4.** $7.80 - 2.03t$ **5.** $0.023x^2 + 2.28x - 0.355$ **7.** 0.19
8. 0.34 **9.** $c_0, c_1, c_2 = -0.42, 0.33, 0.44$ **10.** Any interval from $t = t_1$ to $t = \pm t_1 \pm n/2$
12. (b) $c_1, c_2 = 0, 36/17$ **13.** $c_1, c_2 = 0, 0.253$

16. (a) $2\pi/3$ **(b)** $2/3$ **(c)** $\dfrac{2}{3} + \dfrac{1}{3}\cos\dfrac{2\pi t}{3} - \dfrac{1}{\sqrt{3}}\sin\dfrac{2\pi t}{3}$

19. $\dfrac{4}{\pi}\displaystyle\sum_{n=1}^{\infty}\dfrac{\sin(2n-1)x}{2n-1}$ **20.** $\dfrac{8}{\pi^2}\displaystyle\sum_{n=1}^{\infty}\dfrac{\cos(2n-1)x}{(2n-1)^2}$

21. $\dfrac{2a}{T} + \dfrac{2}{\pi}\displaystyle\sum_{n=1}^{\infty}\dfrac{1}{n}\sin\dfrac{2\pi na}{T}\cos\dfrac{2\pi nt}{T}$ **22.** $\cos x$ **23.** Equation 2.9 for least-squares

polynomial of degree $(M-1)$ and regularly spaced data $f_0, f_1, \ldots, f_{N-1}$ with step size T (all sums from $n=0$ to $n=N-1$):

$$
\begin{bmatrix}
N & T\sum n & \cdots & T^{M-1}\sum n^{M-1} \\[2mm]
T\sum n & T^2\sum n^2 & \cdots & T^M\sum n^M \\[2mm]
T^2\sum n^2 & T^3\sum n^3 & \cdots & T^{M+1}\sum n^{M+1} \\[1mm]
\vdots & \vdots & & \vdots \\[1mm]
T^{M-1}\sum n^{M-1} & T^M\sum n^M & \cdots & T^{2M-2}\sum n^{2M-2}
\end{bmatrix}
\cdot
\begin{bmatrix}
c_0 \\[2mm] c_1 \\[2mm] c_2 \\[1mm] \vdots \\[1mm] c_{M-1}
\end{bmatrix}
=
\begin{bmatrix}
\sum f_n \\[2mm] T\sum nf_n \\[2mm] T^2\sum n^2 f_n \\[1mm] \vdots \\[1mm] T^{M-1}\sum n^{M-1}f_n
\end{bmatrix}
$$

24. Outputs after each row operation:

$$
\begin{bmatrix}
1.0 & 2.0 & 3.0 & 2.0 \\
2.0 & -1.0 & -2.0 & -2.0 \\
3.0 & -2.0 & -3.0 & -3.0
\end{bmatrix}
$$

$$
\begin{bmatrix}
1.0 & 2.0 & -1.0 & -2.0 \\
2.0 & -1.0 & 2.0 & 2.0 \\
3.0 & -2.0 & 1.0 & 1.0
\end{bmatrix}
$$

$$
\begin{bmatrix}
1.0 & 2.0 & -1.0 & -1.0 \\
2.0 & -1.0 & 2.0 & 0.0 \\
3.0 & -2.0 & 1.0 & 1.0
\end{bmatrix}
$$

25. (a) Augmented matrix:

$$
\begin{bmatrix}
8.0 & 56.0 & 560.0 & 6272.0 & 4.0 \\
56.0 & 560.0 & 6{,}272.0 & 74{,}816.0 & 38.0 \\
560.0 & 6{,}272.0 & 74{,}816.0 & 928{,}256.0 & 348.0 \\
6{,}272.0 & 74{,}816.0 & 928{,}256.0 & 11{,}828{,}480.0 & 3{,}752.0
\end{bmatrix}
$$

(b) Optimal c_0, c_1, c_2, c_3:

$$-1.303030, \, 1.267857, \, -0.185877, \, 0.007576$$

26. Intensity curve fitted to data:

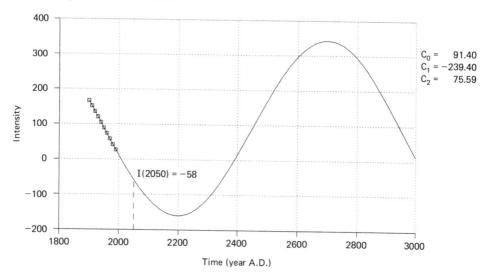

$C_0 = 91.40$
$C_1 = -239.40$
$C_2 = 75.59$

$I(2050) = -58$

Time (year A.D.)

REFERENCES

DANIEL, C., and WOOD, F. S., *Fitting Equations to Data.* New York: Wiley, 1971.

DYM, H., and MCKEAN, H. P., *Fourier Series and Integrals.* New York: Academic Press, 1972.

EDWARDS, R. E., *Fourier Series,* Vols. 1 and 2. New York: Holt, Rinehart and Winston, 1967.

HAMMING, R. W., *Numerical Methods for Scientists and Engineers,* 2nd ed. New York: McGraw-Hill, 1973.

HARMUTH, H. F., *Transmission of Information by Orthogonal Functions.* Berlin: Springer-Verlag, 1969.

HILDEBRAND, F. B., *Introduction to Numerical Analysis,* Chap. 7. New York: McGraw-Hill, 1956.

KELLY, L. G., *Handbook of Numerical Methods and Applications,* Chap. 5. Reading, Mass.: Addison-Wesley, 1967.

KUFNER, A., and KADLEC, J., *Fourier Series,* trans. G. A. Toombs. London: Iliffe Books, 1971.

NIELSON, K. L., *Methods in Numerical Analysis,* Chap. 8. New York: Macmillan, 1967.

SEELEY, R. T., *An Introduction to Fourier Series and Integrals.* New York: W. A. Benjamin, 1966.

3

Review of Continuous Transforms, Transfer Functions, and Convolution

3.1 FOURIER AND LAPLACE TRANSFORMS

The *Fourier transform,* an extension of the Fourier series just discussed, is another process used extensively in engineering analysis. It is defined as

$$F(j\omega) = \int_{-\infty}^{\infty} f(t)e^{-j\omega t}\,dt \; ; \tag{3.1}$$

$F(j\omega)$ is called the Fourier transform of $f(t)$. Because $F(j\omega)$ is a function of ω instead of t, the Fourier transformation is considered an operation that creates from $f(t)$ a function $F(j\omega)$ in the *frequency domain,* where the frequency content of $f(t)$ appears explicitly. To justify associating the word "frequency" with the variable ω, one can demonstrate that $F(j\omega)$ becomes a kind of "continuous coefficient" of the Fourier series when the period of the periodic function $f_p(t)$ increases without bound (and the resulting $f(t)$ becomes aperiodic in the limit). The demonstration proceeds as follows.

First, let a periodic function $f_p(t)$ be expressed using the complex Fourier series as in Section 2.4:

$$f_p(t) = \sum_{n=-\infty}^{\infty} c_n e^{jn\omega_0 t}; \qquad c_n = \frac{\omega_0}{2\pi} \int_{-\pi/\omega_0}^{\pi/\omega_0} f_p(t)e^{-jn\omega_0 t}\,dt \tag{3.2}$$

Here the period of $f_p(t)$ is $2\pi/\omega_0$ seconds with ω_0 measured in radians per second (rad/s), and the harmonic content of $f_p(t)$ is unrestricted. Each c_n is the value of the complex *frequency component* of $f_p(t)$ at the angular frequency $n\omega_0$. As illustrated in Fig. 3.1, the *amplitude spectrum* of $f_p(t)$ can be plotted by plotting each magnitude $|c_n|$ on the ω axis. Since c_n has been shown to be the complex conjugate of c_{-n}, the

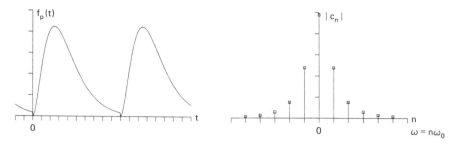

Figure 3.1 Periodic function and its discrete amplitude spectrum.

amplitude spectrum must be even, that is, symmetric about $\omega = 0$ such that $|c_n| = |c_{-n}|$.

The derivation of $F(j\omega)$ now involves letting ω_0, the interval on the ω axis, approach zero so that, in effect, the lines in Fig. 3.1 merge into a continuous spectrum. To do this, the fundamental period, $2\pi/\omega_0$, is allowed to increase without bound by letting $\omega_0 \to 0$. Writing $\Delta\omega$ for ω_0 to emphasize the vanishing frequency interval, Eq. 3.2 can now be written as follows:

$$
\begin{aligned}
f_p(t) &= \sum_{n=-\infty}^{\infty} c_n e^{jn\Delta\omega t} \\
&= \sum_{n=-\infty}^{\infty} \left(\frac{\Delta\omega}{2\pi}\right) \int_{-\pi/\Delta\omega}^{\pi/\Delta\omega} f(\tau)e^{-jn\Delta\omega\tau}\, d\tau \, e^{jn\Delta\omega t} \\
&= \frac{1}{2\pi} \sum_{n=-\infty}^{\infty} \left[\int_{-\pi/\Delta\omega}^{\pi/\Delta\omega} f(\tau)e^{-jn\Delta\omega\tau}\, d\tau\right] e^{jn\Delta\omega t}\, \Delta\omega
\end{aligned}
\tag{3.3}
$$

(The dummy variable τ is used in place of t in the integrand to avoid confusion.) The limit of Eq. 3.3 as $\Delta\omega$ approaches zero can now be taken, recognizing that the product $n\Delta\omega = n\omega_0$ is still equal to ω as in Fig. 3.1. In the limit, the fundamental period of $f_p(t)$ becomes infinite so that $f_p(t)$ is no longer periodic and becomes just $f(t)$. The limit is

$$
\begin{aligned}
f(t) &= \frac{1}{2\pi} \lim_{\Delta\omega \to 0} \sum_{n=-\infty}^{\infty} \left[\int_{-\pi/\Delta\omega}^{\pi/\Delta\omega} f(\tau)e^{-jn\Delta\omega\tau}\, d\tau\right] e^{jn\Delta\omega t}\, \Delta\omega \\
&= \frac{1}{2\pi} \int_{-\infty}^{\infty} \left[\int_{-\infty}^{\infty} f(\tau)e^{-j\omega\tau}\, d\tau\right] e^{j\omega t}\, d\omega
\end{aligned}
$$

Thus

$$
\boxed{f(t) = \frac{1}{2\pi} \int_{-\infty}^{\infty} F(j\omega)e^{j\omega t}\, d\omega; \qquad F(j\omega) = \int_{-\infty}^{\infty} f(t)e^{-j\omega t}\, dt}
\tag{3.4}
$$

where $F(j\omega)$, as given by Eq. 3.1, is used to replace the integral in the square brackets. This then, is the justification for calling $F(j\omega)$ the *spectrum* of $f(t)$: $F(j\omega)$ *replaces the set* $[c_n]$ *of Fourier coefficients* in the limit as $\Delta\omega$ goes to zero. The amplitude spectrum, $|F(j\omega)|$, is now a continuous function as illustrated in Fig. 3.2.

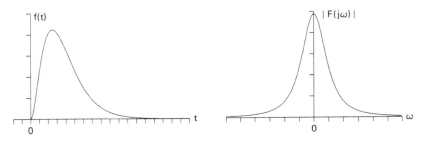

Figure 3.2 Nonperiodic function and its continuous amplitude spectrum.

The result in Eq. 3.4 is the well-known *Fourier transform pair*. The duality of the pair can be seen by letting the frequency variable be $\nu = \omega/2\pi$ (hertz). The formulas are then the same except for the sign of j.

Thus the Fourier transform produces the spectrum of $f(t)$ if the integral in Eq. 3.1 converges. One can easily see that the integral will in fact converge absolutely only if the integral

$$I = \int_{-\infty}^{\infty} |f(t)| \, dt \tag{3.5}$$

converges, because the term $e^{-j\omega t}$ always has a magnitude of 1. Therefore, *if $f(t)$ does not decrease fast enough with increasing* t, F($j\omega$) *does not exist*. Obviously, no periodic function has this property, but all single pulses or transient functions that decay to zero at some finite time do have it, and so do other functions that decay "fast enough." For example, the Fourier transform of $f(t) = e^{-at}$ for $t > 0$ (and $= 0$ for $t \leq 0$) is

$$
\begin{aligned}
F(j\omega) &= \int_{0}^{\infty} e^{-at} e^{-j\omega t} \, dt \\
&= \frac{1}{j\omega + a}
\end{aligned}
\tag{3.6}
$$

Another important property of the Fourier transform is that, like the Fourier series, *it conveys all of the information needed to reconstruct $f(t)$*. This is shown by Eq. 3.4, which gives the reconstruction formula explicitly.

Further interesting properties can be found by expanding the integrand of Eq. 3.1 into real and imaginary parts:

$$
\begin{aligned}
F(j\omega) &= \int_{-\infty}^{\infty} f(t) e^{-j\omega t} \, dt \\
&= \int_{-\infty}^{\infty} f(t) \cos \omega t \, dt - j \int_{-\infty}^{\infty} f(t) \sin \omega t \, dt \\
&= \int_{0}^{\infty} [f(t) + f(-t)] \cos \omega t \, dt - j \int_{0}^{\infty} [f(t) - f(-t)] \sin \omega t \, dt
\end{aligned}
\tag{3.7}
$$

It is now easy to see that:

1. $F(j\omega)$ is *real and even* if, and only if, $f(t)$ is *even* [i.e., $f(t) = f(-t)$].
2. $F(j\omega)$ is *imaginary and odd* if, and only if, $f(t)$ is *odd* [i.e., $f(t) = -f(-t)$].
3. The real part of $F(j\omega)$ is even.
4. The imaginary part of $F(j\omega)$ is odd.

Also, the amplitude spectrum $|F(j\omega)|$ is just the quadrature sum of the real and imaginary components in Eq. 3.7, that is its square is

$$|F(j\omega)|^2 = \left(\int_{-\infty}^{\infty} f(t) \cos \omega t\, dt\right)^2 + \left(\int_{-\infty}^{\infty} f(t) \sin \omega t\, dt\right)^2 \qquad (3.8)$$

from which one can easily show that

$$|F(j\omega)| = |F(-j\omega)| \qquad (3.9)$$

that is, the *amplitude spectrum is an even function* as illustrated in Fig. 3.2, just as in Fig. 3.1 for the discrete spectrum.

The squared amplitude spectrum, $|F(j\omega)|^2$, is called the *energy spectrum* of $f(t)$. This involves a generalization of the physical concept of energy described in Chapter 13, where the energy spectrum is discussed in more detail.

The *Laplace transform* may be considered a modification of the Fourier transform to allow the transformation of a function $f(t)$ that does not necessarily vanish with increasing t. Suppose that the Fourier integral,

$$F(j\omega) = \int_{-\infty}^{\infty} f(t)e^{-j\omega t}\, dt \qquad (3.10)$$

does not converge for a particular $f(t)$. To be able to perform the integration, $f(t)$ can be multiplied by a decaying exponential $e^{-\alpha|t|}$ with $\alpha > 0$, so that the product, $f(t)e^{-\alpha|t|}$, now vanishes with increasing or decreasing t. In addition, although this step is not strictly necessary, most physical problems can be defined so that $f(t)$ "begins" at $t = 0$ and is zero for $t < 0$, so that the magnitude bars can be removed from $e^{-\alpha|t|}$. With these changes, the Fourier transform in Eq. 3.10 becomes

$$F(\alpha + j\omega) = \int_{0}^{\infty} f(t)e^{-(\alpha+j\omega)t}\, dt \qquad (3.11)$$

With the substitution of the complex variable $s = \alpha + j\omega$, the familiar form of the Laplace transform is obtained:

$$\boxed{F(s) = \int_{0}^{\infty} f(t)e^{-st}\, dt} \qquad (3.12)$$

Thus $F(s)$, *the Laplace transform of* f(t), *is also the Fourier transform of* f(t)e^{-\alpha t}, where α is the real part of s. Evidently, Eq. 3.4 is still a valid formula for the inverse transformation, since it provides a means for finding $f(t)e^{-\alpha t}$ from which $f(t)$ can be separated.

Example 3.1

The unit step function at $t = 0$, illustrated in Fig. 3.3, does not have a Fourier transform since it does not vanish with increasing t. It does have a Laplace transform, however:

$$U(s) = \int_0^\infty (1)e^{-st}\,dt$$

$$= \frac{1}{s}$$

(3.13)

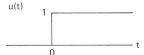

Figure 3.3   Unit step at $t = 0$.

As shown above, this may also be viewed as the *Fourier* transform of $u(t)e^{-\alpha t}$ which, as derived in Eq. 3.6, is obviously the same as in Eq. 3.13 with $s = \alpha + j\omega$.

To summarize the above, the following are important points in comparing the Fourier and Laplace transforms:

1. The Laplace transform $F(s)$ is just the spectrum (Fourier transform) of $f(t)e^{-\alpha t}$, assuming that $f(t) = 0$ for $t < 0$.
2. If $f(t) = 0$ for $t < 0$ and if $F(j\omega)$ exists, then $F(j\omega)$ is equal to $F(s)$ with $s = j\omega$, that is, with $\alpha = 0$. Note that only in this case is the conventional use of the letter "F" to represent both functions strictly justified.
3. The Fourier transform does not require that $f(t) = 0$ for $t < 0$.
4. The Laplace transformation can be performed on a wide variety of practical $f(t)$'s for which the Fourier transform does not exist.

In practice the use of Eq. 3.4 to find the inverse transform of $F(j\omega)$ or $F(s)$ would be tedious. Remembering that $F(s)$ is the Fourier transform of $f(t)e^{-\alpha t}$, Eq. 3.4 gives the *inverse Laplace transformation* as

$$f(t)e^{-\alpha t} = \frac{1}{2\pi} \int_{-\infty}^\infty F(s)e^{j\omega t}\,d\omega$$

$$= \frac{1}{2\pi} \int_{-\infty}^\infty F(\alpha + j\omega)e^{j\omega t}\,d\omega$$

(3.14)

with $s = \alpha + j\omega$. Changing the variable of integration from ω to s and multiplying by $e^{\alpha t}$ in Eq. 3.14, one obtains

$$f(t) = \frac{1}{2\pi} \int_{\alpha - j\infty}^{\alpha + j\infty} F(s)e^{st}\left(\frac{ds}{j}\right)$$

$$= \frac{1}{2\pi j} \int_{\alpha - j\infty}^{\alpha + j\infty} F(s)e^{st}\,ds$$

(3.15)

where the integration now takes place over the region for which α is a constant (i.e., a line parallel to the $j\omega$-axis in the complex s-plane). The residue theorem of complex variable theory often provides a simplified means for finding the integral in Eq. 3.15 [see, e.g., Churchill (1948) and Goldman (1949)] but to save time the "transform pairs" of the form $[f(t), F(j\omega)]$ or $[f(t), F(s)]$ are simply tabulated so that, given $F(s)$ or $F(j\omega)$, $f(t)$ can simply be looked up, and vice versa.

TABLE 3.1 Short Table of Transforms

Line	Function	Laplace Transform*	Fourier Transform
0	$f(t)$	$F(s)$	$F(j\omega)$
1	$Af(t)$	$AF(s)$	$AF(j\omega)$
2	$f(t) + g(t)$	$F(s) + G(s)$	$F(j\omega) + G(j\omega)$
3	$\dfrac{df(t)}{dt}$	$sF(s) - f(0^+)$	$j\omega F(j\omega)$
4	$\displaystyle\int_{-\infty}^{t} f(\tau)\,d\tau$	$\dfrac{F(s)}{s}$	$\dfrac{F(j\omega)}{j\omega}$
5	$tf(t)$	$\dfrac{-dF(s)}{ds}$	$\dfrac{j\,dF(j\omega)}{d\omega}$
6	$e^{-at}f(t);\quad a > 0$	$F(s + a)$	$F(j\omega + a)$
7	$f(t - a);\quad a > 0$	$e^{-as}F(s)$	$e^{-j\omega a}F(j\omega)$
8	$f\left(\dfrac{t}{a}\right);\quad a > 0$	$aF(as)$	$aF(aj\omega)$
9[†]	$f(t) + f(-t)$	—	$F(j\omega) + F(-j\omega)$
A	$\delta(t)$[‡]	1	1
B	$u(t)$[§]	$\dfrac{1}{s}$	—
C	t^n	$\dfrac{n!}{s^{n+1}}$	—
D	e^{-at}	$\dfrac{1}{s + a}$	$\dfrac{1}{j\omega + a}$
E	$\sin \alpha t$	$\dfrac{\alpha}{s^2 + \alpha^2}$	—
F	$\cos \alpha t$	$\dfrac{s}{s^2 + \alpha^2}$	—
G	$e^{-at}\sin \alpha t$	$\dfrac{\alpha}{(s + a)^2 + \alpha^2}$	$\dfrac{\alpha}{(j\omega + a)^2 + \alpha^2}$
H	$Ce^{-at}\sin(\alpha t - \phi);$ $C = -\sqrt{a^2 + \alpha^2}$ $\phi = \tan^{-1}\left(\dfrac{\alpha}{a}\right)$	$\dfrac{\alpha s}{(s + a)^2 + \alpha^2}$	$\dfrac{\alpha j\omega}{(j\omega + a)^2 + \alpha^2}$
I	$e^{-a\lvert t\rvert}$	—	$\dfrac{2a}{\omega^2 + a^2}$

*$f(t) = 0$ for $t < 0$ on all lines where $F(s)$ is given. In particular, on line 7, $f(t - a) = 0$ for $(t - a) < 0$, that is, for $t < a$.

[†]In line 9, if $f(t) = 0$ for $t < 0$, then $f(t) + f(-t)$ is *any even function*.

[‡]The impulse function $\delta(t)$ is the limit as $a \to 0$ of the function at right.

[§]The step function $u(t)$ is defined in Example 3.1.

Table 3.1 is such a tabulation, giving examples of both transforms, which are of course essentially the same when both exist. Appendix A contains a more extensive table of Laplace transforms. Some texts contain tables of the Laplace transform that are still more extensive, and consequently easier to use [see, e.g., Gardner and Barnes (1942), Nixon (1965), Roberts and Kaufman (1966), and Holbrook (1966)]. Note however that a great many transform pairs result immediately from Table 3.1 by using one of the numbered lines in combination with one of the lettered lines. For example, a transform pair obtained by using lines 1 and F is

$$f(t > 0) = 2 \cos \alpha t; \qquad F(s) = \frac{2s}{s^2 + \alpha^2} \tag{3.16}$$

As another example, lines 6 and E might be used to get the Fourier transform on line G, which exists even though the Fourier transform of a sine wave does not:

$$f(t > 0) = e^{-at} \sin \alpha t; \qquad F(j\omega) = \frac{\alpha}{(j\omega + a)^2 + \alpha^2} \tag{3.17}$$

Here the Laplace transform is obtained from lines 6 and E and then $j\omega$ is substituted for s.

3.2 TRANSFER FUNCTIONS

The concept of the transfer function has proved to be very helpful in the analysis of systems that in general operate on a set of "input" functions to produce a set of "output" functions. Here we analyze only systems that have a single input $f(t)$ and a single output $g(t)$, the latter being a linear function of $f(t)$ in the sense described below, but the extension to systems where each output $g_i(t)$ is a function of a set of inputs $f_1(t)$, $f_2(t)$, and so on, is straightforward.

The transfer function concept is most generally applicable to *linear, time-invariant systems*. These are systems described by *linear differential equations* with *constant coefficients*. A general form of such an equation is

$$\left[A_n \frac{d^n}{dt^n} + \cdots + A_1 \frac{d}{dt} + A_0 \right] g(t) = \left[B_m \frac{d^m}{dt^m} + \cdots + B_1 \frac{d}{dt} + B_0 \right] f(t) \tag{3.18}$$

in which each term in the square brackets is a differential operator multiplied by a constant coefficient.

Using line 3 of Table 3.1, the transform of Eq. 3.18 can be obtained by replacing d/dt with $j\omega$:

$$[A_n(j\omega)^n + \cdots + A_1 j\omega + A_0]G(j\omega) = [B_m(j\omega)^m + \cdots + B_1 j\omega + B_0]F(j\omega) \tag{3.19}$$

and the resulting *transfer function* is

$$H(j\omega) = \frac{G(j\omega)}{F(j\omega)} = \frac{\sum\limits_{i=0}^{m} B_i(j\omega)^i}{\sum\limits_{i=0}^{n} A_i(j\omega)^i} \tag{3.20}$$

so that $H(j\omega)$ describes the system in terms of the ratio of the transform of the output to the transform of the input, as illustrated in Fig. 3.4.

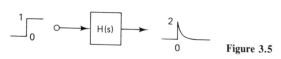

F($j\omega$) ○────▶ H($j\omega$) ────▶ G($j\omega$) = F($j\omega$)H($j\omega$)   **Figure 3.4** System with transfer function $H(j\omega)$.

Note, in line 3 of Table 3.1, that the Laplace transform of Eq. 3.19 is the same as the Fourier transform if initial values are assumed to be zero. Thus a transfer function $H(s)$ can be derived in the same manner by *ignoring the initial values* of $f(t)$, $g(t)$, and their derivatives. Therefore, although as noted above, $F(j\omega)$ and $F(s)$ are functions that are not always the same, in this case when both $H(j\omega)$ and $H(s)$ exist, they are really the same function and $H(j\omega)$ is obtained from $H(s)$ by writing $j\omega$ for s.

Example 3.2

In the system shown in Fig. 3.5, the response to a unit step function at $t = 0$ is $g(t) = 2e^{-at}$. Therefore, for $F(s) = 1/s$, $G(s) = 2/(s + a)$, and the transfer function is $H(s) = 2s/(s + a)$.

1
└┐
 └── ○────▶ H(s) ────▶   2
0   └λ__
                          0   **Figure 3.5**

The most important property of the transfer function of the linear system is that it is a *property only of the system* and does not depend on $f(t)$. This is evident from Eq. 3.20, because the A's and B's clearly do not depend on $f(t)$. In this respect, the transfer function is closely related to the concept of *impedance* in an electrical or mechanical system. The well-known electrical impedance functions, for example, are shown in Table 3.2. Since any linear mechanical system has an electrical analog, the electrical functions will suffice for illustration. The impedance function is the ratio of the transform of the voltage drop across the element to the transform of the current through the element, assuming zero initial values. One need only apply Kirchhoff's law, which states that the algebraic sum of voltage drops in any loop containing these elements along with voltage sources must be zero to derive an appropriate $H(s)$ in the form of Eq. 3.20.

TABLE 3.2. Electrical Impedance Functions

Linear Element		Impedance Function
Resistor	─┤R├─	R
Inductor	─┤L├─	sL
Capacitor	─┤C├─	$\dfrac{1}{sC}$

Example 3.3

The application of Kirchhoff's law to the system shown in Fig. 3.6 yields $(sL + R)I(s) = E_i(s)$, where $I(s)$ is the current transform. Note that the usual convention used here is to

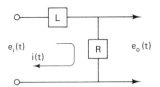

Figure 3.6

give opposite polarity to the input voltage $E_i(s)$. The output voltage transform is $E_o(s) = RI(s)$, and therefore the transfer function is

$$H(s) = \frac{E_o(s)}{E_i(s)} = \frac{R}{sL + R} \qquad (3.21)$$

Example 3.4

In the system illustrated in Fig. 3.7 Kirchhoff's law yields $(sL + 1/(sC) + R)I(s) = E_i(s)$, and in this case the output voltage transform is $E_o(s) = I(s)/(sC)$. Therefore, the transfer function in this case is

$$H(s) = \frac{E_o(s)}{E_i(s)} = \frac{1}{LCs^2 + RCs + 1} \qquad (3.22)$$

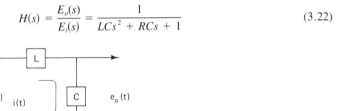

Figure 3.7

The circuits in the examples above are sometimes called *filters* because in general they allow some of the frequency components of $e_i(t)$ to "pass through" while attenuating other components. The characteristics of two common types of filters, the *lowpass filter* and the *bandpass filter,* are illustrated in Fig. 3.8.

The magnitude of the transfer function, $|H(j\omega)|$, is known as the *amplitude response* function of the filter, and the corresponding angle of $H(j\omega)$ is called the *phase shift*. That is, if $R(\omega)$ and $I(\omega)$ are the real and imaginary components of $H(j\omega)$, then

$$H(j\omega) = R(\omega) + jI(\omega) \qquad (3.23)$$

and

$$\text{amplitude response } |H(j\omega)| = [R^2(\omega) + I^2(\omega)]^{1/2} \qquad (3.24)$$

$$\text{phase shift } \theta(j\omega) = \tan^{-1}\left[\frac{I(\omega)}{R(\omega)}\right] \qquad (3.25)$$

Any transfer function $H(s)$ is a *rational* function of s (i.e., a ratio of polynomials in s), as suggested by Eq. 3.20. Therefore, $H(s)$ can in general be factored into various forms and illustrated as such using block diagrams. For example, the case where $H(s)$ is a simple product is illustrated on the left in Fig. 3.9. In this case, $H(s)$ is the product of $H_1(s)$ and $H_2(s)$. The figure implies an intermediate function $x(t)$, which is the output of the first block and the input to the second block.

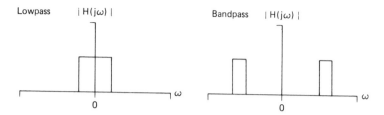

Figure 3.8 Lowpass and bandpass filter characteristics.

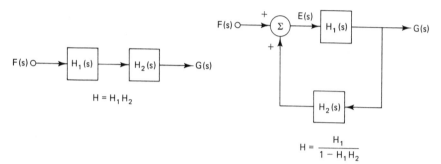

Figure 3.9 Block diagrams of transfer functions.

A simple linear *feedback system* is illustrated on the right in Fig. 3.9. In the feedback system, the output signal $g(t)$ is fed back through a portion of the system and added to the input signal $f(t)$, as suggested by the figure. Given $H_1(s)$ and $H_2(s)$, the overall transfer function is found by noting that

$$G(s) = H_1(s)E(s) \tag{3.26}$$

and

$$E(s) = F(s) + H_2(s)G(s) \tag{3.27}$$

where $E(s)$ is the intermediate signal in Fig. 3.9, sometimes called the "error signal." After eliminating $E(s)$ from Eqs. 3.26 and 3.27, the result can be written

$$H(s) = \frac{G(s)}{F(s)} = \frac{H_1(s)}{1 - H_1(s)H_2(s)} \tag{3.28}$$

As will be shown, the analysis of digital filtering and feedback is closely related to the above, and the transfer function concept remains applicable throughout.

3.3 CONVOLUTION

Closely related to the transfer function concept is the idea of convolution, which is especially useful in understanding the behavior of linear systems. The convolution integral is derived from the inverse transform of the transfer function in the following way.

Let the input to a system with the transfer function $H(s)$ be the impulse function at $t = 0$, $\delta(t)$. According to line A of Table 3.1 the input transform is then $F(s) = 1$. Therefore, the output is

$$G(s) = F(s)H(s) = H(s)$$

and

$$g(t) = h(t) \qquad (3.29)$$

Thus the inverse transform of the transfer function, $h(t)$, is called the *impulse response* of the system, being the response to a unit impulse at $t = 0$.

The *convolution integral* is now defined as follows:

$$f(t) * h(t) = \int_{-\infty}^{t} f(\tau)h(t - \tau)\, d\tau \qquad (3.30)$$

where the star denotes the convolution operation and $h(t)$ is the impulse response function described above. This operation is equivalent to a *multiplication* operation in the frequency or s domain.

A demonstration of the latter statement proceeds as follows: (Fourier transforms are used here; Laplace would do just as well.) First, since $h(t)$ is the response to $\delta(t)$ which occurs at $t = 0$, it is assumed that $h(t) = 0$ for $t < 0$. Therefore, in the integrand in Eq. 3.30, $h(t - \tau)$ must be zero for $\tau > t$, and so infinity can be used in place of t for the upper limit of integration:

$$f(t) * h(t) = \int_{-\infty}^{\infty} f(\tau)h(t - \tau)\, d\tau \qquad (3.31)$$

The *transform of the convolution* is now

$$\int_{-\infty}^{\infty} [f(t) * h(t)]e^{-j\omega t}\, dt = \iint_{-\infty}^{\infty} f(\tau)h(t - \tau)e^{-j\omega t}\, d\tau\, dt \qquad (3.32)$$

When the substitution $x = t - \tau$ is made, Eq. 3.32 becomes the product of two integrals, that is,

$$\int_{-\infty}^{\infty} [f(t) * h(t)]e^{-j\omega t}\, dt = \iint_{-\infty}^{\infty} f(\tau)h(x)e^{-j\omega \tau}e^{-j\omega x}\, d\tau\, dx$$

$$= \int_{-\infty}^{\infty} f(\tau)e^{-j\omega \tau}\, d\tau \int_{-\infty}^{\infty} h(x)e^{-j\omega x}\, dx \qquad (3.33)$$

$$= F(j\omega)H(j\omega)$$

Thus proving that *the convolution* f(t) * h(t) *and the product* F(jω)H(jω) *are a transform pair.*

From the proof above it follows immediately that since the product $F(j\omega)$ $H(j\omega)$ can be commuted, so can the convolution be commuted, and for any linear system,

$$\begin{aligned} g(t) &= h(t) * f(t) \\ &= f(t) * h(t) \end{aligned} \qquad (3.34)$$

Example 3.5

Suppose, in the system illustrated in Fig. 3.10 that the response, $h(t)$, to a unit impulse is a unit step, so that the transfer function is (from Table 3.1) $H(s) = 1/s$. (Such a system is called a "sample-and-hold" system or a "perfect integrator," for obvious reasons.) Sup-

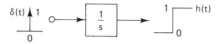

Figure 3.10

pose that the response to an input $f(t) = e^{-at}$ is now desired. Of course, the response is easily obtained from Table 3.1 by noting that $F(s) = 1/(s + a)$, and therefore $G(s) = F(s)H(s) = 1/[s(s + a)]$, and using lines 4 and D of that table gives us

$$g(t) = \int_0^t e^{-a\tau} d\tau = \frac{1}{a}(1 - e^{-at}) \tag{3.35}$$

(Note that the lower limit of integration is zero because the integrand is zero for $\tau < 0$.) The convolution integral in this case is illustrated in Fig. 3.11. As shown, the integrand is the product of $f(\tau)$ and $h(t - \tau)$, the latter beginning at $\tau = t$ and proceeding "backwards" over the τ-axis. The convolution integral is thus equal to the shaded area in the figure, that is,

$$f(t) * h(t) = \int_0^t e^{-a\tau}(1) d\tau$$

$$= \frac{1}{a}(1 - e^{-at}) \tag{3.36}$$

exactly as in Eq. 3.35.

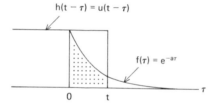

Figure 3.11

3.4 POLE–ZERO PLOTS

Equation 3.20 gives $H(j\omega)$ or $H(s)$ as a rational function, that is, as a ratio of polynomials:

$$H(s) = \frac{B_0 + B_1 s + \cdots + B_n s^n}{A_0 + A_1 s + \cdots + A_m s^m} \tag{3.37}$$

In principle the polynomials can always be factored, and $H(s)$ can be written in the pole–zero form:

$$H(s) = C\frac{(s - q_1)(s - q_2)\cdots(s - q_n)}{(s - p_1)(s - p_2)\cdots(s - p_m)} \tag{3.38}$$

The constant C in front of the latter ratio is equal to B_n/A_m, and the p's and q's depend on the A's and B's, respectively. The roots of the numerator, that is, $q_1, \ldots, q_n$, are the *zeros* of $H(s)$, and $p_1, \ldots, p_m$ are the *poles* of $H(s)$. The poles and zeros of $H(s)$ may be real, complex, or zero, as they may be for any polynomial function.

A plot of the poles and zeros of $H(s)$ on the complex s-plane provides a geometric interpretation of the amplitude response and the phase shift, which were defined in Eqs. 3.24 and 3.25. Given $H(j\omega)$ in the form of Eq. 3.38, the amplitude response must be

$$|H(j\omega)| = \left| C\frac{(j\omega - q_1)\cdots(j\omega - q_n)}{(j\omega - p_1)\cdots(j\omega - p_m)}\right| \tag{3.39}$$

$$= |C|\frac{|j\omega - q_1|\cdots|j\omega - q_n|}{|j\omega - p_1|\cdots|j\omega - p_m|} \tag{3.40}$$

and each component of the form $|j\omega - x|$ can be interpreted as a *distance* on the s-plane from the point given by ω on the imaginary axis to the point x. Thus the amplitude response becomes a ratio of products of distances on the s-plane. Similarly, the total phase shift becomes an algebraic sum of angles. These properties are best illustrated by example.

Example 3.6

Plot the poles and zeros of

$$H(s) = \frac{Cs}{s^2 + 2s + 5} = \frac{Cs}{(s + 1)^2 + 2^2} \tag{3.41}$$

This transfer function has a single zero at $s = 0$ and poles at $s = -1 \pm j2$. These, along with the amplitude response and phase shift, are shown in Fig. 3.12. Note how the amplitude and phase of $H(j\omega)$ change as the operating point moves up the $j\omega$-axis on the s-plane. Computations are given for $\omega = 3$ in the figure. The constant C in Eq. 3.41 does not affect the pole–zero plot, and thus only the normalized amplitude characteristic, $|H(j\omega)|/C$, can be determined from the pole–zero plot.

The s-plane plot is useful in the design as well as the analysis of linear systems. In the design process, one thinks in terms of placing the poles and zeros on the

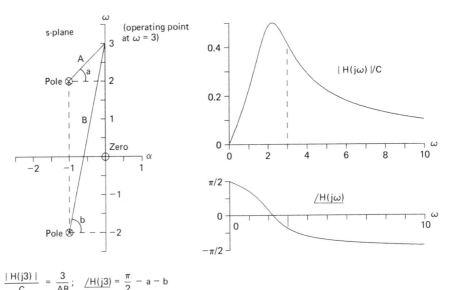

Figure 3.12 Pole–zero plot, gain, and phase shift of $H(s) = Cs/(s^2 + 2s + 5)$.

s-plane to achieve certain objectives, with $H(s)$ then being found from the equivalent of Eq. 3.38. In connection with this, the s-plane plot has several properties worth noting:

1. Poles and zeros must be real or must appear in conjugate pairs.
2. A pole (zero) on the $j\omega$-axis implies that $|H(j\omega)|$ is infinite (zero) at some specific frequency.
3. A pole in the right half plane causes instability in the sense that the response to a transient input signal will increase rather than decay.
4. Poles not on the real axis in general cause oscillation in the output.

The first of these properties follows from the fact that the coefficients in the numerator and denominator of $H(s)$ are real. The p's and q's in Eq. 3.38 must be real or in conjugate pairs for the A's and B's to be real. The second property results from having the distance from the operating point to a pole or zero decrease to zero at a particular operating frequency, as it does at $\omega = 0$ in Fig. 3.6.

The instability suggested by property 3 can be demonstrated by expanding $H(s)$ in Eq. 3.38 into a sum of partial fractions. If $H(s)$ is a proper fraction and there are no multiple poles (i.e., if all p's are different), then the form of the expansion is

$$H(s) = \frac{C_1}{s - p_1} + \frac{C_2}{s - p_2} + \cdots + \frac{C_m}{s - p_m} \tag{3.42}$$

If one of the poles, say $p_n = \alpha_n + j\omega_n$, lies in the right-half plane so that α_n is positive, then its contribution to the impulse response $e^{(\alpha_n + j\omega_n)t}$, will increase without bound, in effect causing the transient response to be unstable. Also, as stated in property 4, if p_n is complex so that $\omega_n \neq 0$, then the impulse response contains the oscillatory factor $e^{j\omega_n t}$.

The main properties of any linear transfer function can thus be observed by finding the locations of the poles and zeros on the s-plane.

EXERCISES

1. Explain the reason for the statement, "If $f(t)$ has a Fourier series, then it has no Fourier transform, and vice versa."
2. Discuss the effects of the differences between the Laplace and Fourier transforms.
3. Prove that if $f(t)$ has a Fourier transform, then $Af(t)$, where A is a constant, also has a Fourier transform.
4. Prove lines 3 and 5 of Table 3.1 for the Laplace transform.
5. Prove lines 6 and 7 of Table 3.1 for the Fourier transform.
6. Derive the Fourier transform in line I of Table 3.1, using lines 9 and D.
7. Using lines 1, 5, and D of Table 3.1, find and sketch the amplitude spectrum of $f(t) = 3te^{-2t}u(t)$.
8. Find the inverse Laplace transform of $F(s) = 1/[(s + a)(s + b)]$ by expanding $F(s)$ into partial fractions so that $F(s) = A/(s + a) + B/(s + b)$.
9. Find the Laplace transform of the infinite train of unit impulse functions at $t = 0, T, 2T, \ldots$. *Hint:* Use lines 7 and A of Table 3.1.

10. Sketch the amplitude spectrum of the decaying sinusoid given by $f(t) = e^{-at} \cos \alpha t \, u(t)$ by letting $(a, \alpha) = (1, 1), (2, 1), (1, 2), (2, 2)$. Discuss the effect of changing a and α.

11. If $g(t) = Ae^{-at}$ is the response (for $t \geq 0$) to a unit impulse function input at $t = 0$, what is the transfer function?

12. What is the output $g(t)$ if the transfer function is $H(j\omega) = 1/(j\omega - \omega^2)$ and a unit impulse is inserted at $t = 0$?

13. Given the transfer function $H(s) = 1/(s^2 + as + b)$, show that the response to an impulse function will oscillate if and only if the function $s^2 + as + b$ has complex roots. *Hint:* Use partial fractions and line D of Table 3.1.

14. Derive the overall transfer function, $G(s)/F(s)$, of the network shown.

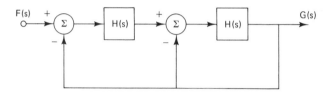

15. What is the form of $H(s)$ if the response to a unit step function is a decaying sine wave?

16. Using the convolution integral, find the output if the input is $f(t) = e^{-2t}$ for $t \geq 0$ and the transfer function is $H(j\omega) = 1/(j\omega + 1)$.

17. Compute the convolution $t^2 u(t) * e^{-2t}u(t)$ directly and then show that the result is correct by taking a product of Laplace transforms.

18. Using lines 3 and E of Table 3.1, derive line F.

19. The rectangular pulse shown in the figure and its Fourier transform are an important transform pair. What is $F(j\omega)$ in this case?

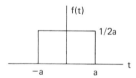

20. (Dual of Exercise 19.) What is $f(t)$, given $F(j\omega)$ in the figure?

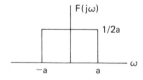

SOME ANSWERS

7. $3/(\omega^2 + 4)$ 8. $(e^{-at} - e^{-bt})/(b - a)$ 9. $1/(1 - e^{-Ts})$
10. $|F(j\omega)| = (\omega^2 + a^2)^{1/2}/[(a^2 + \alpha^2 - \omega^2)^2 + 4a^2\omega^2]^{1/2}$ 11. $A/(s + a)$ 12. $1 - e^{-t}$
14. $H^2/(H^2 + H + 1)$ 15. $H(s) = A\alpha s[(s + a)^2 + \alpha^2]^{-1}$ 16. $e^{-t} - e^{-2t}$
19. $(\sin a\omega)/a\omega$ 20. $(\sin at)/2\pi at$

REFERENCES

CHESTNUT, H. H., and MAYER, R. W., *Servomechanisms and Regulating System Design*. New York: Wiley, 1959.

CHURCHILL, R. V., *Introduction to Complex Variables and Applications*, Chap. 7. New York: McGraw-Hill, 1948.

DYM, H., and MCKEAN, H. P., *Fourier Series and Integrals*. New York: Academic Press, 1972.

GARDNER, M. F., and BARNES, J. L., *Transients in Linear Systems*, Vol. 1. New York: Wiley, 1942.

GOLDMAN, S., *Transformation Calculus and Electrical Transients*, Chap. 7. New York: Prentice-Hall, 1949.

HOLBROOK, J. G., *Laplace Transforms for Electronic Engineers*. Oxford: Pergamon Press, 1966.

MCCOLLUM, P. A., and BROWN, B. F., *Laplace Transform Tables and Theorems*. New York: Holt, Rinehart and Winston, 1965.

NIXON, F. E., *Handbook of Laplace Transformation*. Englewood Cliffs, N.J.: Prentice-Hall, 1965.

PAPOULIS, A., *The Fourier Integral and Its Applications*. New York: McGraw-Hill, 1962.

ROBERTS, G. E., and KAUFMAN, H., *Table of Laplace Transform Pairs*. Philadelphia: W. B. Saunders, 1966.

SEELEY, R. T., *An Introduction to Fourier Series and Integrals*. New York: W. A. Benjamin, 1966.

SPIEGEL, M. R., *Theory and Problems of Laplace Transforms*. New York: McGraw-Hill (Schaum's Outline Series), 1965.

THALER, G. J., and BROWN, R. G., *Servomechanism Analysis*. New York: McGraw-Hill, 1953.

TRUXALL, J. G., *Control System Synthesis*. New York: McGraw-Hill, 1955.

WEBER, E., *Linear Transient Analysis*, Vols. 1 and 2. New York: Wiley, 1954.

4

Sampling and Measurement of Signals

4.1 INTRODUCTION

This chapter deals with the generation or measurement of samples of continuous signals, and serves as an introduction to later chapters in which the processing of digitized signals is discussed. As mentioned in Chapter 1, many types of signal processing systems tend to take the form of Fig. 4.1, due to the speed and versatility of the digital processor. The continuous signal $f(t)$ is sampled to produce the set of numbers $[f_m]$, and the digital processor processes $[f_m]$ and produces an output sample set, $[g_m]$.

Figure 4.1 suggests the actual measurement of a continuous physical quantity $f(t)$, which might be a voltage, current, displacement, shaft angle, or some similar function of time. Figure 4.2 shows an essentially equivalent system that suggests the generation of the sample set $[f_m]$ by some digital process, such as the execution of a computer program. The entire process in Fig. 4.2 could take place within a general-purpose digital computer. The important consideration, mentioned in Chapter 1 and emphasized again here, is that most of the discussion in this chapter, as well as the methods of signal analysis in general, can be applied without regard to the origin of the sample set, or by assuming that $[f_m]$ represents samples of a continuous signal whether it does or not.

Later chapters will show how the digital processor in Figs. 4.1 and 4.2 can effect a linear transfer function just as described in Chapter 3, and how a continuous

Figure 4.1   Digital signal processing.

Figure 4.2   System equivalent to Fig. 4.1.

function $g(t)$ can be reconstructed from the sample set $[g_m]$. Here the emphasis is on some basic properties of the sample set $[f_m]$, and on the information that $[f_m]$ conveys about the original signal $f(t)$. The most important concept concerning the latter is discussed in the next section.

4.2 THE SAMPLING THEOREM

Suppose that the function $f(t)$ is sampled at intervals of T seconds as shown in Fig. 4.3. Assume for now that the samples of $f(t)$ are *impulse samples,* with f_m being equal to $f(mT)$ at regular intervals of T seconds. As the term implies, f_m can be created by multiplying $f(t)$ by the unit impulse function, $\delta(t - mT)$. Figure 4.3 is descriptive of a fairly large class of sampling situations, and the most important question arising from these situations is concerned with the *sampling interval T.*

How small must T be in order to reconstruct $f(t)$, given only the sample set $[f_m]$? Without placing any restrictions on $f(t)$ between the sample points or providing additional information about $f(t)$, the answer is clearly that T must be zero. As suggested by Fig. 4.4, it is always possible to construct different continuous functions (with continuous derivatives also) that pass through all sample points, if T is greater than zero.

In other words, some sort of restriction must be placed on $f(t)$ to describe $f(t)$ completely using a finite set of numbers, $[f_m]$. The *sampling theorem,* which places these restrictions on the *frequency content* of $f(t)$, can be stated as follows:

To be able to recover $f(t)$ exactly, it is necessary to sample $f(t)$ at a rate greater than twice its highest frequency.

If $f(t)$ is a periodic signal, then the validity of this statement follows from Chapter 2, where the lower bound

$$N \geq 2K + 1 \tag{4.1}$$

was established, in which N was the number of samples per fundamental period and K was the highest harmonic in the Fourier series for $f(t)$. If N is the number of

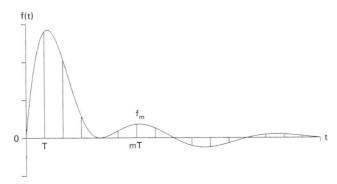

Figure 4.3   Impulse samples of $f(t)$.

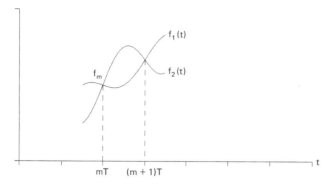

Figure 4.4 Functions with the same samples.

samples per fundamental period and if a period consists of p seconds, then the sampling rate must be N/p samples per second, and the highest frequency component in the Fourier series representation of $f(t)$ must be K/p hertz. Therefore, Eq. 4.1 is really an expression of the sampling theorem for periodic signals.

Note that the sampling rate must be *greater* than twice the highest frequency in $f(t)$ and not equal to the latter. Sometimes the sampling theorem is erroneously stated on this point. Figure 4.5 illustrates a situation in which a sine wave is sampled at a rate equal to exactly twice its (highest) frequency ν, and $f(t)$ obviously cannot be discovered from the sample set, because all sample values are zero! Figure 4.6 illustrates a similar situation, with the sampling rate somewhat higher than 2ν. In this example the sample set repeats at intervals of $8T$ seconds, and there are $N = 8$ samples within each of these intervals. (Note how this repetition period would increase if T were made to approach $1/2\nu$, and how the complete sample set would require increasing values of N.) In the example of Fig. 4.6, there are 3 cycles of $f(t)$ in each interval of $8T$; therefore, $K = 3$, $N = 8$, Eq. 4.1 is satisfied, and the Fourier coefficients of $f(t)$ can be found exactly from the sample set (e.g., see Eq. 2.39). *No other function* of t has this same sample set and is limited to frequencies less than $1/2T$ hertz, where T is the sampling interval.

Figure 4.7, on the other hand, illustrates the ambiguity when the sampling rate is somewhat *lower* than 2ν, that is, T is greater than $1/2\nu$. The samples of $f(t)$ are the same as those of $g(t)$, another sinusoid with a frequency *less than* ν, so, given the sample set as shown, obviously both $f(t)$ and $g(t)$ are possible reconstructions.

This type of ambiguity, caused by violating conditions of the sampling theorem, can be expressed more generally in the following way: As in the figures above,

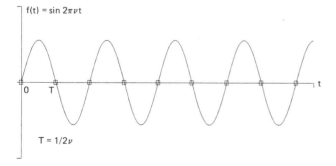

Figure 4.5 Sampling rate = twice highest frequency in $f(t)$.

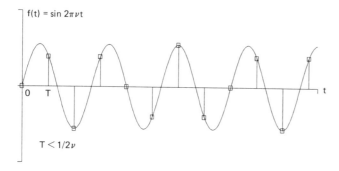

Figure 4.6 Sampling rate > twice highest frequency in $f(t)$.

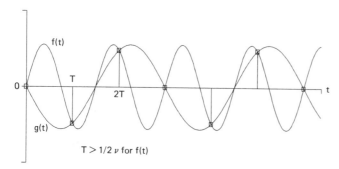

Figure 4.7 Sampling rate < twice highest frequency in $f(t)$.

there is no loss of generality from the present discussion in representing $f(t)$ by a single sinusoid. Therefore, let the sampled function be

$$f(t) = A \sin(2\pi\nu t + \alpha) \tag{4.2}$$

Next, let the sampling interval be T seconds as in Figs. 4.6 and 4.7, so that the sample set of $f(t)$ is given by

$$\text{sample set} = [f_m] = [A \sin(2\pi\nu mT + \alpha)] \tag{4.3}$$

Using the identity $\sin x = \sin(x + 2k\pi)$, where k is any integer, an equivalence between sample sets can be established as follows:

$$[A \sin(2\pi\nu mT + \alpha)] = [A \sin(2\pi\nu mT + \alpha + 2k\pi]$$

$$= [-A \sin(-2\pi\nu mT - \alpha + 2k\pi)] \tag{4.4}$$

$$= \left[\pm A \sin\left(2\pi\left(\pm\nu + \frac{n}{T}\right)mT \pm \alpha\right)\right]$$

where n is any integer so that $mn = k$ is an integer. Equation 4.4, stated in words, is:

Given a sampling interval of T seconds, sinusoidal components at ν and $\pm\nu + n/T$ hertz, n being any integer, are indistinguishable (i.e., have the same sample values).

Sec. 4.2   The Sampling Theorem

Thus Eq. 4.4 amounts to a confirmation of the sampling theorem. Its effect is illustrated in Fig. 4.8, which shows how the frequency domain is in effect "folded" by the sampling process. On the left in Fig. 4.8 is a diagram showing the equivalence between the frequencies $\nu_1 = -1/3T$ and $\nu_2 = 2/3T$, which is the specific situation used for Fig. 4.7. On the right in Fig. 4.8 is a more general diagram that illustrates the folding produced by Eq. 4.4. For each possible signal component at a point marked X, there are components above and below, at other frequencies marked X, having identical sample sets.

One-half of the sampling frequency (i.e., $1/2T$ hertz or π/T rad/s) is called the *folding frequency* for the reason illustrated in Fig. 4.8. It is also often called the *aliasing frequency,* as explained in Chapter 5.

The sampling theorem holds whether or not $f(t)$ is periodic. When $f(t)$ is not periodic (e.g., a transient pulse), its frequency content is expressed in terms of $F(j\omega)$, the Fourier transform of $f(t)$. Since the Fourier transform has already been introduced as a limiting form of the Fourier series, the result above should be expected. The proof of the sampling theorem for aperiodic signals proceeds as follows:

Suppose that the aperiodic signal $f(t)$ is limited to frequencies less than π/T rad/s. Then, starting with Eq. 3.4, $f(t)$ can be expressed as

$$f(t) = \frac{1}{2\pi} \int_{-\infty}^{\infty} F(j\omega)e^{j\omega t}\, d\omega$$

$$= \frac{1}{2\pi} \int_{-\pi/T}^{\pi/T} F(j\omega)e^{j\omega t}\, d\omega \qquad (4.5)$$

The samples of $f(t)$ then take the form

$$f_m = f(mT) = \frac{1}{2\pi} \int_{-\pi/T}^{\pi/T} F(j\omega)e^{jm\omega T}\, d\omega \qquad (4.6)$$

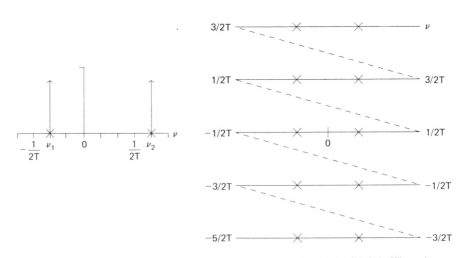

Figure 4.8 Folding of the frequency domain. Components at frequencies labeled "X" can have the same sample sequences.

The latter form is reminiscent of the complex Fourier coefficient formula, Eq. 2.42 in Chapter 2, and is in fact the same if multiplied by T. The quantities ω, T, m, and F here replace t, ω_0, $-k$, and f_p in the Chapter 2 formula. Thus,

> If $F(j\omega)$ is limited to the interval $|\omega| < \pi/T$, then $F(j\omega)$ can be described by a Fourier series within this interval, and Tf_m is the mth coefficient.

The reader might at first question whether a complex function like $F(j\omega)$ can have a Fourier series, since only real functions were treated in Chapter 2. However, there is nothing in the coefficient formula or its derivation to prevent the function from being complex. The only difference is that, where the coefficients c_m and c_{-m} were complex conjugates in Chapter 2, here Tf_m and Tf_{-m} are real and, in general, unequal. In fact, if they are equal, then $F(j\omega)$ is real.

The statement above, that $[Tf_m]$ is the set of Fourier coefficients for $F(j\omega)$ in the interval $|\omega| < \pi/T$, proves the sampling theorem for aperiodic functions. If $f(t)$ can be recovered from $F(j\omega)$, and if $F(j\omega)$ can be constructed using $[Tf_m]$, then clearly $f(t)$ can be recovered from its sample set $[f_m]$.

The next chapter contains further discussion of the sampling theorem and its effects on frequency measurement and waveform reconstruction. Credit for first discovering the theorem and its importance in these areas should be given to E. T. Whittaker, who published the sampling theorem in a remarkable paper (Whittaker, 1915) that lays much of the groundwork for modern digital signal processing. The paper includes, for example, a famous formula for reconstructing the waveform from its sample set, which is discussed in the next chapter.

4.3 SAMPLED-DATA SYSTEMS

The "front-end" of a linear system with a continuous input signal $f(t)$ generally falls into one of the three categories depicted in Fig. 4.9. Both of the two lower schemes include a sampling process to which the sampling theorem is applicable. The digital system in the center is the most widely useful of the three, and the chapters ahead

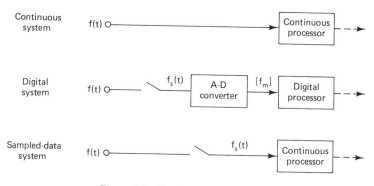

Figure 4.9 Continuous input schemes.

apply mainly to this type of system and its relations to the purely continuous system at the top of the figure.

The *sampled-data system* at the bottom of Fig. 4.9 represents, however, a third category important in the field of automatic control. Note that the sampled-data system is not a digital system. It consists of a continuous system processing samples of $f(t)$. The sampler does not digitize the signals; instead it produces continuous samples, as suggested in Fig. 4.10.

The "continuous processor" in Fig. 4.9 then operates on $f_s(t)$ and produces a continuous output. Thus the sampled-data system is continuous (rather than digital) and can be analyzed using Fourier or Laplace transforms. For comprehensive treatments of the analysis and design of sampled-data systems, see Truxal (1955), Ragazzini and Franklin (1958), Tou (1959), Monroe (1962), or Gibson (1963). In its most basic form, this analysis proceeds as follows.

Suppose that $f_s(t)$ is composed of narrow samples that may be treated as *impulse samples* of $f(t)$. [The case of nonimpulse sampling is discussed in Chapter 5. (see also Section 4.4).] Then, as suggested above, the sample $f_m = f(mT)$ can be obtained by multiplying $f(t)$ by the appropriate unit impulse function, $\delta(t - mT)$. The sample train $f_s(t)$ is then composed of the entire sample set, that is,

$$f_s(t) = \sum_{m=-\infty}^{\infty} f(t)\delta(t - mT) \tag{4.7}$$

Assuming that $f(t)$ starts at $t = 0$, the Laplace transform of $f_s(t)$ can be found as follows:

$$
\begin{aligned}
F_s(s) &= \int_0^\infty f_s(t)e^{-st}\,dt \\
&= \int_0^\infty \sum_{m=-\infty}^{\infty} f(t)\delta(t - mT)e^{-st}\,dt \\
&= \sum_{m=-\infty}^{\infty} \int_0^\infty f(t)e^{-st}\delta(t - mT)\,dt \\
&= \sum_{m=0}^{\infty} f_m e^{-msT}
\end{aligned}
\tag{4.8}
$$

(Each integrand in the third line is nonzero only at $t = mT$, so the final line follows.) The form of this transform has significance beyond sampled-data systems, and is the

Figure 4.10   Signal $f_s(t)$ produced by sampler.

subject of the next chapter. To obtain the output of the continuous processor in Fig. 4.9, $F_s(s)$ is multiplied by the continuous transfer function, just as in the preceding chapter. This process is illustrated in the following simple example.

Example 4.1

Find the response, $g(t)$, of the system illustrated in Fig. 4.11. This particular transfer function is often used to smooth the sampler output $f_s(t)$, and thus make $g(t)$ an approximation to $f(t)$. In this example $f(t)$ is a step function occurring, say, just before $t = 0$, so that the sampled function $f_s(t)$ is as shown. From Eq. 4.8, the transform of the latter is

$$F_s(s) = \sum_{m=0}^{\infty} f_m e^{-msT}$$

$$= \sum_{m=0}^{\infty} e^{-msT} \qquad (4.9)$$

The output transform, $G(s)$, is now

$$G(s) = H(s)F(s) = aT \sum_{m=0}^{\infty} \frac{e^{-msT}}{s + a} \qquad (4.10)$$

Lines 1, 2, 7, and D of Table 3.1 can be used to find the inverse transform of Eq. 4.10. Note that each term in the sum, $e^{-msT}/(s + a)$, inverts to the exponential function e^{-at} for $t > 0$, shifted by the amount mT. Since the unshifted function is zero for $t < 0$, the shifted version must be zero for $t < mT$. Therefore, an expression for $g(t)$ following from Eq. 4.10 is

$$g(t) = aT \sum_{m=0}^{\infty} u(t - mT)e^{-a(t-mT)}$$

$$= aTe^{-at} \sum_{m=0}^{\infty} u(t - mT)e^{maT} \qquad (4.11)$$

To obtain a nicer expression for $g(t)$, assume that t is in the nth sampling interval, between $(n - 1)T$ and nT. Then,

$$g[(n - 1)T < t < nT] = aTe^{-at} \sum_{m=0}^{n-1} e^{mat}$$

$$= aTe^{-at} \frac{1 - e^{nat}}{1 - e^{aT}} \qquad (4.12)$$

When $t = nT^-$ (i.e., just prior to nT),

$$g_n = aTe^{-naT} \frac{1 - e^{naT}}{1 - e^{aT}}$$

$$= \frac{aT}{e^{aT} - 1}(1 - e^{-naT}) \qquad (4.13)$$

Normalized plots of $g(t)$, illustrating the smoothing effect of the transfer function with two different sampling intervals, are shown in Fig. 4.12.

Figure 4.11 Sampled-Data System for Example 4.1.

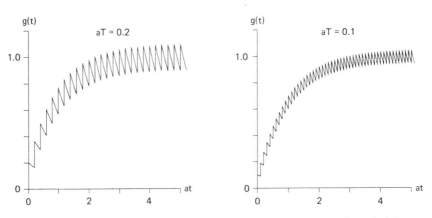

Figure 4.12 Sampled-data system output with $aT = 0.2$ and $aT = 0.1$ (Example 4.1).

4.4 ANALOG-TO-DIGITAL CONVERSION

The fundamental difference between the sampler, discussed above, and the analog-to-digital converter is illustrated in Fig. 4.9. The ADC *digitizes* the waveform samples and provides input to a digital processor as opposed to a continuous processor. A typical digital control system with several inputs and a single output is shown in Fig. 4.13. The system might, for example, be a chemical process control system, an autopilot, or similar system requiring several sensors. The sensor lines are swept in sequence by an analog time-division multiplexer, sampled, and converted to a sequence of encoded numbers by the ADC. As the digitized signals are processed, the DAC produces a continuous control signal from the processor output.

ADC designs take different forms, depending on speed and accuracy requirements. There are textbooks devoted to the design of ADCs and DACs [e.g., Hoeschele (1968) or Schmid (1970)]. The subject of design is beyond the scope of the present discussion; however, Fig. 4.14 gives a simple illustration of one method of analog-to-digital conversion. There is a comparator that causes an n-bit binary counter to count either up or down so that the counter contents, when converted back through the DAC, reflect the current value of $f(t)$. The sample set $[f_m]$ is assembled by reading the contents of the counter every T seconds.

As suggested by Fig. 4.14, each sample of $f(t)$ is a word conveying n bits of information about the value of $f(t)$ at a particular time. Thus, since $f(t)$ is assumed

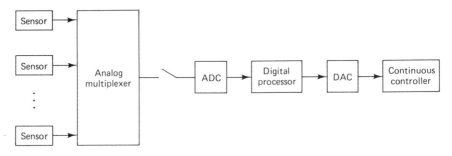

Figure 4.13 Typical digital control system.

Sampling and Measurement of Signals   Chap. 4

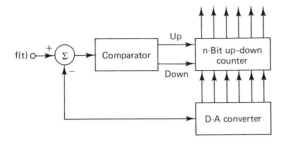

Figure 4.14 Simple A-D converter design.

to vary continuously, there is always a *quantizing error,* as illustrated in Fig. 4.15. As shown, the maximum quantizing error is $q/2$, where q is the value of the least-significant counter bit. (This of course assumes that the ADC comes as close as possible to the correct sample value.) For example, if the range of a 10-bit converter ($n = 10$) represents 0 to 10.24 V, then a single count (q) represents $10.24/2^{10} = 0.01$ V and the maximum quantizing error ($q/2$) is 0.005 V.

Generally, the quantizing error e_m is treated as a random error described in terms of its probability density function. Chapter 13 discusses random functions and shows, for example, that if all values of e_m between $-q/2$ and $q/2$ are equally likely, then

$$\text{average value of } e_m = 0$$

$$\text{rms value of } e_m = \frac{q}{\sqrt{12}} \tag{4.14}$$

When $f(t)$ contains energy at very high frequencies and the sampling rate, $1/T$, is correspondingly high, the *tracking error* becomes a basic limitation to the performance of an ADC. High-speed electronic converters do not generally take the form

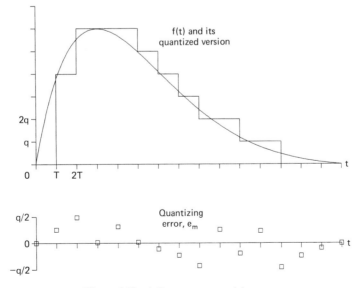

Figure 4.15 A-D converter quantizing error.

of Fig. 4.14 [see Benima (1973), for example], but they must somehow acquire some of the energy in $f(t)$ each time a sample is taken. Therefore, as in Fig. 4.16, an equivalent input series resistance and shunt capacitance are basic to the ADC. A simple estimate of the tracking error in Fig. 4.16 can be obtained as follows: Let $e_i(t)$ be a ramp function with slope e_i' and assume that the sampling switch is closed at $t = 0$, at which time e_i and e_o are equal. Then, from Section 3.2,

$$E_i(s) - E_o(s) = \frac{e_i'}{s^2} - \frac{(e_i'/s^2)(1/sC)}{R + 1/sC}$$

$$= \frac{e_i'}{s(s + 1/RC)}$$

Therefore,

$$
\begin{aligned}
\text{tracking error} &= e_i(t) - e_o(t) \\
&= RCe_i'(1 - e^{-t/RC})
\end{aligned}
\tag{4.15}
$$

That is, if the sampling switch is left closed in order to track e_i to its new value, the tracking error grows toward a steady-state difference equal to RCe_i'.

The question of whether or not the ADC can be treated as an impulse-sampler must also be considered. Sometimes (e.g., in the case of multiplexing) it is convenient to view the converter as a device that examines the continuous function $f(t)$ during a *sampling window* of width w extending from $t = mT - w/2$ to $mT + w/2$, as in Fig. 4.17. Then, if the change in $f(t)$ is insignificant during the window, impulse sampling can be assumed as in Section 4.3. If the pulses in $f_s(t)$ are not almost flat, on the other hand, then the conversion method must be examined more closely. The width w has related effects on the spectrum of the sample set $[f_m]$ and the spectrum of $f_s(t)$. These effects are discussed in Chapter 5.

It is also possible for a *jitter error* to occur in the analog-to-digital process. With this type of error, the sampling window is centered over a time $mT + \mu_m$ instead of mT; that is, the window is offset by the amount μ_m, as in Fig. 4.18. Papoulis (1966) has shown that an independent jitter error is similar to the quantizing error discussed above. From Fig. 4.18, if the rms value of μ_m is σ_μ and if the slope $f'(t)$ is effectively constant over the jitter interval, then

$$\text{rms jitter error} \leq \sigma_\mu \times \text{maximum value of } f'(t) \tag{4.16}$$

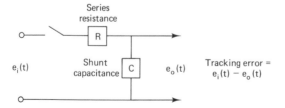

Figure 4.16 Equivalent circuit for estimating the tracking error in an A-D converter.

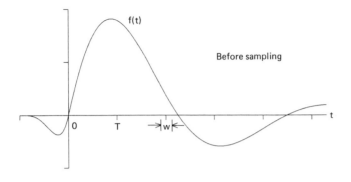

f(t)

Before sampling

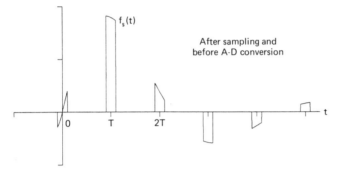

f_s(t)

After sampling and
before A-D conversion

Figure 4.17 Sampling with a window of width w.

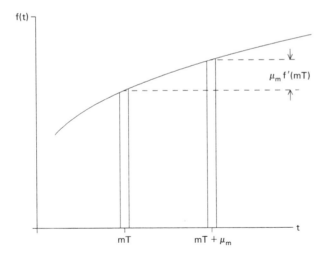

$\mu_m f'(mT)$

mT $mT + \mu_m$

Figure 4.18 Jitter error, $\mu_m f'(mT)$, due to sample offset μ_m.

4.5 DIGITAL-TO-ANALOG CONVERSION

In the digital signal processing system (Fig. 4.13) the DAC functions as a dual of the ADC. It converts an encoded sample set $[f_m]$ into a continuously varying voltage, or shaft position, or similar quantity. In the simplest form of the DAC, illustrated conceptually in Fig. 4.19, the binary sample values $[f_m]$ are placed successively (f_0 at

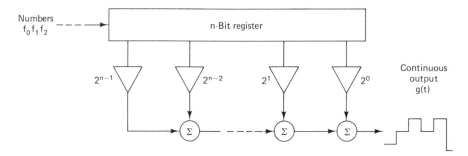

Figure 4.19 D-A conversion concept.

$t = 0$, f_1 at $t = T$, etc.) into an n-bit binary register. The bits in the register are summed according to their weights so that the output, $g(t)$, is a continuous signal that changes value only at the sample points. This is called a *zero-order hold* type of conversion — each sample value is held unchanged in $g(t)$ until the next sample arrives.

The various electrical or mechanical sources of error in DACs [see, e.g., Schmid (1970), Chap. 7] are generally held to a minimum by selecting precision components, using precision power supplies, reducing the operating temperature range, and so on.

The zero-order hold error can (at least in principle) be eliminated by smoothing the output, $g(t)$, as in Fig. 4.20. If $g(t)$ is as described above, then

$$g(t) = \sum_{n=0}^{\infty} f_n[u(t - nT) - u(t - (n + 1)T)] \qquad (4.17)$$

where $u(t - nT)$ is the unit step at $t = nT$. Using the Laplace transform table in Appendix A, the transform $G(s)$ can be written

$$G(s) = \frac{1 - e^{-sT}}{sT} \sum_{n=0}^{\infty} f_n e^{-nsT}(T) \qquad (4.18)$$

If the step size T is considered as the integration step dt, the sum in Eq. 4.18 becomes the transform $F(s)$. (The accuracy of this approximation is discussed further in the next chapter.) Thus,

$$G(s) \approx \frac{1 - e^{-sT}}{sT} F(s) \qquad (4.19)$$

and so the recovery of $f(t)$ from its zero-order hold version involves smoothing with the transfer function $sT/(1 - e^{-sT})$ as illustrated in Fig. 4.20. In practice, various simple approximations to this transfer function (i.e., simple lowpass filters) are used. Papoulis (1966) provides a more comprehensive discussion of the regeneration of $f(t)$ and Davies (1971) discusses the correction of zero-order hold distortion using digital filtering. Tou (1959) discusses the zero-order hold error as an effect of "clamping" the signal $f(t)$.

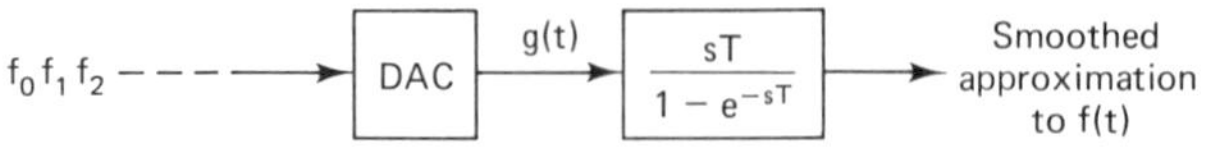

$$f_0 f_1 f_2 \ \text{---} \rightarrow \boxed{\text{DAC}} \xrightarrow{g(t)} \boxed{\dfrac{sT}{1 - e^{-sT}}} \rightarrow \begin{array}{c}\text{Smoothed}\\\text{approximation}\\\text{to } f(t)\end{array}$$

Figure 4.20 Smoothing to correct for zero-order hold.

EXERCISES

1. A periodic signal $f(t)$ contains frequency components up to 1 kHz. At what rate must $f(t)$ be sampled in order to reconstruct it unambiguously?

2. Suppose that $f(t)$, a 1-kHz pure sine wave, is sampled every microsecond to obtain its sample set, $[f_m]$. What other sinusoidal signals have the same $[f_m]$?

3. The Fourier transform of a transient signal $f(t)$ is plotted here.
   (a) What property of $f(t)$ results from $F(j\omega)$ being real?
   (b) What must the sampling interval T be in order to recover $f(t)$ completely?
   (c) What is $f(t)$?

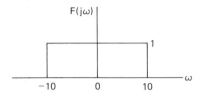

4. Let $f(t)$, the pulse train shown be described by

$$f(t) = \sum_{n=-\infty}^{\infty} \frac{\sin \pi(t - 40n)}{\pi(t - 40n)}$$

How often must $f(t)$ be sampled in order to be able to reconstruct it? *Hint:* Obviously, not much accuracy will be lost if each pulse in the train is treated as if it dies out completely before the next pulse begins.

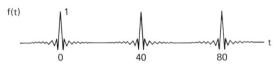

5. Given $f(t)$ in Exercise 4, make a sketch of the sample set for one period of $f(t)$, using a near-maximum value of T.

6. The ramp function $f(t) = 10t$ is sampled from $t = 0$ to $t = kT$ with an impulse sampler. What is the Laplace transform of the sampler output?

7. The sampler output in Exercise 6 is smoothed by passing it through the transfer function, $H(s) = 1/(s + 2)$. Sketch the smoothed output when the sampling interval is $T = 0.05$ s.

8. A 10-bit ADC has a range of ± 10 V. What is the rms quantizing error, assuming that all possible values are equally likely?

9. An ADC has an input capacitance of 10^{-11} F and input series resistance equal to 100 Ω. What is the maximum tracking error if the maximum input rate is $\pm 10^6$ V/s?

SOME ANSWERS

1. Sampling rate > 2000 s^{-1} 2. Pure sine waves at 1.001 MHz, 2.001 MHz, etc.
3. (a) $f(t) = f(-t)$ (b) $T < \pi/10$ (c) $f(t) = (\sin 10t)/\pi t$ 4. $T < 1$
6. $F_s(s) = \sum_{m=0}^{k} 10mTe^{-msT}$ 8. 0.0056 V   9. 0.001 V

REFERENCES

BENIMA, D., and BARGER, J. R., High-Speed, High-Resolution A/D Converters. *Electron. Des. News.* June 5, 1973, p. 62.

DAVIES, A. C., Correction of Zero-Order Hold Distortion in Digital Filters. *IEEE Trans. Audio Electroacoust.*, Vol. AU-19, No. 4, December 1971, p. 289.

ELECCION, M., A/D and D/A Converters. *IEEE Spectrum*, July 1972, p. 63.

GIBSON, J. E., *Nonlinear Automatic Control*, Chap. 3. New York: McGraw-Hill, 1963.

HOESCHELE, D., *Analog-to-Digital/Digital-to-Analog Conversion Techniques*. New York: Wiley, 1968.

JURY, E. I., *Sampled-Data Control Systems*. New York: Wiley, 1958.

MONROE, A. J., *Digital Processes for Sampled Data Systems*, Chap. 6. New York: Wiley, 1962.

PAPOULIS, A., *The Fourier Integral and Its Applications*. New York: McGraw-Hill, 1962.

PAPOULIS, A., Error Analysis in Sampling Theory. *Proc. IEEE*, Vol. 54, No. 7, July 1966, p. 947.

RAGAZZINI, J. R., and FRANKLIN, G. F., *Sampled-Data Control Systems*, Chap. 2. New York: McGraw-Hill, 1958.

SCHMID, H., *Electronic Analog/Digital Conversions*. New York: Van Nostrand Reinhold, 1970.

SCHWARTZ, M., *Information Transmission, Modulation and Noise*, Chap. 3. New York: McGraw-Hill, 1970.

TOU, J. T., *Digital and Sampled-Data Control Systems*, Chaps. 3 and 8. New York: McGraw-Hill, 1959.

TRUXAL, J. G., *Control System Synthesis*, Chap. 9. New York: McGraw-Hill, 1955.

WHITTAKER, E. T., Expansions of the Interpolation-Theory. *Proc. Roy. Soc. Edinburgh*, Vol. 35, 1915, p. 181.

5

The Discrete Fourier
Transform

5.1 INTRODUCTION

In this chapter we are concerned with the spectra of sampled signals. As with continuous signals, there are important differences between the spectra of signals which are periodic and those which are not. We have already seen that the Fourier integral does not converge with periodic signals, whose envelopes do not decrease with time. In addition, we shall see that the spectrum of any type of sampled signal is itself periodic.

The basic entity used to obtain the spectral representation of all sampled signals is the discrete Fourier transform (DFT). Section 5.2 introduces the DFT and its use with aperiodic signals. This discussion is continued in Section 5.3, where a frequency-domain interpretation of the sampling theorem is presented. In Section 5.4 the use of the DFT as it pertains to periodic signals is discussed. The topics that follow include a look at methods for reconstructing sampled waveforms, the effect of nonimpulse or finite-duration sampling, and the relationship between the DFT and discrete convolution. The chapter concludes with some examples of spectra for discrete signals.

5.2 THE DFT AND THE FOURIER TRANSFORM

In this section the DFT is derived from the Fourier transform of a sampled signal. The DFT is itself a sampled complex signal in the sense that it represents a frequency-domain sampling of a Fourier transform, hence the name *discrete* Fourier transform. The Fourier transform of any real function, $f(t)$, with finite energy is (see Chap. 3, Eq. 3.1)

$$F(j\omega) = \int_{-\infty}^{\infty} f(t)e^{-j\omega t}\, dt \qquad (5.1)$$

Here, as before, ω is the frequency variable with units of radians per second. A discrete-time version of $f(t)$, denoted $\bar{f}(t)$, can be written as a sequence of impulses:

$$\bar{f}(t) = f(t) \sum_{n=-\infty}^{\infty} \delta(t - nT)$$

$$= \sum_{n=-\infty}^{\infty} f_n \delta(t - nT) \tag{5.2}$$

where $f_n = f(nT)$ and we have assumed impulse sampling. The second line follows from the first because the product $f(t)\delta(t - nT)$ is zero everywhere except at $t = nT$. The Fourier transform of the sampled signal, denoted $\bar{F}(j\omega)$, is found by substituting $\bar{f}(t)$ into Eq. 5.1,

$$\bar{F}(j\omega) = \int_{-\infty}^{\infty} \bar{f}(t)e^{-j\omega t}\,dt$$

$$= \int_{-\infty}^{\infty} \sum_{n=-\infty}^{\infty} f(t)\delta(t - nT)e^{-j\omega t}\,dt$$

$$= \sum_{n=-\infty}^{\infty} \int_{-\infty}^{\infty} f(t)\delta(t - nT)e^{-j\omega t}\,dt \tag{5.3}$$

$$= \sum_{n=-\infty}^{\infty} f_n e^{-jn\omega T}$$

We usually call this result a DFT, although it is more appropriately called the Fourier transform of the sequence $[f_n]$. The formal definition of the DFT of any real sequence $[f_n]$ is obtained from Eq. 5.3 by placing finite limits on the summation and using samples of the frequency variable ω, as we shall see shortly. The Fourier transform in Eq. 5.3 is a continuous function of ω and is seen to have the following properties:

1. It is a linear transformation, that is, for any two finite-energy sequences $[f_n]$ and $[g_n]$,

$$\mathrm{FT}[k_1 f_n + k_2 g_n] = k_1 \mathrm{FT}[f_n] + k_2 \mathrm{FT}[g_n]$$

where FT denotes the Fourier transform in Eq. 5.3 and k_1 and k_2 are constants.
2. $\bar{F}(j\omega)$ is real and even if, and only if, $f(t)$ is even at the sample points, that is, $f_n = f_{-n}$.
3. $\bar{F}(j\omega)$ is imaginary and odd if, and only if, $f(t)$ is odd at the sample points, that is, $f_n = -f_{-n}$.
4. $\bar{F}(j\omega)$ and $\bar{F}(-j\omega)$ are complex conjugates.
5. $\bar{F}(j\omega)$ is periodic with period $\omega_0 = 2\pi/T$, that is,

$$\bar{F}(j\omega) = \bar{F}(j\omega + jm\omega_0)$$

for all ω and all integer values of m.

The first four properties are also essentially true in general of the Fourier transform, but the fifth is a special property of the summation in Eq. 5.3. To see this, let $\omega = \omega + m\omega_0$ in Eq. 5.3:

$$\overline{F}(j(\omega + m\omega_0)) = \sum_{n=-\infty}^{\infty} f_n e^{-j(\omega + m\omega_0)nT}$$

$$= \sum_{n=-\infty}^{\infty} f_n e^{-jn\omega T} e^{-j2\pi nm} \qquad (5.4)$$

$$= \sum_{n=-\infty}^{\infty} f_n e^{-jn\omega T}$$

$$= \overline{F}(j\omega)$$

The third line follows the second because $e^{-j2\pi nm} = 1$ for all integer values of n and m.

All periodic functions, whether real or complex and whether functions of time or frequency, have Fourier series representations. Such a representation is given by the summation in Eq. 5.3 where the sampled values $[f_n]$ play the role of the Fourier series coefficients. Thus, to recover $[f_n]$ from $\overline{F}(j\omega)$ we use the Fourier series coefficient formula (similar to Eq. 2.42),

$$f_n = \frac{T}{2\pi} \int_{-\pi/T}^{\pi/T} \overline{F}(j\omega) e^{j\omega nT} d\omega \qquad (5.5)$$

Equations 5.3 and 5.5 define the Fourier transform pair for a sequence $[f_n]$. They are of limited practical use when dealing with sampled signals, however, because they cannot be implemented directly on a digital computer.

Any actual computation of the Fourier transform must of course involve a finite summation of terms rather than the infinite sum in Eq. 5.3. In practice, signals are truncated in various ways to obtain a finite sample set. Specifically:

- A transient signal is assumed to equal zero after it decays to a negligible amplitude.
- A periodic signal is sampled over an integral number of periods.
- A random signal is multiplied by a "data window" of finite duration.

In any case, suppose that there are altogether N samples of $f(t)$. If time can be identified so that these samples begin at $t = 0$, then Eq. 5.3 becomes

$$\overline{F}(j\omega) = \sum_{n=0}^{N-1} f_n e^{-jn\omega T} \qquad (5.6)$$

That is, the samples of $f(t)$ are taken at $t = 0, T, 2T, \ldots, (N-1)T$. If time is not so defined and the N samples begin, say, at $t = kT$, then Eq. 5.3 becomes

$$\overline{F}(j\omega) = \sum_{n=k}^{N+k-1} f_n e^{-jn\omega T}$$

$$= e^{-jk\omega T} \sum_{n=0}^{N-1} f_{n+k} e^{-jn\omega T} \qquad (5.7)$$

where k can be any positive or negative integer. Equations 5.5 and 5.6 suggest the same sort of "shifting theorem" given in line 7 of Table 3.1, Chapter 3:

If $f(t)$ is shifted kT seconds to the right (left), then $\overline{F}(j\omega)$, the Fourier transform of $\overline{f}(t)$, must be multiplied by $e^{-jk\omega T}$ ($e^{jk\omega T}$).

The shifting theorem is illustrated in Fig. 5.1, which shows samples of a rectangular pulse beginning at $t = kT$ and ending at $t = (N + k - 1)T$, with $N = 33$. The amplitude spectrum given by $|F(j\omega)|$ does not depend on k, since changing k produces only a phase shift. Nor would it change if sampling were extended beyond the limits of the pulse, as explained in Section 5.7.

Although the frequency ω in Eq. 5.6 or 5.7 is a continuous variable, only N values (real and imaginary) of $\overline{F}(j\omega)$ can actually be independent, because there are only N degrees of freedom in the sample set $[f_n]$. Furthermore, property 5 above suggests that the independent values of $\overline{F}(j\omega)$ should be computed within one of its periods, say from $\omega T = 0$ to 2π. Therefore, letting

$$\omega = \frac{2\pi m}{NT}; \qquad m = 0, 1, \ldots, N - 1 \tag{5.8}$$

the usual formula for computing the independent values of the discrete Fourier transform (DFT) becomes

$$\overline{F}_m = \overline{F}\left(j\frac{2\pi m}{NT}\right) = \sum_{n=0}^{N-1} f_n e^{-j(2\pi mn/N)}; \qquad m = 0, 1, \ldots, N - 1 \tag{5.9}$$

The notation for the spectral sample,

$$\overline{F}_m = \overline{F}\left(j\frac{2\pi m}{NT}\right)$$

parallels the notation used with $f(t)$, that is, $f_n = f(nT)$. Actually, there are N complex values of $\overline{F}_m$, and therefore $2N$ values, real and imaginary, are represented in Eq. 5.9. When the sample sequence $[f_n]$ is real, only N values are independent, however, as seen from the following argument: From property 4 above,

$$\overline{F}_m = \overline{F}_{-m}^* \tag{5.10}$$

where * indicates complex conjugate. Then, from property 5,

$$\overline{F}_{-m} = \overline{F}_{-m+N} \tag{5.11}$$

Therefore,

$$\overline{F}_m = \overline{F}_{N-m}^* \tag{5.12}$$

This result conforms with the sampling theorem in Chapter 4 and proves that only the values of $\overline{F}_m$ for $m = 0$ through $N/2$ really need to be computed. Equation 5.9 is still used, however, because it is valid when $[f_n]$ is complex and is the most useful form for further derivations.

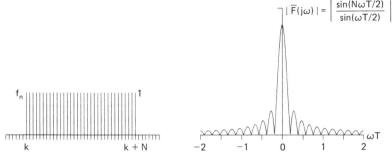

Figure 5.1 Sample set and amplitude spectrum of a sampled rectangular pulse; $N = 33$.

The frequency samples of the DFT in Eq. 5.9 span one period of $\overline{F}(j\omega)$, from $\omega = 0$ to $2\pi/T$ (i.e., from dc to the sampling frequency). Because of the periodic nature of $\overline{F}(j\omega)$ these same samples would be obtained over any period of $\overline{F}(j\omega)$ provided that it begins at frequency $k\pi/(NT)$ for some integer value of k. In Eq. 5.9, we note that $k = 0$. Of course, the position of these samples may shift depending on where the period starts. Often we use $k = -N/2$, so that the range of frequencies of interest is $-\pi/T$ to π/T. This is the range in which the spectrum should fall if the signal is sampled in accordance with the sampling theorem. For the DFT above, with $k = 0$, samples of $\overline{F}(j\omega)$ over the positive frequencies from 0 through π/T are represented by $\overline{F}_m$, $m = 0$ through $N/2$ for N even and $m = 0$ through $(N - 1)/2$ for N odd. Samples of $\overline{F}(j\omega)$ over the negative frequencies $-\pi/T$ to 0 are represented by $\overline{F}_m$ at indices $m = N/2$ through $N - 1$ for N even and $m = (N + 1)/2$ through $N - 1$ for N odd. Thus, when the DFT is computed for the range of indices $m = 0$ to $N - 1$, the samples from $N/2$ to $N - 1$ [or $(N + 1)/2$ to $N - 1$ for N odd] are often associated with the negative frequencies between $-\pi/T$ and 0 rather than the positive frequencies indicated in Eq. 5.8. If N is even, $F_{N/2}$ is the frequency sample for both $-\pi/T$ and π/T (as it is for all frequencies $\pi/T \pm 2\pi l/T$).

The DFT in Eq. 5.9 can be inverted and $[f_n]$ recovered as follows:

$$f_n = \frac{1}{N} \sum_{m=0}^{N-1} \overline{F}_m e^{j(2\pi mn/N)}; \qquad n = 0, 1, \ldots, N - 1 \tag{5.13}$$

This formula is called the inverse DFT. By substituting Eq. 5.9 into Eq. 5.13 one can demonstrate the validity of this inversion and, at the same time, verify that the N independent values of $\overline{F}(j\omega)$ convey all the original information in the sample set. If they did not, $[f_n]$ would not be recoverable from $[\overline{F}_m]$. The substitution gives

$$f_n = \frac{1}{N} \sum_{m=0}^{N-1} \sum_{k=0}^{N-1} f_k e^{-j(2\pi mk/N)} e^{j(2\pi mn/N)}$$

$$= \frac{1}{N} \sum_{k=0}^{N-1} f_k \sum_{m=0}^{N-1} e^{j[2\pi m(n-k)/N]} \tag{5.14}$$

$$= \frac{1}{N} (Nf_n) = f_n$$

Note that the inner sum in the second line is 0 for $k \neq n$ and N for $k = n$.

In summary, for any sample set $[f_n]$, the DFT pair is given by

$$
\begin{aligned}
\overline{F}_m &= \sum_{n=0}^{N-1} f_n e^{-j(2\pi mn/N)}; && m = 0, 1, \ldots, N-1 \\
f_n &= \frac{1}{N} \sum_{m=0}^{N-1} \overline{F}_m e^{j(2\pi mn/N)}; && n = 0, 1, \ldots, N-1 \\
\omega &= \frac{2\pi m}{NT} \quad \text{rad/s}
\end{aligned}
\tag{5.15}
$$

Thus the frequency related to m is $\omega = 2\pi m/NT$, and N and T are, respectively, the number of samples and the time between samples. The DFT has the properties of the Fourier transform listed above and, as mentioned, also has the important property of periodicity. Its relation to the Fourier transform is explored further in the next section.

5.3 RELATION TO THE FOURIER TRANSFORM; ALIASING

In general, the infinite sum for $\overline{F}(j\omega)$ in Eq. 5.3 converges whenever the Fourier transform, $F(j\omega)$, exists. If $\overline{F}(j\omega)$ is viewed as an approximation to $F(j\omega)$, one may be dubious about the accuracy of the approximation. The term $e^{-j\omega t}$ in the integrand changes value rapidly when ω is large, so it appears that the sampling interval T may have to be very small for accuracy. On the other hand, the sampling theorem states that T need only be small enough that $1/T$ hertz is just over twice the highest frequency in $f(t)$.

To reconcile these points, an explicit relationship between $\underline{F}(j\omega)$ and $\overline{F}(j\omega)$ is useful. The inverse Fourier transform (not the inverse DFT) of $\overline{F}(j\omega)$ is useful for this purpose. Let the set $[f_n]$ be, as before, the sample set of $f(t)$, and define a function of the samples of $f(t)$, $\bar{f}(t)$, as in Eq. 5.2 and Fig. 5.2. The arrows in Fig. 5.2 emphasize that $\bar{f}(t)$ is composed of impulse functions, each having zero width, infinite amplitude, and area equal to f_n. Thus, similar to Eq. 5.2, $f(t)$ can be written

$$
\begin{aligned}
\bar{f}(t) &= f(t) \sum_{n=0}^{N-1} \delta(t - nT) \\
&= f(t)d(t)
\end{aligned}
\tag{5.16}
$$

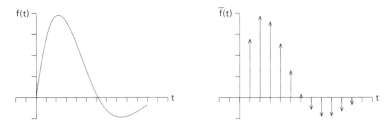

Figure 5.2 $f(t)$ and its impulse train $\bar{f}(t)$, with $N = 12$.

where $d(t)$ now represents the train of N unit impulse functions. Next, since one period of $d(t)$ ranges from $t = 0$ to NT and $f(t)$ can be assumed to be zero outside this range, $d(t)$ can be represented with a Fourier series (not a transform),

$$d(t) = \sum_{n=-\infty}^{\infty} c_n e^{j(2\pi nt/T)} \tag{5.17}$$

The coefficients $[c_n]$ are computed as in Chapter 2, Eq. 2.42:

$$c_n = \frac{1}{T} \int_{-T/2}^{T/2} d(t) e^{-j(2\pi nt/T)} dt$$

$$= \frac{1}{T} \sum_{m=-\infty}^{\infty} \int_{-T/2}^{T/2} \delta(t - mT) e^{-j(2\pi nt/T)} dt \tag{5.18}$$

$$= \frac{1}{T}$$

(The only nonzero term in the sum above is the term for $m = 0$.) Therefore, from Eqs. 5.17 and 5.18,

$$d(t) = \frac{1}{T} \sum_{n=-\infty}^{\infty} e^{j(2\pi nt/T)} \tag{5.19}$$

It is true that the sum in Eq. 5.19 does not converge, but this representation of $d(t)$ is nevertheless adequate for the present discussion. Using this representation of $d(t)$ in Eq. 5.16, we obtain

$$\bar{f}(t) = \frac{1}{T} \sum_{n=-\infty}^{\infty} f(t) e^{j(2\pi nt/T)} \tag{5.20}$$

With $\bar{f}(t)$ in this form, its Fourier transform, which has already been shown to be $\bar{F}(j\omega)$, can be written as follows:

$$\bar{F}(j\omega) = \int_{-\infty}^{\infty} \bar{f}(t) e^{-j\omega t} dt$$

$$= \frac{1}{T} \sum_{n=-\infty}^{\infty} \int_{-\infty}^{\infty} f(t) e^{j(2\pi nt/T)} e^{-j\omega t} dt \tag{5.21}$$

$$= \frac{1}{T} \sum_{n=-\infty}^{\infty} F\left(j\omega - j\frac{2\pi n}{T}\right)$$

The result in the third line is obtained from the second using line 6 of Table 3.1.

This is an important result because it provides an explicit relationship between the Fourier transform of $f(t)$ and that of $\bar{f}(t)$ from which the DFT is derived. It tells us that *the Fourier transform of a sampled sequence is a superposition of an infinite number of shifted Fourier transforms of the unsampled sequence, scaled by $1/T$.*

The result is illustrated for two values of T in Fig. 5.3 for a case where $F(j\omega)$ is a simple real function. $[F(j\omega)$ is in general complex, so that amplitude and phase usually must be plotted separately.] In the first illustration T is small enough for the

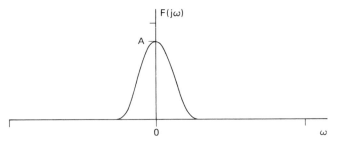

Fourier transform

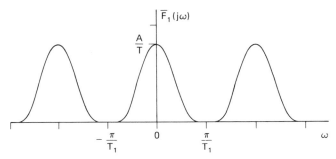

Fourier transform of sampled sequence; T = T₁

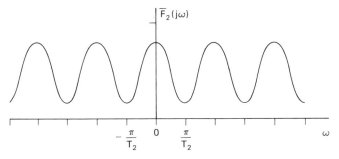

Fourier transform of sampled sequence; T = T₂

Figure 5.3 Fourier transforms of sampled sequences for two values of T, with $T_1 < T_2$. The waveform $f(t)$ has components at frequencies above $1/2T_2$ Hz, but not above $1/2T_1$ Hz.

superimposed functions, that is, the terms in Eq. 5.21, to be distinct. In the second illustration the terms overlap. Figure 5.3 illustrates that the terms will in fact overlap when the nonzero parts of $F(j\omega)$ extend beyond $\omega = \pi/T$. When there is such an overlap, obviously the Fourier transform $F(j\omega)$ cannot be "recovered" from $\overline{F}(j\omega)$ by eliminating portions of $\overline{F}(j\omega)$ outside the interval where $|\omega| < \pi/T$. If $F(j\omega)$ cannot be recovered from $\overline{F}(j\omega)$, then $f(t)$ cannot be recovered from $[f_n]$. Clearly, this is just another statement of the sampling theorem of Chapter 4. The specific effect of aliasing on the sample spectrum is illustrated in the second case of Fig. 5.3, where the spectrum is distorted by the "folding" of the frequency axis described in Chapter 4.

When the amplitude of $\overline{F}(j\omega)$ is of primary interest, the following inequality is obtained from Eq. 5.21:

$$|\overline{F}(j\omega)| = \frac{1}{T}\left|\sum_{n=-\infty}^{\infty} F\left(j\omega - j\frac{2\pi n}{T}\right)\right|$$

$$\leq \frac{1}{T}\sum_{n=-\infty}^{\infty}\left|F\left(j\omega - j\frac{2\pi n}{T}\right)\right| \tag{5.22}$$

Equation 5.22 gives an upper bound on the aliasing effect without reference to the phase of the sampled function $f(t)$.

In practice, sampled functions are never limited completely in frequency content as in Fig. 5.3. Physical spectra, of transients at least, tend to be smooth and to approach zero asymptotically as ω increases to infinity. Thus the designer must choose a sampling interval T so that essentially all rather than all of the spectral content of the waveform is contained below $1/2T$ hertz.

Furthermore, concerning transient signals, the transients themselves generally decay asymptotically to zero, so the number of samples N must also be chosen so that essentially all of the nonzero portions of $f(t)$ are sampled. These practical considerations are illustrated in the following example. The sampling of periodic signals is discussed further in the next section.

Example 5.1

Compute the amplitude of the DFT given

$$f(t) = 2.5e^{-t}\sin 4t; \qquad t \geq 0 \tag{5.23}$$

This function is plotted in Fig. 5.4, along with the DFTs for two different sampling intervals. In both cases the total sampling period NT was 5. The actual amplitude spectrum of $f(t)$, from Table 3.1 of Chapter 3, is

$$|F(j\omega)| = \left|\frac{10}{(j\omega + 1)^2 + 16}\right| = \frac{10}{(\omega^4 - 30\omega^2 + 289)^{1/2}} \tag{5.24}$$

This amplitude spectrum is also plotted in Fig. 5.4 so that the cause of the aliasing in the DFTs can be observed. One might say that with $T = 0.1$ s and $N = 50$, essentially all of the spectral content of $f(t)$ is contained below $\pi/T \approx 31$ rad/s, and thus the DFT is a good approximation to $1/T$ times the Fourier transform. With $T = 0.2$ and $N = 25$, however, some aliasing is noticeable at the higher frequencies.

The amplitude spectra in Fig. 5.4 are found by taking the amplitude of Eq. 5.9, that is,

$$|\overline{F}_m| = \left|\sum_{n=0}^{N-1} f_n e^{-j(2\pi mn/N)}\right|$$

$$= \left\{\left|\sum_{n=0}^{N-1} f_n \cos\left(\frac{2\pi mn}{N}\right)\right|^2 + \left|\sum_{n=0}^{N-1} f_n \sin\left(\frac{2\pi mn}{N}\right)\right|^2\right\}^{1/2} \tag{5.25}$$

The discrete spectra are plotted, as is often the case, for $m = -N/2$ through $N/2$ rather than for $m = 0$ through $N - 1$, that is, over one period of $\overline{F}(j\omega)$ from $\omega = -\pi/T$ to π/T.

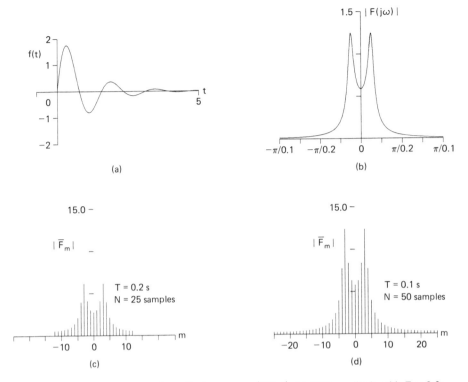

Figure 5.4 (a) $f(t)$; (b) actual amplitude spectrum $|F(j\omega)|$; (c) DFT amplitude with $T = 0.2$; (d) DFT amplitude with $T = 0.1$. Note the $1/T$ scaling factor, that is, $\overline{F} \approx F/T$.

Example 5.2

Suppose that the spectrum of $f(t)$ is real and is given by

$$F(j\omega) = 1 + \cos \omega; \qquad |\omega| \leq \pi$$
$$= 0; \qquad\qquad |\omega| \geq \pi \qquad\qquad (5.26)$$

Find the Fourier transform of $\bar{f}(t)$ for sampling intervals of $T = 0.8$ and 1.25. The result, which can be found using Eq. 5.21, is shown in Fig. 5.5. Since the folding frequency, π/T, must be greater than π to prevent aliasing, aliasing occurs for the larger value of T.

5.4 THE DFT AND THE FOURIER SERIES

Up to now our discussion has been concerned primarily with the spectra of aperiodic signals. In this section we consider the spectra of periodic signals. In the continuous case a periodic signal can be represented by a Fourier series, and its spectrum is nonzero only at the discrete frequencies represented by the Fourier series coefficients. In the preceding section we found that the spectrum of a sampled signal with finite energy is a periodic replication of the continuous signal's spectrum. Here we

The Discrete Fourier Transform   Chap. 5

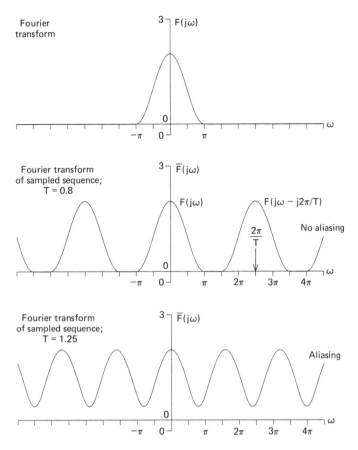

Figure 5.5 Fourier transforms for Example 5.2.

find this same relationship when the sampled signal is periodic. Thus, when the sampled signal is periodic, its spectrum is both discrete and periodic.

To show these properties, we begin by deriving the Fourier series coefficients of a sampled periodic signal. Recall from Chapter 2, Eq. 2.42, that for a periodic signal, $f_p(t)$, the Fourier series coefficients are given by

$$c_m = \frac{1}{P} \int_{-P/2}^{P/2} f_p(t) e^{-jm\omega_0 t}\, dt \qquad (5.27)$$

where P is the fundamental period of the signal with fundamental frequency $\omega_0 = 2\pi/P$. Assume that $f_p(t)$ is sampled (as illustrated in Fig. 2.6) every T seconds to obtain a total of N samples per period, that is, $P = NT$. Also, as in Section 5.3, let $\bar{f}_p(t)$ represent the sampled version of $f_p(t)$ and change the range of t from $(-P/2, P/2)$ to $(0, P)$. Now the Fourier series coefficients of the sampled signal, denoted $\bar{c}_m$, are given by

$$\overline{c}_m = \frac{1}{P} \int_0^P \overline{f}_p(t) e^{-jm\omega_0 t} \, dt$$

$$= \frac{1}{P} \int_0^P f_p(t) \sum_{n=-\infty}^{\infty} \delta(t - nT) e^{-jm\omega_0 t} \, dt$$

$$= \frac{1}{P} \sum_{n=-\infty}^{\infty} \int_0^P f_p(t) \delta(t - nT) e^{-jm\omega_0 t} \, dt \qquad (5.28)$$

$$= \frac{1}{P} \sum_{n=0}^{N-1} f_p(nT) e^{-jnm\omega_0 T}$$

where line 4 follows from line 3 because the only nonzero terms in the sum are those for which the impulse function falls within the limits of the integral. Using $\omega_0 = 2\pi/P$ and $P = NT$ and letting $f_n = f_p(nT)$, this expression can be written

$$\overline{c}_m = \frac{1}{P} \sum_{n=0}^{N-1} f_n e^{-j2\pi mn/N} \qquad (5.29)$$

Note the similarity between this expression and that of the DFT in Eq. 5.9. Specifically, if $\overline{F}_m$ is formed by computing the DFT of the sampled signal over one period, then

$$\overline{c}_m = \frac{1}{P} \overline{F}_m \qquad (5.30)$$

Thus the DFT can be used to compute spectra of periodic signals as well as transients. Like $\overline{F}_m$, $\overline{c}_m$ is periodic with period N, that is,

$$\overline{c}_m = \overline{c}_{m+kN}$$

for any integer value of k. In fact, since $\overline{c}_m$ is just a scaled version of $\overline{F}_m$, it has all the same properties as the DFT mentioned in Section 5.3. To recover f_n from $\overline{c}_m$ the inverse DFT can be used:

$$f_n = P \cdot (\text{inverse DFT of } \overline{c}_m) \qquad (5.31)$$

In practice, it is often preferable to let the DFT, $\overline{F}_m$, represent the signal spectrum rather than $\overline{c}_m$. In this way the spectra of all signals, periodic or not, are computed using the same equation.

It is important here to point out the difference between the use of the DFT in previous sections and its use here. In previous sections the DFT is computed over the entire nonzero portion of the signal. Here, because the signal is periodic and extends indefinitely, the DFT is computed over only one period (or it may be computed over any integral number of periods). Further, in Section 5.3 the DFT was used to compute discrete samples from a continuous spectrum, that is, a finite subset of the entire spectrum. Here these discrete values represent the entire nonzero portion of the signal's spectrum.

We can summarize some findings regarding the spectra of continuous and sampled signals as follows (see Fig. 5.6):

1. The spectrum of a continuous signal with finite energy is continuous and aperiodic.

2. The spectrum of a sampled signal with finite energy is continuous and periodic. It is, in fact, a superposition (replication if no aliasing) of the continuous

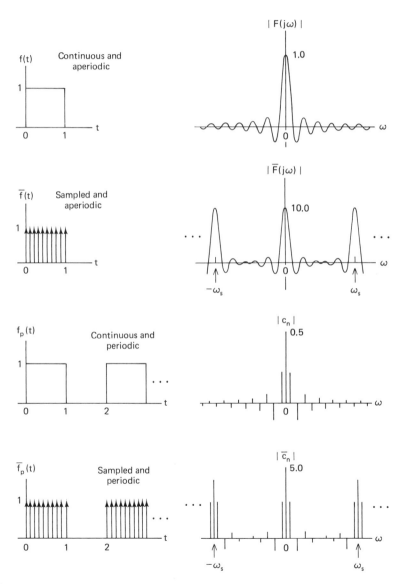

Figure 5.6 Comparable spectra of continuous, sampled, aperiodic, and periodic signals. Frequency corresponding with n is $\omega = 2\pi n/P$.

signal's spectrum at an infinite number of shifted positions, each separated by ω_s, the sampling frequency.

3. The spectrum of a continuous periodic signal is discrete and aperiodic.

4. The spectrum of a sampled periodic signal is discrete and periodic. It is, once again, a superposition of the continuous signal's spectrum at an infinite number of shifted positions each separated by ω_s.

5.5 FOURIER SERIES RECONSTRUCTION

The reader has probably noticed that the DFT, its inverse, and the Fourier series are all similar functions. The DFT in fact is a periodic function, and Eq. 5.9 expresses it as a complex Fourier series with coefficients $[f_n]$.

Furthermore, Eq. 5.15 for f_n suggests a Fourier series reconstruction of $f(t)$. First, to limit the harmonics to frequencies less than the folding frequency, the limits of the sum can be made symmetrical about zero and restricted to $N/2$. Secondly, the continuous variable t can be substituted for nT. The result is

$$\hat{f}(t) = \frac{1}{N} \sum_{|m| \le N/2} \overline{F}'_m e^{j(2\pi mt/NT)}$$

$$\overline{F}'_m = \frac{\overline{F}_{N/2}}{2}; \qquad |m| = \frac{N}{2} \qquad \text{when } N \text{ is even} \qquad (5.32)$$

$$= \overline{F}_m; \qquad |m| < \frac{N}{2}$$

The proof that $\hat{f}(nT)$ is equal to f_n, which is left for an exercise, proceeds essentially as in Eq. 5.14, except that when N is even, the final DFT components must be included at half-strength, so that $\overline{F}'_m$ is required in place of $\overline{F}_m$. Of course when N is odd, $\overline{F}'_m$ and $\overline{F}_m$ are the same, since m is never $N/2$.

Thus the reconstruction $\hat{f}(nT)$ is equal to $f(nT)$, and therefore equal to $f(t)$ at the sample points, but what about between and beyond the sample points? First, beyond the sampling range from $t = 0$ to NT, $\hat{f}(t)$ is seen simply to repeat itself, because Eq. 5.32 is again just a Fourier series. The Fourier coefficients this time are $[\overline{F}'_m/N]$, and the fundamental frequency is $1/NT$ hertz. This suggests that if $f(t)$ is a sampled transient, it is convenient to think of the periodic representation of $f(t)$, denoted $f_p(t)$ as in Fig. 5.7. Then the Fourier series reconstruction, $\hat{f}(t)$, should approximate $f_p(t)$, and approximate $f(t)$ in the interval from 0 to NT.

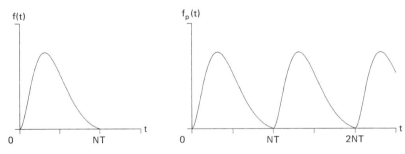

Figure 5.7   Periodic representation of $f(t)$.

The next question is whether $\hat{f}(t)$, being periodic and equal to $f_p(t)$ and $f(t)$ at the sample points, really is equal to $f_p(t)$ at other times. The answer, of course, depends on whether aliasing is present, but it is instructive to arrive at the answer by representing $f_p(t)$ with a Fourier series and then examining the coefficients. Let

$$f_p(t) = \sum_{i=-\infty}^{\infty} c_i e^{j(2\pi i t / NT)} \tag{5.33}$$

This is again the complex form of the Fourier series, and here c_i is the complex conjugate of c_{-i}, assuming that $f_p(t)$ is real-valued.

Equation 5.33 thus expresses $f_p(t)$ and also $f(t)$ in the interval $(0, NT)$, so a component $\overline{F}_m$ of the DFT of $f(t)$ can be expressed as follows:

$$\overline{F}_m = \sum_{n=0}^{N-1} f_n e^{-j(2\pi mn/N)} \tag{5.34}$$

$$= \sum_{n=0}^{N-1} f_p(nT) e^{-j(2\pi mn/N)} \tag{5.35}$$

$$= \sum_{i=-\infty}^{\infty} c_i \sum_{n=0}^{N-1} e^{j[2\pi n(i-m)/N]} \tag{5.36}$$

The inner sum in Eq. 5.36 is seen to be a simple geometric series and therefore has a closed form:

$$\sum_{n=0}^{N-1} e^{j[2\pi n(i-m)/N]} = N \qquad \text{if } i = m + kN$$
$$= 0 \qquad \text{otherwise} \tag{5.37}$$

in which k is any positive or negative integer, including zero. The result in Eq. 5.36 can now be simplified to

$$\overline{F}_m = N \sum_{k=-\infty}^{\infty} c_{m+kN} \tag{5.38}$$

so the DFT is related in this simple manner to the spectral coefficients of $f_p(t)$.

With this result, the question of whether $\hat{f}(t)$ and $f_p(t)$ are equal can be answered. If $f_p(t)$ is limited to frequencies below $1/2T$ hertz, then c_i in Eq. 5.33 will be zero for $|i| \geq N/2$. Therefore, in Eq. 5.38,

$$\overline{F}_m = Nc_m; \qquad m < \frac{N}{2} \qquad \text{if no aliasing} \tag{5.39}$$

Then Eqs. 5.32 and 5.33 become identical; that is, in Eq. 5.32,

$$\hat{f}(t) = \sum_{|m|<N/2} c_m e^{j(2\pi m t / NT)}$$
$$= f_p(t) \qquad \text{if no aliasing} \tag{5.40}$$

so $\hat{f}(t)$ gives an exact reconstruction of $f(t)$ at all times from $t = 0$ to $t = NT$ when there is no aliasing.

If aliasing is present, then Eq. 5.38 gives the effect on the reconstruction: The coefficients of $\hat{f}(t)$ in Eq. 5.32 are sums of the coefficients of $f_p(t)$, so the functions are in general unequal. This result was expected since an accurate reconstruction is not possible if the sampling theorem is violated. Note that

$$
\begin{aligned}
\hat{f}(nT) &= f_p(nT) \\
&= f(nT); \qquad n = 0, 1, \ldots, N - 1
\end{aligned}
\tag{5.41}
$$

as shown above, whether aliasing is present or not. For this reason, $\hat{f}(t)$ provides a practical reconstruction of $f(t)$ in many instances. It of course contains no frequencies above $1/2T$ hertz and often provides a useful smoothing effect.

As a final point on the Fourier series reconstruction, note that $\hat{f}(t)$ is, for the present discussion at least, a real function, yet it is expressed in Eq. 5.32 in terms of complex quantities. By using the fact that $\overline{F}_m$ and $\overline{F}_{-m}$ are conjugates (Eq. 5.10), Eq. 5.32 can be simplified for computing purposes. The result, that is, the preferred Fourier series reconstruction formula, is

$$
\hat{f}(t) = \frac{1}{N} \left\{ \overline{F}_0 + 2 \sum_{m=1}^{m \leq N/2} \left[\mathrm{Re}(\overline{F}_m') \cos\left(\frac{2\pi mt}{NT}\right) - \mathrm{Im}(\overline{F}_m') \sin\left(\frac{2\pi mt}{NT}\right) \right] \right\}
\tag{5.42}
$$

where

$$
\overline{F}_m' = \frac{\overline{F}_{N/2}}{2}; \qquad m = \frac{N}{2}
$$

$$
= \overline{F}_m; \qquad m < \frac{N}{2}
$$

Example 5.3

Suppose that $[f_n]$ is actually the sample set of

$$
f(t) = 2.5e^{-t} \sin 4t
\tag{5.43}
$$

as in Example 5.1. Reconstruct $f(t)$ using a Fourier series. Reconstructions for two values of T using Eq. 5.42 are shown in Fig. 5.8, along with the original function. The effect of aliasing can be noted with the larger sampling interval. In both cases, the effect of the periodicity of $\hat{f}(t)$ (and the sudden change in slope at $t = 0$) can be observed near $t = 5$.

In examples of this type, in which a finite transient beginning at $t = 0$ and decaying to near zero at $t = NT$ is sampled and then reconstructed using the Fourier series, Campbell (1973) proposed a method that generally results in improved reconstruction. The method consists (conceptually) of forming the periodic version $f_p(t)$ out of $f(t)$ plus a shifted version of its inverted mirror image, $-f(NT - t)$, as in the upper waveform in Fig. 5.9. The modified $f_p(t)$ now no longer has the abrupt changes in slope seen in Fig. 5.8, and, in general as well as in the two examples in Fig. 5.9, the reconstruction is more accurate. The modified $f_p(t)$ also is odd-periodic, so the computations in Fig. 5.9 are no more complex than those in Fig. 5.8. From Eq. 5.15

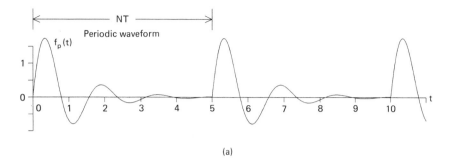

(a)

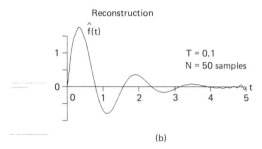

(b)

(c)

Figure 5.8 Fourier series reconstructions. (a) Periodic representation of $f(t)$; (b) reconstruction using 50 samples; (c) reconstruction using only 10 samples.

plus the relationship $f_{N-n} = -f_n$ in Fig. 5.9, and assuming that $f_0 = f_{N/2} = 0$, the DFT in this case is

$$\overline{F}_m = -2j \sum_{n=1}^{(N/2)-1} f_n \sin\left(\frac{2\pi mn}{N}\right) \tag{5.44}$$

and thus the reconstruction formula, Eq. 5.42, has the same number of terms as before, since N is now twice as large but there are now no $\text{Re}(\overline{F}'_m)$ terms.

5.6 WHITTAKER'S RECONSTRUCTION

The work of E. T. Whittaker in connection with the sampling theorem was mentioned in Chapter 4. As part of this work, Whittaker developed a formula that employs the "cardinal function" for interpolating a function between its sample points. It is an important and fundamental formula and can be developed in the following

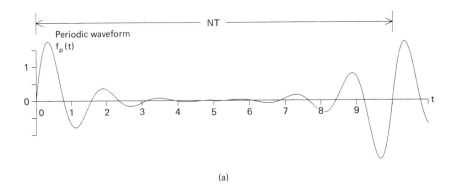

(a)

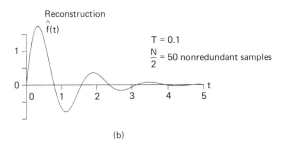

(b)

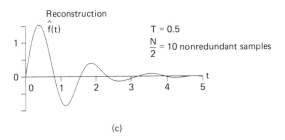

(c)

Figure 5.9 Campbell's Fourier series reconstructions. (a) Odd-periodic representation of $f(t)$; (b) reconstruction using 50 samples; (c) reconstruction using 10 samples.

way based on the foregoing discussion of the Fourier summation and its inverse (Eqs. 5.3 and 5.5).

Refer again to Figs. 5.3 and 5.5, which illustrate how the Fourier summation is a superposition of shifted Fourier transforms of $f(t)$. These figures suggest that if one is interested in reconstructing $f(t)$ from its sample set $[f_n]$, he or she must effectively pass the impulse train, $\bar{f}(t)$, in Eq. 5.2, through a lowpass filter that passes frequencies in the range $|\omega| \leq \pi/T$ to "mask out" all frequencies in $\bar{f}(t)$ that are not in $f(t)$. The suggested operation is illustrated in Fig. 5.10. Note that even if the components of $\bar{F}(j\omega)$ overlap, as when $T = T_2$ in Fig. 5.3, there is still a strong suggestion that the best approximation to $f(t)$ will be obtained if this same filtering function is applied to $\bar{f}(t)$. Note also that in Fig. 5.10, $H(j\omega)$ is equal to T in the passband to account for the difference in scale between $\bar{F}(j\omega)$ and $F(j\omega)$.

The Discrete Fourier Transform   Chap. 5

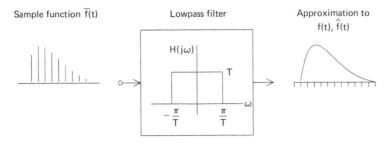

| Sample function $\bar{f}(t)$ | Lowpass filter | Approximation to $f(t)$, $\hat{f}(t)$ |

Figure 5.10   Scheme for obtaining Whittaker's reconstruction.

A general formula for $\hat{f}(t)$, the reconstructed version of $f(t)$, can be derived by carrying out the operation in Fig. 5.10. If $H(j\omega)$ is the transfer function of the filter,

$$\hat{F}(j\omega) = H(j\omega)\overline{F}(j\omega) \tag{5.45}$$

In the time domain, $\hat{f}(t)$ is therefore the convolution (Chapter 3, Section 3.3) of $h(t)$ with $\bar{f}(t)$:

$$\hat{f}(t) = \int_{-\infty}^{\infty} h(\tau)\bar{f}(t - \tau) \, d\tau \tag{5.46}$$

The impulse response $h(t)$ of the filter is found by taking the inverse transform of $H(j\omega)$ as in Exercise 20 of Chapter 3:

$$
\begin{aligned}
h(t) &= \int_{-\infty}^{\infty} H(j\omega)e^{j\omega t} \, d\!\left(\frac{\omega}{2\pi}\right) \\
&= \frac{T}{2\pi} \int_{-\pi/T}^{\pi/T} e^{j\omega t} \, d\omega \\
&= \frac{\sin(\pi t/T)}{\pi t/T}
\end{aligned}
\tag{5.47}
$$

Using this form of $h(t)$ as well as Eq. 5.2 for $\bar{f}(t)$ in the convolution, Eq. 5.46 yields

$$
\begin{aligned}
\hat{f}(t) &= \int_{-\infty}^{\infty} h(\tau) \sum_{n=0}^{N-1} f(t - \tau)\delta(t - \tau - nT) \, d\tau \\
&= \sum_{n=0}^{N-1} f_n h(t - nT) \\
&= \sum_{n=0}^{N-1} f_n \frac{\sin[(\pi/T)(t - nT)]}{(\pi/T)(t - nT)}
\end{aligned}
\tag{5.48}
$$

The sum in Eq. 5.48 is Whittaker's cardinal function and is a general form for reconstructing a function from a set of samples. Whittaker's function has the same important properties discussed above in connection with the Fourier series reconstruction:

First, $\hat{f}(t)$ contains no frequencies higher than one-half the sampling rate, that is, no frequencies higher than $1/2T$ hertz (obvious, since the cutoff frequency of

$H(j\omega)$ is at $\omega = \pi/T$). Since such frequencies inherently cannot be conveyed through the sampling process, any reconstruction that placed them in $\hat{f}(t)$ would clearly be a false form of "data enrichment" unless, of course, other information about $f(t)$ were available.

Second, $\hat{f}(t)$ is equal to $f(t)$ at the sample points. This can be seen by letting $t = nT$ in Eq. 5.48 ($n =$ integer). In this case only the nth term in the sum is nonzero, and the entire sum is equal to f_n, the value of $f(t)$ at $t = nT$.

Finally, when Whittaker's formula is actually being computed, the computation is much more rapid if the sine function is taken from inside the sum in Eq. 5.48. As in Eq. 1.2,

$$\sin[(\pi/T)(t - nT)] = \sin\left(\frac{\pi t}{T}\right)\cos(n\pi) - \cos\left(\frac{\pi t}{T}\right)\sin n\pi$$

$$= (-1)^n \sin\left(\frac{\pi t}{T}\right) \tag{5.49}$$

and therefore the cardinal function becomes

$$\hat{f}(t) = \frac{\sin(\pi t/T)}{\pi/T} \sum_{n=0}^{N-1} f_n \frac{(-1)^n}{t - nT} \tag{5.50}$$

One must be careful not to compute $\hat{f}(t)$ at the sample points using Eq. 5.50, but instead to set $\hat{f}(nT) = f_n$ at these points.

Example 5.4

As in Example 5.1, suppose that $[f_n]$ is actually the sample set of

$$f(t) = 2.5e^{-t}\sin 4t \tag{5.51}$$

Reconstruct $f(t)$ using Whittaker's cardinal function. Reconstructions for two values of T are shown in Fig. 5.11, and can be compared with those in Section 5.5. Again, the effect of aliasing is evident with the larger value of T. The reader should compare these results carefully with those of Example 5.3. Note that here $\hat{f}(t)$ is not periodic but is in fact close to 0 outside the interval from $t = 0$ to 5.

A generalization of Whittaker's cardinal function has been described by Papoulis (1966). Suppose, as illustrated in Fig. 5.12, that the sampling rate is chosen to be considerably greater than twice the highest frequency in $f(t)$, that is, $\pi/T > \omega_1$ in the illustration. Then one obviously has some freedom in choosing ω_0, the cutoff frequency of the lowpass reconstruction filter. In fact, from the figure, if

$$\omega_1 < \omega_0 < \frac{2\pi}{T} - \omega_1 \tag{5.52}$$

then a valid reconstruction should be obtained. Figure 5.10 now becomes a special case of Fig. 5.12 with $\omega_0 = \pi/T$, and the generalized version of Eq. 5.48 becomes

$$\hat{f}(t) = \sum_{n=0}^{N-1} f_n \frac{\sin[\omega_0(t - nT)]}{(\pi/T)(t - nT)} \tag{5.53}$$

Still more generally, the reconstruction filter can have any shape between ω_1 and ω_0 where $\overline{F}(j\omega)$ is zero and, as long as $H(j\omega) = T$ for $|\omega| \le \omega_1$, the reconstruction

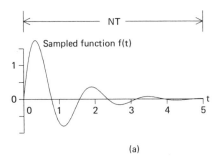

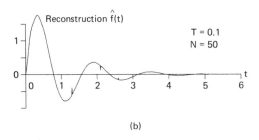

(a)

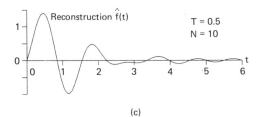

(b)

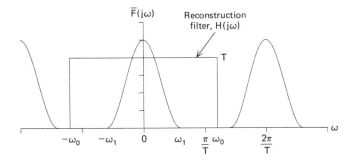

Figure 5.11 Whittaker reconstructions.
(a) Sampled function $f(t)$;
(b) reconstruction using $T = 0.1$;
(c) reconstruction using $T = 0.5$.

(c)

Figure 5.12 Reconstruction filter with cutoff at $\omega_0 \neq \pi/T$.

must be valid. As in the derivation of Eq. 5.48, the reconstruction in its most general form is

$$\hat{f}(t) = \sum_{n=0}^{N-1} f_n h(t - nT) \tag{5.54}$$

where $h(t)$ is the inverse transform of $H(j\omega)$ having the properties just described.

5.7 RECONSTRUCTION USING DIGITAL INTERPOLATION

In this section we develop an interpolation method for reconstructing the signal between the samples of the original sequence. The interpolation method is similar to the methods discussed in the previous two sections, but differs from them in that both of the previous methods were capable of reconstructing $f(t)$ for any continuous range of t, while the method discussed here reconstructs values only at discrete intervals. Furthermore, the interpolation method is easier to implement than either of the methods discussed in the previous sections and has both time- and frequency-domain implementations. We focus first on the time-domain implementation, which is suitable to real-time applications.

There are two steps in time-domain interpolation. The first is the insertion of zeros between the original sampled values. The second is lowpass filtering of the resulting sequence. The output of the filter is the interpolated sequence. It contains the original samples, unmodified, with the appropriate values interpolated in between. The interpolated sequence, because it has more samples over the same time span, has a higher effective sampling frequency, and consequently a higher folding frequency. Assuming the initial sequence was sampled in accordance with the sampling theorem (i.e., at greater than twice its highest frequency), the spectrum of the interpolated sequence is identical to the spectrum that would result from sampling the original signal at a faster rate. Consequently, the spectrum of the interpolated sequence is zero from one-half the original sampling rate to the new folding frequency.

To understand how this method works, let us first determine the effect of zero insertion on the spectrum of a sample sequence. Let $[x_n]$ be the original sequence of length N, and $[y_n]$ be the same sequence with Z zeros inserted after each element.

$$[y_n] = [x_0 \quad \underbrace{0 \cdots 0}_{Z \text{ zeros}} \quad x_1 \quad \underbrace{0 \cdots 0}_{Z \text{ zeros}} \quad \cdots \quad x_{N-1} \quad \underbrace{0 \cdots 0}_{Z \text{ zeros}}]$$

(5.55)

The DFT of $[y_n]$ is given by

$$\overline{Y}_k = \sum_{n=0}^{N(Z+1)-1} y_n e^{-j[2\pi nk/N(Z+1)]}; \qquad k = 0, 1, \ldots, N(Z+1) - 1 \qquad (5.56)$$

From Eq. 5.55, we note that the only nonzero terms in the sum in Eq. 5.56 occur every $Z + 1$ elements. The sum can be rewritten in terms of only these elements by using a new index i chosen such that $n = (Z + 1)i$. Using this substitution and the fact that $y_{(Z+1)i} = x_i$, Eq. 5.56 can be rewritten

$$\overline{Y}_k = \sum_{i=0}^{N-1} x_i e^{-j[2\pi k(Z+1)i/N(Z+1)]}$$

$$= \sum_{i=0}^{N-1} x_i e^{-j(2\pi ki/N)} \qquad (5.57)$$

$$= \overline{X}_k; \qquad k = 0, 1, \ldots, N(Z + 1) - 1$$

we see that the frequency index of $\overline{Y}_k$ extends from 0 to $(Z + 1)N - 1$, that is, over $Z + 1$ periods of $\overline{X}_k$ (recall that $\overline{X}_k$ is periodic with period N). Thus, the DFT of $[y_n]$

is equal to the DFT of $[x_n]$ repeated an additional Z times. In general, then, we conclude that the spectrum of any sequence formed by inserting Z zeros after each element is identical to the original spectrum replicated Z times.

As an example, let $[x_n]$ be the exponential sequence shown (along with its DFT) in Fig. 5.13. Note that the DFT values from $N/2$ to $N - 1$ are interpreted as negative frequencies as described in Section 5.2. Suppose that $[y_n]$ is formed from $[x_n]$ by inserting four zeros after each element. The resulting sequence and its DFT are shown in Fig. 5.14. As in Eq. 5.57, $[\overline{Y}_k]$ is identical to $[\overline{X}_k]$ from $k = -N/2$ to $k = N/2$ (i.e., from $\omega = -\pi/T$ to π/T), and is then repeated another four times, twice at positive frequencies and twice at negative frequencies. The frequency samples in Fig. 5.14 lie between the new folding frequencies, $-5\pi/T$ and $5\pi/T$.

Now that we know the effect of zero insertion, the purpose of the lowpass filter becomes clear. To obtain a signal with the desired spectrum we need only to pass $[y_n]$ through an ideal lowpass filter with a cutoff frequency at $\omega = \pi/T$.

As mentioned at the beginning of this section, there is also a *frequency-domain* implementation of digital interpolation. In the frequency-domain implementation, one first forms the DFT of the N-sample sequence $[x_n]$, then pads the DFT with zeros to form the filtered version of $[\overline{Y}_k]$ as illustrated in Fig. 5.14, and finally takes the inverse DFT and forms an interpolated sequence with $(Z + 1)N$ samples. More specifically, when $[x_n]$ is a real sequence and N is even, the *filtered* version of $[\overline{Y}_k]$ in Fig. 5.14 becomes the DFT, $[\overline{X}_k]$ for $0 \le k \le N/2$, extended on the right with $Z(N/2)$ zeros. The interpolated sequence is then obtained as the scaled inverse of

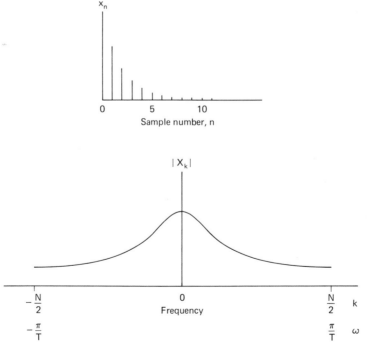

Figure 5.13 Sampled exponential sequence and its DFT.

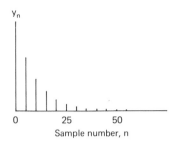

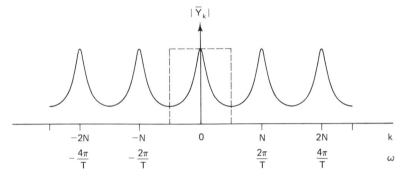

Figure 5.14 Sampled exponential with four zeros inserted after each sample, and DFT.

this extended DFT. Thus, the frequency-domain and time-domain implementations produce essentially the same interpolated sequences.

In either case, let us use $[v_n]$ to designate the real sequence produced by low-passing $[y_n]$ or, equivalently, by taking the inverse DFT of $[\overline{V_k}]$, which is $[\overline{X_k}]$ padded with $Z(N/2)$ zeros. We now demonstrate that $[v_n]$ must be scaled by $(Z + 1)$ to produce the correct interpolated sequence, $[w_n]$. The demonstration proceeds by showing that every $(Z + 1)$st element of v_n, so scaled, is equal to x_n. We begin by expressing v_n in terms of its inverse DFT:

$$v_n = \frac{1}{N(Z + 1)} \sum_{k=0}^{N(Z+1)-1} \overline{V_k} e^{j[2\pi nk/N(Z+1)]}; \qquad n = 0, 1, \ldots, Z(N + 1) - 1$$

(5.58)

Let us assume for simplicity here that N is odd. (The following argument can be easily modified when N is even.) Because of the lowpass filter, $[\overline{V_k}]$ is equal to $[X_k]$ at frequencies between $-\pi/T$ and π/T, that is, at indices from 0 through $(N - 1)/2$ and from $ZN + (N + 1)/2$ through $N(Z + 1) - 1$, and is zero elsewhere. Thus expression 5.58 can be written

$$v_n = \left[\frac{1}{Z + 1}\right] \frac{1}{N} \left[\sum_{k=0}^{(N-1)/2} \overline{X_k} e^{j[2\pi nk/N(Z+1)]} + \sum_{k=NZ+(N+1)/2}^{N(Z+1)-1} \overline{X_k} e^{j[2\pi nk/N(Z+1)]} \right]$$

This expression can be simplified by noting that the samples of $[\overline{X_k}]$ from $k = ZN + (N + 1)/2$ to $N(Z + 1) - 1$ are identical to those from $k = (N + 1)/2$ to

$N - 1$, due to the periodicity of $[X_k]$ noted above and illustrated in Fig. 5.14. Thus, v_n can be written

$$v_n = \left[\frac{1}{Z+1}\right] \frac{1}{N} \sum_{k=0}^{N-1} \overline{X}_k e^{j[2\pi nk/N(Z+1)]} \qquad (5.59)$$

We wish to evaluate this expression at every $(Z + 1)$st element, that is, for $n = i(Z + 1), i = 0, 1, \dots, N$. Making this substitution we find that

$$
\begin{aligned}
v_{i(Z+1)} &= \left[\frac{1}{Z+1}\right] \frac{1}{N} \sum_{k=0}^{N-1} \overline{X}_k e^{j[2\pi ki(Z+1)/N(Z+1)]} \\
&= \left[\frac{1}{Z+1}\right] x_i; \qquad i = 0, 1, \dots, N
\end{aligned} \qquad (5.60)
$$

Thus, if $w_n = (Z + 1)v_n$, w_n will be equal to x_n as desired.

The zero insertion (time domain) method is summarized in Fig. 5.15. The corresponding frequency-domain approach is pictured in Fig. 5.16. The frequency-domain approach is more general than the time-domain approach in that the DFT of x_n can be padded with any number (K) of zeros (not just NZ zeros). The resulting interpolated sequence will always have the desired spectrum. Only when the number of zeros is an integer multiple of N, however, will the original samples, $[x_n]$, appear in the interpolated sequence. When this is the case, we have $K = ZN$ in Fig. 5.16, and we see that the scaling in Fig. 5.16 is the same as in Fig. 5.15, that is, $Z + 1$.

We have seen that zero padding in the frequency domain can be used to increase the resolution in the time domain (i.e., padding the DFT produces more samples within a fixed time interval). One might wonder if the reverse is also true, that is, that zero padding in the time domain can be used to increase resolution in the frequency domain. The answer depends on the nature of the signal being padded. In the above discussion, zero padding in the frequency domain can be viewed as an extended sampling of the continuous signal's spectrum. Because the signal is assumed to be bandlimited, the additional samples are naturally all zeros. Similarly, zero padding is justified in the time domain when the continuous signal itself is zero beyond the original sampling range. Thus, zero padding is justified for signals of finite duration that have already been sampled over all nonzero portions. Clearly, the most general form of the DFT in Eq. 5.3 is unchanged whether the additional zeros are included in the sum or not. The advantage of zero padding is that it can be used to obtain a larger number of *discrete* frequency values when using the DFT defined in Eq. 5.9 on a digital computer.

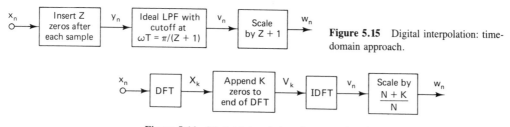

Figure 5.15   Digital interpolation: time-domain approach.

Figure 5.16   Digital interpolation: frequency-domain approach.

Examples of time- and frequency-domain interpolation are provided in Figs. 5.17 and 5.18. In Fig. 5.17 we see an interpolation of a time sequence $[f_n]$ that places three samples between each original pair of samples. The interpolation is obtained by extending the DFT, $[F_m]$, with zeros, scaling, and then taking the inverse DFT. In Fig. 5.18, we see an interpolation of the DFT accomplished by extending the time sequence, $[f_n]$, with zeros and then taking the DFT. In both Figs. 5.17 and 5.18, we note that the samples of the original sequence agree with those of the interpolated sequence whenever the two coincide.

When the signal is not of finite duration, periodic for example, zero padding can lead to erroneous results. For example, if a periodic signal is sampled over one period and then extended with zeros, the resulting DFT is that of the zero-extended periodic sequence, not the original periodic sequence, as illustrated in Fig. 5.19. Clearly the DFT of the zero-extended sequence is not simply a higher resolution version of the original signal's DFT. In conclusion, one must be careful about the use of zero padding with sampled signals. It is justified only when a signal is of finite duration and has already been sampled over the range where it is nonzero.

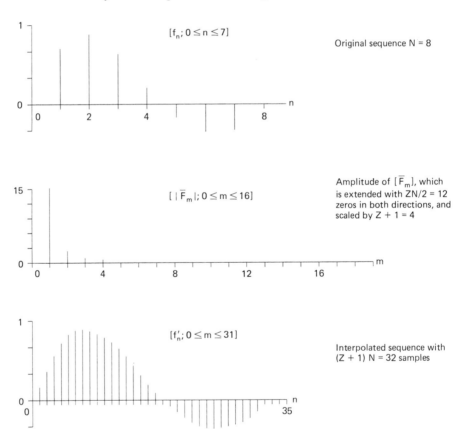

Figure 5.17 Digital interpolation of a time sequence using zero padding in the frequency domain with $Z = 3$.

The Discrete Fourier Transform   Chap. 5

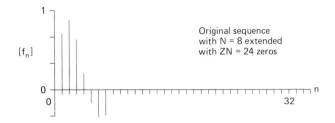

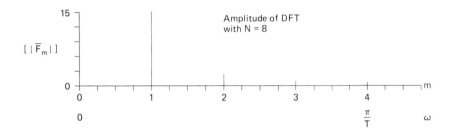

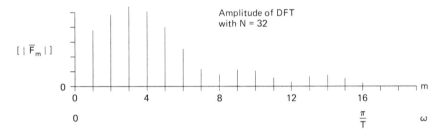

Figure 5.18 Digital interpolation of the DFT using zero padding in the time domain with $Z = 3$.

5.8 FINITE SAMPLE DURATION

As mentioned in Section 4.4, cases sometime arise where samples of the function $f(t)$ are not impulse functions and cannot be treated as such. When the sampling process involves measuring a physical quantity and performing an analog-to-digital conversion, energy usually must be absorbed by the measurer. Unless only the absorption of a single quantum of energy is involved, the absorption takes time, so the sample is of nonzero duration.

In many cases, samples can be treated as though they were impulse samples even though they are not. The judgment of whether or not they can be so treated can be based on whether or not $f(t)$ is relatively constant while each sample is being taken. Here it is assumed that variations in $f(t)$ during the sampling interval must be taken into account.

In a sense, two types of spectral effects are caused by finite sampling. The first is the effect on the spectrum of the sample set itself and is most relevant where

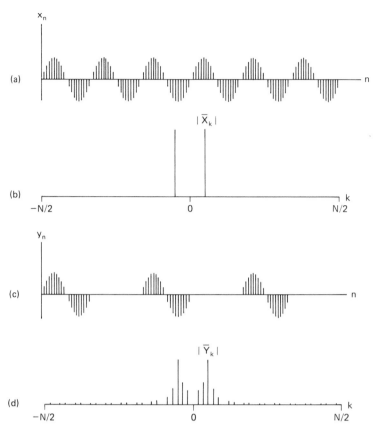

Figure 5.19 Example of padding a sampled periodic sequence with zeros: (a) sampled periodic sequence $[x_n]$; (b) $[X_k]$, the DFT of one period of $[x_n]$; (c) sampled periodic signal with zero insertion, $[y_n]$; (d) $[Y_k]$, the DFT of one period of $[y_n]$.

sampled-data systems are concerned. The second is the effect on the computed spectrum of $f(t)$, that is, on the DFT.

First, concerning the spectrum of the sample set itself, recall that $\overline{F}_s(j\omega)$, the transform of the sampler output in Chapter 4, was shown to be the Fourier transform (not the DFT) of the train of impulse samples, so that Eq. 5.21 gives

$$\overline{F}_s(j\omega) = \frac{1}{T} \sum_{n=-\infty}^{\infty} F\left(j\omega - j\frac{2\pi n}{T}\right) \tag{5.61}$$

This result applies with impulse sampling. What happens to $F_s(j\omega)$, viewed now as the Fourier transform of the sample train, when the samples become finite in duration? This question is answered, following Ragazzini and Franklin (1958), as in Section 5.3 for the impulse case, by letting $p(t)$ be a train of pulses as illustrated in Fig. 5.20. The duration of each pulse is w seconds, and the amplitude is shown as T/w. [For any other constant amplitude, $p(t)$ can simply be multiplied by a constant in the following discussion.] The product $f_s(t) = f(t)p(t)$ now represents the train of finite samples of $f(t)$.

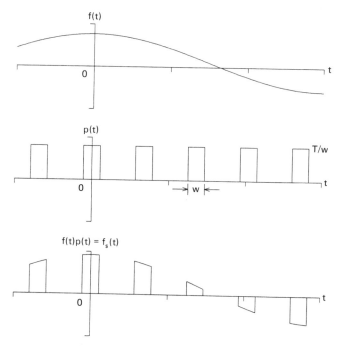

Figure 5.20 Operation of the finite-duration sampler.

The pulse train $p(t)$, being a periodic function, can be expanded in a Fourier series as follows:

$$p(t) = \sum_{n=-\infty}^{\infty} c_n e^{j(2\pi nt/T)}$$

$$c_n = \frac{1}{T} \int_{-T/2}^{T/2} p(t) e^{-j(2\pi nt/T)} dt$$

$$= \frac{1}{T} \int_{-w/2}^{w/2} \frac{T}{w} e^{-j(2\pi nt/T)} dt$$

$$= \frac{T}{n\pi w} \sin\left(\frac{n\pi w}{T}\right)$$

Therefore,

$$p(t) = \sum_{n=-\infty}^{\infty} \frac{T}{nw\pi} \sin\left(\frac{n\pi w}{T}\right) e^{j(2\pi nt/T)} \tag{5.62}$$

so the sample pulse train of $f(t)$ is

$$f_s(t) = f(t)p(t)$$

$$= \sum_{n=-\infty}^{\infty} \frac{Tf(t)}{nw\pi} \sin\left(\frac{n\pi w}{T}\right) e^{j(2\pi nt/T)} \tag{5.63}$$

and its spectrum, obtained from line 6 of Table 3.1 (Chapter 3) is

$$F_s(j\omega) = \sum_{n=-\infty}^{\infty} \frac{T}{nw\pi} \sin\left(\frac{n\pi w}{T}\right) F\left(j\omega - j\frac{2\pi n}{T}\right) \tag{5.64}$$

Thus, in contrast with Eq. 5.61, an envelope of the form $\sin(x)/x$ is imposed on the superposition of the original spectra.

Two limiting cases of Eq. 5.64 can be noted immediately. The first is where the sample duration w approaches zero, and

$$\lim_{w \to 0}\left[\frac{T}{nw\pi} \sin\left(\frac{nw\pi}{T}\right)\right] = 1 \tag{5.65}$$

so that Eqs. 5.64 and 5.61 become identical except for a constant, that is,

$$F_s(j\omega) = \sum_{n=-\infty}^{\infty} F\left(j\omega - j\frac{2\pi n}{T}\right) \quad \text{if } w \to 0 \tag{5.66}$$

The missing constant, $1/T$, in Eq. 5.66 results because the area of each pulse in $p(t)$ was assumed to be T instead of one (see Fig. 5.20).

The second limiting case in Eq. 5.64 is where $w = T$, so that there is no time between samples. Then only the term for $n = 0$ is nonzero, and

$$F_s(j\omega) = F(j\omega) \quad \text{if } w \to T \tag{5.67}$$

This limiting case is also reasonable, since the output of the sampler is just $f(t)$.

Turning now to the second effect of finite sampling mentioned at the beginning of this section, what changes occur in the computed spectrum of $f(t)$, as opposed to the spectrum of the sampled pulse train discussed above?

Here, of course, it is assumed that the spectral computation is based on the output of an ADC and naturally the characteristics of the converter affect the outcome. Many converters detect the peak value of the finite sample, thus causing a jitter in the sequence of samples. The effect of this jitter is difficult to assess in general terms, because it depends on the details of the waveform (see Chapter 4, Section 4.4).

On the other hand, if it is assumed that the average value of each finite-duration sample is recorded, the general effect on the spectral computations can be found. Suppose that a segment of $f(t)$ from $t = 0$ to $t = (N - 1)T$ is sampled as in Fig. 5.21 with the objective of computing the DFT or, equivalently, a Fourier component, at a set of frequencies including the frequency ω. For this purpose $f(t)$ can be said to contain an unknown component at the frequency ω plus components at other frequencies:

$$f(t) = A \sin(\omega t + \alpha) + g(t) \tag{5.68}$$

In this expression for $f(t)$, A is the amplitude and α the phase of the unknown component and $g(t)$ contains no components at ω.

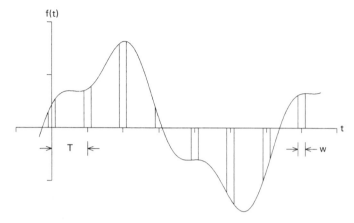

Figure 5.21 Finite-width samples for A-D conversion.

If the ADC averages each sample of $f(t)$, the value of the nth sample of $f(t)$ in Eq. 5.68 must be

$$f_n = \frac{1}{w} \int_{nT-w/2}^{nT+w/2} f(t)\, dt$$

$$= \frac{1}{w} \int_{nT-w/2}^{nT+w/2} A\ \sin(\omega t + \alpha)\, dt + g_n \qquad (5.69)$$

$$= \left[\frac{\sin(\omega w/2)}{\omega w/2} \right] A\ \sin(\omega nT + \alpha) + g_n$$

where g_n is the nth sample of $g(t)$.

Thus Eq. 5.69 indicates that a sampling interval of width w has the effect of scaling the computed component of $f(t)$ at the frequency ω by the amount in the square brackets. That is, finite-width sampling imposes an envelope of the form

$$E = \frac{\sin(\omega w/2)}{\omega w/2} \qquad (5.70)$$

on the computed spectrum of $f(t)$ when $f(t)$ is averaged over the pulse duration. Note the singular case where $w = 2\pi/\omega$ and E is therefore zero. This value of E is reasonable since here a sample is an average of $f(t)$ over a complete period of the component in question. The other singular case is where $w = 0$ and there is impulse sampling, with $E = 1$.

To summarize this section, the two following formulas give the effect of finite-duration sampling on the spectrum of the pulse train and on the computed spectrum. The first is Eq. 5.64, and the second is Eq. 5.61 times its envelope E in Eq. 5.70.

The spectrum of the finite pulse train is

$$F_s(j\omega) = \sum_{n=-\infty}^{\infty} \frac{T}{nw\pi} \sin\left(\frac{nw\pi}{T}\right) F\left(j\omega - j\frac{2\pi n}{T}\right) \qquad (5.71)$$

Assuming that the ADC averages $f(t)$ over each sample duration, the computed spectrum of the finite sample set is

$$\overline{F}(j\omega) = \frac{1}{T} \sum_{n=-\infty}^{\infty} \frac{\sin((\omega - 2\pi n/T)w)}{(\omega - 2\pi n/T)w} F\left(j\omega - j\frac{2\pi n}{T}\right) \qquad (5.72)$$

Example 5.5

A function $f(t)$ having the spectrum

$$F(j\omega) = 1 + \cos \omega; \qquad |\omega| \leq \pi$$
$$= 0; \qquad\qquad |\omega| \geq \pi \qquad (5.73)$$

is sampled at 0.8-s intervals (fast enough so that no aliasing is involved), with sample width $w = 0.4$ s. What is the spectrum $F_s(j\omega)$ of the sample pulse train, and what is the computed DFT, $\overline{F}(j\omega)$, if the ADC averages $f(t)$ over each sample pulse? The answer is shown in Fig. 5.22. In both cases, the finite sample duration imposes an envelope on the superposition of Fourier transforms. The plot of $\overline{F}(j\omega)$ should be compared with the impulse sampling case in Example 5.2.

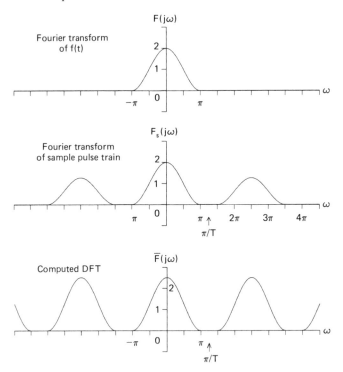

Figure 5.22 Spectra for Example 5.5; $T = 0.8$ s, $w = 0.4$ s, showing the effect of finite-width sampling on the sample pulse train and on the computed DFT.

5.9 THE DFT IN CONVOLUTION AND CORRELATION

Just as in the case of continuous convolution described in Section 3.3, discrete convolution in the time domain often involves the convolution of an input sequence with an impulse response to produce an output sequence and implies a corresponding

The Discrete Fourier Transform   Chap. 5

product of DFTs. Similarly, a frequency-domain convolution of DFTs implies a product of functions in the time domain. The two convolution theorems, being closely related, are presented together in this section. Since convolution and correlation are similar operations, we also present a correlation theorem.

The theorem for discrete convolution in the time domain (Gold and Rader, 1969; Bergland, 1969) emphasizes an important difference between continuous transform products and DFT products: In Chapter 3 the inverse transform of the product $F(j\omega)H(j\omega)$ was shown to be the convolution of $f(t)$ with $h(t)$, but here the inverse transform of the DFT product $\overline{F}_m\overline{H}_m$ is shown to be the "circular" convolution of one sample set, say $[h_n]$, with the periodic extension of the other sample set, say $[\bar{f}_n]$, where the bar over f_n indicates periodicity and

$$\text{periodic } \bar{f}_{n+kN} = f_n; \qquad \begin{aligned} n &= 0, 1, \ldots, N-1 \\ k &= 0, \pm 1, \ldots, \pm\infty \end{aligned} \qquad (5.74)$$

The theorem itself can be stated as follows:

Discrete Convolution in Time
The inverse DFT of the product of two DFTs is a periodic or circular convolution, that is,

$$\text{if } \overline{G}_m = \overline{F}_m\overline{H}_m; \qquad m = 0, 1, \ldots, N-1$$

$$\text{then } g_n = \sum_{m=0}^{N-1} f_m\overline{h}_{n-m} = \sum_{m=0}^{N-1} h_m\bar{f}_{n-m} \qquad (5.75)$$

The proof of the theorem proceeds from the inverse DFT formula, Eq. 5.13:

$$g_n = \frac{1}{N} \sum_{i=0}^{N-1} \overline{G}_i e^{j(2\pi ni/N)}$$

$$= \frac{1}{N} \sum_{i=0}^{N-1} \overline{F}_i \overline{H}_i W_N^{-ni} \qquad (5.76)$$

where $W_N = e^{-j(2\pi/N)}$ is used to simplify the notation. Next the DFT formulas are substituted for $\overline{F}_i$ and $\overline{H}_i$ and the sums are reordered:

$$g_n = \frac{1}{N} \sum_{i=0}^{N-1} \left[\sum_{m=0}^{N-1} f_m W_N^{mi} \sum_{k=0}^{N-1} h_k W_N^{ki} \right] W_N^{-ni}$$

$$= \frac{1}{N} \sum_{m=0}^{N-1} \sum_{k=0}^{N-1} f_m h_k \sum_{i=0}^{N-1} W_N^{(m+k-n)i} \qquad (5.77)$$

Using the geometric series formula in Chapter 1, Eq. 1.19, the sum on the right in Eq. 5.77 is seen to be zero unless $(m + k - n)$ is zero or a multiple of N, in which case each term is one and the sum equals N. Therefore, the double sum may have nonzero terms only when $k = n - m +$ (multiple of N). Thus the double sum is really a single sum with h_k replaced by $\overline{h}_{m-n}$, and g_n becomes

$$g_n = \sum_{m=0}^{N-1} f_m \overline{h}_{n-m} \qquad (5.78)$$

so the theorem is proved. (The reversed form in Eq. 5.75 follows simply by reversing the roles of $\overline{F}$ and $\overline{H}$.)

The periodic convolution is illustrated in Fig. 5.23. On the left are two sample sets, $[f_n]$ and $[h_n]$, and on the right are the terms f_m and $\overline{h}_{m-n}$ in the periodic convolution, Eq. 5.75. The convolution is the product of the two waveforms on the right, summed over m.

The form of the theorem above suggests a possible problem: Suppose that one wishes to approximate the continuous convolution of $f(t)$ and $h(t)$ by taking a product of DFTs. How can the error introduced by the periodicity illustrated in Fig. 5.23 be avoided? A solution (Bergland, 1969) is illustrated in Fig. 5.24, where N is doubled by padding the sample sets with zeros. The revised form of Eq. 5.75 is shown

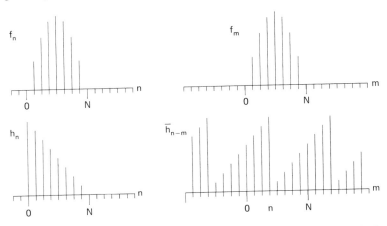

Figure 5.23 Illustrating the periodic or circular convolution, Eq. 5.75. Convolution is the sum (over m) of f_m times $\overline{h}_{n-m}$ on the right.

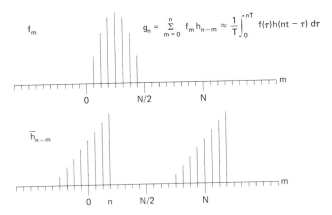

$$g_n = \sum_{m=0}^{n} f_m h_{n-m} \approx \frac{1}{T} \int_0^{nT} f(\tau)h(nt - \tau)\, d\tau$$

Figure 5.24 Approximation to continuous convolution obtained by adding zeros to sample sets.

The Discrete Fourier Transform   Chap. 5

in the figure, and is seen to be the "correct" approximation to the convolution integral. This procedure, in which the ordinary convolution is obtained via a product of specially chosen DFTs is useful in digital filtering applications. The input sample set is $[f_n]$, the filter impulse response is $[h_n]$ padded with zeros as in Fig. 5.24, and filtering is actually accomplished by taking the inverse transform of the product, $[\overline{F}_m \overline{H}_m]$.

The theorem for convolution in the frequency domain, which is similar to the preceding theorem, can be stated as follows:

Discrete Convolution in Frequency
The DFT of the product of two sample sets is a periodic convolution of DFTs, that is,

$$\text{if } g_n = f_n h_n; \qquad n = 0, 1, \ldots, N - 1$$

$$\text{then } \overline{G}_m = \frac{1}{N} \sum_{n=0}^{N-1} \overline{F}_n \overline{H}_{m-n} \tag{5.79}$$

$$= \frac{1}{N} \sum_{n=0}^{N-1} \overline{H}_n \overline{F}_{m-n}$$

The proof of this theorem proceeds as in Eqs. 5.76 through 5.78 for time convolution, that is,

$$\begin{aligned}
\overline{G}_m &= \sum_{i=0}^{N-1} g_i W_N^{mi} \\
&= \sum_{i=0}^{N-1} \left[\frac{1}{N^2} \sum_{n=0}^{N-1} \sum_{k=0}^{N-1} \overline{F}_n \overline{H}_k \right] W_N^{(m-n-k)i} \\
&= \frac{1}{N^2} \sum_{n=0}^{N-1} \sum_{k=0}^{N-1} \overline{F}_n \overline{H}_k \sum_{i=0}^{N-1} W_N^{(m-n-k)i} \\
&= \frac{1}{N} \sum_{n=0}^{N-1} \overline{F}_n \overline{H}_{m-n}
\end{aligned} \tag{5.80}$$

As in Eq. 5.78, the last line follows because the sum of W_N-terms equals N only when k equals $m - n$ plus a multiple of N, and $[\overline{H}_k]$ is of course periodic with $\overline{H}_k = \overline{H}_{k+N}$, and so on.

To this point we have discussed the relationships between convolution in time, multiplication in frequency, and vice versa. Since convolution and correlation are similar operations, it is not surprising that similar relationships exist for correlation. We define a *periodic correlation function* similar to Eq. 5.75:

$$g_n = \sum_{m=0}^{N-1} f_m \overline{h}_{m+n} = \sum_{m=0}^{N-1} h_m \overline{f}_{m-n} \tag{5.81}$$

Here we have a total product similar to Eq. 5.75, except that the periodic time series is not reversed in this case. In fact, if we use the reversed sequence $\bar{f}'_k = \bar{f}_{-k}$ in the second version in Eq. 5.81, we have a convolution:

$$g_n = \sum_{m=0}^{N-1} h_m \bar{f}'_{n-m} \tag{5.82}$$

But Eqs. 5.9 and 5.10 give the DFT of any reversed sequence as

$$\mathrm{DFT}[f_{-n}] = [\overline{F}_{-m}] = [\overline{F}^*_m] \tag{5.83}$$

Thus, given the theorem for discrete convolution in time, we have proved a similar theorem for

---

Discrete Correlation in Time
The DFT of the correlation function is a product of two DFTs, that is,

$$\text{if } g_n = \sum_{m=0}^{N-1} f_m \bar{h}_{m+n} = \sum_{m=0}^{N-1} h_m \bar{f}_{m-n}$$

$$\text{then } \overline{G}_m = \overline{F}^*_m \overline{H}_m; \qquad m = 0, 1, \ldots, N-1 \tag{5.84}$$

The correlation function is defined in more detail in Chapter 13 and used in Chapters 14 and 15. The correlation theorem given here is of considerable practical interest because, using the fast Fourier transform (FFT) discussed in Chapter 6, it provides a means for reducing the computation necessary to obtain the correlation function.

5.10 EXAMPLES OF DISCRETE SPECTRA

Some examples of digital spectral computations made with real physical data are shown in the following figures. In each case the computations were made on a general-purpose computer using the "SPFFT" routine in Appendix B. In the first example the signal is a transient pulse, but in the remaining cases the signals extend beyond the limits of observation and have varying amounts of periodic content.

The first example in Fig. 5.25 gives the amplitude spectrum of a detected radar pulse, courtesy of Sandia National Laboratories, digitized with a 100-ns sampling interval. The folding frequency is thus 5 MHz in this example. To prevent aliasing, the pulse was passed through a lowpass analog filter cutting off at 2 MHz before being digitized. The lower-frequency components in the amplitude spectrum reflect the overall triangular shape of the pulse, while the higher components convey mostly information about its fine structure, which in this case represents the "signature" of the radar target. The components above $m = 25$ are lower due to the effect of the lowpass analog filter.

The second spectral illustration using physical data is in Fig. 5.26. The signal in this case is a seismic waveform, courtesy of Professor Jon Berger, University of California at San Diego, representing about 4 minutes of earth strain resulting from an underground explosion. The recording was made at the earth's surface, several

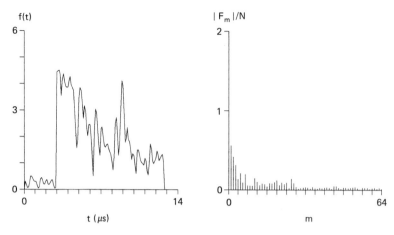

Figure 5.25 Radar echo pulse and its amplitude spectrum. Sampling interval $T = 100$ ns; $N = 128$.

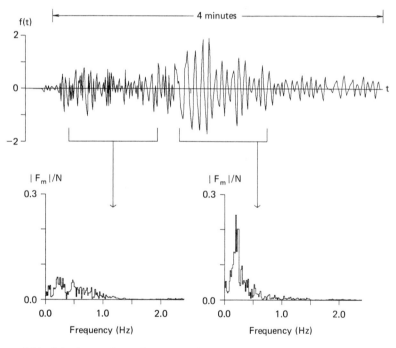

Figure 5.26 Seismic waveform with spectra of two different segments. Time step $T = 0.125$ s, $N = 512$ for each spectrum.

hundred miles distant from the site of the explosion. The digitizing rate in this case was eight samples per second. Amplitude (strain) units are arbitrary.

The two spectra in Fig. 5.26 tend to confirm one's observation of the waveform itself: The leading portions are varying at frequencies above those found in the later portions. Seismologists use this type of information to deduce facts about the origin of the seismic signal and its modes of propagation.

Sec. 5.10   Examples of Discrete Spectra

95

The third example of computed spectra in Fig. 5.27 is between the first two in its range of frequencies. The figure shows two speech waveforms, namely two hard E (as in bee) sounds, and their amplitude spectra. The upper sound was made by a male speaker and the lower by a female. Note how the two waveforms and spectra differ. How amazing it is that the human ear is so easily able to discern the gestalt of the E sound from waveforms such as these.

Finally, Fig. 5.28 again shows 32 ms of a sound wave and its spectrum, but this time the sound of a guitar string is substituted for the voiced sound. A western guitar of traditional style was used. The guitar was not tuned to an absolute musical scale. Here, as well as generally in the case of musical string sounds, the spectrum is richer than the voice spectrum in harmonics of the fundamental string frequency. The fundamental or "pitch" frequency in this case is approximately 290 Hz, and the other major components in the spectrum are at multiples of this frequency.

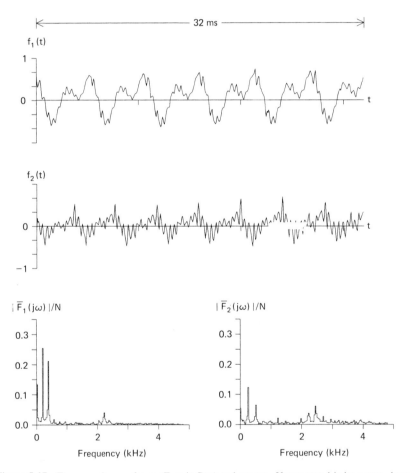

Figure 5.27 Two speech waveforms (E as in Bee) and spectra. Upper sound is by a man, lower by a woman. Sampling interval $T = 62.5$ μs; $N = 512$.

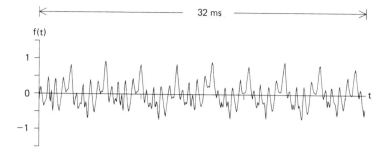

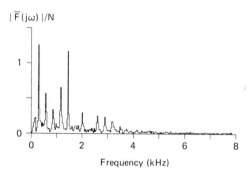

Figure 5.28 Sound and computed amplitude spectrum of a guitar E string. $T = 62.5\ \mu\text{sec}$; $N = 512$.

EXERCISES

1. Verify properties 1 through 4 in Section 5.2.
2. A waveform $f(t)$ is shown below.
   (a) Derive the Fourier transform, $F(j\omega)$, and sketch its magnitude.
   (b) Suppose that $a = 1$ and $f(t)$ is sampled every $T = 0.2$ s to form an impulse sequence, $\bar{f}(t)$. Derive the Fourier transform, $\bar{F}(j\omega)$, and sketch its magnitude.
   (c) Suppose that $a = 1$ and $f(t)$ is replicated every 2 s to form a periodic signal, $f_p(t)$. Derive the Fourier series coefficients $[c_n]$ and sketch the amplitude spectrum.
   (d) Suppose that $a = 1$ and the periodic signal, $f_p(t)$, in part (c) is sampled every $T = 0.2$ s. Derive the Fourier series coefficients $[c_n]$ and sketch the amplitude spectrum.

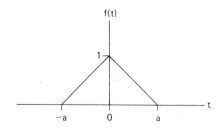

3. A waveform $s(t)$ has the following spectrum:

$$S(j\omega) = 1 - \left|\frac{\omega}{4\pi}\right|; \qquad |\omega| < 4\pi$$

$$= 0; \qquad |\omega| \geq 4\pi$$

and is sampled over its entire nonzero range at intervals of T seconds. Sketch the Fourier transform, $\bar{S}(j\omega)$, as well as the DFT for $N = 1$.
(a) If $T = 0.5$ s
(b) If $T = 0.3$ s
(c) If $T = 0.2$ s

4. A waveform $u(t)$ having the spectrum

$$U(j\omega) = \cos(10^{-3}\pi\omega); \qquad |\omega| < 500$$

$$= 0; \qquad |\omega| \geq 500$$

is sampled every millisecond over its nonzero range.
(a) Sketch the Fourier transform, $\bar{U}(j\omega)$ in Eq. 5.21.
(b) Sketch the spectrum of the Whittaker reconstruction, $u^*(t)$.

5. Given the sampled function shown, find the DFT in Eq. 5.3 **(a)** using $N = 5$ nonzero samples, and **(b)** with $N = 11$. Comment on the difference between the DFTs, if any.

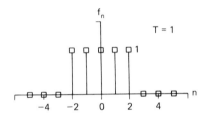

6. Given the function $f(t) = 200te^{-10t}$ for $t \geq 0$ and a sampling interval of $T = \pi/30$ s:
(a) Derive $F(j\omega)$ and sketch its magnitude.
(b) Derive a formula for $\bar{F}(j\omega)$.
(c) Using the sketch of $|F(j\omega)|$, make an approximate sketch of $|\bar{F}(j\omega)|$.

7. If $f(t)$ has the spectrum shown, sketch the DFT in Eq. 5.3 for sampling rates of r, $r/2$, and $r/4$ samples per second.

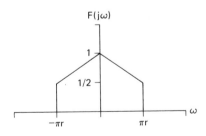

8. Given the samples of $f(t)$ shown, compute the DFT in Eq. 5.9 for **(a)** $N = 4$ samples, and **(b)** $N = 6$.

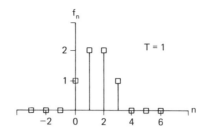

9. If $f(t) = e^{-t} \cos(t + \gamma)$ is sampled in the range $(0, 10)$ often enough to convey essentially all of its frequency content, give a bound on the error in the computed amplitude spectrum, $|F(j\omega)|$. Assume that the entire error is due to truncation of $f(t)$.

10. For $f(t)$ in Exercise 8, state a formula for the reconstruction, $\hat{f}(t)$, for **(a)** $N = 4$ samples, and **(b)** $N = 6$.

11. Let $f(t) = 3e^{-t/10}$ for $t \geq 0$ be sampled at $t = 0, 1, \ldots, 63$.
    (a) Find the Fourier transform, $F(j\omega)$.
    (b) Comment on the adequacy of the sampling rate.
    (c) Comment on the adequacy of the number of samples.
    (d) Find the Fourier transform, $\overline{F}(j\omega)$.
    (e) Plot the magnitudes of $F(j\omega)$ and $\overline{F}(j\omega)$ and observe the relative error.

12. Given $f(t) = te^{-t/10}$ for $t \geq 0$, sampled at $t = 0, 1, \ldots, 127$.
    (a) Find the Fourier transform, $F(j\omega)$.
    (b) Find the Fourier transform, $\overline{F}(j\omega)$.
    (c) Plot and compare $|F(j\omega)|$ and $|\overline{F}(j\omega)|$.
    (d) Compute the reconstruction, $\hat{f}(t)$, at and halfway between the sample points.
    (e) Plot and compare $f(t)$ and $\hat{f}(t)$.

13. Samples are taken over effectively all of the nonzero portion of the function

$$f(t) = e^{-t}; \quad t \geq 0$$

starting at $t = 0$ and proceeding at intervals of T s. Plot the amplitude spectrum $|F(j\omega)|$, along with the magnitude $|\overline{F}(j\omega)|$, for $T = 0.1$ and $T = 0.2$, and compare the three plots.

14. Derive the simplified form of

$$\sum_{n=0}^{N-1} e^{j2\pi n(i-m)/N}$$

in Eq. 5.37.

15. Given the DFT shown with $N = 5$, find the Fourier series reconstruction of $f(t)$.

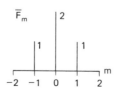

16. In Eq. 5.32, prove that $\hat{f}(nT) = f_n$ for the case where N is even.

17. Letting R_m and I_m be the real and imaginary parts of $\overline{F}_m$, express the reconstruction formula in Eq. 5.42 as a real Fourier series of the form $\hat{f}(t) = a_0/2 + \sum_m (a_m \cos m\omega_0 t + b_m \sin m\omega_0 t)$.

18. A function $f(t)$ having the spectrum $F(j\omega) = \cos(\omega/2)$ for $|\omega| \le 3\pi$ and zero elsewhere is sampled every half-second and then reconstructed using Whittaker's formula. Sketch the spectrum of the reconstruction.

19. Given the sample set as shown here, with t in seconds,

t	0	1	2	3	4	5
$f(t)$	0	3	5	2	1	0

find a function $\hat{f}(t)$ having this same sample set and having no frequencies greater than $\frac{1}{2}$ Hz. Plot $\hat{f}(t)$ for $0 \le t \le 5$.

20. Start with Eq. 5.58 and derive Eq. 5.60 for N even. (*Note:* Assume that the ideal lowpass filter reduces the frequency samples at π/T and $-\pi/T$ to half strength.)

21. The spectrum of a signal $x(t)$, sampled at 10 samples per second, is shown below. Sketch the spectrum of $[y_n]$ that is formed from $[x_n]$ by inserting three zeros after each sample. What is the effective folding frequency? Suppose that we wish to form an interpolated version of $[x_n]$ by passing $[y_n]$ through a lowpass filter. What can you say about the specific requirements of the filter in this case?

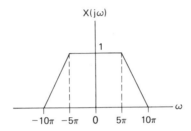

22. Given an ADC with sampling interval $= 10~\mu s$ that averages each sample of $f(t)$ over a window of width w, choose w so that aliasing due to a 1-MHz noise component in $f(t)$ will be effectively eliminated.

23. Show that the inverse DFT of the product of $\overline{F}_m^*$ and $\overline{H}_m$ (where * denotes complex conjugate) is the periodic correlation

$$g_n = \sum_{m=0}^{N-1} f_m \overline{h}_{n+m}$$

24. Given the sampled sequence

$$f(n) = \sin(0.4n); \qquad n = 0, 1, \ldots, 127$$

(a) Use the time-domain implementation of digital interpolation to construct an interpolated version of $f(n)$ with four times as many samples. Plot your results.

(b) Perform the same interpolation as in part (a) using the frequency-domain approach. Are the results identical? Explain.

SOME ANSWERS

2. (a) $F(j\omega) = a\left[\dfrac{\sin(\omega a/2)}{\omega a/2}\right]^2$ **5.** $\overline{F}(j\omega) = 1 + 2\cos\omega + 2\cos 2\omega$

6. $|F(j\omega)| = 200/(\omega^2 + 100)$

8. (a)

Real	Imaginary
6.00	0.00
−1.00	−1.00
0.00	0.00
−1.00	1.00

(b)

Real	Imaginary
6.00	0.00
0.00	−3.46
0.00	0.00
0.00	0.00
0.00	0.00
0.00	3.46

11. (a) $F(j\omega) = 3/(j\omega + 0.1)$ **(d)** $\overline{F}(j\omega) = 3(1 - e^{-64(j\omega + 0.1)})/(1 - e^{-(j\omega + 0.1)})$

15. $\hat{f}(t) = \frac{2}{5}[1 + \cos(2\pi t/5T)], \ 0 \le t \le 5T$

17. $\omega_0 = 2\pi/NT; \ a_0 = 2\overline{F}_0/N; \ a_m = 2R_m/N; \ b_m = -2I_m/N$

19. $\hat{f}(t) = \sum_{n-1}^{4} f_n \sin[(t - n)\pi]/[(t - n)\pi]$ **22.** $w = 1 \ \mu s$

24. The results should be identical since the processes (a) and (b) are equivalent.

REFERENCES

BERGLAND, G. D., A Guided Tour of the Fast Fourier Transform. *IEEE Spectrum,* July 1969, p. 41.

BRACEWELL, R., *The Fourier Transform and Its Applications,* Chap. 10. New York: McGraw-Hill, 1965.

CAMPBELL, A. B., *A New Sampling Theorem for Causal (Non-bandlimited) Functions,* Ph.D. dissertation. University of New Mexico, Albuquerque, July 1973.

COOLEY. J. W., LEWIS, P. A. W., and WELCH, P. D., The Finite Fourier Transform. *IEEE Trans. Audio Electroacoust.,* Vol. AU-17, No. 2, June 1969, p. 77.

GOLD, B., and RADER, C. M., *Digital Processing of Signals,* Chap. 6. New York: McGraw-Hill, 1969.

HOVANESSIAN, S. A., and PIPES, L. A., *Digital Computer Methods in Engineering,* Chap. 4. New York: McGraw-Hill, 1969.

PAPOULIS, A., Error Analysis in Sampling Theory. *Proc. IEEE,* Vol. 54, No. 7, July 1966, p. 947.

RAGAZZINI, J. R., and FRANKLIN, G. F., *Sampled-Data Control Systems.* New York: McGraw-Hill, 1958.

TOU, J. T., *Digital and Sampled-Data Control Systems,* Chap. 3. New York: McGraw-Hill, 1959.

WHITTAKER, E. T., Expansions of the Interpolation-Theory. *Proc. Roy. Soc. Edinburgh,* Vol. 35, 1915, p. 181.

6

The Fast Fourier

Transform

6.1 INTRODUCTION

The fast Fourier transform (FFT) is not a new type of transformation. It is instead an algorithm for computing the finite DFT just described in Chapter 5. It is important because, by eliminating most of the repetition in the DFT formula, it allows a much more rapid computation of the DFT. The FFT also generally allows a more accurate computation of the DFT by reducing round-off errors.

The FFT has a great variety of applications in digital signal processing. The methods of spectral analysis in such fields as speech communications, crystallography, seismology, vibration mechanics, and others often require the direct computation of spectra from large sample sets, or repeated computations of spectra from large numbers of sample sets. In these cases the FFT may provide the only possible means for spectral computation within the limits of time and computing cost.

The FFT algorithm can be realized as a program in a general-purpose computer or as a piece of special-purpose hardware. Examples of the former, in the form of Fortran subroutines, are presented in Appendix B. Examples of special-purpose hardware can be found in the FFT "black boxes" in modern spectrum analyzers.

The purpose of this chapter is to illustrate the redundancy encountered in computing the DFT, to show how the FFT algorithm eliminates this redundancy, and to illustrate different versions of the FFT. The discussion is generally applicable to hardware as well as to software implementations of the FFT.

6.2 REDUNDANCY IN THE DFT

As in Chapter 5 a complete set of values of the DFT is given by

$$\overline{F}_m = \sum_{n=0}^{N-1} f_n e^{-j(2\pi mn/N)}; \qquad m = 0, 1, \ldots, N - 1 \qquad (6.1)$$

in which m is the frequency index (i.e., $\omega = 2\pi m/NT$ rad/s), f_n is the nth sample of $f(t)$, and N is the total number of samples. The range of m in Eq. 6.1 need only be from 0 to $N/2$ when $f(t)$ is real, but as in Chapter 5 the discussion is simplified by assuming that the full set of values $\overline{F}_0, \overline{F}_1, \ldots, \overline{F}_{n-1}$ is to be computed.

A simple measure of the amount of computation in Eq. 6.1 is the *number of complex products* implied by the equation and the range of m. It is easy to see that there are N sums each with N products, or N^2 products in the total computation of the set $[\overline{F}_m]$. Because of the cyclic property of the exponential function, one might guess that some of these products are redundant, that is, duplicated once or more in the course of the total computation.

To simplify the notation in Eq. 6.1, let W_N represent the invariant part of the exponential term, that is,

$$W_N = e^{-j(2\pi/N)} \tag{6.2}$$

so that the DFT formula in Eq. 6.1 now becomes

$$\overline{F}_m = \sum_{n=0}^{N-1} W_N^{mn} f_n; \qquad m = 0, 1, \ldots, N-1 \tag{6.3}$$

Each $\overline{F}_m$ can be viewed as a linear combination of the sample set $[f_n]$, with $[W_N^{mn}]$ being the set of coefficients.

The cyclic character of the coefficients, which is easily seen from the definition of W_N in Eq. 6.2, is illustrated for $N = 8$ in Fig. 6.1. Note that N must be specified in order to state exactly the equivalence of different powers of W_N. This fact creates problems in any general discussion of the FFT, and the specific case with $N = 8$ will therefore be used for most of the illustrations in this chapter.

How can one take advantage of the cyclic property of W_N^{mn} to eliminate some of the N^2 products in Eq. 6.3? A partial but simple answer to this question can be found by assuming first that N is not prime, but has a factor, say 2, so that $N = 2P$ and P is also an integer.[†] One could then decompose the sample set $[f_n]$ into

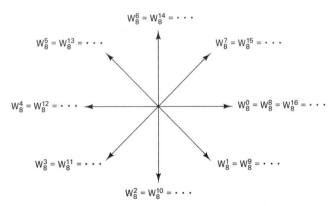

Figure 6.1   Illustrating the equivalence of different powers of W_N with $N = 8$.

[†]The development here is based on Cochran, Cooley, et al. (1967).

two subsets, one containing even-numbered samples and the second containing odd-numbered samples, and rewrite Eq. 6.3 as follows:

$$\overline{F}_m = \sum_{n=0}^{P-1} f_{2n} W_N^{2mn} + W_N^m \sum_{n=0}^{P-1} f_{2n+1} W_N^{2mn} \tag{6.4}$$

The range of m here is still 0 to $N - 1$, and P is $N/2$ by definition; thus each sum in Eq. 6.4 has $N/2$ products, and so it appears at first that there are now $N + 1$ products for each m (including the product of W_N^m times the second sum), or $N^2 + N$ products in all.

However, upon closer examination, each sum in Eq. 6.4 can be written in the form of a DFT. Let $[a_n] = [f_{2n}]$ and $[b_n] = [f_{2n+1}]$ represent the set of even- and odd-numbered samples, respectively. The DFTs of $[a_n]$ and $[b_n]$ are

$$\overline{A}_m = \sum_{n=0}^{P-1} a_n W_P^{mn}; \qquad m = 0, 1, \ldots, P - 1$$

$$\overline{B}_m = \sum_{n=0}^{P-1} b_n W_P^{mn}; \qquad m = 0, 1, \ldots, P - 1 \tag{6.5}$$

But from Eq. 6.2 it is also obvious that

$$W_P = e^{-j(2\pi/P)} = e^{-j(2\pi \cdot 2/N)} = (e^{-j(2\pi/N)})^2 = W_N^2 \tag{6.6}$$

Therefore, $W_N^{2mn} = W_P^{mn}$ and, for $m < P$, $\overline{A}_m$ and $\overline{B}_m$ can be substituted into Eq. 6.4 to obtain

$$\overline{F}_m = \overline{A}_m + W_N^m \overline{B}_m; \qquad m = 0, 1, \ldots, P - 1 \tag{6.7}$$

But also, when $m \geq P$, DFTs in Eq. 6.5 simply repeat, and indeed, Eq. 5.11 of Chapter 5 gives

$$\overline{A}_m = \overline{A}_{m+P} \quad \text{and} \quad \overline{B}_m = \overline{B}_{m+P} \tag{6.8}$$

Thus m in Eq. 6.5 can be extended through $N - 1$ and the complete DFT of N samples can be expressed as

$$\overline{F}_m = \overline{A}_m + W_N^m \overline{B}_m; \qquad m = 0, 1, \ldots, N - 1 \tag{6.9}$$

The important thing about this result is that *the DFT of N samples has become a linear combination of two smaller DFTs,* each of $N/2$ samples. Each smaller DFT requires $(N/2)^2$ products, so Eq. 6.9 requires altogether $2(N/2)^2 + N = N(N/2 + 1)$ products, which is a considerable savings from the original N^2 products when N is large.

Note that, if N were divisible by 4 (i.e., P divisible by 2), the two DFTs $[A_m]$ and $[\overline{B}_m]$ could then be further decomposed into two smaller DFTs, and so on, eliminating more and more redundant products from the original DFT. The smaller DFTs result when the original sample set is decomposed into subsets. Decomposition, which is the subject of the next section, is thus used to develop the general forms of the FFT.

6.3 SAMPLE SET DECOMPOSITIONS

In the development of Eq. 6.5 the original sample set $[f_n]$ was decomposed into two smaller sets by taking every other sample. This decomposition is illustrated in Fig. 6.2, in which integers are used in place of the samples to simplify the notation. The most straightforward derivation of the FFT is based on having N be a power (rather than a multiple) of 2 so that the decomposition can be repeated as explained below. However, the reduction in products described above does not depend on this, and $[f_n]$ can be decomposed in other ways as well.

For example, suppose that N is a multiple of 3 so that $N = 3Q$. Then a decomposition could be accomplished by taking every third sample as illustrated in Fig. 6.3. The original DFT could then be decomposed into a combination of three smaller DFTs in a manner essentially the same as that of the preceding section. In fact, when $N = 3Q$, the DFT formula may be written

$$
\begin{aligned}
\overline{F}_m &= \sum_{n=0}^{N-1} f_n W_N^{mn} \\
&= \sum_{n=0}^{Q-1} f_{3n} W_N^{3mn} + W_N^m \sum_{n=0}^{Q-1} f_{3n+1} W_N^{3mn} + W_N^{2m} \sum_{n=0}^{Q-1} f_{3n+2} W_N^{3mn}
\end{aligned}
\tag{6.10}
$$

The procedure in Section 6.2 could now be used to express the DFT as a linear combination of three smaller DFTs, again with a reduction in the required number of complex products.

Comparing Eqs. 6.4 and 6.10, the extension to cases where N has a factor of 4, 5, and so on, is made easily. However, the FFT is based on repeated decompositions rather than the single decompositions discussed so far. It is most convenient now to assume that N is a power of 2, that is,

$$
N = 2^q
\tag{6.11}
$$

(It is important to note that some base other than 2 could have been selected for the discussion that follows, and that 2 is just the most convenient base to use in illustrating the FFT.)

With $N = 2^q$ each successive decomposition of the original sample set $[f_n]$ can be further decomposed until there are altogether q decompositions. For example,

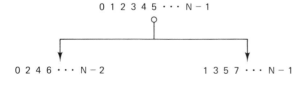

0 1 2 3 4 5 $\cdots$ N − 1

0 2 4 6 $\cdots$ N − 2   1 3 5 7 $\cdots$ N − 1

Figure 6.2 Decomposition of the sample set by taking every other sample.

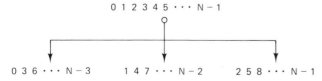

0 1 2 3 4 5 $\cdots$ N − 1

0 3 6 $\cdots$ N − 3   1 4 7 $\cdots$ N − 2   2 5 8 $\cdots$ N − 1

Figure 6.3 Decomposition of the sample set into three subsets by taking every third sample.

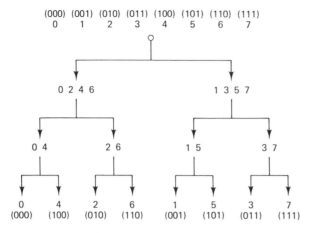

(000) (001) (010) (011) (100) (101) (110) (111)
  0   1   2   3   4   5   6   7

0 2 4 6

1 3 5 7

0 4

2 6

1 5

3 7

0   4   2   6   1   5   3   7
(000)   (100)   (010)   (110)   (001)   (101)   (011)   (111)

Figure 6.4 Complete decomposition of the sample set with $N = 8$ showing bit reversal.

Fig. 6.4 shows the complete decomposition with $N = 8$, or $q = 3$. Each level in the diagram is simply a repetition of Fig. 6.2. In general, the set of q decompositions accomplishes a phenomenon called *bit reversal*, which is also illustrated. The ordering of samples is kept intact within subsets during the decomposition, but at the final level the order is as if each subscript were written in binary notation and then reversed. For example, the subscript at position zero (000) remains unchanged, the subscript at position one (001) becomes four (100), and so on. Bit reversal occurs similarly for all values of q and is thus a convenient way to remember the effect of the complete decomposition on the ordering of samples.

Each stage in the complete decomposition of the sample set suggests a corresponding decomposition of the DFT into a set of smaller DFTs, with the FFT being the final result. Thus Eq. 6.9 could be used recursively to express the FFT, but the notation would become complicated. A better vehicle for expressing the FFT is the signal-flow diagram, originated by S. J. Mason (Mason, 1953), which is the subject of the next section.

6.4 SIGNAL-FLOW DIAGRAMS

The signal-flow diagram is a network of nodes connected by line segments and is used generally to describe how a set of output signals is formed by combining a set of input signals. A rather special form of the general signal-flow diagram (Truxal, 1955) is used here to illustrate the FFT. All of the essential properties of this special form are illustrated in Fig. 6.5. Note that the flow is from left to right, that n represents an exponent of W_N, and that an unbroken line represents a zero exponent (i.e., a coefficient equal to 1). To illustrate further, Fig. 6.6 shows a complete DFT (or FFT, since no simplification is possible in this case) for $N = 2$. The diagram has the useful effect of showing at a glance how the samples are combined to form the DFT

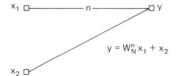

$$y = W_N^n x_1 + x_2$$

Figure 6.5 Simple signal-flow diagram.

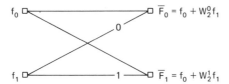

$$\overline{F}_0 = f_0 + W_2^0 f_1$$

$$\overline{F}_1 = f_0 + W_2^1 f_1$$

Figure 6.6 Signal-flow diagram for the DFT (or FFT) with $N = 2$.

values. It also illustrates the number of complex products (two in the present example), because each complex product is represented by a broken line, that is, a single exponent[†] in the diagram.

6.5 THE FFT USING TIME DECOMPOSITION

The process discussed in Section 6.2 and illustrated in Section 6.3 is known as *time decomposition* because the time series, that is, the sample set $[f_n]$ is decomposed as described above. The signal-flow diagram introduced in Section 6.4 will now be used to illustrate the time-decomposition FFT.

Consider the specific case with $N = 8$. Figure 6.4 illustrates the decomposition of the sample set and, as mentioned, implies three applications of Eq. 6.5, with the FFT being the final result. This result is now given in Fig. 6.7 using the signal-flow diagram. The diagram shows explicitly all of the sums and products that are implied after Eq. 6.5 has been applied three times to the original DFT formula. The samples are listed vertically in bit-reversed order because, as in Fig. 6.4, this is the order that results after the complete decomposition.

To verify that Fig. 6.7 is indeed correct, one can also trace the signal paths and see that each DFT sum is correct (e.g., for $m = 1$, $\overline{F}_1 = f_0 + f_1 W_8^1 + \cdots + f_7 W_8^7$, etc.).

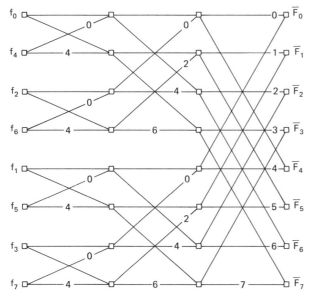

Figure 6.7 FFT for $N = 8$ using time decomposition with input bit reversal.

[†]Since $W_N^0 = 1$, a broken line with $n = 0$ is of course the same as an unbroken line in the diagram. Broken lines with $n = 0$ are added in Fig. 6.6 and in subsequent diagrams for the sake of symmetry and ease of programming.

Starting from Fig. 6.7 it is easy to go to other values of q (with $N = 2^q$). To increase q from 3 to 4, for example, one would double each of the exponents in Fig. 6.7, repeat the entire diagram below itself, and add a fourth section on the right that would tie the two diagrams together. To decrease q from 3 to 2, simply take the upper left-hand network having three columns of four nodes each and divide each of the exponents by 2. (In all cases the inputs must of course be bit-reversed.)

Since each complex product in the FFT is represented by an exponent in the diagram, it is easy to see that there are q columns of products with N products in each column. But also, half of the products in each column are redundant, because, from Fig. 6.1, $W_N^n = -W_N^{n-N/2}$ when $n \geq N/2$. Thus there are q columns each with $N/2$ unique products, and

$$\text{number of complex products} = \frac{Nq}{2} = \frac{N}{2} \log_2 N \qquad (6.12)$$

Comparing this number with N^2, the number of products in the DFT, gives

$$\frac{\text{number of FFT products}}{\text{number of DFT products}} = \frac{1}{2N} \log_2 N \qquad (6.13)$$

This fraction becomes impressively small as N increases. For example, it is less than 1% for $N = 512$. Thus the FFT produces a significant savings in the computing effort.

Figure 6.7 of course specifies a computing algorithm, and in fact Appendix B contains Fortran subroutines based on time decomposition with input bit reversal. Other time-decomposition algorithms can be obtained by reordering the nodes in Fig. 6.7. For example, if the rows of nodes (each row having four nodes) are exchanged, carrying the branches along in each exchange so that the inputs are in correct order, Fig. 6.8 results. In this version the inputs are taken in correct order but now the outputs are in bit-reversed order. Thus Fig. 6.7 requires a shuffling of the input sample set, while Fig. 6.8 requires a shuffling of computed DFT values.

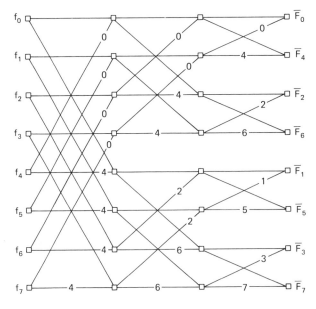

Figure 6.8 FFT for $N = 8$ using time decomposition with output bit reversal.

The Fast Fourier Transform   Chap. 6

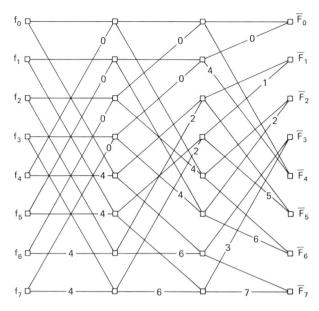

Figure 6.9 FFT for $N = 8$ using time decomposition without bit reversal.

The third time-decomposition form of the FFT (Cochran, Cooley, et al., 1967) is illustrated for $N = 8$ in Fig. 6.9. Here the inputs and outputs are in natural order but the simple pattern of the network is lost.

Again regarding the computing algorithms implied by these diagrams, notice that the diagrams imply how the algorithms must be composed. Each algorithm must proceed from left to right, beginning with the sample set and ending with the DFT.

Furthermore, each diagram has implications regarding temporary storage requirements during the computation. Note that Figs. 6.7 through 6.9 are superpositions of two-element DFT forms (i.e., repetitions of the pattern of Fig. 6.6). Therefore, the computation in these cases can be done *in place* with the use of only two auxiliary storage cells for complex products. For example, in Fig. 6.7, the sums $f_0 + W_8^0 f_4$ and $f_0 + W_8^4 f_4$ could be computed and then stored in place of f_0 and f_4, and so on.

6.6 THE FFT USING FREQUENCY DECOMPOSITION

In addition to the three different forms of the FFT just discussed, there are three analogous forms[†] based on decomposing the set $[\overline{F}_m]$ of DFT values instead of the sample set $[f_n]$. To introduce these forms there is a decomposition formula similar to Eq. 6.9.

To derive the decomposition formula, begin again with the sample set $[f_n]$ and let $P = N/2$ as before, but this time simply divide $[f_n]$ at the halfway point. Let

$$
\begin{aligned}
a_n &= f_n; & n &= 0, 1, \ldots, P - 1 \\
b_n &= f_{n+P}; & n &= 0, 1, \ldots, P - 1
\end{aligned}
\tag{6.14}
$$

[†]Here again the development is based on Cochran, Cooley, et al. (1967), and also on Gentleman and Sande (1966).

Then write the DFT formula as follows

$$\overline{F}_m = \sum_{n=0}^{N-1} f_n W_N^{mn}$$

$$= \sum_{n=0}^{P-1} (a_n + W_N^{mN/2} b_n) W_N^{mn} \tag{6.15}$$

Now again we use Eq. 6.6, that is, $W_P = W_N^2$, and also note from Fig. 6.1 that $W_N^{N/2} = -1$. The DFT is now

$$\overline{F}_m = \sum_{n=0}^{P-1} [a_n + (-1)^m b_n] W_P^{mn/2} \tag{6.16}$$

This form of the DFT suggests *frequency decomposition* because it has different forms for even and odd values of m. First, when m is even, let $k = m/2$, and

$$\overline{F}_{2k} = \sum_{n=0}^{P-1} (a_n + b_n) W_P^{nk}; \qquad k = \frac{m}{2} = 0, 1, \ldots, P - 1 \tag{6.17}$$

Second, when m is odd, let $k = (m - 1)/2$ and, again using $W_P = W_N^2$,

$$\overline{F}_{2k+1} = \sum_{n=0}^{P-1} (a_n - b_n) W_P^{(2k+1)n/2}$$

$$= \sum_{n=0}^{P-1} [(a_n - b_n) W_N^n] W_P^{nk}; \qquad k = \frac{m-1}{2} = 0, 1, \ldots, P - 1 \tag{6.18}$$

Taken together, Eqs. 6.17 and 6.18 give all N of the DFT values. Taken separately, however, each represents a DFT of order P instead of N. The sample sets are $[a_n + b_n]$ for Eq. 6.17 and $[(a_n - b_n) W_N^n]$ for Eq. 6.18.

Thus, in Eqs. 6.17 and 6.18, the DFT is decomposed into sets of even and odd frequency values, each set resulting from a smaller DFT. Just as in the case of Eq. 6.9, these equations can be applied iteratively to decompose the DFT completely into the FFT.

Figure 6.10 illustrates the decomposition of the sample set into the sets $[a_n]$ and $[b_n]$ when $N = 8$, that is, a single application of Eq. 6.14. There is no bit reversal in this case; the samples remain in their original order. Because of the combination of terms in Eqs. 6.17 and 6.18, however, Fig. 6.10 cannot be extended in the simple manner of Fig. 6.4.

Again for $N = 8$, the signal-flow diagram in Fig. 6.11 illustrates the three applications of Eqs. 6.17 and 6.18. Proceeding from left to right, at each level the sums of the form $(a_n + b_n)$ and $(a_n - b_n)$ are multiplied by the appropriate coefficients and combined. As suggested by Fig. 6.10 there is no bit reversal at the input; however, the DFT values, being now decomposed as in Fig. 6.4, come out in bit-

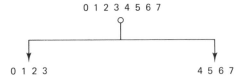

Figure 6.10 Decomposition of the sample set with $N = 8$ and without bit reversal.

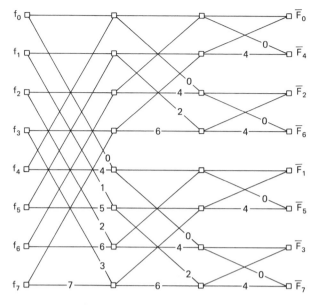

Figure 6.11 FFT for $N = 8$ using frequency decomposition with output bit reversal.

reversed order. As in all of the diagrams, the correctness of the DFT can be verified by tracing each of the complete signal paths from input to output to obtain, for example, $\overline{F}_1 = f_0 + f_1 W_8^1 + f_2 W_8^2 + \cdots$.

As in the case of time decomposition, other versions of the frequency-decomposition FFT can be obtained by reordering the nodes (and carrying along the branches as before) in Fig. 6.11. Figures 6.12 and 6.13 show the other two versions resulting first from input bit reversal, and second, from no bit reversal.

Taken together, the three frequency-decomposition diagrams in Figs. 6.11 through 6.13 plus the three time-decomposition diagrams in Figs. 6.7 through 6.9

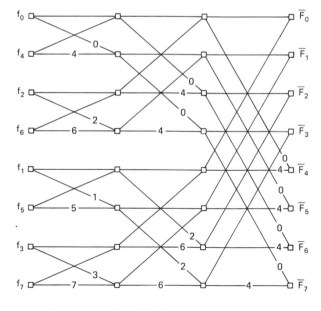

Figure 6.12 FFT for $N = 8$ using frequency decomposition with input bit reversal.

Sec. 6.6   The FFT Using Frequency Decomposition

111

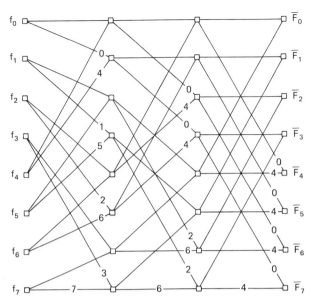

Figure 6.13 FFT for $N = 8$ using frequency decomposition without bit reversal.

comprise a basic set of FFTs. The four diagrams with bit reversal are all essentially superpositions of two-point DFTs and thus, as discussed above, allow in-place computation. It is also interesting to note that to change from time decomposition to frequency decomposition, or vice versa, one can simply reverse the direction of flow and interchange the roles of f and $\overline{F}$ in any of the six diagrams (e.g., Fig. 6.11 is a reversal of Fig. 6.7, etc.).

6.7 MATRIX FACTORING

The derivation of the FFT from the original DFT formula, Eq. 6.1, can also be accomplished by factoring the coefficient matrix in the set of linear equations representing the DFT, as described by Good (1958), Andrews and Caspari (1970), and Kahaner (1970). In fact the FFT or any decomposition can be expressed as a product of matrix factors.

To illustrate, let $N = 8$ again so that matrix factorings can be given for the signal-flow diagrams above. Let the DFT, Eq. 6.1, be expressed as follows:

$$
\begin{bmatrix}
\overline{F}_0 \\
\overline{F}_1 \\
\overline{F}_2 \\
\overline{F}_3 \\
\overline{F}_4 \\
\overline{F}_5 \\
\overline{F}_6 \\
\overline{F}_7
\end{bmatrix}
=
\begin{bmatrix}
0 & 0 & 0 & 0 & 0 & 0 & 0 & 0 \\
0 & 1 & 2 & 3 & 4 & 5 & 6 & 7 \\
0 & 2 & 4 & 6 & 0 & 2 & 4 & 6 \\
0 & 3 & 6 & 1 & 4 & 7 & 2 & 5 \\
0 & 4 & 0 & 4 & 0 & 4 & 0 & 4 \\
0 & 5 & 2 & 7 & 4 & 1 & 6 & 3 \\
0 & 6 & 4 & 2 & 0 & 6 & 4 & 2 \\
0 & 7 & 6 & 5 & 4 & 3 & 2 & 1
\end{bmatrix}
\cdot
\begin{bmatrix}
f_0 \\
f_1 \\
f_2 \\
f_3 \\
f_4 \\
f_5 \\
f_6 \\
f_7
\end{bmatrix}
\qquad (6.19)
$$

Here, just as in the diagrams, each element of the 8×8 coefficient matrix is an *exponent* (n) in W_8^n in order to simplify the notation. Also, in accordance with Fig. 6.1, exponents greater than 7 are taken mod 8.

Now using the exponent notation in Eq. 6.19, the three FFT factorings corresponding to the time-decomposition FFT diagrams are as follows (since a zero in the coefficient matrix represents W_8^0, a dot is now used to represent a coefficient equal to zero):

Time decomposition with input bit reversal:

$$
\begin{bmatrix} \bar{F}_0 \\ \bar{F}_1 \\ \bar{F}_2 \\ \bar{F}_3 \\ \bar{F}_4 \\ \bar{F}_5 \\ \bar{F}_6 \\ \bar{F}_7 \end{bmatrix}
=
\begin{bmatrix}
0 & \cdot & \cdot & \cdot & 0 & \cdot & \cdot & \cdot \\
\cdot & 0 & \cdot & \cdot & \cdot & 1 & \cdot & \cdot \\
\cdot & \cdot & 0 & \cdot & \cdot & \cdot & 2 & \cdot \\
\cdot & \cdot & \cdot & 0 & \cdot & \cdot & \cdot & 3 \\
0 & \cdot & \cdot & \cdot & 4 & \cdot & \cdot & \cdot \\
\cdot & 0 & \cdot & \cdot & \cdot & 5 & \cdot & \cdot \\
\cdot & \cdot & 0 & \cdot & \cdot & \cdot & 6 & \cdot \\
\cdot & \cdot & \cdot & 0 & \cdot & \cdot & \cdot & 7
\end{bmatrix}
\cdot
\begin{bmatrix}
0 & \cdot & 0 & \cdot & \cdot & \cdot & \cdot & \cdot \\
\cdot & 0 & \cdot & 2 & \cdot & \cdot & \cdot & \cdot \\
0 & \cdot & 4 & \cdot & \cdot & \cdot & \cdot & \cdot \\
\cdot & 0 & \cdot & 6 & \cdot & \cdot & \cdot & \cdot \\
\cdot & \cdot & \cdot & \cdot & 0 & \cdot & 0 & \cdot \\
\cdot & \cdot & \cdot & \cdot & \cdot & 0 & \cdot & 2 \\
\cdot & \cdot & \cdot & \cdot & 0 & \cdot & 4 & \cdot \\
\cdot & \cdot & \cdot & \cdot & \cdot & 0 & \cdot & 6
\end{bmatrix}
$$

$$
\cdot
\begin{bmatrix}
0 & 0 & \cdot & \cdot & \cdot & \cdot & \cdot & \cdot \\
0 & 4 & \cdot & \cdot & \cdot & \cdot & \cdot & \cdot \\
\cdot & \cdot & 0 & 0 & \cdot & \cdot & \cdot & \cdot \\
\cdot & \cdot & 0 & 4 & \cdot & \cdot & \cdot & \cdot \\
\cdot & \cdot & \cdot & \cdot & 0 & 0 & \cdot & \cdot \\
\cdot & \cdot & \cdot & \cdot & 0 & 4 & \cdot & \cdot \\
\cdot & \cdot & \cdot & \cdot & \cdot & \cdot & 0 & 0 \\
\cdot & \cdot & \cdot & \cdot & \cdot & \cdot & 0 & 4
\end{bmatrix}
\cdot
\begin{bmatrix} f_0 \\ f_4 \\ f_2 \\ f_6 \\ f_1 \\ f_5 \\ f_3 \\ f_7 \end{bmatrix}
\qquad (6.20)
$$

Time decomposition with output bit reversal:

$$
\begin{bmatrix} \bar{F}_0 \\ \bar{F}_4 \\ \bar{F}_2 \\ \bar{F}_6 \\ \bar{F}_1 \\ \bar{F}_5 \\ \bar{F}_3 \\ \bar{F}_7 \end{bmatrix}
=
\begin{bmatrix}
0 & 0 & \cdot & \cdot & \cdot & \cdot & \cdot & \cdot \\
0 & 4 & \cdot & \cdot & \cdot & \cdot & \cdot & \cdot \\
\cdot & \cdot & 0 & 2 & \cdot & \cdot & \cdot & \cdot \\
\cdot & \cdot & 0 & 6 & \cdot & \cdot & \cdot & \cdot \\
\cdot & \cdot & \cdot & \cdot & 0 & 1 & \cdot & \cdot \\
\cdot & \cdot & \cdot & \cdot & 0 & 5 & \cdot & \cdot \\
\cdot & \cdot & \cdot & \cdot & \cdot & \cdot & 0 & 3 \\
\cdot & \cdot & \cdot & \cdot & \cdot & \cdot & 0 & 7
\end{bmatrix}
\cdot
\begin{bmatrix}
0 & \cdot & 0 & \cdot & \cdot & \cdot & \cdot & \cdot \\
\cdot & 0 & \cdot & 0 & \cdot & \cdot & \cdot & \cdot \\
0 & \cdot & 4 & \cdot & \cdot & \cdot & \cdot & \cdot \\
\cdot & 0 & \cdot & 4 & \cdot & \cdot & \cdot & \cdot \\
\cdot & \cdot & \cdot & \cdot & 0 & \cdot & 2 & \cdot \\
\cdot & \cdot & \cdot & \cdot & \cdot & 0 & \cdot & 2 \\
\cdot & \cdot & \cdot & \cdot & 0 & \cdot & 6 & \cdot \\
\cdot & \cdot & \cdot & \cdot & \cdot & 0 & \cdot & 6
\end{bmatrix}
$$

$$
\cdot
\begin{bmatrix}
0 & \cdot & \cdot & \cdot & 0 & \cdot & \cdot & \cdot \\
\cdot & 0 & \cdot & \cdot & \cdot & 0 & \cdot & \cdot \\
\cdot & \cdot & 0 & \cdot & \cdot & \cdot & 0 & \cdot \\
\cdot & \cdot & \cdot & 0 & \cdot & \cdot & \cdot & 0 \\
0 & \cdot & \cdot & \cdot & 4 & \cdot & \cdot & \cdot \\
\cdot & 0 & \cdot & \cdot & \cdot & 4 & \cdot & \cdot \\
\cdot & \cdot & 0 & \cdot & \cdot & \cdot & 4 & \cdot \\
\cdot & \cdot & \cdot & 0 & \cdot & \cdot & \cdot & 4
\end{bmatrix}
\cdot
\begin{bmatrix} f_0 \\ f_1 \\ f_2 \\ f_3 \\ f_4 \\ f_5 \\ f_6 \\ f_7 \end{bmatrix}
\qquad (6.21)
$$

Time decomposition without bit reversal:

$$
\begin{bmatrix} \overline{F}_0 \\ \overline{F}_1 \\ \overline{F}_2 \\ \overline{F}_3 \\ \overline{F}_4 \\ \overline{F}_5 \\ \overline{F}_6 \\ \overline{F}_7 \end{bmatrix}
=
\begin{bmatrix}
0 & 0 & \cdot & \cdot & \cdot & \cdot & \cdot & \cdot \\
\cdot & \cdot & 0 & 1 & \cdot & \cdot & \cdot & \cdot \\
\cdot & \cdot & \cdot & \cdot & 0 & 2 & \cdot & \cdot \\
\cdot & \cdot & \cdot & \cdot & \cdot & \cdot & 0 & 3 \\
0 & 4 & \cdot & \cdot & \cdot & \cdot & \cdot & \cdot \\
\cdot & \cdot & 0 & 5 & \cdot & \cdot & \cdot & \cdot \\
\cdot & \cdot & \cdot & \cdot & 0 & 6 & \cdot & \cdot \\
\cdot & \cdot & \cdot & \cdot & \cdot & \cdot & 0 & 7
\end{bmatrix}
\cdot
\begin{bmatrix}
0 & \cdot & 0 & \cdot & \cdot & \cdot & \cdot & \cdot \\
\cdot & 0 & \cdot & 0 & \cdot & \cdot & \cdot & \cdot \\
\cdot & \cdot & \cdot & \cdot & 0 & \cdot & 2 & \cdot \\
\cdot & \cdot & \cdot & \cdot & \cdot & 0 & \cdot & 2 \\
0 & \cdot & 4 & \cdot & \cdot & \cdot & \cdot & \cdot \\
\cdot & 0 & \cdot & 4 & \cdot & \cdot & \cdot & \cdot \\
\cdot & \cdot & \cdot & \cdot & 0 & \cdot & 6 & \cdot \\
\cdot & \cdot & \cdot & \cdot & \cdot & 0 & \cdot & 6
\end{bmatrix}
$$

$$
\cdot
\begin{bmatrix}
0 & \cdot & \cdot & \cdot & 0 & \cdot & \cdot & \cdot \\
\cdot & 0 & \cdot & \cdot & \cdot & 0 & \cdot & \cdot \\
\cdot & \cdot & 0 & \cdot & \cdot & \cdot & 0 & \cdot \\
\cdot & \cdot & \cdot & 0 & \cdot & \cdot & \cdot & 0 \\
0 & \cdot & \cdot & \cdot & 4 & \cdot & \cdot & \cdot \\
\cdot & 0 & \cdot & \cdot & \cdot & 4 & \cdot & \cdot \\
\cdot & \cdot & 0 & \cdot & \cdot & \cdot & 4 & \cdot \\
\cdot & \cdot & \cdot & 0 & \cdot & \cdot & \cdot & 4
\end{bmatrix}
\cdot
\begin{bmatrix} f_0 \\ f_1 \\ f_2 \\ f_3 \\ f_4 \\ f_5 \\ f_6 \\ f_7 \end{bmatrix}
\qquad (6.22)
$$

Notice how these three equations correspond with Figs. 6.7, 6.8, and 6.9, respectively. The first matrix product on the *right* (i.e., the right-hand coefficient matrix times the sample vector) corresponds with the first stage of the FFT diagram on the *left,* and so on. Thus the pattern of the time-decomposition FFT is seen in each of these equations. Derivation of the frequency-decomposition equations is left as an exercise.

6.8 THE FFT IN PRACTICE

Some of the details encountered in programming the FFT or constructing FFT hardware are worth noting. First, concerning the number of complex products, there are nominally $(N/2) \log_2 N$ as in Eq. 6.12, after redundancies of the form $W_N^n = -W_N^{n-N/2}$ have been taken into account in the programming. If it is considered worthwhile, this number can be further reduced by recognizing that $W_N^0 = 1$ in the algorithm.

Furthermore, the number of products as well as the storage requirements can be reduced again if the sampled function $f(t)$ has special properties. If $f(t)$ is real, only half as much storage is needed (only $\overline{F}_0$ through $\overline{F}_{N/2}$ need be found).

Since $f(t)$ is in fact real in the majority of engineering applications, the FFT routines in Appendix B assume that $[f_n]$ is a real sequence and utilize this property. Thus, in the forward transform, the complex spectrum $[\overline{F}_m; m = 0, \ldots, N/2 - 1]$ replaces the time sequence $[f_n; n = 0, \ldots, N - 1]$. Only two extra storage locations are required for the final spectral value, $\overline{F}_{N/2}$. Moreover, the FFT routines in Appendix B may be used when $[f_n]$ is a complex sequence, say

$$[f_n] = [r_n + ji_n] \qquad (6.23)$$

In this case the FFT of the complex sequence is obtained easily from two FFTs of real sequences:

$$[\overline{F}_m] = [\overline{R}_m + j\overline{I}_m] \tag{6.24}$$

(see Eq. 6.28 in Section 6.10). Note that both $\overline{R}_m$ and $\overline{I}_m$ are in general complex.

If $f(t)$ is an even function of t then only the real part of $[\overline{F}_m]$ is nonzero, and if $f(t)$ is odd, only the imaginary part is nonzero (Chapter 5, Section 5.2). When a large percentage of the samples in $[f_n]$ is zero, a process called "FFT pruning" (Markel, 1971), in which operations on zeros are effectively eliminated, can be used.

The computation of a sine or a cosine usually takes longer than the computation of a complex product. In most general-purpose FFT programs, the sine and cosine calculations are embedded in the program for convenience, but particularly if N is fixed, it is preferable for the sake of speed to tabulate the sine and cosine values so that they may be looked up during computation. Then, when the FFT is performed repeatedly with N constant, the trigonometric computations need not be repeated. Such a tabulation is possible with a modification of the FFT routines in Appendix B.

With clever scaling and checking for overflow and underflow, the FFT can be programmed using integer (rather than floating-point) arithmetic (DeJongh and DeBoer, 1971). This technique can produce significant savings in computing time, especially in small computers without floating-point hardware.

6.9 USING THE FFT FOR DISCRETE CONVOLUTION

In Section 5.9 of Chapter 5 we found that the DFT can be used to perform *periodic convolution,* that is, a convolution of one sequence with the periodic extension of the other. We also found that this convolution can be modified to perform the "correct" approximation to a continuous convolution, which, with sampled sequences, is called *linear convolution.* It is linear convolution that is of interest here. We now extend the discussion of Chapter 5, showing how the FFT can be used to implement linear convolution in a computationally efficient manner. The procedure that results is called *fast convolution.* Compared to a direct implementation of the convolution sum, fast convolution provides a significant reduction in computational requirements. Fast convolution is often used to implement finite impulse response (FIR) filtering, which is discussed in Chapter 8. FIR filtering requires the input to be convolved with an impulse response which is finite in length, but often quite long. In such cases, where both sequences in the convolution are of considerable length, the fast convolution method often provides the only method for obtaining a cost-efficient real-time implementation.

Linear convolution, also known as aperiodic or noncyclic convolution, of two causal sequences f_n and h_n is defined as follows:

$$g_n = \sum_{m=0}^{\infty} f_m h_{n-m} = \sum_{m=0}^{\infty} h_m f_{n-m} \tag{6.25}$$

If the causal sequences f_n and h_n are of length N_1 and N_2, respectively, then the result, g_n, is in general of length $N_3 = N_1 + N_2 - 1$. Without loss of generality let us assume $N_1 \geq N_2$. In this case Eq. 6.25 above can be rewritten

$$g_n = \sum_{m=0}^{N_1-1} f_m h_{n-m} = \sum_{m=0}^{N_1-1} h_m f_{n-m} \tag{6.26}$$

The operation implied by this equation involves flipping one of the sequences about the vertical axis and then shifting it with respect to the other, forming a sum of products at each shifted position as illustrated in Fig. 6.14. A brute-force implementation of the equation requires that N_1 product terms be summed at each of $N_1 + N_2 - 1$ shifted positions, resulting in a total of $N_1(N_1 + N_2 - 1)$ multiplications and $(N_1 - 1)(N_1 + N_2 - 1)$ additions. If, in the implementation, we are clever enough to avoid computing the zero product terms at positions where the sequences do not overlap then the total number of computations can be reduced to $N_1 N_2$ multiplications and $(N_1 - 1)(N_2 - 1)$ additions.

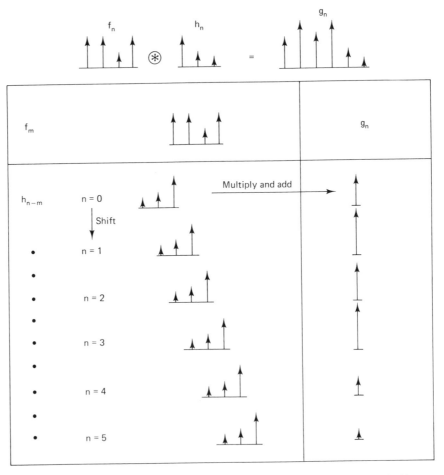

Figure 6.14 Linear convolution: flip about vertical axis, then shift, multiply, and add.

The DFT approach to convolution discussed in Chapter 5 is based on the fact that convolution in the time domain can be performed using multiplication in the frequency domain. We found there that the inverse DFT of the product of two DFTs results in a periodic convolution, as illustrated in Fig. 6.15. Note that to compute the product, the DFTs must be of equal length. This means that the shorter of the two original sequences must be extended with zeros to the length of the other before its DFT is formed. Even with DFTs of the same length, the resulting convolution is not in general the same as the linear convolution shown in Fig. 6.14. If we are to use DFT products to perform linear convolution, the sequences must be modified using an approach similar to that discussed earlier in Chapter 5. Since the product of DFTs always results in a periodic convolution, we must modify the data so that a periodic convolution and a linear convolution are equivalent. The key lies in the realization that when performing the first (and last) $N_2 - 1$ summations of the linear convolu-

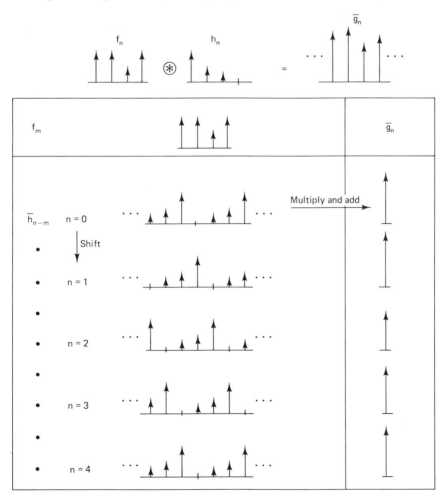

Figure 6.15 Periodic convolution: form periodic extension, flip about vertical axis, then shift, multiply, and add.

Sec. 6.9   Using the FFT for Discrete Convolution   **117**

tion, the two sequences are not completely overlapping, and many of the product terms in Eq. 6.25 are zero (refer to Fig. 6.14). In periodic convolution, the two sequences are always completely overlapping because as the end of one period is shifted out the beginning of the next is shifted in (refer to Fig. 6.15). To achieve linear convolution we must insert enough zeros into the data so that the product terms from the end of the period being shifted out are zero. Thus, we pad the original sequence with zeros before forming its periodic extension. For linear convolution, the sequence must be extended to a length of $N_1 + N_2 - 1$ or greater, as illustrated in Fig. 6.16.

To implement linear convolution using FFTs, the DFT must conform to a size supported by the FFT algorithm. Therefore, the sequences must be padded with additional zeros (beyond length $N_1 + N_2 - 1$) so that their length conforms to an allowable FFT size. These additional zeros produce extra zeros in the result at the end of the desired convolution which can simply be ignored. The fast convolution method is summarized schematically in Fig. 6.17. The first step is to pad both input se-

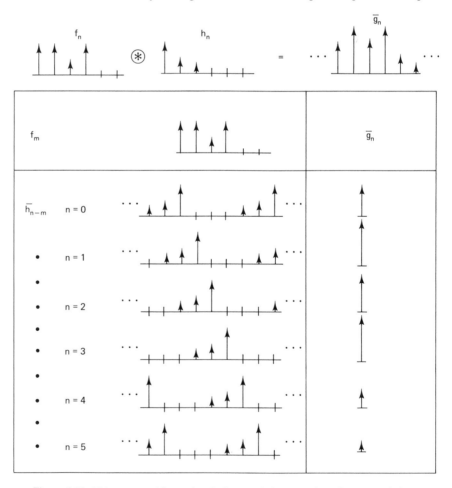

Figure 6.16 Using zero padding and periodic convolution to perform linear convolution.

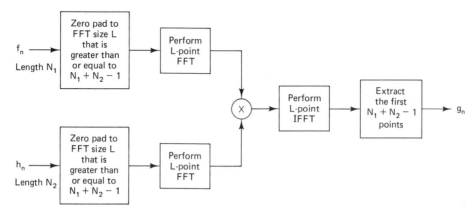

Figure 6.17 Block diagram of fast convolution method.

quences with zeros to a length corresponding to an allowable FFT size, L, that is greater than or equal to $N_1 + N_2 - 1$. Next the FFT is computed for both sequences; the complex product is formed; and the inverse FFT is used to obtain the result. The desired linear convolution is contained in the first $N_1 + N_2 - 1$ terms of this result.

The computational requirements of this method can be summarized as follows. Let L represent the FFT size (typically, the first integer power of 2 that is greater than or equal to $N_1 + N_2 - 1$). This method requires three L-point FFTs, each requiring $L \log_2 L$ complex additions and $(L/2) \log_2 L$ complex multiplications. In addition, L complex multiplications are required to perform the frequency-domain multiplication. Since one complex multiplication requires four regular multiplications and two additions, and one complex addition requires two regular additions, the total number of computations is

$$\text{total multiplications: } 6L \log_2 L + 4L$$

$$\text{total additions: } 9L \log_2 L + 2L$$

It is difficult to make quantitative comparisons between these results and those for the direct implementation in Eq. 6.26 without placing restrictions on the sequence lengths. For example, let us assume that the sequences are of equal length, that is, $N_1 = N_2$. For this case a plot of the computational requirements for the two implementations is shown in Fig. 6.18. Clearly, for larger sequence lengths, the fast convolution method provides a significant reduction in computational requirements.

6.10 FFT ROUTINES

Two FFT subroutines, SPFFT and SPIFFT, are included in Appendix B. These routines perform the forward and inverse discrete Fourier transforms, respectively, on a sequence that is real in the time domain. The length of the time-domain sequence, N, must be a power of 2. The calling sequences are described as follows:

```
              CALL SPFFT(X,N)
    X(0:N+1) = REAL array containing the sequence
               X₀, X₁, . . . , X_{N-1} plus two extra
               locations which need not be
```

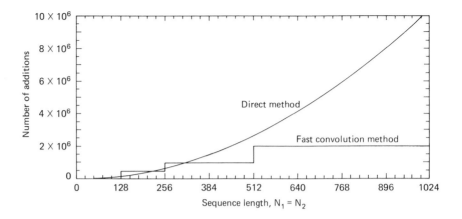

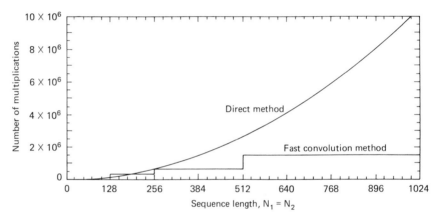

Figure 6.18 Computational comparisons for direct versus fast convolution methods when both sequences are of equal length.

initialized. After execution, the FFT
components $Re[X_0]$, $Im[\overline{X}_0]$, $Re[\overline{X}_1]$,
$Im[\overline{X}_1]$, . . . , $Re[\overline{X}_{N/2}]$, $Im[\overline{X}_{N/2}]$ are
contained in order in $X(0)$ through
$X(N+1)$.

N = Number of time-sequence samples. Must
be a power of 2 greater than 2.

CALL SPIFFT(X,N)

$X(0:N+1)$ = REAL array containing the complex
spectral components $Re[\overline{X}_0]$, $Im[\overline{X}_0]$,
$Re[\overline{X}_1]$, $Im[\overline{X}_1]$, . . . , $Re[\overline{X}_{N/2}]$,
$Im[\overline{X}_{N/2}]$. After execution, $X(0:N-1)$
contains the time sequence scaled by N.

N = Number of time-sequence samples. Must
be a power of 2 greater than 2.

The Fast Fourier Transform   Chap. 6

Thus, SPIFFT produces a scaled version of the original time sequence from the complex transform produced by SPFFT. That is, if SPFFT and SPIFFT are called in sequence, the result is a scaled version of the original time sequence, as illustrated in Fig. 6.19.

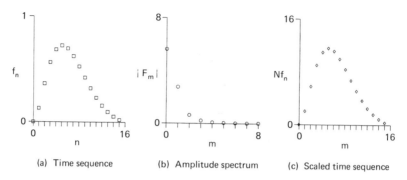

(a) Time sequence   (b) Amplitude spectrum   (c) Scaled time sequence

Figure 6.19 Using SPFFT to transform (a) into (b), then SPIFFT to transform (b) into (c).

Regarding the application of SPFFT, we note that the power-of-2 requirement on N can always be satisfied by extending the time sequence with zeros as described previously. Furthermore, suppose that the time sequence is complex, say

$$x_n = r_n + ji_n; \qquad n = 0, \ldots, N - 1 \qquad (6.27)$$

We can still use SPFFT to find the transform, $[\overline{X}_m]$, simply by transforming $[r_n]$ and $[i_n]$ individually and adding the transforms:

$$\overline{X}_m = \overline{R}_m + j\overline{I}_m; \qquad m = 0, 1, \ldots, \frac{N}{2}$$

$$= \overline{R}_{N-m}^* + j\overline{I}_{N-m}^*; \qquad m = \frac{N}{2} + 1, \ldots, N - 1 \qquad (6.28)$$

A technique that amounts essentially to an inversion of this procedure has been described by Brigham (1974) and is discussed in Stearns and David (1988). The technique allows one to obtain the transform of a real N-point sequence from the transform of a complex $N/2$-point sequence, and is used in the SPFFT routine.

EXERCISES

1. Using Eq. 6.1, write the four equations for the complete DFT with $N = 4$.
2. Construct a diagram showing the equivalent powers of W_N with $N = 4$.
3. Using diagrams like Fig. 6.3 show all possible time decompositions when $N = 10$ samples.
4. Express the DFT in the form of Eq. 6.10 **(a)** with five sums and **(b)** with three sums, when $N = 15$.
5. Show the complete time decomposition with input bit reversal, **(a)** when $N = 4$, and **(b)** when $N = 16$.
6. Construct the signal-flow diagram of the FFT for $N = 4$ using time decomposition with output bit reversal.

Chap. 6   Exercises

7. Construct the signal-flow diagram of the FFT for $N = 4$ using time decomposition with no bit reversal.

8. Describe how to modify Fig. 6.7 to obtain the signal-flow diagram of the FFT for $N = 16$ using time decomposition with input bit reversal.

9. How many complex products are required for the complete FFT of 4096 samples?

10. What fraction of the N^2 DFT products are redundant when $N = 4096$?

11. Discuss the use of an auxiliary storage unit in the time-decomposition FFT without bit reversal.

12. Write a Fortran subroutine to accomplish bit reversal by shuffling the samples in $[f_n]$. Let $k = \log_2 N$ as well as the array $[f_n]$ be inputs in the calling sequence.

13. Construct the signal-flow diagram of the FFT for $N = 4$ using frequency decomposition with output bit reversal.

14. Construct the signal-flow diagram of the FFT for $N = 16$ using frequency decomposition with input bit reversal.

15. Construct the signal-flow diagram of the FFT for $N = 4$ using frequency decomposition without bit reversal.

16. Give the matrix factoring for the frequency-decomposition FFT with output bit reversal; $N = 8$.

17. Give the matrix factoring for the frequency-decomposition FFT with input bit reversal; $N = 8$.

18. Give the matrix factoring for the frequency-decomposition FFT without bit reversal; $N = 8$.

19. Show the matrix-factored form of the time-decomposition FFT; $N = 2$.

20. Show the matrix-factored form of the time-decomposition FFT without bit reversal; $N = 4$.

21. Show the matrix-factored form of the frequency-decomposition FFT without bit reversal; $N = 4$.

22. Use the routine SPFFT to compute the complex spectra of the following sequences:
   (a) $[0 \quad 1 \quad -1 \quad 0]$
   (b) $[0 \quad 2 \quad 4 \quad 6 \quad 4 \quad 2 \quad 0 \quad 0]$
   (c) $[0 \quad 0 \quad 2 \quad -2 \quad 0 \quad 0 \quad 0 \quad 0]$
   (d) $[1 \quad -1 \quad 1 \quad -1 \quad 1 \quad -1 \quad 1 \quad -1]$
   (e) $[1 \quad 1 \quad 1 \quad 1 \quad 1 \quad 1 \quad 1 \quad 1]$

23. Plot the amplitude spectra in Exercise 22.

24. Using the fast transform routines, compute the linear convolution of the following sequences in Exercise 22:
   (a) Sequences (a) and (c)
   (b) Sequences (b) and (c)
   (c) Sequences (b) and (a)
   (d) Sequences (d) and (e)

25. Using SPFFT, compute and plot the amplitude spectra of:
   (a) $[1 + j0 \quad 2 - j1 \quad 1 - j2 \quad 1 + j0]$
   (b) $[0 + j0 \quad 1 + j1 \quad 2 + j2 \quad 3 + j3]$

SOME ANSWERS

1. $\overline{F}_0 = f_0 + f_1 + f_2 + f_3;\ \overline{F}_1 = f_0 + f_1 W_4 - f_2 - f_3 W_4;$
   $\overline{F}_2 = f_0 - f_1 + f_2 - f_3;\ \overline{F}_3 = f_0 - f_1 W_4 - f_2 + f_3 W_4$

4. (b) $\overline{F}_m = \sum_{n=0}^{4} f_{3n} W_{15}^{3mn} + W_{15}^{m} \sum_{n=0}^{4} f_{3n+1} W_{15}^{3mn} + W_{15}^{2m} \sum_{n=0}^{4} f_{3n+2} W_{15}^{3mn}$

9. 24,576   **10.** 99.9%

22. (a) $0 + j0, 1 - j1, -2 + j0$ **(e)** $8 + j0, 0 + j0, 0 + j0, 0 + j0, 0 + j0$

24. (a) $[0, 0, 0, 2, -4, 2, 0, 0, 0, 0, 0]$ **(c)** $[0, 0, 2, 2, 2, -2, -2, -2, 0, 0, 0, 0, 0, 0, 0]$

25. (a) $5.831, 1.414, 1.414, 3.162$

REFERENCES

ANDREWS, H. C., and CASPARI, K. L., A Generalized Technique for Spectral Analysis. *IEEE Trans. Audio Electroacoust.,* Vol. C-19, No. 1, January 1970, p. 16.

BERGLAND, G. D., A Guided Tour of the Fast Fourier Transform. *IEEE Spectrum,* July 1969, p. 41.

BRIGHAM, E. O., *The Fast Fourier Transform,* Chap. 10. Englewood Cliffs, N.J.: Prentice Hall, 1974.

COCHRAN, W. T., COOLEY, J. W., ET AL., What Is the Fast Fourier Transform? *IEEE Trans. Audio Electroacoust.,* Vol. AU-15, No. 2, June 1967, p. 45.

COOLEY, J. W., and TUKEY, J. W., An Algorithm for the Machine Calculation of Complex Fourier Series. *Math Comput.,* Vol. 19, April 1965, p. 297.

DEJONGH, H. R., and DEBOER, E., The Fast Fourier Transform and Its Use. *1971 DECUS Proceedings,* Maynard, Mass.: Digital Equipment Corp.

GENTLEMAN, W. M., and SANDE, G., Fast Fourier Transforms—For Fun and Profit. *1966 Fall Joint Computer Conf. AFIPS Proc.,* Vol. 29, p. 563. Washington, D.C.: Spartan, 1966.

GLISSON, T. H., BLACK, C. I., and SAGE, A. P., The Digital Computation of Discrete Spectra Using the Fast Fourier Transform. *IEEE Trans. Audio Electroacoust.,* Vol. AU-18, No. 3, September 1970, p. 271.

GOLD, B., and RADER, C. M., *Digital Processing of Signals,* Chap. 6. New York: McGraw-Hill, 1969.

GOOD, I. J., The Interaction Algorithm and Practical Fourier Series. *J. Roy. Statist. Soc. Ser. B,* Vol. 20, 1958, p. 361; Vol. 22, 1960, p. 372.

HOVANESSIAN, S. A., and PIPES, L. A., *Digital Computer Methods in Engineering,* Chap. 4. New York: McGraw-Hill, 1969.

IEEE Transactions on Audio and Electroacoustics (Special Issues on the Fast Fourier Transform), Vol. AU-15, No. 2, June 1967; Vol. AU-17, No. 2, June 1969.

KAHANER, D. K., Matrix Description of the Fast Fourier Transform. *IEEE Trans. Audio Electroacoust.,* Vol. AU-18, No. 4, December 1970, p. 442.

MARKEL, J. D., FFT Pruning. *IEEE Trans. Audio Electroacoust.,* Vol. AU-19, No. 4, December 1971, p. 305.

MASON, S. J., Feedback Theory—Some Properties of Signal Flow Graphs. *Proc. IRE,* Vol. 41, No. 9, September 1953, p. 1144.

STEARNS, S. D., and DAVID, R. A., *Signal Processing Algorithms,* Chap. 3. Englewood Cliffs, N.J.: Prentice Hall, 1988.

TRUXAL, J. G., *Control System Synthesis,* Chap. 2. New York: McGraw-Hill, 1955.

7

The z-Transform

7.1 INTRODUCTION

The z-transform plays a vital role in the analysis and representation of linear shift-invariant discrete-time systems. It is to discrete-time systems what the Laplace transform is to continuous-time systems. It can be viewed as a generalization of the DFT just as the Laplace transform can be viewed as a generalization of the Fourier transform. In this chapter we introduce the z-transform and discuss some of its properties.

7.2 DEFINITION OF THE z-TRANSFORM

The z-transform of a sampled sequence $[x_n]$, denoted $\tilde{X}(z)$ or $\mathscr{Z}[x_n]$, is defined

$$\tilde{X}(z) = \mathscr{Z}[x_n] = \sum_{n=-\infty}^{\infty} x_n z^{-n} \tag{7.1}$$

where z is complex. The symbol $\tilde{X}$ is used to emphasize that $\tilde{X}(z)$ is a new function, different from $X(j\omega)$ and $\bar{X}(j\omega)$. The definition above is commonly called the two-sided z-transform. Occasionally, a one-sided transform is used. The one-sided transform is

$$\tilde{X}(z) = \sum_{n=0}^{\infty} x_n z^{-n} \tag{7.2}$$

For causal sequences where $x_n = 0$ for $n < 0$, the two-sided z-transform reduces to a one-sided z-transform. In this book our formal definition is Eq. 7.1. However, since we often work with causal sequences we will find the one-sided definition useful.

From Eq. 7.1, the z-transform is seen to have the following important features:

1. $\tilde{X}(z)$ is a polynomial in z and is determined by the complete sample set $[x_n]$.
2. Each factor z^{-n} serves to isolate x_n from the rest of the sample set so that the entire sample set can be recovered if $\tilde{X}(z)$ is given.

3. $\tilde{X}(z)$ is formally independent of the sampling interval T, but

4. The factor z^{-n}, when associated with x_n, corresponds with the time nT, and in this sense z^{-n} implies a delay of nT seconds from the time $t = 0$.

5. If the argument of $\tilde{X}(z)$ is changed from z to $e^{j\omega T}$, the result is the DFT; that is, by definition,

$$\overline{X}(j\omega) = \tilde{X}(e^{j\omega T})$$

Thus, the DFT is found by evaluating the z-transform on the unit circle in the z-plane.

The z-transform defined in Eq. 7.1 does not necessarily exist for all values of z in the complex plane. For the z-transform to exist, the summation in Eq. 7.1 must converge. The domain of the z-plane within which the summation converges is called the region of convergence. The location of this region depends on the sequence $[x_n]$ as illustrated in the following examples.

Example 7.1

Let $[x_n]$ be the exponential sequence

$$x_n = a^n; \qquad n \geq 0$$
$$= 0; \qquad n < 0$$

The z-transform is given by

$$\tilde{X}(z) = \sum_{n=-\infty}^{\infty} x_n z^{-n} = \sum_{n=0}^{\infty} a^n z^{-n} = \sum_{n=0}^{\infty} (az^{-1})^n$$

This is simply a geometric series which converges to (see Eq. 1.18)

$$\tilde{X}(z) = \frac{1}{1 - az^{-1}} = \frac{z}{z - a}$$

for $|az^{-1}| < 1$, that is,

$$|z| > |a|$$

Thus the region of convergence includes all values of z with magnitude greater than $|a|$, that is, all values outside a circle of radius $|a|$ in the z-plane, as illustrated in Fig. 7.1.

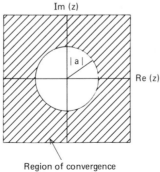

Region of convergence

Figure 7.1 Region of convergence for Example 7.1.

Example 7.2

Let $[x_n]$ be the left-sided sequence

$$x_n = b^n; \qquad n \le 0$$
$$= 0; \qquad n > 0$$

The z-transform is

$$\tilde{X}(z) = \sum_{n=-\infty}^{0} b^n z^{-n}$$

Performing the substitution $m = -n$, this sum can be written

$$\tilde{X}(z) = \sum_{m=0}^{\infty} b^{-m} z^m = \sum_{m=0}^{\infty} (b^{-1} z)^m$$

This again is a geometric series that converges to

$$\tilde{X}(z) = \frac{1}{1 - b^{-1} z} = \frac{b}{b - z}$$

for $|b^{-1} z| < 1$ or

$$|z| < |b|$$

In this example the region of convergence lies inside a circle of radius $|b|$ in the complex plane, as illustrated in Fig. 7.2.

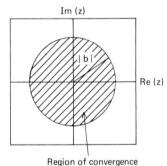

Region of convergence

Figure 7.2 Region of convergence for Example 7.2.

Example 7.3

Now consider the two-sided sequence

$$x_n = a^n; \qquad n \ge 0$$
$$= b^n; \qquad n < 0$$

The z-transform in this case can be written in terms of the two one-sided transforms in the previous examples:

$$\tilde{X}(z) = \sum_{n=-\infty}^{0} b^n z^{-n} + \sum_{n=0}^{\infty} a^n z^{-n} - 1$$

This expression converges to

$$\tilde{X}(z) = \frac{b}{b - z} + \frac{z}{z - a} - 1 = \frac{z(a - b)}{(z - a)(z - b)}$$

for $|z| > |a|$ and $|z| < |b|$, that is,

$$|a| < |z| < |b|$$

Thus, the region of convergence in this case is the intersection of the two regions in Examples 7.1 and 7.2 (see Fig. 7.3). For the regions to overlap, $|a|$ must be less than $|b|$. If the sequence $[x_n]$ has finite energy then, for the right segment, $|a| < 1$, and for the left segment $|b| > 1$. In this case an overlap is guaranteed.

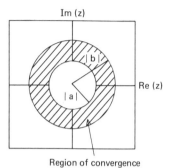

Region of convergence

Figure 7.3 Region of convergence for Example 7.3.

Example 7.4

Let $[x_n]$ be the damped sinusoid defined by

$$x_n = e^{-anT} \cos(bnT); \qquad n \geq 0$$
$$= 0; \qquad n < 0$$

Using the z-transform in Eq. 7.2 yields

$$\tilde{X}(z) = \sum_{n=0}^{\infty} e^{-anT} \cos(bnT) z^{-n}$$

It is easiest at this point to write the cosine in terms of complex exponentials using Euler's identity in Eq. 1.8:

$$\tilde{X}(z) = \sum_{n=0}^{\infty} \frac{e^{-anT}}{2} (e^{jbnT} + e^{-jbnT}) z^{-n}$$
$$= \frac{1}{2} \left[\sum_{n=0}^{\infty} (e^{-aT} e^{jbT} z^{-1})^n + \sum_{n=0}^{\infty} (e^{-aT} e^{-jbT} z^{-1})^n \right]$$

which converges to

$$\tilde{X}(z) = \frac{1}{2} \left(\frac{z}{z - e^{-aT} e^{jbT}} + \frac{z}{z - e^{-aT} e^{-jbT}} \right)$$
$$= \frac{z(z - e^{-aT} \cos(bT))}{z^2 - 2e^{-aT} \cos(bT) z + e^{-2aT}}$$

for

$$|z| > e^{-aT}$$

To obtain this result, we have used $|e^{-aT} e^{jbT}| = |e^{-aT}||e^{jbT}| = e^{-aT}$. Again, the region of convergence is the area outside a circle in the z-plane, this time with radius e^{-aT}.

Sec. 7.2   Definition of the z-Transform

It is important to note that with finite energy sequences, the region of convergence in all four examples includes the unit circle. We have already stated that the z-transform evaluated on the unit circle is the DFT. If the region of convergence does not include the unit circle, then the DFT is not defined for the sequence.

7.3 PROPERTIES OF THE z-TRANSFORM

The following are some important properties of the z-transform that can be easily verified using the definition in Eq. 7.1. As in Eq. 7.1, we use $\mathcal{Z}[\cdot]$ to stand for "z-transform of."

1. The z-transform is a *linear* transformation, that is,

$$\mathcal{Z}[a_1 x_n + a_2 y_n] = a_1 \mathcal{Z}[x_n] + a_2 \mathcal{Z}[y_n]$$

   for arbitrary constants a_1 and a_2.

2. The effect of shifting a sequence on its z-transform is given by the

---

Shifting Theorem
Let $\tilde{X}(z)$ be the z-transform of x_n. Then the z-transform of the shifted sequence x_{n+m} is

$$\mathcal{Z}[x_{n+m}] = z^m \tilde{X}(z) \tag{7.3}$$

---

3. Differentiation of the z-transform has the effect of multiplying the sequence by the negative time index. Thus,

$$\mathcal{Z}[n x_n] = -z \frac{d\tilde{X}(z)}{dz} \tag{7.4}$$

4. Multiplication of a sequence x_n by an exponential a^n results in a scaling of the z variable such that

$$\mathcal{Z}[a^n x_n] = \tilde{X}(a^{-1} z) \tag{7.5}$$

5. The z-transform of the convolution of two sequences is the product of their z-transforms, that is,

$$\mathcal{Z}\left(\sum_{m=-\infty}^{\infty} x_m y_{n-m} \right) = \tilde{X}(z)\tilde{Y}(z) \tag{7.6}$$

A proof of this relation proceeds as follows. By definition the z-transform of the convolution is

$$\tilde{W}(z) = \sum_{n=-\infty}^{\infty} \sum_{m=-\infty}^{\infty} x_m y_{n-m} z^{-n}$$

$$= \sum_{m=-\infty}^{\infty} x_m \sum_{n=-\infty}^{\infty} y_{n-m} z^{-n}$$

Now substitute k for $n - m$ and the expression above reduces to

$$\tilde{W}(z) = \sum_{m=-\infty}^{\infty} x_m \sum_{k=-\infty}^{\infty} y_k z^{-k-m}$$

$$= \sum_{m=-\infty}^{\infty} x_m z^{-m} \sum_{k=-\infty}^{\infty} y_k z^{-k} \qquad (7.7)$$

$$= \tilde{X}(z)\tilde{Y}(z)$$

which completes the proof.

6. The z-transform of a sampled sequence that is flipped about the zero axis is found by substituting z^{-1} for z, that is,

$$\mathscr{Z}[x_{-n}] = \tilde{X}(z^{-1}) \qquad (7.8)$$

7.4 TABLE OF z-TRANSFORMS

A short listing of z-transform relationships is provided in Table 7.1 on pages 130–131. As in Table 3.1 the relationships on the numbered lines in the upper part of the table can be used to generalize the specific transforms on the lettered lines in the lower part of the table.

As noted in the table, $f(t) = 0$ for $t < 0$ is assumed on the lettered lines, so the z-transform formula for these lines might as well be the one-sided definition in Eq. 7.2. Unless carefully understood, this fact can lead to an erroneous conclusion when the shifting theorem, line 5 of the table, is applied to one of the lettered lines. The source of the error can be seen in the following example.

Example 7.5

Using lines 5 and C of Table 7.1, find the inverse transform of

$$\tilde{G}(z) = \frac{z^{-1}}{z - a}$$

Observing lines 5 and C, we set $n = 2$ in line 5, so m must be replaced with $m - 2$ in line C to give $g_m = a^{m-2}$. This delays the original function a^m by two intervals, as illustrated in Fig. 7.4. Note, however, that the original (undelayed) function is 0 for $m < 0$; therefore,

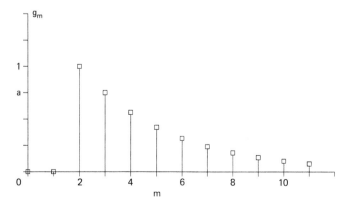

Figure 7.4 Inverse transform of $z^{-1}/(z - a)$, using $a = 0.75$.

TABLE 7.1. Short Table of Transforms

Line	Function	mth Sample
0	$f(t)$	$f_m = f(mT)$
1	$Af(t)$	Af_m
2	$f(t) + g(t)$	$f_m + g_m$
3	$tf(t)$	mTf_m
4	$e^{-at}f(t); \quad a > 0$	$e^{-maT}f_m$
5	$f(t - nT); \quad n > 0$	f_{m-n}
6	$f\left(\dfrac{t}{a}\right); \quad a > 0$	f_m
	(*Note:* In the following lines, $f(t) = 0$ for $t < 0$.)	
A	$d(t)$	$d_0 = \dfrac{1}{T}$
B	$u(t)$	$u_m = 1; \quad m \geq 0$
C	$a^{t/T}$	a^m
D	$\left(\dfrac{a}{a - b}\right)a^{t/T} + \left(\dfrac{b}{b - a}\right)b^{t/T}$	$\dfrac{a^{m+1} - b^{m+1}}{a - b}$
E	$e^{-at}; \quad a > 0$	e^{-maT}
F	$\sin at$	$\sin maT$
G	$\cos at$	$\cos maT$
H	$1 - e^{-aT}; \quad a > 0$	$1 - e^{-maT}$
I	$e^{-at} \sin bt; \quad a > 0$	$e^{-maT} \sin mbT$
J	$e^{-at} \cos bt; \quad a > 0$	$e^{-maT} \cos mbT$
K	$\dfrac{1}{b} R^{(t/T)+1} \sin\left[\left(\dfrac{t}{T} + 1\right)\theta\right];$ $R = \sqrt{a^2 + b^2} < 1;$ $\theta = \tan^{-1}\left(\dfrac{b}{a}\right)$	$\dfrac{1}{b} R^{m+1} \sin[(m + 1)\theta]$

z-Transform	DFT
$\tilde{F}(z) = \displaystyle\sum_{m=-\infty}^{\infty} f_m z^{-m}$	$\overline{F}(j\omega) = \tilde{F}(e^{j\omega T})$
$A\tilde{F}(z)$	$A\overline{F}(j\omega)$
$\tilde{F}(z) + \tilde{G}(z)$	$\overline{F}(j\omega) + \overline{G}(j\omega)$
$-Tz\dfrac{d}{dz}[\tilde{F}(z)]$	$j\dfrac{d}{d\omega}[\overline{F}(j\omega)]$
$\tilde{F}(ze^{aT})$	$\overline{F}(j\omega + a)$
$z^{-n}\tilde{F}(z)$	$e^{-jn\omega T}\overline{F}(j\omega)$
$\tilde{F}(z)$ with $\dfrac{T}{a} \to T$	$a\overline{F}(aj\omega)$
$\dfrac{1}{T}$	$\dfrac{1}{T}$
$\dfrac{z}{z-1}$	—
$\dfrac{z}{z-a}$	—
$\dfrac{z^2}{(z-a)(z-b)}$	—
$\dfrac{z}{z-e^{-aT}}$	$\dfrac{1}{1-e^{-(a+j\omega)T}}$
$\dfrac{z \sin aT}{(z^2 - 2z \cos aT + 1)}$	—
$\dfrac{z(z - \cos aT)}{(z^2 - 2z \cos aT + 1)}$	—
$\dfrac{(1 - e^{-aT})z}{(z-1)(z-e^{-aT})}$	—
$\dfrac{(e^{-aT} \sin bT)z}{z^2 - (2e^{-aT} \cos bT)z + e^{-2aT}}$	$\dfrac{e^{-aT} \sin bT}{e^{j\omega T} + e^{-(2a+j\omega)T} - 2e^{-aT} \cos bT}$
$\dfrac{z(z - e^{-aT} \cos bT)}{z^2 - (2e^{-aT} \cos bT)z + e^{-2aT}}$	$\dfrac{e^{j\omega T} - e^{-aT}\cos bT}{e^{j\omega T} + e^{-(2a+j\omega)T} - 2e^{-aT} \cos bT}$
$\dfrac{z^2}{(z-a)^2 + b^2}$	$\dfrac{1}{1 - 2ae^{-j\omega T} + (a^2 + b^2)e^{-2j\omega T}}$

$g_m = a^{m-2}$ is not correct for $m < 2$ in the delayed function. The correct inverse transform is

$$g_m = 0; \qquad m < 2$$
$$= a^{m-2}; \qquad m \geq 2$$

as illustrated in Fig. 7.4.

7.5 THE INVERSE z-TRANSFORM

Now let us assume that the z-transform $\tilde{X}(z)$ is given and we wish to recover the sampled sequence, $[x_n]$. There are two approaches to computing the inverse z-transform. The first is essentially a table lookup method and the second uses an integral relationship obtained from the realm of complex variable theory. In either case the z-transform is assumed to be given as a ratio of polynomials in z,

$$\tilde{X}(z) = \frac{\tilde{B}(z)}{\tilde{A}(z)} = K\left(\frac{z^M + b_{M-1}z^{M-1} + \cdots + b_1 z + b_0}{z^N + a_{N-1}z^{N-1} + \cdots + a_1 z + a_0}\right) \tag{7.9}$$

By factoring the numerator and denominator, $\tilde{X}(z)$ can be written

$$\tilde{X}(z) = K\left(\frac{(z - q_0)(z - q_1)\cdots(z - q_{M-1})}{(z - p_0)(z - p_1)\cdots(z - p_{N-1})}\right) \tag{7.10}$$

We assume for the present that there are no repeated roots in the numerator or denominator. When expressed in this form, the q_i's are called the *zeros* of $\tilde{X}(z)$ because at these values $\tilde{X}(z)$ is zero. Similarly, the p_i's are called the *poles* of $\tilde{X}(z)$. At these values $\tilde{X}(z)$ becomes infinite. Both methods for computing the inverse z-transform usually require that the denominator be factored so that the poles are known. This in itself can be difficult if the order of the polynomial is greater than 2. Nevertheless, assuming that the poles can be found, the inverse can be computed using one of the two following methods.

In the first method the objective is to manipulate the z-transform so that it is expressed as the sum of one or more of the forms in column 3 of Table 7.1. Then the inverse can be obtained directly from the relationships in the table. In the simplest case the required manipulation involves the following steps:

1. Find the poles of $\tilde{X}(z)$ so that the denominator can be expressed as a product of first-order terms,

$$\tilde{X}(z) = \frac{\tilde{B}(z)}{(z - p_0)\cdots(z - p_{N-1})} \tag{7.11}$$

2. Perform a partial-fraction expansion into a sum of first-order terms,

$$\frac{\tilde{X}(z)}{z} = \frac{b_0}{z - p_0} + \frac{b_1}{z - p_1} + \cdots + \frac{b_{N-1}}{z - p_{N-1}}, \quad \text{or}$$

$$\tilde{X}(z) = \frac{b_0 z}{z - p_0} + \frac{b_1 z}{z - p_1} + \cdots + \frac{b_{N-1} z}{z - p_{N-1}} \tag{7.12}$$

This requires that we solve for the numerator coefficients $b_0, b_1, \ldots, b_N$. Note that when the poles are complex they will always appear in conjugate pairs.

That is, if p_i is a complex pole, then there is another pole at the conjugate, p_i^*. Assuming that $p_i^* = p_{i+1}$, we can show that the numerator coefficients b_i and b_{i+1} are also complex conjugates (see Exercise 4). Thus, for conjugate pole pairs we need only solve for one of the numerator coefficients; the other is given automatically as the conjugate. Once the numerator coefficients are found, the conjugate pairs should be combined into a single second-order term before proceeding to the next step.

3. Once $\tilde{X}(z)$ has been expressed as the sum of first and/or second-order terms, we simply look in column 3 of Table 7.1 for the forms that best match the terms in $\tilde{X}(z)$. Most first-order terms will match line C, while the second-order terms can match either F, G, I, J, or K. The inverse z-transforms are read directly from column 2 of the table.

This method is illustrated in the following examples.

Example 7.6

Find the inverse z-transform of

$$\tilde{X}(z) = \frac{0.9z}{z^2 - 0.1z - 0.2}$$

The poles are easily found to be at $z = 0.5$ and -0.4, so we write

$$\tilde{X}(z) = \frac{0.9z}{(z - 0.5)(z + 0.4)}$$

The partial-fraction expansion requires that we solve for b_0 and b_1 in the following expression:

$$\tilde{X}(z) = \frac{b_0 z}{z - 0.5} + \frac{b_1 z}{z + 0.4}$$

b_0 and b_1 are found by cross multiplying and substituting appropriate values for z, for example,

$$b_0 = (z - 0.5) \frac{\tilde{X}(z)}{z} \bigg|_{z=0.5}$$

Using this approach, b_0 and b_1 are found to be 1 and -1, respectively. With this, $\tilde{X}(z)$ is given by

$$\tilde{X}(z) = \frac{z}{z - 0.5} - \frac{z}{z + 0.4}$$

Now using line C of Table 7.1, the inverse z-transform is given by

$$x_n = 0.5^n - (-0.4)^n; \qquad n \geq 0$$
$$= 0; \qquad\qquad n < 0$$

Example 7.7

Find the inverse z-transform of

$$\tilde{X}(z) = \frac{z^3 - 0.566z^2 + 0.527z}{z^3 - 1.33z^2 + 0.866z - 0.128}$$

The poles in this case are 0.2, $0.8e^{j\pi/4}$, and $0.8e^{-j\pi/4}$. The required partial fraction expansion has the form

$$\tilde{X}(z) = \frac{b_0 z}{z - 0.2} + \frac{b_1 z}{z - 0.8e^{j\pi/4}} + \frac{b_1^* z}{z - 0.8e^{-j\pi/4}}$$

Solving for b_0 gives 1. The solution for b_1 is $-0.5j$, implying that b_1^* is $0.5j$. Using these results and combining the complex conjugate terms, $\tilde{X}(z)$ can be written

$$\tilde{X}(z) = \frac{z}{z - 0.2} + \frac{0.564z}{z^2 - 1.13z + 0.64}$$

The inverse z-transform for the first-order term is once again found in line C of Table 7.1. For the second-order term we use line I with $aT = 0.223$ and $bT = \pi/4$. The inverse z-transform is then

$$x_n = 0.2^n + e^{-0.223n} \sin(n\pi/4); \qquad n \geq 0$$

$$= 0; \qquad\qquad\qquad\qquad n < 0$$

The partial-fraction method just described above can be used to invert nearly any z-transform. It must be modified, however, to handle transforms with shifted sequences and transforms with repeated poles. Rather than pursue these modifications, we proceed to the second method, which handles these situations automatically.

The second method for computing the inverse z-transform relies on some fundamental results from complex variable theory. It is not our intent here to derive these results; rather, the emphasis is placed on their use to obtain the inverse transform. We hasten to add that the reader need not have an extensive background in complex analysis to understand the method.

The formal definition of the inverse z-transform is given by the integral equation,

$$x_n = \frac{1}{2\pi j} \oint_C \tilde{X}(z) z^{n-1} dz \qquad (7.13)$$

where the closed path of integration is counterclockwise along a circular contour (C) that lies within the region of convergence of $\tilde{X}(z)$ and is centered at the origin. The evaluation of this integral is greatly simplified through the use of the residue theorem. [See Churchill et al. (1976), or any text on complex variables.] The residue theorem states that

$$x_n = \sum_k \text{Res}[z^{n-1}\tilde{X}(z) \text{ at pole } p_k] \qquad (7.14)$$

That is, the integral is equal to a finite sum of entities called residues evaluated at the poles of $z^{n-1}\tilde{X}(z)$ that lie within the contour of integration. Residues are found as follows.

Let $\tilde{Y}(z) = z^{n-1}\tilde{X}(z)$ be a rational function of z, and p_k a pole of $\tilde{Y}(z)$ which is repeated m times. Also, let $\tilde{W}(z)$ be formed as follows:

$$\tilde{W}(z) = (z - p_k)^m \tilde{Y}(z) \qquad (7.15)$$

Then the residue of $\tilde{Y}(z)$ at the pole p_k is given by the following expression:

Pole of order m at p_k

$$\text{Res}[\tilde{Y}(z) \text{ at } p_k] = \frac{1}{(m-1)!} \left[\frac{d^{m-1}\tilde{W}(z)}{dz^{m-1}} \right]_{z=p_k}$$

(7.16)

For simple poles where $m = 1$, Eq. 7.16 reduces to

Simple pole at p_k

$$\text{Res}[\tilde{Y}(z) \text{ at } p_k] = \tilde{W}(p_k)$$

(7.17)

The residue theorem thus states that the inverse z-transform, $[x_n]$, can be formed by summing the residues of $\tilde{Y}(z) = z^{n-1}\tilde{X}(z)$ at those pole locations that lie within the contour of integration. (Some authors use the factor $2\pi j$ in Eq. 7.14 with the corresponding factor $1/2\pi j$ in Eqs. 7.16 and 7.17.) The residue method is illustrated in the following examples.

Example 7.8

Use the residue method to find the inverse z-transform of $\tilde{X}(z)$ in Example 7.6. In this case $\tilde{Y}(z)$ is

$$\tilde{Y}(z) = z^{n-1}\tilde{X}(z)$$

$$= \frac{0.9z^n}{(z-0.5)(z+0.4)}$$

and the region of convergence is $|z| > 0.5$. The choice of integration path is arbitrary as long as it lies on a circular contour with radius greater than 0.5. (It is often convenient to choose the unit circle as the path of integration. The reason for this choice will become clear shortly.)

For $n \geq 0$ there are only two poles within the contour of integration, 0.5 and -0.4. Thus, the inverse z-transform is given by

$$x_n = \text{Res}(0.5) + \text{Res}(-0.4)$$

where the residues are computed as follows:

$$\text{Res}(0.5) = \left. \frac{0.9z^n}{z+0.4} \right|_{z=0.5} = 0.5^n$$

$$\text{Res}(-0.4) = \left. \frac{0.9z^n}{z-0.5} \right|_{z=-0.4} = -(-0.4)^n$$

So, for $n \geq 0$, x_n is the sum of these residues, that is,

$$x_n = 0.5^n - (-0.4)^n; \qquad n \geq 0$$

For $n < 0$, $\tilde{Y}(z)$ has an additional pole of order n at the origin. For example, at $n = -1$, $\tilde{Y}(z)$ is given by

$$\tilde{Y}(z) = \frac{0.9}{z(z - 0.5)(z + 0.4)}$$

and the inverse z-transform is

$$y_{-1} = \text{Res}(0.0) + \text{Res}(0.5) + \text{Res}(-0.4)$$

These residues are found as follows:

$$\text{Res}(0.0) = \frac{0.9}{(z - 0.5)(z + 0.4)}\bigg|_{z=0} = -4.5$$

$$\text{Res}(0.5) = \frac{0.9}{z(z + 0.4)}\bigg|_{z=0.5} = 2.0$$

$$\text{Res}(-0.4) = \frac{0.9}{z(z - 0.5)}\bigg|_{z=-0.4} = 2.5$$

Summing these results, we have

$$y_{-1} = 0.0$$

In a similar fashion we could evaluate y_n for $n = -2, -3$, and so on. This, however, would be a long and tedious process. Fortunately, there is an easier method for evaluating the inverse z-transform for $n < 0$. Through a simple series of manipulations one can show that the integral in Eq. 7.13 is identical to the following integral:

$$\boxed{x_n = \frac{1}{2\pi j} \oint_{C'} \tilde{X}(z^{-1}) z^{-n-1} \, dz} \qquad (7.18)$$

where C' is a new contour defined as follows. If C is the contour $|z| = r$, then C' is the contour $|z| = 1/r$. Note that if C is the unit circle, then C' is identical to C. In the integrand, the argument z^{-1} of $\tilde{X}$ has the effect of relocating the poles. Specifically, if a pole of $\tilde{X}(z)$ is at $z = p_i$, then the corresponding pole of $\tilde{X}(z^{-1})$ is at $z = 1/p_i$. Thus, all poles that were inside the unit circle are reflected outside the unit circle, and vice versa. The residue theorem can be used to evaluate Eq. 7.18 just as it was used to evaluate Eq. 7.13. To illustrate, let us complete Example 7.8 by finding x_n for $n < 0$. To do so we must evaluate the integral

$$x_n = \frac{1}{2\pi j} \oint_{C'} \frac{0.9 z^{-1}}{(z^{-1} - 0.5)(z^{-1} + 0.4)} z^{-n-1} \, dz$$

$$= \frac{1}{2\pi j} \oint_{C'} \frac{(5)(0.9) z^{-n}}{(2 - z)(2.5 + z)} \, dz$$

where C' is the unit circle. For $n \leq 0$ there are two poles, one at $z = 2$ and one at $z = -2.5$. Both are outside the contour of integration, implying that there are no nonzero residues for this integral. Thus, $x_n = 0$ for all $n \leq 0$. Note that for $n > 0$ the integrand has

The z-Transform   Chap. 7

a pole of order n at $z = 0$. Thus, to evaluate this integral for $n > 0$, we would require the same tedious procedure that was needed to evaluate x_n for $n < 0$ using the first integral.

Example 7.9

Use the residue theorem to find the inverse z-transform of

$$\tilde{X}(z) = \frac{3.8z}{(z - 0.2)(z - 4)}$$

where the region of convergence is $0.2 < |z| < 4$. Starting with the integral in Eq. 7.13 we have

$$x_n = \frac{1}{2\pi j} \oint_C \frac{3.8z^n}{(z - 0.2)(z - 4)} \, dz$$

Once again, we let C be the unit circle. For $n \geq 0$ the integrand has only one pole inside the contour at $z = 0.2$. Thus,

$$x_n = \text{Res}(0.2)$$

$$= \frac{3.8z^n}{z - 4}\bigg|_{z=0.2}$$

$$= -(0.2)^n; \quad n \geq 0$$

For $n < 0$ we use the integral in Eq. 7.18,

$$x_n = \frac{1}{2\pi j} \oint_{C'} \frac{3.8z^{-1}z^{-n-1}}{(z^{-1} - 0.2)(z^{-1} - 4)} \, dz$$

$$= \frac{1}{2\pi j} \oint_{C'} \frac{(3.8)(1.25)z^{-n}}{(z - 5)(z - 0.25)} \, dz$$

where C' is again the unit circle. For $n \leq 0$ the only pole inside C' is at $z = 0.25$. Thus,

$$x_n = \text{Res}(0.25)$$

$$= \frac{4.75z^{-n}}{(z - 5)}\bigg|_{z=0.25}$$

$$= -(0.25)^{-n}; \quad n \leq 0$$

Example 7.10

Use the residue theorem to find the inverse transform of $\tilde{X}(z)$ in Example 7.7. Using the integral in Eq. 7.13 the integrand, $\tilde{Y}(z)$, takes on the form

$$\tilde{Y}(z) = z^{n-1}\tilde{X}(z) = \frac{z^{n+2} - 0.566z^{n+1} + 0.527z^n}{(z - 0.2)(z - 0.8e^{j\pi/4})(z - 0.8e^{-j\pi/4})}$$

The region of convergence is $|z| > 0.8$. For $n \geq 0$ all three poles lie within the contour of integration, so that x_n is the sum of residues

$$x_n = \text{Res}(0.2) + \text{Res}(0.8e^{j\pi/4}) + \text{Res}(0.8e^{-j\pi/4})$$

which are given by

$$\text{Res}(0.2) = \frac{z^n(z^2 - 0.566z + 0.527)}{z^2 - 1.13z + 0.64}\bigg|_{z=0.2} = 0.2^n$$

$$\text{Res}(0.8e^{j\pi/4}) = \frac{z^n(z^2 - 0.566z + 0.527)}{(z - 0.2)(z - 0.8e^{-j\pi/4})}\Bigg|_{z=0.8e^{j\pi/4}} = \frac{j}{2}0.8^n e^{jn\pi/4}$$

$$\text{Res}(0.8e^{-j\pi/4}) = \frac{z^n(z^2 - 0.566z + 0.527)}{(z - 0.2)(z - 0.8e^{j\pi/4})}\Bigg|_{z=0.8e^{-j\pi/4}} = \frac{j}{2}0.8^n e^{-jn\pi/4}$$

Combining these values gives

$$x_n = 0.2^n + 0.8^n \sin\left(\frac{n\pi}{4}\right); \quad n \geq 0$$

In this result, the complex residues have been combined into a single sinusoidal term. Using $e^{-0.223} = 0.8$, we observe that this result is identical to that in Example 7.7. Using the integral in Eq. 7.18, we find that all the poles are reflected outside the contour of integration, so that x_n is zero for $n < 0$.

Example 7.11

Use the residue theorem to find the inverse z-transform of

$$\tilde{X}(z) = \frac{0.6}{z^2 - 1.2z + 0.36}$$

Solving for the poles of $\tilde{X}(z)$, we see that the denominator has a repeated root at $z = 0.6$. Thus $\tilde{Y}(z)$ can be expressed

$$\tilde{Y}(z) = z^{n-1}\tilde{X}(z) = \frac{0.6z^{n-1}}{(z - 0.6)^2}$$

For a stable solution we take the region of convergence to be $|z| > 0.6$. Thus, for $n \geq 1$, x_n is given by a single residue:

$$x_n = \text{Res}(0.6) = \frac{1}{(2 - 1)!}\left[\frac{d}{dz}(0.6z^{n-1})\right]_{z = 0.6}$$

$$= 0.6(n - 1)z^{n-2}|_{z=0.6}$$

$$= (n - 1)0.6^{n-1}$$

Using the integral in Eq. 7.18 the solution for $n \leq 0$ is found to be 0. Therefore,

$$x_n = (n - 1)0.6^{n-1}; \quad n \geq 1$$

$$= 0; \quad n < 1$$

7.6 THE CHIRP z-TRANSFORM

In Section 5.2 we derived the DFT by sampling the Fourier transform at equally spaced frequencies over one of its periods, typically taken to be the period from $-\pi/T$ to π/T. (The Fourier transform is given in Eq. 5.6 and the DFT in Eq. 5.9.) In Chapter 6 the FFT was developed as an efficient algorithm for computing the DFT. In many applications only the DFT samples are required. In some applications, however, we are not interested in the entire range of frequencies from $-\pi/T$ to π/T, but only in a small band within that range. For example, we may wish to resolve two closely spaced components in a small region of the spectrum, necessitating a higher-frequency resolution than that provided by the DFT. The problem then becomes one of computing a large number of spectral components over a small range of frequencies. One approach is to evaluate the samples directly using a range of values of ω in Eq. 5.6. This approach, however, suffers from the same type of computational burden as a direct evaluation of the DFT. A second approach might be to pad the original sequence with zeros and obtain a larger FFT with the desired resolution. Spectral components outside the band of interest would then be discarded. This method can lead to large FFT sizes which in turn require as many or more computations than the direct method, since most of the effort goes into computing samples which are to be discarded. The chirp z-transform described in this section provides a solution to the problem. It is an efficient algorithm for computing samples of the Fourier transform over a small band of frequencies.

More specifically, the chirp z-transform (CZT) is an efficient algorithm for evaluating the z-transform of a finite-length sequence at equally spaced samples along a generalized contour in the z-plane. It uses the FFT to perform the major computational step of the transformation. Since we are interested in acquiring samples of the Fourier transform, the z-plane contours are restricted to arcs along the unit circle. The algorithm, however, is not limited to such contours. A more general class of contours is the subject of Exercise 10 at the end of the chapter.

Let $[x_n]$ be the N-point sequence whose spectrum is of interest. The z-transform of $[x_n]$ is given by

$$\tilde{X}(z) = \sum_{n=0}^{N-1} x_n z^{-n} \tag{7.19}$$

The Fourier transform of $[x_n]$ is obtained by substituting $z = e^{j\omega T} = e^{j\theta}$, where $\theta = \omega T$ is the angular frequency in radians. Now suppose that we wish to sample the Fourier transform at M discrete frequencies, beginning at angular frequency θ_0, and with sampling resolution ϕ_0. This corresponds to sampling of the z-transform along an arc on the unit circle as shown in Fig. 7.5.

The samples along this arc are given by

$$z_k = AB^{-k}; \qquad k = 0, 1, \dots, M - 1 \tag{7.20}$$

where

$$A = e^{j\theta_0}; \qquad B = e^{-j\phi_0}$$

Substituting these samples of z into Eq. 7.19 above gives

$$\begin{aligned} \tilde{X}(z_k) &= \sum_{n=0}^{N-1} x_n (AB^{-k})^{-n} \\ &= \sum_{n=0}^{N-1} x_n A^{-n} B^{nk}; \qquad k = 0, 1, \dots, M - 1 \end{aligned} \tag{7.21}$$

The next step in our development, which is due to Bluestein (1970), appears to complicate rather than simplify this expression, but is the key step in arriving at an efficient implementation. This step uses the following identity:

$$nk = \tfrac{1}{2}[n^2 + k^2 - (k - n)^2]$$

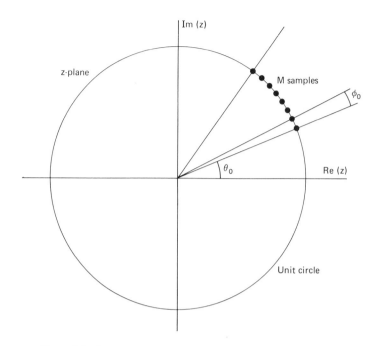

Figure 7.5 Sampling the z-transform along an arc on the unit circle.

With this substitution, Eq. 7.21 becomes

$$\tilde{X}(z_k) = \sum_{n=0}^{N-1} x_n A^{-n} B^{n2/2} B^{k2/2} B^{-(k-n)2/2}$$

$$= B^{k2/2} \sum_{n=0}^{N-1} (x_n A^{-n} B^{n2/2}) B^{-(k-n)2/2}$$

Now if we let

$$g_n = x_n A^{-n} B^{n2/2} \tag{7.22}$$

and

$$b_n = B^{-n2/2} \tag{7.23}$$

then Eq. 7.21 can be written

$$\tilde{X}(z_k) = B^{k2/2} \sum_{n=0}^{N-1} g_n b_{k-n}; \qquad k = 0, 1, \ldots, M - 1 \tag{7.24}$$

Thus, using Bluestein's substitution, we have transformed the sum of products expression in Eq. 7.21 into the linear convolution in Eq. 7.24. The key to an efficient implementation is to perform this linear convolution using the fast convolution method developed in Section 6.9. What results is called the chirp z-transform (CZT), which is illustrated in Fig. 7.6.

The name "chirp" stems from the nature of b_n. This sequence is of the form

$$b_n = e^{jn2\phi_0/2} = e^{jn(n\phi_0/2)}$$

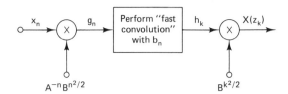

Figure 7.6 Chirp z-transform algorithm block diagram.

That is, b_n is a complex sinusoid with linearly increasing frequency, often called a chirp signal. An example of a chirp is shown in Fig. 7.7. (This type of signal is used in radar systems for frequency compression, similar to the application here.) It is important to note that the b_n sequence in Eq. 7.24 is noncausal; that is, it is nonzero for $n < 0$.

The steps required to carry out the CZT algorithm are the following:

1. Choose the FFT size, L, to be the first compatible size (usually, a power of 2) that is greater than or equal to $N + M - 1$.
2. Form the L-point sequence $[g_n]$:

$$g_n = A^{-n}B^{n^2/2}x_n; \qquad n = 0, 1, \ldots, N - 1$$
$$= 0; \qquad n = N, N + 1, \ldots, L - 1$$

[Note that the product $(A^{-n}B^{n^2/2})$ can often be precomputed.]

3. Use the FFT to compute the L-point DFT of $[g_n]$, denoted $[\overline{G}_m]$.
4. Form the L-point sequence $[b_n]$:

$$b_n = B^{-n^2/2}; \qquad 0 \le n \le M - 1$$
$$= 0; \qquad M \le n \le L - N$$
$$= B^{-(L-n)^2/2}; \qquad L - N + 1 \le n \le L - 1$$

The final term here places values in the last N locations of the sequence, which in the fast convolution method represent samples of $[b_n]$ for $n < 0$.

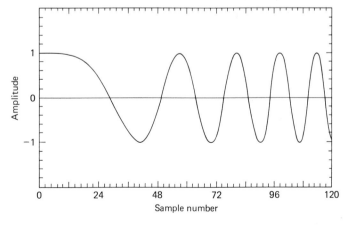

Figure 7.7 "Chirp" signal; sinusoid with linearly increasing frequency.

5. Use the FFT to compute the L-point DFT of $[b_n]$, denoted $[\overline{B}_m]$. $[\overline{B}_m]$, like the product in step 2, can often be computed ahead of time.
6. Multiply the DFTs point by point to form $[\overline{H}_m] = [\overline{G}_m \overline{B}_m]$.
7. Compute the L-point IFFT to give $[h_k]$.
8. Multiply h_k by $B^{k^2/2}$ for $k = 0, 1, \ldots, M - 1$ to give $[\tilde{X}(z_k)]$, the desired transform.

The computational requirements of the various steps are:

1. N complex multiplications to form $[g_n]$ (assuming that $A^{-n}B^{n^2/2}$ is precomputed)
2. $L \log_2 L$ complex multiplications, and $2L \log_2 L$ complex additions in the FFT and IFFT
3. L complex multiplications to compute the frequency-domain product $\overline{B}_m \overline{G}_m$
4. M complex multiplications to form the final result, $\tilde{X}(z_k) = h_k b_k$

The total computational requirements of this method are $(N + M + L + L \log_2 L)$ complex multiplications and $2L \log_2 L$ complex additions. On the other hand, using the direct method to evaluate $\tilde{X}(z_k)$ at M different frequencies requires a total of NM complex multiplications and $(N - 1)M$ complex additions. As an example, let the length of x_n be $N = 100$, and suppose that we wish to evaluate $\tilde{X}(z_k)$ at $M = 100$ different frequencies. The CZT (with $L = 256$) would require 2504 complex multiplications and 4096 complex additions, while the direct method would require 10,000 complex multiplications and 9900 complex additions. The savings are even greater for larger values of N and/or M. The computational savings of the CZT not only enhance the execution time of the algorithm, but also tend to reduce numerical round-off errors.

The CZT, like the FFT, is an efficient algorithm for computing components of the Fourier transform of a finite-length sequence. In many ways the CZT overcomes some of the inherent limitations of the FFT. For this reason it is instructive to review the following comparative aspects of the CZT and the FFT.

1. By appropriate selection of θ_0, ϕ_0, and M, the CZT can be used to compute the DFT. The FFT, however, is far more efficient in this case.
2. In the CZT the number of points (N) in the input sequence need not equal the number of frequency samples (M) at the output. This is not true for the FFT.
3. In the CZT there are no restrictions on N and M; that is, they are not required to be a power of 2 as is often the case with the FFT. Thus the CZT can be used to compute any size DFT in a relatively efficient manner.
4. The frequency range of the CZT is arbitrary; for the FFT it is always one period of the Fourier transform, typically $-\pi/T$ to π/T.
5. The frequency resolution of the CZT is arbitrary; for the FFT it is always a function of the FFT size, that is, $\Delta\omega = 2\pi/NT$.

EXERCISES

1. Find the z-transform of the following sequences, and give the region of convergence for each.
   (a) $x_n = 3e^{-4n}; \quad n \geq 0$
   (b) $x_n = 2a^{-3|n|}$
   (c) $x_n = a^{-2n} \cos(0.2\pi n); \quad n \geq 0$
   (d) $x_n = na^{-4n}; \quad n \geq 0$
   (e) $[x_n] = [1, 2, -3, -2]; \quad n = 0, 1, 2, 3$

2. Verify properties 1–4 and 6 in Section 7.3.

3. Show that the z-transform of the correlation of two sequences $[x_n]$ and $[y_n]$ is the product $\tilde{X}(z^{-1})\tilde{Y}(z)$, that is,

$$\mathscr{Z}\left[\sum_{n=-\infty}^{\infty} x_n y_{n+m}\right] = \tilde{X}(z^{-1})\tilde{Y}(z)$$

4. Show that in the partial-fraction expansion

$$\tilde{X}(z) = \frac{Kz}{(z - p_0)(z - p_1)} = \frac{b_0 z}{z - p_0} + \frac{b_1 z}{z - p_1}$$

   if K is real and >0, and if (p_0, p_1) are complex conjugates, then so are (b_0, b_1).

5. Use the table lookup (or partial-fraction expansion) method discussed in Section 7.5 to find the inverse z-transform of the following:

   (a) $\tilde{X}(z) = \dfrac{z}{z^2 - 0.2z - 0.24}$

   (b) $\tilde{X}(z) = \dfrac{z}{(z - e^{-2})(z - e^{-0.4})}$

   (c) $\tilde{X}(z) = \dfrac{1}{(z - a)(z - b)}$ with $|a| < 1$ and $|b| < 1$

   (d) $\tilde{X}(z) = \dfrac{0.647z}{z^2 - 0.94z + 0.64}$

   (e) $\tilde{X}(z) = \dfrac{z(z + 0.185)}{z^2 + 0.37z + 0.36}$

6. Use the residue theorem to find the inverse z-transforms in Exercise 5. (Assume that all sequences are causal.)

7. Use the residue method to find the inverse z-transform of

$$\tilde{H}(z) = \frac{z}{(z - 0.5)(z - 4)}$$

   given that the region of convergence is
   (a) $|z| > 4$
   (b) $0.5 < |z| < 4$
   (c) $|z| < 0.5$
   Which of these results produces a sequence with finite energy?

8. Use the residue method to find the inverse z-transform of each of the following assuming that the region of convergence includes the unit circle.

   (a) $\tilde{H}(z) = \dfrac{z}{z^2 - 0.6z + 0.09}$

(b) $\tilde{H}(z) = \dfrac{z^{-1}}{z^2 - 3.2z + 0.6}$

(c) $\tilde{H}(z) = \dfrac{z^4 - 0.707z^3 + 1.283z^2}{(z + 0.4)(z^2 - 1.414z + 1)}$

(d) $\tilde{H}(z) = 1 + 4z^{-1} - 0.7z^{-2} + 3.2z^{-3} - 4z^{-5}$

9. (Computer) Use the CZT algorithm in Section 7.6 to compute 40 samples of the Fourier transform of $[x_n]$ given below, starting at $\theta_0 = 0.05\pi$ and with sampling resolution $\phi_0 = 0.01\pi$.

$$x_n = e^{-n/50} \sin\left(\frac{2\pi n}{10}\right); \qquad 0 \le n \le 199$$

10. In this exercise you are to derive a more general form of the Chirp z-transform algorithm. The z-transform of an N-length sequence is to be evaluated at M points along an arc in the z-plane given by (see the figure)

$$z_k = A_0 B_0^k e^{j(\theta_0 + k\phi_0)}; \qquad k = 0, 1, 2, \ldots, M - 1$$

where A_0 is the initial radius of the arc, θ_0 is the initial angle, ϕ_0 is the angular step size from one point to the next on the arc, and B_0 determines the change in radius from one point to the next. Show that the algorithm derived in Section 7.6 is valid for this problem if we let $A = A_0 e^{j\theta_0}$ and $B = B_0 e^{-j\phi_0}$.

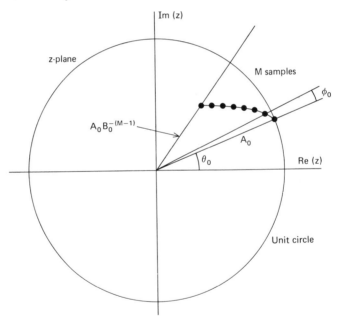

SOME ANSWERS

1. **(a)** $\dfrac{3z}{z - e^{-4}}; \quad |z| > e^{-4}$ **(b)** $\dfrac{2z(a^{-3} - a^3)}{z^2 - (a^3 + a^{-3})z + 1}; \quad a^{-3} < |z| < a^3$

(c) $\dfrac{z^2 - a^{-2}\cos(0.2\pi)z}{z^2 - 2a^{-2}\cos(0.2\pi)z + a^{-4}}; \quad |z| > a^{-2}$ **(d)** $\dfrac{a^{-4}z}{z^2 - 2a^{-4}z + a^{-8}}; \quad |z| > a^{-4}$

(e) $1 + 2z^{-1} - 3z^{-2} - 2z^{-3}; \quad |z| > 0$

6. **(a)** $0.6^n - (-0.4)^n$; $\quad n \geq 0$ **(b)** $\left(\dfrac{1}{e^{-0.4} - e^{-2.0}}\right)(e^{-0.4n} - e^{-2n})$; $\quad n \geq 0$

 (c) $\left(\dfrac{1}{a - b}\right)(a^{n-1} - b^{n-1})$; $\quad n \geq 1$ **(d)** $(0.8)^n \sin(0.3\pi n)$; $\quad n \geq 0$

 (e) $(0.6)^n \cos(0.6\pi n)$; $\quad n \geq 0$

7. **(a)** $\left(\dfrac{1}{3.5}\right)(4^n - 0.5^n)$; $\quad n \geq 0$ **(b)** $-\left(\dfrac{1}{3.5}\right)(0.5)^n$; $\quad n \geq 0$

$\qquad\quad 0$; $\hspace{3.5cm} n < 0$ $\qquad\quad -\left(\dfrac{1}{3.5}\right)4^n$; $\hspace{1cm} n < 0$

 (c) $\left(\dfrac{1}{3.5}\right)(0.5^n - 4^n)$; $\quad n \leq 0$

$\qquad\quad 0$; $\hspace{3.5cm} n > 0$

8. **(a)** $n(0.3)^{n-1}$; $\quad n \geq 0$ **(b)** $-\left(\dfrac{1}{2.8}\right)(0.2)^{n-2}$; $\quad n \geq 2$

$\qquad\quad 0$; $\hspace{2cm} n < 0$ $\qquad\quad -\left(\dfrac{1}{2.8}\right)3^{n-2}$; $\hspace{1cm} n \leq 1$

REFERENCES

AHMED, N., and NATARAJAN, T., *Discrete-Time Signals and Systems*. Reston, Va.: Reston, 1983.

BLUESTEIN, L. I., A Linear Filtering Approach to the Computation of the Discrete Fourier Transform. *IEEE Trans. Audio Electroacoust.*, Vol. AU-18, December 1970, p. 451.

CHURCHILL, R. V., BROWN, J. W., and VERHEY, R. F., *Complex Variables and Applications*. New York: McGraw-Hill, 1976.

JURY, E. I., *Theory and Application of the z-Transform Method*. Melbourne, Fla.: R. E. Krieger, 1984.

OPPENHEIM, A. V., and SCHAFER, R. W., *Digital Signal Processing*. Englewood Cliffs, N.J.: Prentice-Hall, 1975.

RABINER, L. R., and GOLD, B., *Theory and Application of Digital Signal Processing*. Englewood Cliffs, N.J.: Prentice-Hall, 1975.

8

Nonrecursive Digital Systems

8.1 DIGITAL FILTERING

The concept of "filtering," that is, passing certain frequencies in a signal and rejecting others, comes to digital signal analysis from its origins in the theory of linear continuous systems. Figure 8.1 illustrates the analogy. In the analog or continuous case $f(t)$ is a continuous function of time and $g(t)$ is the output, or filtered version of $f(t)$. In the digital case $f(t)$ is replaced by the set $[f_m]$ of discrete values, which may be considered a set of *samples* of $f(t)$, and a digital computation is then made to produce the set $[g_m]$.

In broadest terms, the concept of "digital filtering" refers to the assessment of the frequency-domain effects of any digital system or processing algorithm for which there is an "input" and an "output" — a rather inclusive concept in these terms. Sometimes the goal is the simulation of an analog system — a circuit, a control system, or some other continuous process. At other times no continuous system is involved, and only the spectral properties of a digital processing scheme are of interest.

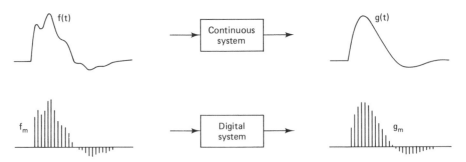

Figure 8.1 Filtering analogy.

In any case, digital filtering as discussed here involves a computing algorithm in which the set $[g_m]$ is produced as the result of operating on the set $[f_m]$. Each value g_m is thus some function of the set $[f_m]$ as suggested by Fig. 8.1. If this function is *linear,* then the equivalent of the linear transfer function exists, as shown below.

The related problems of analyzing, synthesizing, and realizing digital filters are discussed in this and subsequent chapters. This chapter concentrates on *nonrecursive* filters and begins with the definition of this type of filter and its transfer function in Sections 8.2 and 8.3. The discussion then proceeds in Sections 8.4 through 8.6 to the design of lowpass nonrecursive filters and a definition of the digital impulse response. Finally, the realization and synthesis of nonrecursive filters is discussed in Sections 8.7 through 8.10.

8.2 THE NONRECURSIVE ALGORITHM

Digital filters can be divided into two broad classes: those in which the formula for g_m contains explicitly the "past" values, g_{m-1}, g_{m-2}, and so on, as well as the set $[f_m]$, and those for which g_m is given explicitly in terms only of $[f_m]$. The former are called *recursive* filters because the samples of $g(t)$ are given recursively in terms of past values of $g(t)$, and the latter are *nonrecursive* filters because samples of $g(t)$ are computed directly in terms only of the samples of $f(t)$. Here the emphasis is on nonrecursive filters, with recursive filters being considered in the next chapter.

A general form of the linear nonrecursive algorithm is

$$g_m = \sum_{n=-N}^{N} b_n f_{m-n} \tag{8.1}$$

so that each value of the output is a linear function of the input sample set as just described. The limit N could theoretically be any value from zero to infinity, but must obviously be finite for any realizable nonrecursive filter. Each coefficient b_n can be assumed to be any real value including zero. Equation 8.1 thus allows one to consider $g(t)$ as a "moving weighted average" of $f(t)$, and the analysis problem becomes simply that of finding the frequency-domain effect of this averaging process. Note that, if any of the b_n's in Eq. 8.1 are nonzero for n negative, $g(t)$ appears to be given in terms of *future* values of $f(t)$. In analog systems this usually produces a realizability problem, but in digital processing the values of $f(t)$ needed to compute g_m can, in many practical cases, be stored in advance. If "real-time computing" or other constraints demand that g_m be given only in terms of the present and past values f_m, f_{m-1}, and so on, then b_n must be zero for $n < 0$.

Notice also that the nonrecursive formula, Eq. 8.1, is in the form of a discrete convolution. Therefore, a set $[g_m]$ may be computed rapidly by taking the inverse FFT of the product of the two FFTs, $[\overline{F}_m]$ times $[\overline{B}_m]$, as shown in Chapter 6. When computing Eq. 8.1 in this manner one must often address the following issue. The length of the data segment, $[f_m]$ in Eq. 8.1 is often very long, so long in fact that it becomes impractical to compute the FFT of the entire segment. The solution is to partition $[f_m]$ into blocks of reasonable size and then compute the convolution of each block using the FFT approach. When using this approach either the blocks must be overlapping or the result of each block be overlapped and combined in the appropriate manner, so that the result is the same as a single large convolution. Two methods that perform the desired overlap are the *overlap add* and *overlap save* methods. A detailed description of these methods can be found in Rabiner and Gold (1975).

8.3 TRANSFER FUNCTION

When the sample set $[f_m]$ is processed as in Eq. 8.1 to produce the set $[g_m]$, a transfer function is involved in the sense illustrated in Fig. 8.1 — an underlying function $f(t)$ is filtered and samples of its output, $g(t)$, are being computed.

A linear system in general alters only the magnitude and phase of a sinusoidal input. Therefore the linear transfer function can always be determined if one can ascertain the amplitude and phase angle of the output when the input is a known sinusoid. The output is, of course, also a sinusoid at the same frequency. Letting $e^{j\omega t}$ represent the known sinusoid at the frequency ω, one can say that

$$\text{if } f(t) = e^{j\omega t}, \quad \text{then } g(t) = Ae^{j(\omega t + a)}$$

Therefore,

$$\text{transfer function } \overline{H}(j\omega) = Ae^{ja} \tag{8.2}$$

(The bar is used on the transfer function because $\overline{H}(j\omega)$ is the Fourier transform of a sample sequence, as seen below.) Letting t take on the discrete values where the samples of $f(t)$ and $g(t)$ exist (i.e., $t = mT$, where m is an integer and T is the sampling interval), Eq. 8.2 can be stated in terms of the sample values:

$$\text{if } f_m = e^{jm\omega T}, \quad \text{then } g_m = \overline{H}(j\omega)e^{jm\omega T} \tag{8.3}$$

These sample values, if substituted into Eq. 8.1, provide a solution for $\overline{H}(j\omega)$:

$$\overline{H}(j\omega)e^{jm\omega T} = \sum_{n=-N}^{N} b_n e^{j(m-n)\omega T} \tag{8.4}$$

Therefore,

$$\overline{H}(j\omega) = \sum_{n=-N}^{N} b_n e^{-jn\omega T} \tag{8.5}$$

Equation 8.5 gives $\overline{H}(j\omega)$ in terms of the sampling interval T and the set $[b_n]$ of smoothing weights; furthermore, as shown in Chapter 5, Eq. 8.5 expresses $\overline{H}(j\omega)$ as the *Fourier transform of* the sequence of impulses $[b_n]$, and also as a complex Fourier series with real coefficients $[b_{-n}]$.

With the transfer function thus expressed as a Fourier transform, several of its properties are evident:

1. $\overline{H}(j\omega)$ is a *periodic* function of ω with the period being $2\pi/T$; that is, the sampling frequency.
2. $\overline{H}(j\omega)$ and $\overline{H}(-j\omega)$ are complex conjugates, and therefore the amplitude spectrum $|\overline{H}(j\omega)|$ is an *even* function of ω.

3. If $b_n = b_{-n}$ for all n, then $\overline{H}(j\omega)$ is *real* and in this case

$$\overline{H}(j\omega) = b_0 + 2 \sum_{n=1}^{N} b_n \cos n\omega T = \overline{H}(-j\omega) \qquad (8.6)$$

4. The complex Fourier coefficient formula (Eq. 2.42 in Chapter 2) gives each b_n in terms of $\overline{H}(j\omega)$ and thus provides a *filter synthesis formula:*[†]

$$b_n = \frac{T}{2\pi} \int_{-\pi/T}^{\pi/T} \overline{H}(j\omega) e^{jn\omega T}\, d\omega; \qquad -N \le n \le N \qquad (8.7)$$

Again, when $\overline{H}(j\omega)$ is *real,* Eq. 8.7 simplifies to

$$b_n = \frac{T}{\pi} \int_0^{\pi/T} \overline{H}(j\omega) \cos n\omega T\, d\omega = b_{-n} \qquad (8.8)$$

As shown previously, b_n must be real and must vanish beyond some maximum magnitude of n in order for $\overline{H}(j\omega)$ to be realizable in nonrecursive form. That is, $\overline{H}(j\omega)$ must be expressible as a finite Fourier series with real coefficients as in Eq. 8.5. On the other hand, suppose that a "desired" transfer function, $H_d(j\omega)$, has sharp corners or other features that necessitate an infinite Fourier series. Then, as shown in Chapter 2, $\overline{H}(j\omega)$ can be made a *least-squares approximation* to $H_d(j\omega)$ simply by using Eq. 8.7 and finding b_n for n in the interval $[-N, N]$. The approximation of course improves as N increases; but the complexity of the digital computation in Eq. 8.1 also increases with N.

These properties of the nonrecursive filter and its transfer function are illustrated in the following examples.

Example 8.1

A sample set $[f_m]$ is smoothed by averaging each sample with its two nearest neighbors, as illustrated in Fig. 8.2, so that

$$g_m = \tfrac{1}{3}(f_{m-1} + f_m + f_{m+1}) \qquad (8.9)$$

What is the transfer function $\overline{H}(j\omega)$? Here, Eq. 8.9 is an example of Eq. 8.1 with $b_{-1} = b_0 = b_1 = \tfrac{1}{3}$ and $N = 1$. Since $b_1 = b_{-1}$, Eq. 8.6 can be used in place of Eq. 8.5, and therefore

$$\overline{H}(j\omega) = \tfrac{1}{3}(1 + 2 \cos \omega T) \qquad (8.10)$$

and is a real function in this case. $\overline{H}(j\omega)$ is plotted versus ωT in Fig. 8.3. Note that, although Eq. 8.9 does accomplish smoothing in the sense that higher frequencies in the interval $(-\pi/T \le \omega \le \pi/T)$ are somewhat attenuated, $\overline{H}(j\omega)$ is not a particularly good lowpass transfer function. As illustrated in the next example, one can do somewhat better with a nonrecursive filter just as simple as Eq. 8.9. Outside the interval $(-\pi/T \le \omega \le \pi/T)$, $\overline{H}(j\omega)$ is of course periodic and, as property 1 above indicates, the period can be increased *only* by decreasing the sampling interval T.

[†]Equation 8.7 is also easy to prove by substituting Eq. 8.5 for $\overline{H}(j\omega)$.

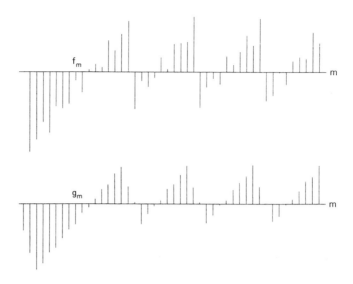

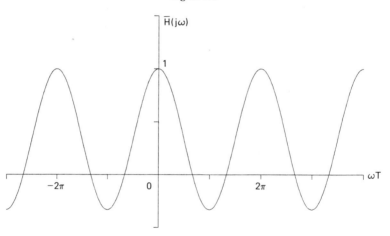

Figure 8.2

Figure 8.3

Example 8.2

Suppose that the desired effect of a nonrecursive filter is given by the transfer function $H_d(j\omega)$ shown in Fig. 8.4, and the sampling interval is $T = \pi/2$. [Note that $\overline{H}(j\omega)$ in Example 8.1 provides a crude approximation to $H_d(j\omega)$ here.] Design the nonrecursive fil-

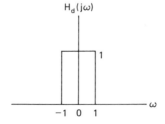

Figure 8.4

ter that provides a least-squares approximation to $H_d(j\omega)$ with $N = 1$, so that there are only three terms in the algorithm as in Example 8.1. Since $H_d(j\omega)$ is real, Eq. 8.8 provides the solution for this example:

$$b_n = b_{-n} = \frac{T}{\pi} \int_0^{\pi/T} H_d(j\omega) \cos n\omega T \, d\omega$$

$$= \frac{1}{2} \int_0^1 \cos\left(\frac{n\omega\pi}{2}\right) d\omega \tag{8.11}$$

$$= \frac{1}{2} \quad \text{for } n = 0; \qquad \frac{\sin(n\pi/2)}{n\pi} \quad \text{for } n \neq 0$$

As indicated by the solution, the sharp corners of $H_d(j\omega)$ can be obtained only by letting N, the maximum value of n in the filter algorithm, increase without bound. With $N = 1$, as specified in this example, the algorithm and transfer function are

$$g_m = \frac{f_m}{2} + \frac{1}{\pi}(f_{m-1} + f_{m+1}) \tag{8.12}$$

$$\overline{H}(j\omega) = \frac{1}{2} + \frac{2}{\pi} \cos\left(\frac{\omega\pi}{2}\right) \tag{8.13}$$

The transfer function $\overline{H}(j\omega)$ is of course composed of the first two terms of the Fourier series for $H_d(j\omega)$. The approximation, shown in Fig. 8.5, is better than that of Example 8.1 but still not very good, suggesting that N should be increased to a value greater than 1 to obtain a better approximation to $H_d(j\omega)$.

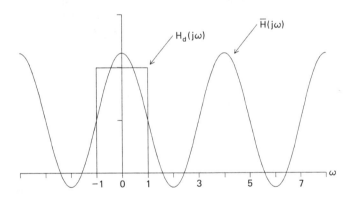

Figure 8.5

8.4 LOWPASS FILTER WITH ZERO PHASE SHIFT

Example 8.2 can be generalized to obtain the design of a lowpass nonrecursive digital filter with no phase shift at any frequency—a filter that might be used, for example, to eliminate high-frequency noise while passing a low-frequency signal without shifting its phase. Let the desired transfer function be as shown, this time in periodic form, in Fig. 8.6, with gain equal to 1 out to the cutoff frequency ω_c and zero phase shift everywhere. Obviously, the spectral content of both the signal *and the noise* must enter into the choice of the sampling interval T in Fig. 8.6.

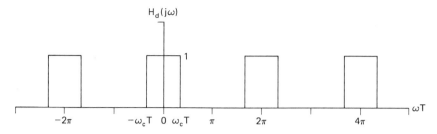

Figure 8.6 Lowpass transfer function.

Equation 8.8 now gives the filter coefficients (i.e., the generalized version of Eq. 8.11):

$$b_n = b_{-n} = \frac{T}{\pi} \int_0^{\pi/T} H_d(j\omega) \cos n\omega T \, d\omega$$

$$= \frac{T}{\pi} \int_0^{\omega_c} \cos n\omega T \, d\omega \qquad (8.14)$$

$$= \frac{\omega_c T}{\pi} \frac{\sin n\omega_c T}{n\omega_c T}$$

(The coefficient b_0 is understood to be $b_0 = \omega_c T/\pi$ in this result.) With these coefficients, the filter output and transfer function are now given by Eqs. 8.1 and 8.6:

$$g_m = \frac{\omega_c T}{\pi} \sum_{n=-N}^{N} \frac{\sin n\omega_c T}{n\omega_c T} f_{m-n}, \qquad (8.15)$$

and

$$\overline{H}(j\omega) = \frac{\omega_c T}{\pi} \left(1 + 2 \sum_{n=1}^{N} \frac{\sin n\omega_c T}{n\omega_c T} \cos n\omega T \right) \qquad (8.16)$$

With N equal to infinity the latter is of course just the Fourier series for $H_d(j\omega)$, as expected. With N finite so that the filter becomes realizable, $\overline{H}(j\omega)$ is the least-squares approximation to $H_d(j\omega)$, and the goodness of fit depends on the ratio of ω_c to the aliasing frequency π/T as well as on N. That is, the forms of both $g(t)$ and $\overline{H}(j\omega)$, as seen in Eqs. 8.15 and 8.16, are determined by the product $\omega_c T$ as well as by N.

Illustrations of one period of $\overline{H}(j\omega)$ for two values of N and two relative values of the cutoff frequency ω_c are given in Fig. 8.7. As seen in this illustration, the approximation to $H_d(j\omega)$ for either value of ω_c is improved when N is increased.

The lowpass transfer functions expressed in Eq. 8.16 and illustrated in Fig. 8.7 exhibit the property known as the *Gibbs phenomenon* (Stockham, 1969; Wait, 1970): Truncation of the Fourier series for a rectangular function leads to a fixed-percentage overshoot in the approximation. Note that the degree of overshoot in Fig. 8.7 does not change when N is changed from 20 to 30.

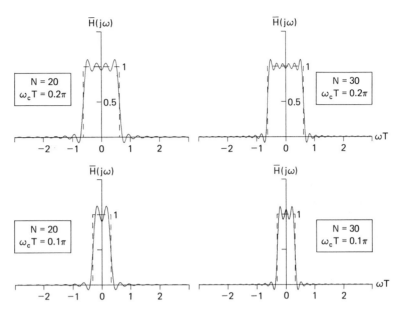

Figure 8.7   First period of the nonrecursive lowpass transfer function for two values of N and $\omega_c T$.

Gibbs phenomenon can be explained as follows. Let the truncated version of $[b_n]$ be denoted $[h_n]$. $[h_n]$ can be viewed as the product of $[b_n]$ times a rectangular window function $[w_n]$, that is, $[h_n] = [b_n w_n]$, where

$$
\begin{aligned}
w_n &= 1; \qquad |n| \le N \\
&= 0; \qquad \text{elsewhere}
\end{aligned}
\tag{8.17}
$$

As a result, the transfer function $\overline{H}(j\omega)$ is the convolution of the desired transfer function $H_d(j\omega)$ with the DFT of the window function. To verify this, consider the Fourier transform of $[h_n]$,

$$
\begin{aligned}
\overline{H}(j\omega) &= \sum_{n=-\infty}^{\infty} h_n e^{-jn\omega T} \\
&= \sum_{n=-\infty}^{\infty} b_n w_n e^{-jn\omega T}
\end{aligned}
$$

Substituting the inverse transform from Eq. 5.5 in Chapter 5 for $[b_n]$ gives

$$
\begin{aligned}
\overline{H}(j\omega) &= \sum_{n=-\infty}^{\infty} \left(\frac{T}{2\pi} \int_{-\pi/T}^{\pi/T} \overline{H}_d(j\theta) e^{jn\theta T}\, d\theta \right) w_n e^{-jn\omega T} \\
&= \frac{T}{2\pi} \int_{-\pi/T}^{\pi/T} \overline{H}_d(j\theta) \left(\sum_{n=-\infty}^{\infty} w_n e^{-jn(\omega-\theta)T} \right) d\theta \\
&= \frac{T}{2\pi} \int_{-\pi/T}^{\pi/T} \overline{H}_d(j\theta) \overline{W}(j(\omega-\theta))\, d\theta \\
&= \overline{H}_d(j\omega) * \overline{W}(j\omega)
\end{aligned}
\tag{8.18}
$$

Thus, the amount by which the actual frequency response differs from the desired response depends on the window spectrum, $\overline{W}(j\omega)$. For the rectangular window function above, the spectrum is given by

$$\overline{W}_R(j\omega) = \sum_{n=-N}^{N} e^{-jn\omega T} \qquad (8.19)$$

which can be simplified using the geometric series formula, Eq. 1.17, to

$$\overline{W}_R(j\omega) = \frac{\sin[(\omega T/2)(2N+1)]}{\sin(\omega T/2)} \qquad (8.20)$$

At $\omega = 0$ this expression is equal to $2N + 1$ (see Eq. 8.19). As ω increases to the Nyquist frequency, π/T, the denominator grows larger. This attenuates the higher-frequency numerator term, resulting in the damped sinusoidal function shown in Fig. 8.8. Because $[w_n]$ is sampled, $\overline{W}(j\omega)$ is repeated every ω_s rad/s. Within one

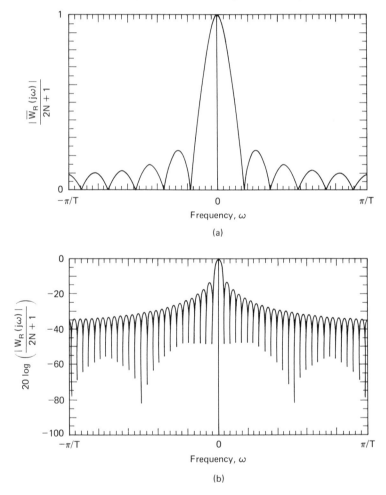

Figure 8.8   Rectangular window spectrum: (a) Linear plot with $N = 5$; (b) dB plot with $N = 25$.

period this function closely resembles the sinc function, although the two are not identical. When this function is convolved with the ideal spectral response in Fig. 8.6, the sidelobes of the window spectrum produce the rippling effect (Gibbs phenomenon) seen in the actual filter responses in Fig. 8.7. The sidelobes can be attributed largely to the sharp transition of the window sequence at the ends. As we shall see, the sidelobes can be greatly reduced by allowing the window to taper to zero in a more gradual manner. This is the topic of the next section.

8.5 DISCRETE-TIME WINDOW FUNCTIONS AND THEIR PROPERTIES

By inspection, Eq. 8.18 tells us that the ideal window spectrum is an impulse, that is, $\overline{W}(j\omega) = \delta(j\omega)$, which would give $\overline{H}(j\omega) = H_d(j\omega)$. However, an impulse-shaped spectrum would require a window sequence of infinite length. Since an infinite window is unrealizable, we seek finite-length window sequences with spectra that resemble the impulse function. In this section we describe several such windows and their properties.

For comparison with other windows, we first discuss the rectangular window, whose spectrum is given in Eq. 8.20. The first null in this spectrum occurs at $\omega T = 2\pi/(2N + 1)$. Thus, the width of the main lobe (distance between nulls on either side of zero) is $4\pi/(2N + 1)$. As we increase the length of the window the main lobe narrows and its peak value rises. In this sense, as N gets large, the window spectrum approaches an impulse. However, the amplitudes of the side lobes also increase. In fact, they grow at approximately the same rate as the main lobe, resulting in a constant peak-to-sidelobe ratio (PSLR). The ratio of the main lobe amplitude to the first sidelobe amplitude is

$$\text{PSLR} = \left| \frac{\overline{W}(0)}{\overline{W}(j\omega_1)} \right| \tag{8.21}$$

where ω_1 is the frequency at which first sidelobe reaches its maximum value. For the window spectrum in Eq. 8.20, $\omega_1 T = 3\pi/(2N + 1)$. Substituting this expression into Eq. 8.20 and simplifying yields

$$\text{PSLR} = (2N + 1) \sin\left(\frac{3\pi}{2(2N + 1)} \right) \tag{8.22}$$

For large N,

$$\sin\left(\frac{3\pi}{2(2N + 1)} \right) \approx \frac{3\pi}{2(2N + 1)}$$

so the ratio becomes

$$\text{PSLR} \approx \frac{3\pi}{2} = 13.5 \text{ dB} \quad \text{(rectangular window)} \tag{8.23}$$

Thus, while increasing the window length succeeds in narrowing the width of the main lobe, there is little effect on the peak-to-sidelobe ratio. The result is a fixed overshoot, regardless of N, in the actual filter responses seen in Fig. 8.7. The peak-to-sidelobe ratio can be improved, however, by using a window function that tapers to zero in a smooth fashion. The price for this improvement is an increased main lobe width. Thus, there is a trade-off between the main lobe width and the sidelobe amplitude.

Some of the more common window functions are shown in Table 8.1. The algorithms in Table 8.1 are encoded in FUNCTION SPWIND, which is included in Appendix B. Note that all of the windows are symmetric. This is a requirement if we are to retain the linear phase property of the filter.

The *Bartlett window*, which has a triangular shape, is well known because of the way it evolves from standard methods of spectral estimation, which is the subject of Chapter 15. Although it is not a particularly good window for filter design, it clearly has a smoother transition to zero than the rectangular window. It can, in fact, be formed approximately by convolving two rectangular windows of length $N + 1$. Therefore, its spectrum is approximately the square of the spectrum of a rectangular window of length $N + 1$ (see Eq. 5.75). That is, from Eq. 8.20,

$$\overline{W}_{BT}(j\omega) \approx \frac{\sin^2\left[\left(\dfrac{N+1}{2}\right)\omega T\right]}{\sin^2(\omega T/2)} \tag{8.24}$$

The first null in this spectrum occurs at $\omega = 2\pi/(N + 1) \approx 2\pi/N$. Thus, the width of the main lobe is approximately twice that of the rectangular window. On the other hand one can show that the peak-to-sidelobe ratio for the Bartlett window is approximately 27 dB, or twice that of the rectangular window (see Fig. 8.9).

The *Hanning window* is basically one period of a raised cosine function. By breaking the cosine term into two complex exponentials (see Eq. 1.8) one can easily show that its window spectrum is of the form

$$\overline{W}_{HN}(j\omega) = 0.5\overline{W}_R(j\omega) + 0.25\overline{W}_R\left(j\left(\omega - \frac{\pi}{NT}\right)\right)$$
$$+ 0.25\overline{W}_R\left(j\left(\omega + \frac{\pi}{NT}\right)\right) \tag{8.25}$$

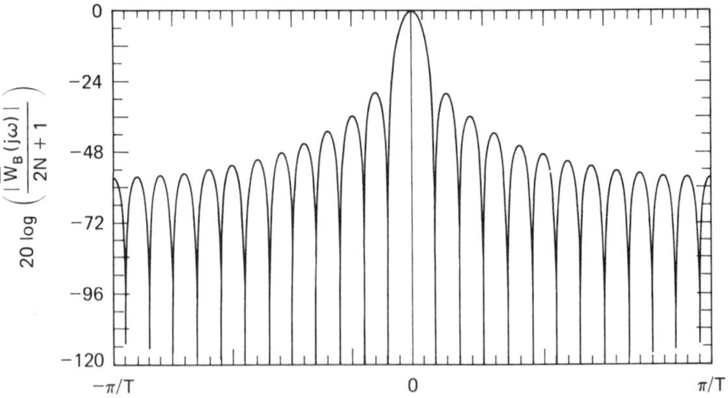

Figure 8.9 Bartlett window spectrum, $N = 25$.

TABLE 8.1. Common Window Functions[a, b]

1. Rectangular

$$w_n = 1; \quad |n| \le N$$
$$ = 0; \quad \text{elsewhere}$$

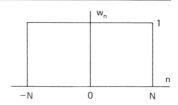

2. Bartlett

$$w_n = 1 - \frac{|n|}{N}; \quad |n| \le N$$
$$ = 0; \quad \text{elsewhere}$$

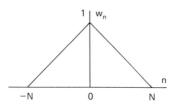

3. Hanning

$$w_n = 0.5\left(1.0 + \cos\left(\frac{\pi n}{N}\right)\right); \quad |n| \le N$$
$$ = 0; \quad \text{elsewhere}$$

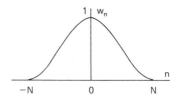

4. Hamming

$$w_n = 0.54 + 0.46 \cos\left(\frac{\pi n}{N}\right); \quad |n| \le N$$
$$ = 0; \quad \text{elsewhere}$$

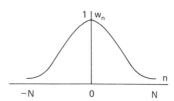

5. Blackman

$$w_n = 0.42 + 0.5 \cos\left(\frac{\pi n}{N}\right) + 0.08 \cos\left(\frac{2\pi n}{N}\right); \quad |n| \le N$$
$$ = 0; \quad \text{elsewhere}$$

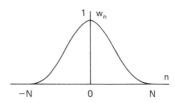

6. Kaiser

$$w_n = \frac{I_0(\beta\sqrt{1 - (n/N)^2})}{I_0(\beta)}; \quad |n| \le N$$
$$ = 0; \quad \text{elsewhere}$$

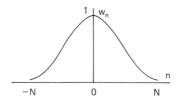

[a]These functions are encoded in FUNCTION SPWIND in Appendix B.
[b]Some authors define windows as shown, but with N replaced by $N + 1$ to keep from eliminating the Nth data samples with windows 2, 3, and 5, for which $w_N = 0$.

where $\overline{W}_R(j\omega)$ is the spectrum of the rectangular window in Eq. 8.20. The first null of $\overline{W}_{HN}(j\omega)$ occurs at approximately $\omega T = 2\pi/N$, resulting in a main lobe width of approximately $4\pi/N$. Thus, like the Bartlett spectrum, the width of the main lobe is approximately twice that of the rectangular window. However, for large N, the peak-to-sidelobe ratio is approximately 31 dB (see Fig. 8.10), an improvement of 4 dB over the Bartlett spectrum.

The *Hamming window* is similar to the Hanning except that it does not taper to zero at the ends. Using the same approach as above, the Hamming window spectrum is found to be

$$
\begin{aligned}
\overline{W}_{HM}(j\omega) = 0.54\overline{W}_R(j\omega) &+ 0.23\overline{W}_R\left(j\left(\omega - \frac{\pi}{NT}\right)\right) \\
&+ 0.23\overline{W}_R\left(j\left(\omega + \frac{\pi}{NT}\right)\right)
\end{aligned}
\tag{8.26}
$$

The Hamming and Hanning main lobe widths are approximately the same. The peak-to-sidelobe ratio of the Hamming window is approximately 41 dB, a significant improvement over that of the Hanning window (see Fig. 8.11).

The fifth window function in Table 8.1 is the *Blackman window*. It is similar to the Hanning and Hamming windows except for the addition of a higher-frequency cosine term. Like the Hanning window, the Blackman window tapers to zero. Again, using the same approach as for the Hamming window, the Blackman window spectrum can be shown to be

$$
\begin{aligned}
\overline{W}_{BK}(j\omega) = 0.42\overline{W}_R(j\omega) \\
+ 0.25\left[\overline{W}_R\left(j\left(\omega - \frac{\pi}{NT}\right)\right) + \overline{W}_R\left(j\left(\omega + \frac{\pi}{NT}\right)\right)\right] \\
+ 0.04\left[\overline{W}_R\left(j\left(\omega - \frac{2\pi}{NT}\right)\right) + \overline{W}_R\left(j\left(\omega + \frac{2\pi}{NT}\right)\right)\right]
\end{aligned}
\tag{8.27}
$$

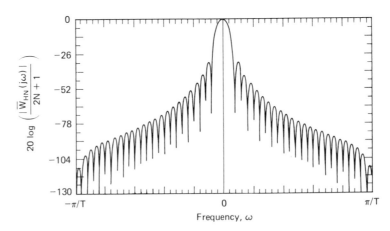

Figure 8.10 Hanning window spectrum, $N = 25$.

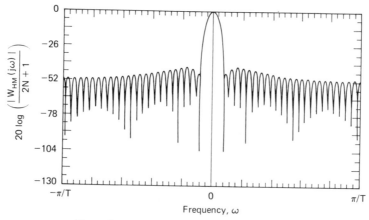

Figure 8.11 Hamming window spectrum, $N = 25$.

The addition of the second cosine term has the effect of increasing the width of the main spectral lobe to $\omega T \approx 6\pi/N$, but at the same time improving the peak-to-sidelobe ratio to approximately 57 dB (see Fig. 8.12).

Which window is best? The answer depends on the design specifications of the filter. The rectangular window is best in the sense that, for a fixed length N, its spectrum has the most narrow main lobe, providing the sharpest transition in the filter response. However, the large sidelobes that lead to Gibbs phenomenon are generally unacceptable. For this reason, a window function is generally selected on the basis of its ability to meet certain peak-to-sidelobe requirements. The main lobe can always be narrowed by increasing the window length, that is, the filter length. Thus, one might choose the "best" window function to be the one that provides the largest energy in the main lobe for a given sidelobe amplitude. For continuous functions, the optimal window in this sense involves a class of functions called *prolate spher-*

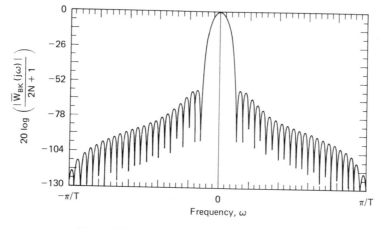

Figure 8.12 Blackman window spectrum, $N = 25$.

Sec. 8.5   Discrete-Time Window Functions and Their Properties   **159**

oidal wave functions. The *Kaiser window* (Hamming, 1977) uses discrete-time approximations to these functions. The Kaiser window has the form

$$w_n = \frac{I_0(\beta[1 - (n/N)^2]^{1/2})}{I_0(\beta)}; \qquad |n| < N \tag{8.28}$$

$$= 0; \qquad\qquad\qquad\qquad \text{elsewhere}$$

where $I_0(\beta)$ is the zeroth-order modified Bessel function of the first kind given by

$$I_0(\beta) = 1 + \sum_{k=1}^{\infty} \left(\frac{(\beta/2)^k}{k!}\right)^2 \tag{8.29}$$

The argument β in the Kaiser window is typically between 4 and 9 and is varied to trade main lobe energy with side lobe amplitude. An example of the Kaiser window and its spectrum for $N = 25$ is shown in Fig. 8.13. Similar to the Hamming window, the Kaiser window tapers to a nonzero value. In fact, by substituting $n = N$ in the definition above, the end value is found to be $1/I_0(\beta)$. When $\beta = \sqrt{3}\,\pi$ the Kaiser window spectrum has approximately the same main lobe width as the Hamming window spectrum and, even though the peak-to-sidelobe ratio of the two are about the same, the energy in the sidelobes is far less for the Kaiser window, as shown in Fig. 8.14. Similarly, when $\beta = 2\sqrt{2}\,\pi$, the Kaiser window spectrum has approximately the same main lobe width as Blackman's window spectrum. In this case the Kaiser peak-to-sidelobe ratio is much better than the Blackman ratio, as shown in Fig. 8.15. As a general rule the Kaiser window is the best window in the sense that it optimizes the trade-off between main lobe energy and sidelobe amplitude.

The features of window functions pertaining to the design of FIR filters can be summarized as follows. The coefficients of the desired filter, given by Eq. 8.14, are generally infinite in extent. Any practical filter must have a finite number of coefficients. Window functions provide a means of truncating the desired filter coefficients in a manner that allows the resulting filter response to retain a desired likeness to the ideal response. Truncation by any window leads to two types of distortion. The first is Gibbs phenomenon, a rippling effect in the filter response brought on by

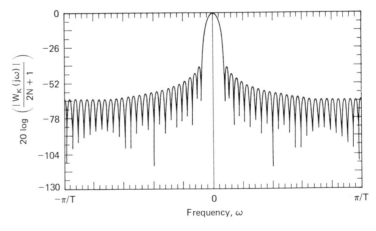

Figure 8.13 Kaiser window spectrum, $N = 25$ ($\beta = 5.44$).

Nonrecursive Digital Systems   Chap. 8

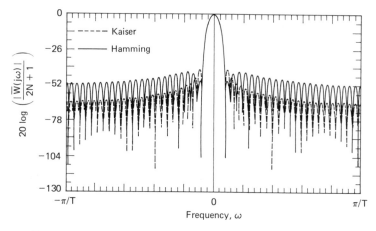

Figure 8.14 Kaiser and Hamming window spectra, $N = 25$ ($\beta = 5.44$).

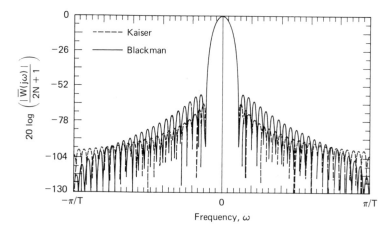

Figure 8.15 Kaiser and Blackman window spectra, $N = 25$ ($\beta = 8.89$).

the sidelobes of the window spectrum. The second is a broadening of the transition between passband and stopband due to the width of window's main spectral lobe. A window is generally chosen mainly on the basis of its ability to reduce Gibbs phenomenon (see Fig. 8.16 for a comparison). The sharpness of the transition from passband to stopband is controlled by the length of the filter.

However, an optimal window function does not necessarily result in an optimal filter design. That is, the filter characteristic obtained by truncating the desired filter coefficients with an optimal window function is not necessarily the best approximation to the desired response.

Direct optimization techniques for FIR filter design that do achieve such an approximation are generally quite complex. They are often formulated into a constrained optimization problem that is solved using linear or nonlinear programming techniques. Although such techniques are beyond the scope of this book, it is instructive to discuss the basic formulation of these solutions.

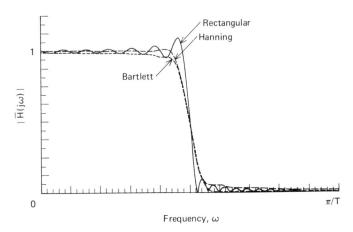

Figure 8.16 Lowpass filter responses using different window functions: rectangular, Bartlett, and Hanning. Results for Hamming, Blackman, and Kaiser are similar to Hanning.

As mentioned above, the best filter in terms of minimizing the mean squared error uses the rectangular window. The minimized function is

$$\text{MSE} = \frac{T}{2\pi} \int_{-\pi/T}^{\pi/T} [\overline{H}_d(j\omega) - \overline{H}(j\omega)]^2 \, d\omega \tag{8.30}$$

Unfortunately, this approach places no restrictions on how the error is distributed over frequency. As a result, it allows large errors near the transition region (Gibbs phenomenon) in order to keep the errors elsewhere at a minimum. One way to compensate for this behavior is to choose the filter coefficients to minimize a *weighted* squared error,

$$\text{WMSE} = \frac{T}{2\pi} \int_{-\pi/T}^{\pi/T} (\overline{W}(j\omega)[\overline{H}_d(j\omega) - \overline{H}(j\omega)])^2 \, d\omega \tag{8.31}$$

where $\overline{W}(j\omega)$ is a frequency-dependent weighting function used to place more emphasis on the error in the transition region. While this approach has been used with some success, a more popular approach is to minimize the maximum absolute error,

$$\text{AE}_{\max} = \underset{\omega \in R}{\text{maximum}} |\overline{W}(j\omega)[\overline{H}_d(j\omega) - \overline{H}(j\omega)]| \tag{8.32}$$

where once again $\overline{W}(j\omega)$ is a weighting function. Here the filter coefficients are selected to minimize the maximum absolute error within a frequency range (R) of interest. This method is much better at distributing the error more equally among all frequencies. For a more detailed discussion on these methods, see Rabiner and Gold, 1975.

8.6 IMPULSE RESPONSE

The concept of the inverse transform of the transfer function being the response to a unit impulse input can, with some revisions, be applied to digital filters. For continuous filters the unit impulse input is $\delta(t)$, defined in Chapter 3. The Fourier transform, $\Delta(j\omega)$, is equal to one at all frequencies; that is, an input transient with constant energy density at all frequencies is used to obtain the impulse response of a continuous filter.

162   Nonrecursive Digital Systems   Chap. 8

As in Eq. 8.5, the digital filter transfer function $\overline{H}(j\omega)$ is a periodic function of ω—a fact that turns out to be true for recursive as well as nonrecursive digital filters. Therefore, the integral of $|\overline{H}(j\omega)|$ does not in general converge, and the inverse transform of $\overline{H}(j\omega)$ does not usually exist as it does for realizable continuous filters. That is, the response of a digital filter to $\delta(t)$ is not in general defined. Note that this is a consequence not of $\delta(t)$ being infinite at $t = 0$, but rather of the constant energy density of $\delta(t)$ at all frequencies.

Consider the two impulse functions shown, each with its Fourier transform, in Fig. 8.17. The second impulse, $d(t)$, is the inverse Fourier transform of a spectrum, $D(j\omega)$, that is equal to T in the frequency interval $-\pi/T \le \omega \le \pi/T$. That is, as in Chapter 3, Exercise 20,

$$
\begin{aligned}
d(t) &= \frac{1}{2\pi} \int_{-\infty}^{\infty} D(j\omega) e^{j\omega t}\, d\omega \\
&= \frac{T}{2\pi} \int_{-\pi/T}^{\pi/T} e^{j\omega t}\, d\omega = \frac{\sin(\pi t/T)}{\pi t/T}
\end{aligned}
\tag{8.33}
$$

As the time step T decreases toward zero, the spectrum $D(j\omega)$ approaches $T\Delta(j\omega)$, and $d(t)$ approaches an impulse function with diminishing amplitude T.

We define the *digital impulse function* to be a sequence with a single unit sample at $t = 0$, that is,

$$
\begin{aligned}
d_n &= 1; & n &= 0 \\
&= 0; & n &\ne 0
\end{aligned}
\tag{8.34}
$$

Thus, d_n is seen to be the sequence of samples of $d(t)$ in Fig. 8.13, and in this sense the digital and continuous impulse functions are related through a scaling factor

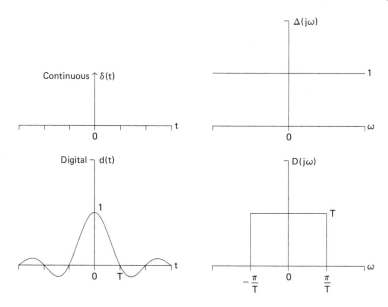

Figure 8.17 Impulse functions and spectra for continuous and digital filters.

equal to T. A "sample set" of $\delta(t)$ is defined to be the single sample $1/T$ at $n = 0$. See Appendix A, line 100.

In other words, the following two filters have outputs that are equal at the sample points $t = 0, T, 2T, \ldots$:

1. A digital filter with transfer function $\overline{H}(j\omega)$ and with input d_n
2. A continuous filter with transfer function equal to the first period of $T\overline{H}(j\omega)$ and with input $\delta(t)$

If the time step T approaches zero so that the sample points draw closer together, the continuous filter (2) becomes essentially a limiting case of the digital filter (1).

Restricting the discussion again to nonrecursive digital filters, it is easy to see how the set $[b_n]$ of filter coefficients is related to the impulse response. Equation 8.7, rewritten here, gives each filter coefficient b_n in terms of the first period of $\overline{H}(j\omega)$:

$$b_n = \frac{T}{2\pi} \int_{-\pi/T}^{\pi/T} \overline{H}(j\omega)e^{jn\omega T}\,d\omega \qquad (8.35)$$

If b_n is now viewed as $b(nT)$, a sample of a function $b(t)$ at $t = nT$, then $b(t)$ is in fact the inverse transform of the first period of $T\overline{H}(j\omega)$, so the set $[b_n]$ represents the impulse response of the nonrecursive digital filter, as illustrated in Fig. 8.18. It follows from Eq. 8.5 that the first period of $T\overline{H}(j\omega)$ is the Fourier transform of the impulse response envelope $b(t)$, which also corresponds with the continuous case. Note also that Eq. 8.1 gives the impulse response directly. If the input f_m is 1 for $m = 0$ and zero elsewhere, the output g_m must be equal to b_m.

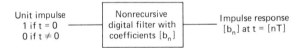

Figure 8.18 Impulse response of nonrecursive digital filter.

Example 8.3

Find the impulse response of the lowpass nonrecursive filter. Using the unmodified filter coefficients $[b_n]$ in Eq. 8.14, the impulse response is seen to be

$$b_n = \frac{\omega_c T}{\pi}\,\frac{\sin n\omega_c T}{n\omega_c T}; \qquad |n| \le N$$
$$= 0; \qquad\qquad\quad |n| > N \qquad (8.36)$$

The envelope of this response, found by letting $t = nT$ as above, is shown in Fig. 8.19. The impulse response is found by taking values of $b(t)$ at $t = nT$ for $|n| < N$, as indicated in the figure. The fact that the impulse response occurs for negative as well as positive values of t does not affect the practical realizability of the lowpass filter in the sense described above. Finally, if ω_c is made equal to π/T in Fig. 8.19, then the entire impulse spectrum $D(j\omega)$ in Fig. 8.17 is passed unaltered by the filter, and $b(t)$ in Fig. 8.19 equals $Td(t)$ in Fig. 8.13.

8.7 FREQUENCY RESPONSE AND BLOCK DIAGRAMS

As in Section 7.2, if we substitute $z = e^{j\omega T}$ into the transfer-function formula, Eq. 8.5, we obtain the nonrecursive transfer function in the form of a z-transform. Thus,

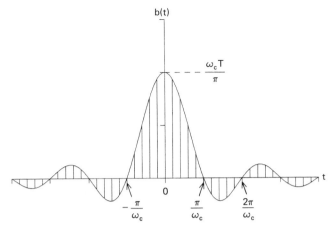

Figure 8.19 Impulse response of the lowpass nonrecursive filter.

$$\tilde{H}(z) = \overline{H}(j\omega)\big|_{z=e^{j\omega T}} = \sum_{n=-N}^{N} b_n z^{-n} \tag{8.37}$$

Using $\tilde{H}(z)$, we can derive a digital transfer-function relationship similar to the continuous relationships given in Chapter 3. We take the z-transform of Eq. 8.1 and apply the convolution-product relationship in Eq. 7.6 to obtain

$$\tilde{G}(z) = \mathscr{Z}\left[\sum_{n=-N}^{N} b_n f_{m-n}\right]$$
$$= \tilde{B}(z)\tilde{F}(z) = \tilde{H}(z)\tilde{F}(z) \tag{8.38}$$

That is, the transfer function $\tilde{H}(z)$ is the transform of the coefficient sequence, $[b_n]$, and is also the ratio of input and output transforms.

A Fourier transform version of Eq. 8.38 is easily obtained via the substitution $z = e^{j\omega T}$. Furthermore, we will see that, although the form of $\tilde{H}(z)$ is different, the input–output relationship $\tilde{G}(z) = \tilde{H}(z)\tilde{F}(z)$ holds for recursive as well as nonrecursive systems. Thus we have the same kind of transfer-function characterization for both continuous and digital linear systems, suggesting a close duality between the two, which is explored further in Chapters 10 and 11. Using the transform notations in Chapter 3 as well as above, we can summarize:

CONTINUOUS SYSTEMS	DIGITAL SYSTEMS	
$H(j\omega) = \dfrac{G(j\omega)}{F(j\omega)}$	$\overline{H}(j\omega) = \tilde{H}(e^{j\omega T}) = \dfrac{\overline{G}(j\omega)}{\overline{F}(j\omega)}$	
$H(s) = \dfrac{G(s)}{F(s)}$	$\tilde{H}(z) = \dfrac{\tilde{G}(z)}{\tilde{F}(z)}$	(8.39)

The notation here again serves to emphasize that whereas $H(j\omega)$ and $H(s)$ are the same function with different arguments, $\overline{H}(j\omega)$ and $\tilde{H}(z)$ are different functions, as seen in Eq. 8.37.

The *amplitude response* and *phase shift* of any linear digital system are also similar to the continuous amplitude response and phase shift described in Chapter 3, Section 3.2. Thus,

	CONTINUOUS SYSTEMS	DIGITAL SYSTEMS	
Amplitude response:	$|H(j\omega)|$	$|\tilde{H}(e^{j\omega T})|$	(8.40)
Phase shift:	$\arg[H(j\omega)]$	$\arg[\tilde{H}(e^{j\omega T})]$	

The simple representation in Eq. 8.39 of the digital transfer function is useful for many purposes, and leads directly to a block-diagram representation of the filter. As with the z-transform itself, the principal uses of the diagram are with recursive filters, but it is introduced here to show its form in a simple manner.

Using property 4 in Section 7.2, which states that z^{-1} is equivalent to a unit delay of T seconds, one can diagram the nonrecursive transfer function, $\tilde{H}(z)$ in Eq. 8.37. The result is shown for $N = 3$ in Fig. 8.20, with the unit delays, the coefficients $[b_n]$, and the summation all illustrated explicitly. The diagram is suggestive of a realization of the filter in either hardware or algorithmic form, with the "future" sample f_{m+N} being available at time $f = mT$ and the samples $f_{m+N-1}, \ldots, f_m$, $\ldots, f_{m-N}$ all being stored, amplified, and summed in the filter to produce g_m at time t.

An equivalent realization of the nonrecursive filter is illustrated in Fig. 8.21. For each path from input to output in Fig. 8.20, there is an equivalent path in Fig. 8.21. Figure 8.21, however, suggests a slight difference in the realization of the filter: The samples of $f(t)$ are amplified before being stored and delayed in the filter, instead of being stored first as in Fig. 8.20. Both diagrams imply the same amount of storage, but in the next chapter there are cases where different realizations of the same transfer function involve different amounts of storage.

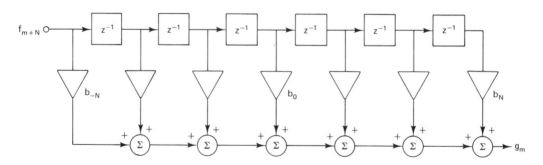

Figure 8.20 Nonrecursive filter with $N = 3$.

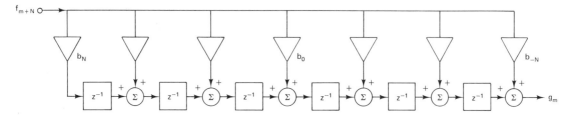

Figure 8.21 Block diagram equivalent to Fig. 8.20.

8.8 SYNTHESIS OF NONRECURSIVE FILTERS

As in Section 8.4, the synthesis of nonrecursive filters involves the selection of a set of filter coefficients, $[b_n]$ or $[w_n b_n]$, to achieve a desired frequency response in the filter. Two basic procedures are discussed in this section.

The first procedure is essentially that of Section 8.3, in which the coefficients are derived such that $\overline{H}(j\omega)$, the digital transfer function, is a least-squares approximation to $H_d(j\omega)$, the desired transfer function. (If a nonrectangular window is used, the approximation is smoothed as described in Section 8.5). As in Section 8.3, the general coefficient formula is

$$b_n = \frac{T}{2\pi} \int_{-\pi/T}^{\pi/T} H_d(j\omega)e^{jn\omega T}\, d\omega; \qquad -N \le n \le N \tag{8.41}$$

A modification to this approach, described by Helms (1968) and Rabiner (1971), can be used when the integral in Eq. 8.41 is difficult to evaluate. By using a set of samples of $H_d(j\omega)$, $[H_{dm}]$, one can effectively substitute the inverse DFT for the integral in Eq. 8.41 to obtain a zero-order approximation to the latter. Let the sample set $[H_{dm}]$ be obtained at $M + 1$ regularly spaced points over the frequency range from $\omega = 0$ to π/T (including the endpoints), so that the spacing between points is $d\omega = \pi/MT$. (Since $H_d[-j\omega]$ is the complex conjugate of $H_d[j\omega]$, the sample set is of course known for negative ω.) The zero-order approximation for b_n can now be derived from Eq. 8.41 as follows:

$$b_n = \frac{T}{2\pi} \int_{-\pi/T}^{\pi/T} H_d(j\omega)e^{jn\omega T}\, d\omega; \qquad -N \le n \le N$$

$$\approx \frac{T}{2\pi} \sum_{m=-M}^{M} H_{dm}e^{j(mn\pi/M)}\left(\frac{\pi}{MT}\right) \tag{8.42}$$

$$\approx \frac{1}{2M} \sum_{m=-M}^{M} H_{dm}e^{j(mn\pi/M)}; \qquad -N \le n \le N$$

(The endpoints, H_{dm} for $m = \pm M$, should be included at half-strength if nonzero.) In this approximation, M should be larger than N and moreover large enough so that $H_d(j\omega)$ is sampled at an adequate rate.

The second synthesis procedure is suggested by the equivalence of the analog and digital impulse response function described in Section 8.6. Given a desired transfer function $H_d(j\omega)$, the digital impulse response envelope, $b(t)$, could be set equal to T times the desired impulse response, $h_d(t)$. Then the digital transfer function, $\overline{H}(j\omega)$, should approximate $H_d(j\omega)$.

The two synthesis procedures are summarized in Table 8.2. They are quite similar, and in fact the two coefficient solutions are identical if $H_d(j\omega)$ is zero outside the interval $|\omega| < \pi/T$. [In this case $1/2\pi$ times the integral in procedure 1 becomes just the inverse transform of $H_d(j\omega)$, or $h_d(t)$ in procedure 2.] If $H_d(j\omega)$ is *not* zero outside of this interval, however, different coefficients result from the two procedures. In procedure 1, when the rectangular window is used, the coefficients become such that $\overline{H}(j\omega)$ is a least-squares approximation to $H_d(j\omega)$ in the above interval, but in procedure 2 the coefficients are such that $\overline{H}(j\omega)$ is the zero-order approximation to $H_d(j\omega)$, that is, $T\overline{H}_d(j\omega)$. The examples below illustrate the two procedures.

TABLE 8.2. Nonrecursive Filter Synthesis Procedures

Starting point	Procedure 1: Desired Transfer Function, $H_d(j\omega)$	Procedure 2: Desired Impulse Response, $h_d(t)$
Solution for coefficients	$b_n = \dfrac{T}{2\pi} \displaystyle\int_{-\pi/T}^{\pi/T} H_d(j\omega)e^{jn\omega T}\,d\omega$ $\approx \dfrac{1}{2M}\displaystyle\sum_{m=-M}^{M} H_{dm}e^{j(mn\pi/M)}$	$b_n = Th_d(nT)$
Digital filter formula[a]	$g_m = \displaystyle\sum_{n=-N}^{N} w_n b_n f_{m-n}$	Same
Digital filter transfer function	$\overline{H}(j\omega) = \displaystyle\sum_{n=-N}^{N} w_n b_n e^{-jn\omega T}$	Same
Properties of $\overline{H}(j\omega)$ when $[w_n]$ is the rectangular window	Least-squares approximation to $H_d(j\omega)$ for $|\omega| < \pi/T$	Fouier transform of $Th_d(t)$ [i.e., $T\overline{H}_d(j\omega)$]

[a]Use of the data window $[w_n]$ is discussed in Section 8.5.

Example 8.4

Design the general lowpass filter with zero phase shift, Eqs. 8.14 through 8.16, using procedure 2. [Note that procedure 1 was used in Section 8.4 (see Eq. 8.14).] To use procedure 2, the desired impulse response $h_d(t)$ must be determined. As illustrated in Fig. 8.22, $Th_d(t)$ is exactly the envelope of b_n in Eq. 8.14, that is, $b(nT) = Th_d(nT)$ (see also Fig. 8.19).

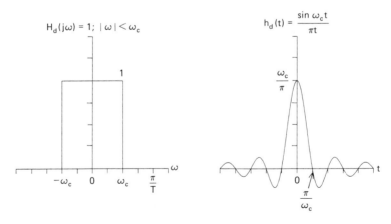

Figure 8.22 Lowpass transfer function and impulse response.

Therefore, procedure 2 gives the same result as procedure 1 for the lowpass filter, that is, the result in Eqs. 8.14 through 8.16. This is as expected, since the desired transfer function $H_d(j\omega)$ in this case is zero for $|\omega| \geq \pi/T$. Thus, in this particular example according to Table 8.2, the resulting transfer function $\overline{H}(j\omega)$ is $T\overline{H}_d(j\omega)$ as well as a least-squares approximation to $H_d(j\omega)$.

8.9 A NONRECURSIVE DIFFERENTIATOR

As a further example of the filter synthesis procedure discussed in the preceding section, we design here a nonrecursive differentiating filter, that is, a filter that takes the derivative of the input. The effect of taking the time derivative is to multiply the Fourier transform by $j\omega$ (see Table 3.1). Thus, the desired transfer function of the differentiator is

$$H_d(j\omega) = j\omega \tag{8.43}$$

In this case the desired transfer function is purely imaginary. Using procedure 1 in Table 8.2, the desired filter coefficients are given by

$$b_n = \frac{T}{2\pi} \int_{-\pi/T}^{\pi/T} j\omega e^{jn\omega T} d\omega: \qquad -N \leq n \leq N$$

$$= 0; \qquad\qquad n = 0 \tag{8.44}$$

$$= \frac{(-1)^n}{nT}; \qquad\qquad n \neq 0$$

These coefficients have odd symmetry, that is, $b_{-n} = -b_n$. Thus, the filter output $[y_k]$, a function of the input $[x_k]$, can be written

$$
\begin{aligned}
y_k &= \sum_{n=1}^{\infty} b_n(x_{k-n} - x_{k+n}) \\
&= \sum_{n=1}^{\infty} \frac{(-1)^n}{nT}(x_{k-n} - x_{k+n})
\end{aligned}
\tag{8.45}
$$

This is called a sum of weighted symmetric differences. The weighting is less for differences that are farther from the center of the filter. Intuitively, this structure seems quite reasonable for a differentiator.

To implement the filter, the desired coefficients must be truncated using the appropriate window function. From Table 8.2. using a window sequence $[w_n]$ with $2N + 1$ samples, we have

$$
\begin{aligned}
\overline{H}(j\omega) &= \sum_{n=-N}^{N} w_n b_n e^{-jn\omega T} \\
&= \sum_{n=1}^{N} w_n \left[\frac{(-1)^n}{nT} e^{-jn\omega T} - \frac{(-1)^n}{nT} e^{jn\omega T} \right] \\
&= \frac{-2j}{T} \sum_{n=1}^{N} w_n \frac{(-1)^n}{n} \sin(n\omega T)
\end{aligned}
\tag{8.46}
$$

Using $N = 10$, plots of $|\overline{H}(j\omega)|$ are presented in Fig. 8.23. These plots show the magnitude response using the rectangular and Blackman windows. The response us-

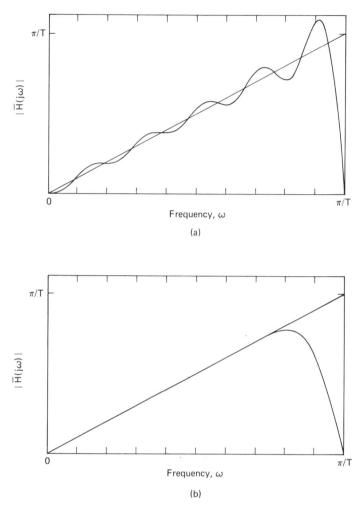

Figure 8.23 Amplitude response of differentiating digital filter, $N = 10$, using (a) rectangular window, and (b) Blackman window.

ing the Blackman window is smoother, while the rectangular window provides the best least-squares fit to $H_d(j\omega)$.

By definition, the differentiating filter amplifies higher frequencies more than lower frequencies. If the signal of interest is oversampled and is known to lie in the range of frequencies $\omega = 0$ to ω_c, then it is undesirable to amplify any noise that might be present in the range from ω_c to π/T. In this case the desired frequency response of the differentiator would be

$$H_d(j\omega) = j\omega; \qquad |\omega| \le \omega_c$$
$$= 0; \qquad |\omega| > \omega_c$$

Again using procedure 1 in Table 8.2, the desired coefficients are now given by

$$b_n = \frac{T}{2\pi} \int_{-\omega_c}^{\omega_c} j\omega e^{jn\omega T} \, d\omega$$

$$= \frac{T}{\pi} \left[\frac{\omega_c \cos(n\omega_c T)}{nT} - \frac{\sin(n\omega_c T)}{(nT)^2} \right]; \qquad -N \le n \le N \tag{8.47}$$

One can easily verify that this expression is equivalent to Eq. 8.44 when $\omega_c = \pi/T$. After truncating this sequence with rectangular and Blackman windows of length 21 (i.e., $N = 10$), $|\overline{H}(j\omega)|$ takes the forms shown in Fig. 8.24. In effect, we have a wideband differentiator in cascade with a lowpass filter.

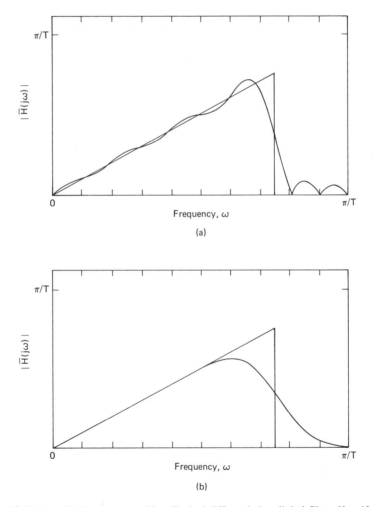

Figure 8.24 Amplitude response of bandlimited differentiating digital filter, $N = 10$, with (a) rectangular window and (b) Blackman window; $\omega_c = 0.75\pi/T$.

8.10 NONRECURSIVE HILBERT TRANSFORMER

In this section we present a further example of the filter synthesis procedure in Section 8.8. The *Hilbert transform* is a transform that expresses the functional dependency, when it exists, between the real and imaginary parts of a complex function. For complex signals whose spectra are one-sided, such a relationship is known to exist, and in fact the relationship is quite simple. The imaginary part is simply shifted by $\pm 90°$ with respect to the real part ($-90°$ if the nonzero part of the spectrum is greater than zero, and $+90°$ if it is less than zero). Consider for example the complex exponential signal $e^{j\omega_0 t} = \cos\omega_0 t + j\sin\omega_0 t$. Clearly the imaginary part, $\sin\omega_0 t$, is shifted $-90°$ with respect to the real part, $\cos\omega_0 t$. Also, the spectrum of this signal is one-sided in that it contains only the positive frequency component at ω_0. Similarly, the signal $e^{-j\omega_0 t} = \cos\omega_0 t - j\sin\omega_0 t$ whose imaginary part is shifted by $+90°$ with respect to the real part contains only the negative frequency component at $-\omega_0$. It is instructive to note that the spectra of purely real signals are always two-sided. This can be seen from the property in equation 3.9 which states that for real signals $|F(j\omega)| = |F(-j\omega)|$. With this property the only way that one half of the spectrum can be zero is if the other half is also zero. In addition we should note that not all complex signals have one-sided spectra, only those whose real and imaginary parts are related by the Hilbert transform described above.

The Hilbert transform is defined for *complex sequences* when the DFT is zero for $-\pi/T \le \omega < 0$ or for $0 < \omega \le \pi/T$. In this section we will design a filter that performs the Hilbert transform for such sequences; that is, a filter that, given the real part of a sequence, generates the imaginary part. This transform is useful for generating the quadrature component of narrowband signals like those used in radar systems, and is useful in general for producing a 90° phase shift.

The transfer function of the ideal Hilbert transformer is

$$H_d(j\omega) = -j; \qquad \omega \ge 0$$
$$= j; \qquad \omega < 0 \tag{8.48}$$

It is easy to see that this transfer function represents a $\pm 90°$ phase shift as described above, with unity gain at all frequencies.

Using procedure 1 in Table 8.2, the coefficients of the ideal Hilbert transformer are

$$b_n = \frac{T}{2\pi} \int_{-\pi/T}^{\pi/T} H_d(j\omega) e^{jn\omega T} \, d\omega; \qquad -N \le n \le N$$

$$= \frac{Tj}{2\pi} \int_{-\pi/T}^{0} e^{jn\omega T} \, d\omega - \frac{Tj}{2\pi} \int_{0}^{\pi/T} e^{jn\omega T} \, d\omega$$

$$= \frac{1}{\pi n} [1 - \cos(n\pi)] \tag{8.49}$$

$$= 0; \qquad n \text{ even}$$

$$= \frac{2}{\pi n}; \qquad n \text{ odd}$$

After truncating these coefficients with rectangular and Blackman windows of lengths 31 and 63, the actual filter responses are shown in Fig. 8.25. The approximation is seen to improve when the filter size is increased.

To provide an example of transformer operation, we let the input to the filter be $\cos(n\omega_0 T)$, where $\omega_0 T = 0.1\pi$. Using a Hilbert transformer that has been truncated to length 51 ($N = 25$) using the Blackman window, we produced the output shown in Fig. 8.26. Note, after the time span of half the filter, or 25 samples, that

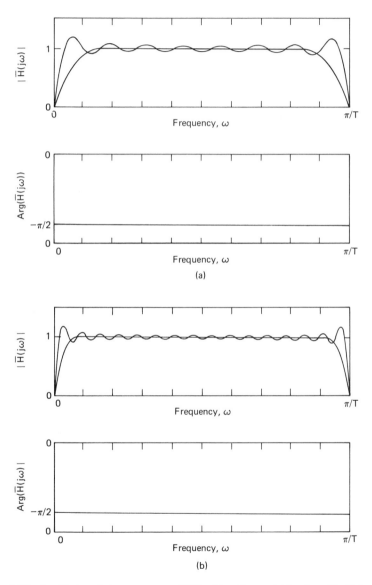

Figure 8.25 Amplitude and phase responses of Hilbert transformer: (a) length = 31, (b) length = 63; Blackman window used to obtain smoother responses.

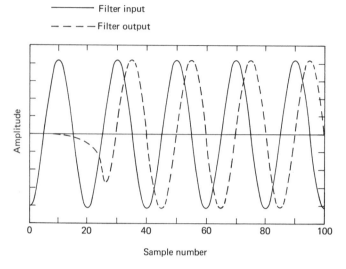

Figure 8.26 Input and output signals for the Hilbert transformer.

the output has approximately the same magnitude and frequency as the input, but is delayed by a quarter cycle, that is, shifted $-90°$ in phase.

The results in Fig. 8.26 are produced using a noncausal filter, with coefficients from $n = -25$ to $n = 25$ as in Eq. 8.49. If the filter were made causal, the output waveform in Fig. 8.26 would be delayed an additional 25 samples, that is, half the filter length.

EXERCISES

1. Given Eq. 8.5 and the formula in Chapter 2 for c_k, the complex Fourier series coefficient, derive Eq. 8.7.

2. A set $[x_m]$ of regular samples is smoothed by computing $[y_m] = [2x_{m-1} + x_m + 2x_{m+1}]$. Sketch the transfer function $\overline{H}(j\omega)$ for this operation.

3. What is the gain and phase shift effected by the smoothing formula $v_m = u_m - \frac{1}{2}u_{m-1} + \frac{1}{6}u_{m-2}$?

4. What is the complex gain accomplished by the smoothing formula

$$y_m = \frac{1}{N} \sum_{n=0}^{N-1} x_{m-n} ?$$

5. What gain and phase shift are accomplished by

$$y_m = \frac{1}{2N + 1} \sum_{n=-N}^{N} x_{m-n} ?$$

6. A "desired" bandpass characteristic with zero phase shift is shown here. Show the approximation, $\overline{H}(j\omega)$, that can be obtained in a nonrecursive filter, and derive the filter algorithm. Use a rectangular data window.

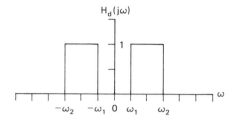

$H_d(j\omega)$

$-\omega_2 \quad -\omega_1 \ 0 \ \omega_1 \quad \omega_2$

ω

7. (Computer) Plot the optimum nonrecursive lowpass filter gain functions for $\omega_c T = 0.2\pi$ with $N = 10$ and $N = 100$, and compare with Fig. 8.7. Use the Hanning data window.

8. Give the impulse response of the filter in Exercise 2.

9. What is the impulse response of the filter in Exercise 3?

10. Give the impulse response for Exercise 4.

11. Give $\tilde{H}(z)$ for the filter in Exercise 2.

12. Give $\tilde{H}(z)$ for Exercise 3 and draw a diagram similar to Fig. 8.20.

13. Draw a diagram like Fig. 8.21 for Exercise 4.

14. Give $\tilde{H}(z)$ for Exercise 5.

15. Prove that multiplying a transfer function $\tilde{H}(z)$ by z^k, k being any integer, does not affect the amplitude gain.

16. Letting $\theta(j\omega)$ represent the phase shift of a digital filter in radians, show the effect, if any, on $\theta(j\omega)$ of multiplying $\tilde{H}(z)$ by z^{-1}.

17. A digital filter with transfer function $\tilde{H}(z)$ is described by

$$y_m = \sum_{n=-N}^{N} b_n x_{m-n}$$

Give a similar algorithm for the filter below whose transfer function is $z^k \tilde{H}(z)$.

$x_m \circ \longrightarrow \boxed{z^k} \longrightarrow \boxed{\tilde{H}(z)} \longrightarrow y'_m$

18. Two different nonrecursive filters have the following algorithms:

$$u_m = \sum_{n=-N}^{N} b_n x_{m-n}; \qquad v_m = \sum_{n=0}^{2N} b'_{n-N} x_{m-n}$$

The b's are the same in the sense that $b_1 = b'_1$, $b_2 = b'_2$, and so on. Using the result in Exercise 17,

(a) Find the function $\tilde{H}_v(z)$ in terms of $\tilde{H}_u(z)$,

(b) Find $\overline{H}_v(j\omega)$ in terms of $\overline{H}_u(j\omega)$ and compare the result with Exercises 15 and 16.

(c) Discuss the realizability of the two filters.

19. (Computer) The desired lowpass response below cuts off at 125 kHz or 785 krad/s, with zero phase shift at all frequencies. The sampling interval is 1 μs.
Using the rectangular data window,
(a) Write a general expression for the filter coefficients $[b_n]$.
(b) Plot the actual response functions for filters with $N = 4$, 8, and 32.
(c) Plot the response $g(t)$ when the input consists of 100 samples of $f(t) = \sin(5 \times 10^5 t)$ and N is 32.
(d) Plot $g(t)$ when $f(t) = \sin(8 \times 10^5 t)$ and $N = 32$, and compare with part c.
(e) Plot $g(t)$ when $f(t)$ is a 50-kHz square wave and N is 32.

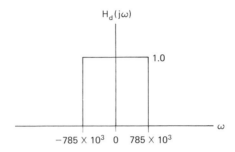

$H_d(j\omega)$

1.0

-785×10^3 0 785×10^3

ω

20. (Computer) Use the results in Exercise 6 and the window routine SPWIND in Appendix B to obtain the coefficients of a 21-weight ($N = 10$) FIR bandpass filter with $\omega_1 T = 0.2\pi$ and $\omega_2 T = 0.4\pi$ using each of the windows; the rectangular, Bartlett, Hanning, Hamming, Blackman, and Kaiser ($\beta = 5.44$). Use the SPFFT routine in Appendix B to compute samples of the actual amplitude responses of these filters. Plot and compare your results.

21. (Computer) Repeat Exercise 20 using a filter of length 41 ($N = 20$). How do these results compare with those in Exercise 20?

22. Derive Eq. 8.25. Verify that the first null in $\overline{W}_{HN}(j\omega)$ occurs at approximately $\omega T = 2\pi/N$.

23. Derive Eq. 8.27. Verify that the first null in $\overline{W}_{BK}(j\omega)$ occurs at approximately $\omega T = 3\pi/N$.

24. (Computer) Evaluate and plot the Kaiser window function $[w_n]$ for $\beta = 5$ and $N = 10$. Compare this with Blackman's window function.

25. (Computer) Use the subroutines in Appendix B to help derive the coefficients of a nonrecursive differentiating filter of length 31 using the Kaiser window function with $\beta = 5.44$. Use these coefficients to filter the sequence

$$x_n = 0.99^n; \qquad n = 0, 1, 2, \ldots, 999$$

Does the result approximate the derivative of x_n?

26. (Computer) Repeat Exercise 25 using a differentiator with a cutoff frequency $\omega_c T = 0.8\pi$. Compare your result with that in Exercise 25.

27. (Computer) Compute the filter coefficients of a nonrecursive Hilbert transformer of length 15 using a rectangular window. Use these coefficients to filter the following sequences.

(a) $x_n = \cos(0.12\pi n); \qquad n = 0, 1, 2, \ldots, 200$

(b) $x_n = \cos(0.24\pi n); \qquad n = 0, 1, 2, \ldots, 200$

Compare your results and explain. *Hint:* Plot the frequency response of the filter.

28. (Computer) Check the validity of the differentiator

$$y_k = \sum_{n=1}^{9} w_n b_n (x_{k-n} - x_{k+n})$$

with $[b_n]$ given in Eq. 8.44 and $[w_n]$ representing the Hanning window by differentiating a unit ramp function with $x_k = k$.

SOME ANSWERS

2. $\overline{H}(j\omega) = 1 + 4 \cos \omega T$ **3.** $|\overline{H}(j\omega)| = \frac{1}{6}(34 - 42 \cos \omega T + 24 \cos^2 \omega T)^{1/2}$

4. $\overline{H}(j\omega) = \dfrac{1}{N} \dfrac{\sin(N\omega T/2)}{\sin(\omega T/2)} e^{j(1-N)\omega T/2}$ **5.** $\overline{H}(j\omega) = \dfrac{1}{2N+1}\left[\dfrac{\sin[(N+1/2)\omega T]}{\sin(\omega T/2)}\right]$

6. $g_m = \displaystyle\sum_{n=-N}^{N} \dfrac{1}{n\pi}(\sin n\omega_2 T - \sin n\omega_1 T) f_{m-n}$ **8.** $[2, 1, 2]$ at $t = -T, 0, T$

10. $1/N$ at $t = 0, T, \ldots, (N-1)T$

11. $\tilde{H}(z) = 2z + 1 + 2z^{-1}$ **14.** $\tilde{H}(z) = \dfrac{1}{2N+1}\displaystyle\sum_{n=-N}^{N} z^{-n} = \dfrac{1}{2N+1}\left[\dfrac{z^{-N} - z^{N+1}}{1-z}\right]$

16. $\theta(j\omega)$ is decreased by ωT radians   **17.** $y'_m = \displaystyle\sum_{n=-N-k}^{N-k} b_{n+k} x_{m-n}$

18. (a) $\tilde{H}_v(z) = z^{-N}\tilde{H}_u(z)$

REFERENCES

BRUCE, J. D., Discrete Fourier Transforms, Linear Filters, and Spectrum Weighting. *IEEE Trans. Audio Electroacoust.,* Vol. AU-16, No. 4, December 1968, p. 495.

CROOKE, A. W., and CRAIG, J. W., Digital Filters for Sample-Rate Reduction. *IEEE Trans. Audio Electroacoust.,* Vol. AU-20, No. 4, October 1972, p. 308.

GOLD, B., and JORDAN, K. L., JR., A Direct Search Procedure for Designing Finite Duration Impulse Response Filters. *IEEE Trans. Audio Electroacoust.,* Vol. AU-17, No. 1, March 1969, p. 33.

HADDAD, R. A., A Class of Orthogonal Nonrecursive Binomial Filters. *IEEE Trans. Audio Electroacoust.,* Vol. AU-19, No. 4, December 1971, p. 296.

HAMMING, R. W., *Digital Filters.* Englewood Cliffs, N.J.: Prentice-Hall, 1977.

HARRIS, F. J., On the Use of Windows for Harmonic Analysis with the Discrete Fourier Transform. *Proc. IEEE,* Vol. 66, No. 1, January 1978, p. 51.

HELMS, H. D., Nonrecursive Digital Filters: Design Methods for Achieving Specifications on Frequency Response. *IEEE Trans. Audio Electroacoust.,* Vol. AU-16, No. 3, September 1968, p. 336.

HOWARD, R. A., *Dynamic Programming and Markov Processes,* Chap. 1. Cambridge, Mass.: MIT Press, 1960.

KAISER, J. F., and KUO, F. F., *System Analysis by Digital Computer,* Chap. 7. New York: Wiley, 1966.

KELLOGG, W. C., Time Domain Design of Nonrecursive Least Mean-Square Digital Filters. *IEEE Trans. Audio Electroacoust.,* Vol. AU-20, No. 2, June 1972, p. 155.

NOWAK, D. J., and SCHMID, P. E., A Nonrecursive Digital Filter for Data Transmission. *IEEE Trans. Audio Electroacoust.,* Vol. AU-16, No. 3, September 1968, p. 343.

OPPENHEIM, A. V., and SCHAFER, R. W., *Digital Signal Processing.* Englewood Cliffs, N.J.: Prentice-Hall, 1975.

ORMSBY, J. F. A., Design of Numerical Filters with Applications to Missile Data Processing. *J. Assoc. Compute Mach.,* Vol. 8, No. 3, July 1961, p. 440.

RABINER, L. R., The Design of Wide-Band Recursive and Nonrecursive Digital Differentiators. *IEEE Trans. Audio Electroacoust.,* Vol. AU-18, June 1970, p. 204.

RABINER, L. R., Techniques for Designing Finite-Duration Impulse-Response Digital Filters. *IEEE Trans. Communication Tech.,* Vol. COM-19, April 1971, p. 188.

RABINER, L. R., and GOLD, B., *Theory and Application of Digital Signal Processing.* Englewood Cliffs, N.J.: Prentice-Hall, 1975.

REQUICHA, A. A. G., and VOELCKER, H. B., Design of Nonrecursive Filters by Specification of Frequency-Domain Zeros. *IEEE Trans. Audio Electroacoust.,* Vol. AU-18, No. 4, December 1970, p. 464.

STOCKHAM, T. G., High-Speed Convolution and Correlation with Applications to Digital Filtering, Chap. 7 in *Digital Processing of Signals,* B. Gold, C. M. Rader, et al. New York: McGraw-Hill, 1969.

WAIT, J. V., Digital Filters, Chap. 5 in *Active Filters: Lumped, Distributed, Integrated, Digital, and Parametric,* ed. L. P. Huelsman. New York: McGraw-Hill, 1970.

9

Recursive Digital Systems

9.1 INTRODUCTION

The various concepts applied to the analysis of nonrecursive filtering algorithms will now be applied to recursive algorithms. Nonrecursive filters turn out to be a subset of recursive filters, so that the transfer function, impulse response, and other properties of recursive filters become generalizations of the results in the last chapter.

Recursive algorithms, being more general in nature, offer a greater variety of transfer functions than nonrecursive algorithms. With a fixed number of filter coefficients, one can usually attain a better approximation to a "desired" transfer function (e.g., one with sharp corners) using a recursive algorithm. Also, as shown in the next chapter, the digital simulation of continuous systems in general requires the recursive algorithm.

Linear recursive systems require the same sort of hardware or software "components" as nonrecursive systems, that is, unit delays, stored numerical coefficients, multiplications, and additions. As explained in the preceding chapter, the unique feature of recursive systems is the involvement of past values of the output in the computation of its own present value. This characteristic results generally in the presence of feedback loops in the filter diagram. For example, look again at Fig. 1.7 of Chapter 1, which contains a diagram of a simple recursive filter.

In this chapter the linear recursive algorithm and its transfer function are discussed in Sections 9.2 and 9.3. Using the z-transform, systems are then "solved" for specific inputs in Section 9.4, and realization in terms of schematic diagrams is discussed in Sections 9.5 and 9.6. In Section 9.7 the transfer function is discussed again in terms of pole and zero locations on the z-plane. Finally, Section 9.8 is on the attainment of linear phase shift in recursive filters.

9.2 THE RECURSIVE ALGORITHM

In the nonrecursive formula, Eq. 8.1, a sample of the filter output, g_m, was given in terms of past as well as future samples of the input $f(t)$. To include the recursive property, a weighted sum of the past samples of $g(t)$ can be added, so that

$$g_m = \sum_{n=-N}^{N} b_n f_{m-n} - \sum_{n=1}^{N} a_n g_{m-n} \qquad (9.1)$$

[The second sum is subtracted to give Eq. 9.1 a more symmetrical form (see Eq. 9.6).] Note that it is practical to compute g_m only in terms of the *past* samples of $g(t)$ (i.e., g_{m-1}, g_{m-2}, etc.), so in the second sum n has only positive nonzero values.

As in Chapter 8, N can be as large as desired but must be finite for the filter to be realizable, and a coefficient (a_n or b_n) can have any real value including zero. Also, the remarks in the last chapter on the realizability of the filter when b_n is nonzero for $n < 0$ still apply. The effect of excluding this case, that is, limiting n to positive values in Eq. 9.1, is discussed below.

The nonrecursive algorithm is obviously a subset of Eq. 9.1, as shown by letting $[a_n] = [0]$ in Eq. 9.1; hence Eq. 9.1 is a general linear algorithm. Moreover, Eq. 9.1 can be written in nonrecursive form if N is allowed to increase without bound—a property common to all linear recursion formulas. To demonstrate this property, use Eq. 9.1 itself to substitute for g_{m-n} in the second sum, changing the index in the substitution from n to k:

$$g_m = \sum_{n=-N}^{N} b_n f_{m-n} - \sum_{n=1}^{N} a_n \left(\sum_{k=-N}^{N} b_k f_{m-n-k} - \sum_{k=1}^{N} a_k g_{m-n-k} \right) \qquad (9.2)$$

In this formulation, the most recent sample of $g(t)$ on the right has changed from g_{m-1} to g_{m-2}. If Eq. 9.1 is again substituted, this time for g_{m-n-k} in Eq. 9.2, the most recent sample becomes g_{m-3}, and so on. In most cases of interest it can be assumed that $g(t)$ is zero prior to some value of t, so that by substituting Eq. 9.1 repeatedly one could eventually eliminate the past samples of $g(t)$ from the right side, and g_m would be of the form

$$g_m = \sum_{n=-N}^{\infty} c_n f_{m-n} \qquad (9.3)$$

with each c_n being a function of the a's and b's.

Since the recursive algorithm in Eq. 9.1 has the nonrecursive form in Eq. 9.3, the general nonrecursive transfer function properties given in Eqs. 8.39 and 8.40 must hold as well for recursive filters, because these formulas did not require that N be finite. The properties are repeated here:

$$\tilde{H}(z) = \frac{\tilde{G}(z)}{\tilde{F}(z)} \qquad (9.4)$$

$$\overline{H}(j\omega) = \frac{\overline{G}(j\omega)}{\overline{F}(j\omega)} \qquad (9.5)$$

Given these relationships, other properties of the recursive transfer function can easily be determined.

9.3 TRANSFER FUNCTION

Equation 9.5, which was shown to be valid for any linear digital filter, gives the transfer function as a DFT. Thus, if the filter coefficients $[a_n]$ and $[b_n]$ are real, the transfer function has the properties of any DFT of a real-time sequence.

1. If $e^{j\omega T}$ is substituted for z in $\tilde{H}(z)$, the result is $\overline{H}(j\omega)$.
2. $\overline{H}(j\omega)$ is periodic with period $2\pi/T$.
3. $\overline{H}(j\omega)$ and $\overline{H}(-j\omega)$ are complex conjugates.
4. $|\overline{H}(j\omega)| = |\overline{H}(-j\omega)|$ — the amplitude gain is an even function of ω.

A formula for the transfer function in terms of the filter coefficients can be obtained by first writing Eq. 9.1 in symmetric form:

$$\sum_{n=0}^{N} a_n g_{m-n} = \sum_{n=-N}^{N} b_n f_{m-n}; \qquad a_0 = 1 \tag{9.6}$$

The coefficient a_0 is used in this expression and hereafter for convenience. It is always understood to be 1. We take the z-transform of Eq. 9.6 and, as in Chapter 8, Eq. 8.38, apply Eq. 7.6 to obtain

$$\tilde{A}(z)\tilde{G}(z) = \tilde{B}(z)\tilde{F}(z) \tag{9.7}$$

The transfer function $\tilde{H}(z)$ is then

$$\tilde{H}(z) = \frac{\tilde{G}(z)}{\tilde{F}(z)} = \frac{\tilde{B}(z)}{\tilde{A}(z)} \tag{9.8}$$

$$= \frac{b_{-N}z^N + \cdots + b_0 z^0 + \cdots + b_N z^{-N}}{1 + a_1 z^{-1} + \cdots + a_N z^{-N}} \tag{9.9}$$

We again note that for *causal* systems, that is, systems that are *realizable in real time*, the b's with negative subscripts must all be zero so that the system does not require future values of the input. The essential transfer-function relationships are summarized in Table 9.1.

Again, the nonrecursive transfer functions in Chapter 8 are obtained by setting $a_n = 0$ for $n > 0$ and remembering that $a_0 = 1$. A specific recursive transfer function is illustrated in the following example.

Example 9.1

Given the algorithm $g_m = f_m + 0.5g_{m-1}$, find the transfer function. Here the coefficients are $a_0 = 1$, $b_0 = 1$, and $a_1 = -0.5$. Therefore $\tilde{A}(z) = 1 - 0.5z^{-1}$ and $\tilde{B}(z) = 1$, so

$$\tilde{H}(z) = \frac{1}{1 - 0.5z^{-1}} = \frac{z}{z - 0.5} \tag{9.12}$$

and $\overline{H}(j\omega)$ is found simply by substituting $e^{j\omega T}$ for z in $\tilde{H}(z)$:

$$\overline{H}(j\omega) = \frac{1}{1 - 0.5e^{-j\omega T}} \tag{9.13}$$

The amplitude response is plotted in Fig. 9.1, and both $|\overline{H}(j\omega)|$ and $\overline{H}(j\omega)$ are seen to have properties 1 through 4 listed above.

TABLE 9.1. Linear Digital Transfer Functions Corresponding to Eqs. 9.1 and 9.19

Transfer Function	In Terms of Output/Input	In Terms of Coefficients	
$\tilde{H}(z)$	$= \dfrac{\tilde{G}(z)}{\tilde{F}(z)}$	$= \dfrac{\tilde{B}(z)}{\tilde{A}(z)}$	(9.10)
$\overline{H}(j\omega) = \tilde{H}(e^{j\omega T})$	$= \dfrac{\overline{G}(j\omega)}{\overline{F}(j\omega)}$	$= \dfrac{\overline{B}(j\omega)}{\overline{A}(j\omega)}$	(9.11)

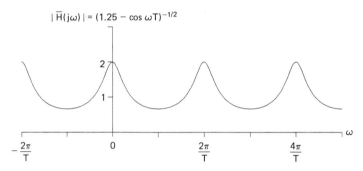

$$|\overline{H}(j\omega)| = (1.25 - \cos \omega T)^{-1/2}$$

Figure 9.1 Amplitude response for $g_m = f_m + 0.5g_{m-1}$.

Having the forms of the transfer function summarized in Table 9-1, we can again address the realizability question mentioned previously, that is, the question of including "future" samples of $f(t)$ in the filtering algorithm. Observing Eq. 9.11 in Table 9.1, it is obvious that having $b_n = 0$ for $n < 0$ is going to have some limiting effect on the form of the transfer function $\overline{H}(j\omega)$. The exact nature of this effect is given in the following restatement of the shifting theorem:

SHIFTING THEOREM

Let k be any integer. The following operations are equivalent in any linear digital system:

1. Replacing the input sample set $[f_m]$ with the set $[f_{m-k}]$
2. Multiplying $\tilde{H}(z)$ by z^{-k}
3. Multiplying $\overline{H}(j\omega)$ by $e^{-jk\omega T}$

The shifting theorem is proved as follows: Assume that operation 1 is done so that $[g'_m]$ is the revised output sequence and Eq. 9.6 becomes

$$\sum_{n=0}^{N} a_n g'_{m-n} = \sum_{n=-N}^{N} b_n f_{m-n-k} \tag{9.14}$$

Equation 9.7, which followed from Eq. 9.6, is also changed:

$$\tilde{A}(z)\tilde{G}'(z) = z^{-k}\tilde{B}(z)\tilde{F}(z) \tag{9.15}$$

and therefore the new transfer function, $\tilde{H}'(z)$, is

$$\tilde{H}'(z) = \frac{\tilde{G}'(z)}{\tilde{F}(z)} = \frac{z^{-k}\tilde{B}(z)}{\tilde{A}(z)} = z^{-k}\tilde{H}(z) \qquad (9.16)$$

Therefore operation 2 is equivalent. Operation 3 then follows by substituting $e^{j\omega T}$ for z in Eq. 9.16, that is, from Eq. 8.39, and the theorem is proved (see also Chapter 8, Exercises 15–18.)

Since a unit delay of T seconds is easy to achieve in digital filters, the shifting theorem is of considerable practical interest. It shows, for example, that the exclusion of the future samples of $f(t)$ can only result in a phase shift, and that *any amplitude response function that can be achieved using future samples of* f(t) *can also be achieved using only past values.*

Example 9-2

Let the input in Example 9.1 be delayed by two sampling intervals, so that $g'_m = f_{m-2} + 0.5g'_{m-1}$, and find the resulting transfer function. The shifting theorem gives the result here in terms of the result in Example 9.1:

$$\tilde{H}'(z) = z^{-2}\tilde{H}(z) = \frac{1}{z(z - 0.5)} \qquad (9.17)$$

and

$$\overline{H}'(j\omega) = e^{-2j\omega T}\overline{H}(j\omega) = \frac{e^{-2j\omega T}}{1 - 0.5e^{-j\omega T}} \qquad (9.18)$$

The amplitude response, $|\overline{H}'(j\omega)|$, is of course the same as in Fig. 9.1.

With the shifting theorem in mind, it will be convenient in the remainder of this chapter to exclude future samples of $f(t)$ from the algorithm, realizing that only a phase shift is involved in doing so. We will use the following causal formulas:

Causal formulas

$$g_m = \sum_{n=0}^{N} b_n f_{m-n} - \sum_{n=1}^{N} a_n g_{m-n}$$

$$\tilde{H}(z) = \frac{b_0 + b_1 z^{-1} + \cdots + b_N z^{-N}}{1 + a_1 z^{-1} + \cdots + a_N z^{-N}} \qquad (9.19)$$

The transfer functions in Table 9.1 of course remain applicable.

9.4 SOLUTION OF DIFFERENCE EQUATIONS: IMPULSE AND STEP FUNCTION RESPONSE

When the input set $[f_m]$ is given, Eq. 9.19 becomes a *difference equation* that is, theoretically at least, solvable for $[g_m]$ as indicated by Eq. 9.3. Two special solutions are of particular interest in the analysis of continuous linear systems, namely the solutions for the responses to the *unit impulse* and *step function* inputs.

Sec. 9.4   Solution of Difference Equations

183

Considering first the impulse response, the impulse function $d(t)$ having unit spectral content in the frequency interval $(-\pi/T, \pi/T)$ was seen in the preceding chapter to have the single nonzero sample at $t = 0$, $d_0 = 1$. Its z-transform $\tilde{D}(z)$ is therefore also 1, so Eq. 9.10 in Table 9.1 gives

$$\tilde{G}(z) = \frac{\tilde{B}(z)}{\tilde{A}(z)} \tilde{D}(z) = \frac{\tilde{B}(z)}{\tilde{A}(z)}, \tag{9.20}$$

or

$$\sum_{m=-\infty}^{\infty} g_m z^{-m} = \frac{\sum_{n=0}^{N} b_n z^{-n}}{\sum_{n=0}^{N} a_n z^{-n}} = \tilde{H}(z) \tag{9.21}$$

The version in Eq. 9.21 can be "solved" for g_m if the right side, which is a ratio of polynomials in z, can be written as a geometric series in z, for then the coefficient of z^{-m} in the latter can be equated to g_m. A trivial example occurs when $a_n = 0$ for $n > 0$ and the filter is nonrecursive. In this case Eq. 9.21 gives

$$\sum_{m=-\infty}^{\infty} g_m z^{-m} = \sum_{n=0}^{N} b_n z^{-n} \tag{9.22}$$

Therefore,

$$g_m = b_m \tag{9.23}$$

This result, obtained by equating coefficients of like power of z in Eq. 9.22, is the same as in Chapter 8.

In general, the solution for g_m in a difference equation can be obtained by:

1. Factoring the denominator of $\tilde{G}(z)$
2. Expanding $\tilde{G}(z)$ using partial fractions
3. Using the geometric series formula,

$$\frac{z}{z - \alpha} = \sum_{m=0}^{\infty} \alpha^m z^{-m} \tag{9.24}$$

   on the result [assuming that $\tilde{G}(z)$ has only first-order poles],
4. Equating the coefficients of like powers of z

Steps 1 and 4 are illustrated in the following example, and partial fractions are used in Example 9.4.

Example 9.3

Give the impulse response of the filter used in Example 9.1, with $g_m = f_m + 0.5 g_{m-1}$. As in Eq. 9.12, $\tilde{H}(z)$ in this case is $z/(z - 0.5)$, so Eq. 9.21 becomes

$$\sum_{m=-\infty}^{\infty} g_m z^{-m} = \frac{z}{z - 0.5} \tag{9.25}$$

$$= \sum_{m=0}^{\infty} (0.5)^m z^{-m} \tag{9.26}$$

Therefore,

$$g_m = (0.5)^m; \quad m \geq 0 \tag{9.27}$$

with Eq. 9.26 following from step 3 above and Eq. 9.27 resulting from step 4. The result in Eq. 9.27 is plotted in Fig. 9.2.

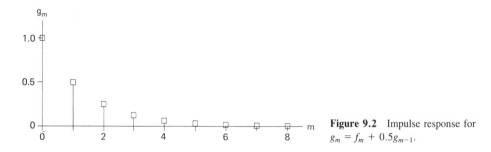

Figure 9.2 Impulse response for $g_m = f_m + 0.5g_{m-1}$.

The difference equation for the response of a digital filter to a step function produces a form similar to Eq. 9.21, and steps 1–4 above are again applicable. As in Chapter 2, the unit step function $u(t)$ is equal to zero for $t < 0$ and equal to one for $t \geq 0$. Therefore, the input sample set is

$$
\begin{aligned}
u_m &= 0; \quad m < 0 \\
&= 1; \quad m \geq 0
\end{aligned}
\tag{9.28}
$$

(The sample at $t = 0$ may require redefinition for some purposes, but $u_0 = 1$ can be assumed here.) Thus, as in the case of the unit impulse, the input sample set has a simple z-transform. Using Eq. 9.24,

$$\tilde{U}(z) = \sum_{m=0}^{\infty} z^{-m} = \frac{z}{z - 1} \tag{9.29}$$

Using $\tilde{U}(z)$ for the step function in place of $\tilde{D}(z)$ for the impulse function, results similar to Eqs. 9.20 and 9.21 are obtained:

$$\tilde{G}(z) = \frac{\tilde{B}(z)}{\tilde{A}(z)} \tilde{U}(z) \tag{9.30}$$

$$= \frac{z\tilde{H}(z)}{z - 1} \tag{9.31}$$

Since the right side of Eq. 9.31 is again a ratio of polynomials in z, steps 1–4 above can again be applied to obtain a solution for g_m.

Example 9.4

Give the step-function response of the filter used above, with $g_m = f_m + 0.5g_{m-1}$. Again, for this filter, $\tilde{H}(z) = z/(z - 0.5)$, so Eq. 9.31 gives

$$\tilde{G}(z) = \frac{z\tilde{H}(z)}{z - 1} = \frac{z^2}{(z - 1)(z - 0.5)} \tag{9.32}$$

Applying step 2 to expand Eq. 9.32 into partial fractions gives

$$\sum_{m=-\infty}^{\infty} g_m z^{-m} = \frac{Az}{z-1} + \frac{Bz}{z-0.5} \tag{9.33}$$

$$= \frac{2z}{z-1} - \frac{z}{z-0.5} \tag{9.34}$$

with the solutions for A and B being obtained by cross-multiplication in Eq. 9.33. Next, steps 3 and 4 are applied as before:

$$\sum_{m=-\infty}^{\infty} g_m z^{-m} = 2\sum_{m=0}^{\infty} z^{-m} - \sum_{m=0}^{\infty} (0.5)^m z^{-m}$$

$$= \sum_{m=0}^{\infty} (2 - 0.5^m) z^{-m} \tag{9.35}$$

Therefore,

$$g_m = 2 - 0.5^m; \quad m \geq 0 \tag{9.36}$$

and thus the step-function response is obtained. It is plotted in Fig. 9.3.

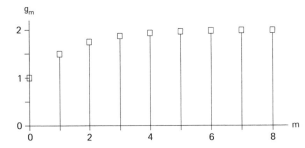

Figure 9.3 Step-function response for $g_m = f_m + 0.5 g_{m-1}$.

A fairly large class of difference equations can be solved by the method used in this section [see also Gold and Rader (1969), Chap. 2]. The steps taken in the solution, steps 1–4 above, really involve finding the *inverse z-transform* of $\tilde{G}(z)$, after $\tilde{G}(z)$ has been written as a ratio of polynomials in z. Therefore, if one is armed with a table of z-transforms as in Table 7.1 or Appendix A, he can usually avoid steps 3 and 4, and possibly even step 2. Note, for instance, that the solutions to Eqs. 9.25 and 9.32 can be found in Table 7.1 and written down directly.

9.5 BLOCK DIAGRAMS

Just as with continuous filters, block diagrams of digital filters help one to visualize the filter function and effect, and also suggest different realizations in either hardware or algorithmic form. The nonrecursive diagrams in the preceding chapter are now extended to linear filters in general, with interesting results.

The general linear filter is first diagrammed from Eq. 9.19 in Fig. 9.4, and is seen to be an extension of Fig. 8.16, except that f_m is used in place of f_{m+N} because Eq. 9.19 excludes future values of the input. Again, as shown by the Shifting Theorem, the transfer function would be unchanged if

$$f_{m+k} \longrightarrow \boxed{z^{-k}} \longrightarrow f_m$$

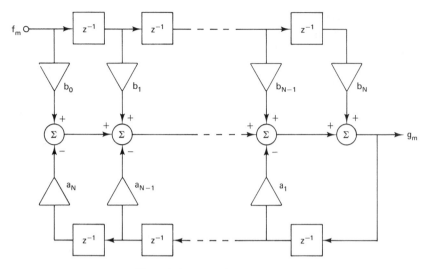

Figure 9.4 Linear filter diagram 1.

were attached to the input in Fig. 9.4. Future values are excluded only for convenience, to give the diagram a more symmetrical form.

Note how the upper half of Fig. 9.4 represents the first sum in Eq. 9.19, and the lower half the second sum. All of the paths need not exist, of course — any of the a's or b's in the figure can be zero.

Example 9.5

The filter with $g_m = f_m + 0.5g_{m-1}$, used in the examples above starting with Example 9.1, is shown in Fig. 9.5.

Since each delay element $\boxed{z^{-1}}$ in a filter diagram implies the storage of a quantity at $t = (m - 1)T$ so that this same quantity will be available at $t = mT$, the *number of z^{-1}'s in the diagram can in general be equated to the number of storage cells* required by the filter algorithm or hardware. Each "cell" of course provides storage for a complete real number — not just a single bit or digit.

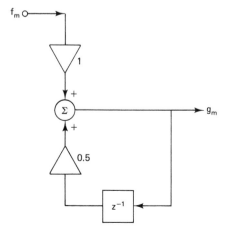

Figure 9.5

A second linear diagram, Fig. 9.6, can be derived from Fig. 9.4 just as Fig. 8.21 was derived from Fig. 8.20. For each path in Fig. 9.4 from input to output a similar path can be traced in Fig. 9.6, although the order of operations along the path may differ. There is, however, an important difference between the two diagrams: *Diagram 1 contains 2N delays (storage cells), whereas diagram 2 contains only N.* Diagram 2 implies an algorithm different from Eq. 9.19—an algorithm involving only N past values instead of $2N$. The algorithm can be written by labeling the input to each delay in diagram 2, just as each delay input in Fig. 9.4 is implicitly labeled with a past value of $f(t)$ or $g(t)$. In diagram 2, let each intermediate sum be labeled

$$x_m^n = b_n f_m - a_n g_m + x_{m-1}^{n+1} \tag{9.37}$$

so that m represents the time $t = mT$ as usual and the superscript n (*not* an exponent) designates the delay-unit input point in the diagram. For example, x_m^N is the input to the leftmost delay unit. The algorithm corresponding to diagram 2 then follows:

$$\left. \begin{array}{l} g_m = b_0 f_m + x_{m-1}^1 \\ x_m^n = b_n f_m + x_{m-1}^{n+1} - a_n g_m; \quad n = 1, 2, \ldots, N-1 \\ x_m^N = b_N f_m - a_N g_m \end{array} \right\} \tag{9.38}$$

If the computations are made in this order, with values at $t = mT$ replacing values at $t = (m-1)T$ as they are computed, then only N storage cells are needed as implied in diagram 2. Equations 9.38 and 9.19 are of course completely equivalent, and as a matter of fact, Eq. 9.19 can be obtained by eliminating the x's in Eq. 9.38.

To illustrate that other N-cell realizations of the linear filter are possible, diagram 3 is presented in Fig. 9.7. Again there is an input–output path in diagram 3 equivalent to each such path in diagram 1 or 2. (As before, the order of operations along a path may be altered, but the resulting contribution to g_m is the same.) Again

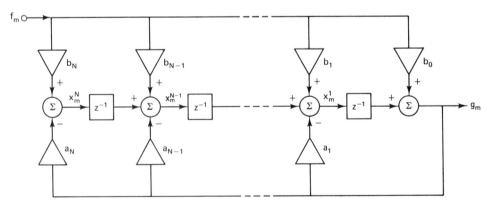

Figure 9.6 Linear filter diagram 2.

Recursive Digital Systems   Chap. 9

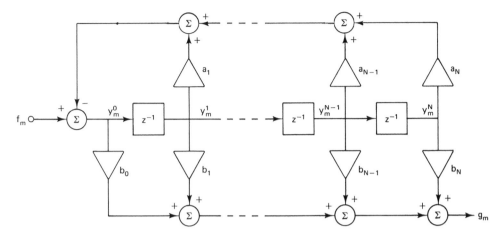

Figure 9.7 Linear filter diagram 3.

using an intermediate variable y_m^n, this time defined as shown in diagram 3, the algorithm for diagram 3 is

$$
\left.
\begin{aligned}
y_m^n &= y_{m-1}^{n-1}, \qquad n = 1, 2, \ldots, N \\
y_m^0 &= f_m - \sum_{n=1}^{N} a_n y_m^n \\
g_m &= \sum_{n=0}^{N} b_n y_m^n
\end{aligned}
\right\}
\tag{9.39}
$$

Again, for the storage of past values, the algorithm requires only N cells for y_m^0 through y_m^{N-1} (i.e., only y_{m-1}^0 through y_{m-1}^{N-1} appear in Eq. 9.39).

Example 9.6

Given the recursive algorithm $g_m = 2f_m + 3f_{m-1} - 4g_{m-1}$, draw diagrams 1–3 shown in Fig. 9.8. (In the second diagram, the second line in Eq. 9.38 is nonexistent because there is only one "x" here.) By eliminating the x's and y's, the three algorithms can be made identical.

In Chap. 3 the linear, continuous transfer function was shown in block-diagram form, and of course the linear digital system can be represented in similar form. Figure 9.9 illustrates two basic decompositions of the digital transfer function, $\tilde{H}(z)$ [see Gold and Rader (1969), Chap. 2; also Wait (1970)]. These forms are in contrast with the forms in Figs. 9.4, 9.6, and 9.7 because the latter do not involve the decomposition of $\tilde{H}(z)$ into simpler transfer functions. Here the numerator and denominator polynomials in $\tilde{H}(z)$ must be factorable into products of simpler polynomials (e.g., as in Eq. 9.74 below).

The *cascade form* expresses $\tilde{H}(z)$ as a product of its factors and is simple to construct once $\tilde{H}(z)$ has been factored. The *parallel form* expresses $\tilde{H}(z)$ as a sum and can be constructed by expanding $\tilde{H}(z)$ into a sum of partial fractions. The following example illustrates these two basic forms.

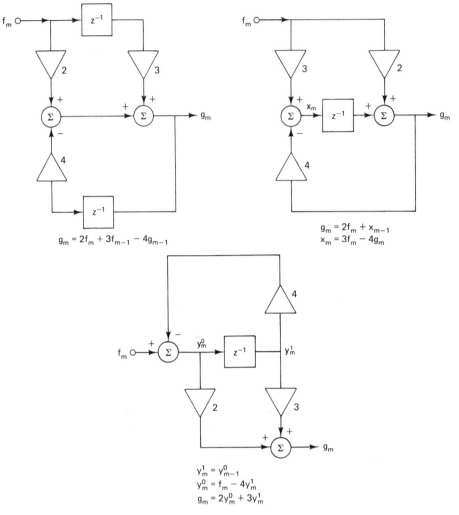

$$g_m = 2f_m + 3f_{m-1} - 4g_{m-1}$$

$$g_m = 2f_m + x_{m-1}$$
$$x_m = 3f_m - 4g_m$$

$$y_m^1 = y_{m-1}^0$$
$$y_m^0 = f_m - 4y_m^1$$
$$g_m = 2y_m^0 + 3y_m^1$$

Figure 9.8 Three equivalent forms in Example 9.6.

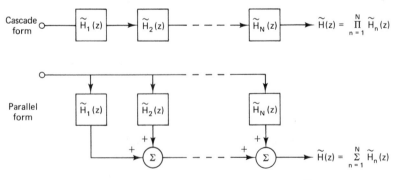

Cascade form $\widetilde{H}_1(z) \rightarrow \widetilde{H}_2(z) - - - - \rightarrow \widetilde{H}_N(z) \rightarrow$ $\widetilde{H}(z) = \prod_{n=1}^{N} \widetilde{H}_n(z)$

Parallel form $\widetilde{H}(z) = \sum_{n=1}^{N} \widetilde{H}_n(z)$

Figure 9.9 Decompositions of the transfer function $\widetilde{H}(z)$.

Recursive Digital Systems   Chap. 9

Example 9.7

Show how to decompose

$$\tilde{H}(z) = \frac{3z^3 - 5z^2 + 10z}{z^3 - 3z^2 + 7z - 5} \tag{9.40}$$

into cascade and parallel forms. The factored form of $\tilde{H}(z)$ is

$$\tilde{H}(z) = \frac{z(3z^2 - 5z + 10)}{(z - 1)(z^2 - 2z + 5)} \tag{9.41}$$

If $\tilde{H}(z)$ is factored further imaginary coefficients will result, so Eq. 9.41 is the preferred factored form. The cascade and parallel forms of $\tilde{H}(z)$ are

$$\text{cascade:} \quad \tilde{H}(z) = \tilde{H}_1(z)\tilde{H}_2(z) = \left(\frac{z}{z - 1}\right)\left(\frac{3z^2 - 5z + 10}{z^2 - 2z + 5}\right) \tag{9.42}$$

$$\text{parallel:} \quad \tilde{H}(z) = \tilde{H}_1(z) + \tilde{H}_2(z) = \left(\frac{2z}{z - 1}\right) + \left(\frac{z^2}{z^2 - 2z + 5}\right) \tag{9.43}$$

In the parallel form, Eq. 9.43 is the partial-fraction expansion. Equations 9.42 and 9.43 are in the form of Fig. 9.9, and of course the individual factors can be in the form of Figs. 9.4, 9.6, and 9.7. Using 9.5, complete constructions are given in Fig. 9.10.

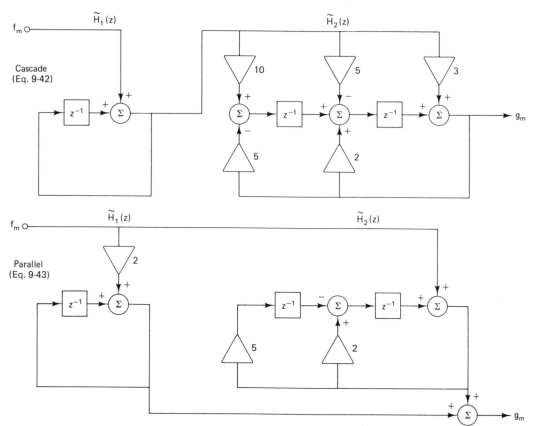

Figure 9.10   Cascade and parallel versions of $H(z) = (3z^3 - 5z^2 + 10z)/(z^3 - 3z^2 + 7z - 5)$.

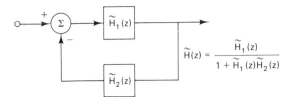

$$\tilde{H}(z) = \frac{\tilde{H}_1(z)}{1 + \tilde{H}_1(z)\tilde{H}_2(z)}$$

Figure 9.11 Digital feedback diagram.

Besides the cascade and parallel forms, the digital *feedback* diagram illustrated in Fig. 9.11 is important, particularly in closed-loop digital control systems. The overall transfer function $\tilde{H}(z)$ is found in terms of $\tilde{H}_1(z)$ and $\tilde{H}_2(z)$ exactly as in the analog case (Chapter 3, Section 3.2).

9.6 LATTICE STRUCTURES

Another interesting digital structure that we introduce here is the *lattice structure*. Two lattice elements, called *symmetric two-multiplier lattice stages*, are shown in Fig. 9.12. As in the previous diagrams, the superscripts on the signals $[x_m^n]$ and

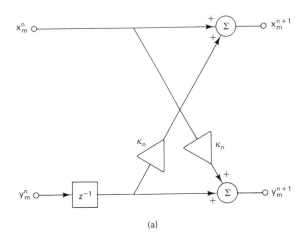

(a)

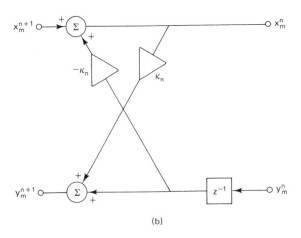

(b)

Figure 9.12 Symmetric two-multiplier lattice stages.

Recursive Digital Systems   Chap. 9

$[y_m^n]$ denote the location of a signal in the lattice. Using subscripts on the z-transforms to correspond with these superscripts, we have for stage (a) in Fig. 9.12,

$$\begin{bmatrix} \tilde{X}_{n+1}(z) \\ \tilde{Y}_{n+1}(z) \end{bmatrix} = \begin{bmatrix} 1 & z^{-1}\kappa_n \\ \kappa_n & z^{-1} \end{bmatrix} \begin{bmatrix} \tilde{X}_n(z) \\ \tilde{Y}_n(z) \end{bmatrix} \tag{9.44}$$

For stage (b) in Fig. 9.12, we note that y_m^{n+1} is an output signal, and

$$\tilde{X}_n(z) = \tilde{X}_{n+1} - z^{-1}\kappa_n \tilde{Y}_n(z)$$
$$\tilde{Y}_{n+1}(z) = \kappa_n \tilde{X}_n(z) + z^{-1}\tilde{Y}_n(z) \tag{9.45}$$

Rewriting these equations, we have the same result as in Eq. 9.44:

$$\begin{bmatrix} \tilde{X}_{n+1}(z) \\ \tilde{Y}_{n+1}(z) \end{bmatrix} = \begin{bmatrix} 1 & z^{-1}\kappa_n \\ \kappa_n & z^{-1} \end{bmatrix} \begin{bmatrix} \tilde{X}_n(z) \\ \tilde{Y}_n(z) \end{bmatrix} \tag{9.46}$$

Thus, observing the different signal labels in Fig. 9.12 and noting that Eqs. 9.44 and 9.46 are the same, we would say that the two stages in Fig. 9.12 are closely related.

A lattice structure can be formed by connecting stages in cascade. To construct the lattice equivalent of a general recursive digital filter, we use stage (b) from Fig. 9.12 in the structure shown in Fig. 9.13, following Gray and Markel (1973). We note that the output, g_m, is formed as a weighted sum of the signals y_m^n, that is,

$$G(z) = \sum_{n=0}^{N} \nu_n \tilde{Y}_n(z) \tag{9.47}$$

First we wish to show that $\tilde{G}(z)/\tilde{F}(z)$ in Fig. 9.13 is equal to a rational transfer function of order N in the form of Eq. 9.19, and at the same time to show how to convert the lattice to a direct form. We note in Fig. 9.13 that $\tilde{F}(z) = \tilde{X}_N(z)$ so, from Eq. 9.47, the transfer function is

$$\tilde{H}(z) = \frac{\tilde{G}(z)}{\tilde{F}(z)} = \sum_{n=0}^{N} \frac{\nu_n \tilde{Y}_n(z)}{\tilde{X}_N(z)} \tag{9.48}$$

Now, to find the terms in Eq. 9.48, we first define transfer functions from $\tilde{X}_0(z)$ and $\tilde{Y}_0(z)$, which are equal in Fig. 9.13, to $\tilde{X}_n(z)$ and $\tilde{Y}_n(z)$, as follows:

$$\tilde{P}_n(z) = \frac{\tilde{X}_n(z)}{\tilde{X}_0(z)} \qquad \tilde{Q}_n(z) = \frac{\tilde{Y}_n(z)}{\tilde{Y}_0(z)} \tag{9.49}$$

Since $\tilde{X}_0(z) = \tilde{Y}_0(z)$, we can divide Eq. 9.46 by either term and write Eq. 9.46 in the form

$$\begin{bmatrix} \tilde{P}_n(z) \\ \tilde{Q}_n(z) \end{bmatrix} = \begin{bmatrix} 1 & z^{-1}\kappa_{n-1} \\ \kappa_{n-1} & z^{-1} \end{bmatrix} \cdots \begin{bmatrix} 1 & z^{-1}\kappa_0 \\ \kappa_0 & z^{-1} \end{bmatrix} \begin{bmatrix} 1 \\ 1 \end{bmatrix} \tag{9.50}$$

In the first of these two equations, we now substitute z^{-1} for z and multiply by z^{-n} to obtain

$$\begin{bmatrix} z^{-n}\tilde{P}_n(z^{-1}) \\ \tilde{Q}_n(z) \end{bmatrix} = \begin{bmatrix} z^{-1} & \kappa_{n-1} \\ \kappa_{n-1} & z^{-1} \end{bmatrix} \cdots \begin{bmatrix} z^{-1} & \kappa_0 \\ \kappa_0 & z^{-1} \end{bmatrix} \begin{bmatrix} 1 \\ 1 \end{bmatrix} \tag{9.51}$$

194

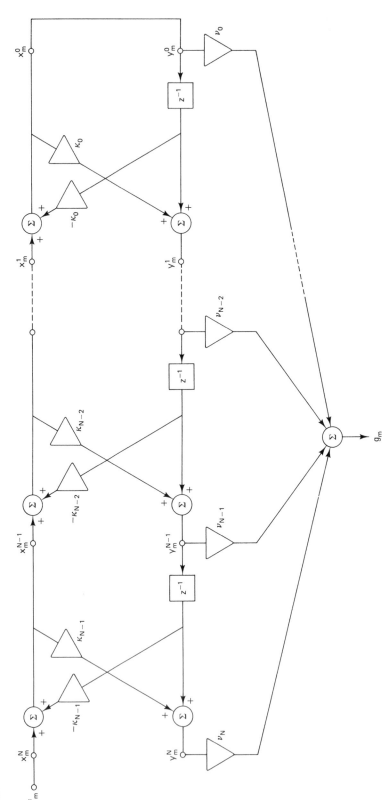

Figure 9.13 Symmetric lattice form of a linear recursive digital filter using stage (b) in Fig. 9.12.

If we think of carrying out this product from right to left, we see that the two equations are the same, so

$$\tilde{Q}_n(z) = z^{-n}\tilde{P}_n(z^{-1})$$ (9.52)

This result is important in lattice theory because it shows that $\tilde{P}_n(z)$ and $\tilde{Q}_n(z)$, and therefore $\tilde{X}_n(z)$ and $\tilde{Y}_n(z)$, are related in a simple manner in symmetric lattices.

Using the definitions in Eq. 9.49, we can rewrite Eqs. 9.48 and 9.50 and obtain the following:

SYMMETRIC LATTICE-TO-DIRECT CONVERSION

$$\tilde{H}(z) = \frac{\displaystyle\sum_{n=0}^{N} \nu_n \tilde{Q}_n(z)}{\tilde{P}_N(z)}$$ (9.53)

with

$$\begin{bmatrix} \tilde{P}_n(z) \\ \tilde{Q}_n(z) \end{bmatrix} = \begin{bmatrix} 1 & z^{-1}\kappa_{n-1} \\ \kappa_{n-1} & z^{-1} \end{bmatrix} \begin{bmatrix} \tilde{P}_{n-1}(z) \\ \tilde{Q}_{n-1}(z) \end{bmatrix}$$ (9.54)

and

$$\tilde{P}_0(z) = \tilde{Q}_0(z) = 1$$ (9.55)

Since each coefficient matrix in Eq. 9.50 has a one in the upper left corner, we see that p_0^N, the coefficient of z^0 in $\tilde{P}_N(z)$, must be 1, so $\tilde{P}_N(z)$ in Eq. 9.53 becomes the denominator, $\tilde{A}(z)$, of $\tilde{H}(z)$ in Eq. 9.19. Similarly, the numerator in Eq. 9.53 is $\tilde{B}(z)$ in Eq. 9.19.

Thus, in Eqs. 9.53 through 9.55, we have an algorithm for converting from the lattice form in Fig. 9.13 to the direct form in Fig. 9.4. A Fortran implementation of this lattice-to-direct algorithm is included in Appendix B. Its calling sequence is

```
CALL SPLTOD(KAPPA,NU,N,B,A)
```

```
KAPPA(0:N - 1)    = REAL lattice coefficients in Fig. 9.13
NU(0:N)           = REAL lattice coefficients in Fig. 9.13
N                 = Number of lattice stages (order of filter)
B(0:N)            = Numerator coefficients in direct form, Eq. 9.19
A(1:N)            = Denominator coefficients in direct form, Eq. 9.19
```

Note carefully that the first index of A is 1, not 0. The routine converts the lattice coefficients stored in KAPPA and NU into the direct-form coefficients stored in B and A in accordance with Eqs. 9.53 through 9.55. It is used in Example 9.8 below and in the exercises at the end of this chapter.

We turn now to the problem of converting from direct to lattice form. From Eq. 9.53 we have the numerator and denominator polynomials in Eq. 9.19, that is,

$$\tilde{H}(z) = \frac{B(z)}{A(z)} = \frac{\sum\limits_{n=0}^{N} \nu_n \tilde{Q}_n(z)}{\tilde{P}_N(z)} \tag{9.56}$$

So in this case we must begin with $\tilde{P}_N(z) = A(z)$ and decompose $\tilde{P}_N(z)$ to obtain $\tilde{P}_{N-1}(z)$, $\tilde{P}_{N-2}(z)$, and so on. At each stage (n) of the decomposition, we can find the corresponding lattice coefficient, κ_{n-1} and, also, having Eq. 9.52 to relate $\tilde{Q}_n(z)$ to $\tilde{P}_n(z)$, we can find the numerator coefficient, ν_N in Eq. 9.56. This procedure is stated more clearly below, but first we need a recursion algorithm for obtaining $\tilde{P}_{n-1}(z)$ from $\tilde{P}_n(z)$. We get this from Eq. 9.54, which may be written

$$\begin{bmatrix} \tilde{P}_n(z) \\ \tilde{Q}_n(z) \end{bmatrix} = \begin{bmatrix} 1 & \kappa_{n-1} \\ \kappa_{n-1} & 1 \end{bmatrix} \begin{bmatrix} \tilde{P}_{n-1}(z) \\ z^{-1}\tilde{Q}_{n-1}(z) \end{bmatrix} \tag{9.57}$$

Using Eq. 1.30, we invert Eq. 9.57 and obtain

$$\begin{bmatrix} \tilde{P}_{n-1}(z) \\ z^{-1}\tilde{Q}_{n-1}(z) \end{bmatrix} = \frac{1}{1 - \kappa_{n-1}^2} \begin{bmatrix} 1 & -\kappa_{n-1} \\ -\kappa_{n-1} & 1 \end{bmatrix} \begin{bmatrix} \tilde{P}_n(z) \\ \tilde{Q}_n(z) \end{bmatrix} \tag{9.58}$$

(We will see shortly that κ_{n-1}^2 must be less than 1 for stability, so the denominator in Eq. 9.58 is acceptable.) From this form we obtain the desired recursion algorithm:

$$\tilde{P}_{n-1}(z) = \frac{\tilde{P}_n(z) - \kappa_{n-1}\tilde{Q}_n(z)}{1 - \kappa_{n-1}^2}; \qquad n = N, \dots, 1 \tag{9.59}$$

Using this algorithm, the procedure for converting from direct to lattice form is as follows:

SYMMETRIC DIRECT-TO-LATTICE CONVERSION

Initially,

$$\tilde{P}_N(z) = \tilde{A}(z) = 1 + a_1 z^{-1} + \cdots + a_N z^{-N}$$

$$\tilde{S}_N(z) = \tilde{B}(z) = b_0 + b_1 z^{-1} + \cdots + b_N z^{-N}$$

For $n = N, N-1, \dots, 1$:

$$\tilde{Q}_n(z) = z^{-n}\tilde{P}_n(z^{-1}) \tag{9.60}$$

$$\nu_n = \text{coef. } (s_n) \text{ of } z^{-n} \text{ in } \tilde{S}_n(z) \tag{9.61}$$

$$\tilde{S}_{n-1}(z) = \tilde{S}_n(z) - \nu_n \tilde{Q}_n(z) \tag{9.62}$$

$$\kappa_{n-1} = \text{coef. } (p_n) \text{ of } z^{-n} \text{ in } \tilde{P}_n(z) \tag{9.63}$$

$$\tilde{P}_{n-1}(z) = \frac{\tilde{P}_n(z) - \kappa_{n-1}\tilde{Q}_n(z)}{1 - \kappa_{n-1}^2} \tag{9.64}$$

Finally,

$$\nu_0 = s_0 \tag{9.65}$$

This procedure works as described above, by decomposing the direct-form polynomials. Initially, $\tilde{P}_N(z) = \tilde{A}(z)$ and $\tilde{S}_N(z) = \tilde{B}(z)$. Then, as the procedure continues, Eq. 9.60 is the same as Eq. 9.52, and $\nu_n = s_n$ in Eq. 9.61 is the leading coefficient in the nth numerator term in Eq. 9.53, which is then removed from the sum $\tilde{S}_n(z)$ in Eq. 9.62. That Eq. 9.63 is correct may be seen in Eq. 9.50, where κ_{n-1} is seen to be the coefficient of z^{-n}. (The reader may wish to verify this result with $n = 1$ and then $n = 2$ in Eq. 9.50.) Finally, Eq. 9.64 is the same as Eq. 9.59.

A Fortran routine implementing the direct-to-lattice procedure just described is included in Appendix B. Its calling sequence is

```
CALL SPDTOL(B,A,N,KAPPA,NU,IERROR)
```

```
B(0:N)        = Numerator coefficients in Eq. 9.21
A(1:N)        = Denominator coefficients in Eq. 9.21
N             = Order of filter = number of lattice stages
KAPPA(0:N-1)  = REAL lattice coefficients in Fig. 9.13
NU(0:N)       = REAL lattice coefficients in Fig. 9.13
IERROR        = 0: no errors
              = 1: unstable filter
```

As before, we note that the first index of A is 1, not 0. The routine converts the direct-form coefficients stored in A and B into lattice coefficients in KAPPA and NU. It does so without destroying A and B, and without using an extra work array. It is used in the exercises at the end of this chapter. The following example illustrates the use of the two routines, SPDTOL and SPLTOD.

Example 9.8

Find the symmetric lattice version of the recursive filter given by

$$\tilde{H}(z) = \frac{4 + 3z^{-1} + 2z^{-2} + z^{-3}}{1 - 0.3z^{-1} + 0.2z^{-2} + 0.3z^{-3}}$$

Then convert the lattice back to the direct form. The program DSA0908, shown here, accomplishes both tasks. The lattice coefficients printed during execution are κ_0, κ_1, and κ_2 and ν_0, ν_1, ν_2, and ν_3 in Fig. 9.13, and the direct-form coefficients obtained from SPLTOD are of course those in $\tilde{H}(z)$ above.

```
      PROGRAM DSA0908
C-DSA EXAMPLE 9.8 -- DIRECT TO LATTICE, THEN BACK TO DIRECT.
      REAL B(0:3),A(1:3),KAPPA(0:2),NU(0:3)
      DATA B/4.,3.,2.,1./, A/-.3,.2,.3/
      CALL SPDTOL(B,A,3,KAPPA,NU,IERR)
      IF(IERR.NE.0) STOP
      PRINT '(/'' KAPPA:'',3F10.6)', KAPPA
      PRINT '('' NU:    '',4F10.6)', NU
      CALL SPLTOD(KAPPA,NU,3,B,A)
      PRINT '(/'' B:    '',4F10.6)', B
      PRINT '('' A:    '',3F10.6/)', A
      STOP
      END
$ RUN DSA0908
```

```
KAPPA:  -0.300000 0.318681 0.300000
NU:      4.080000 3.709890 2.300000 1.000000

B:       4.000000 3.000000 2.000000 1.000000
A:      -0.300000 0.200000 0.300000

FORTRAN STOP
$
```

Another important general property of lattices, proved by Jury (1964), can be observed in principle from the recursion implied in Eq. 9.50 or, equivalently, from Eq. 9.46. If the right-hand terminals of the lattice stage in Fig. 9.12(b) are connected, the transfer function from the right to the upper left terminal is given by $(1 - z^{-1}\kappa_n)$. Thus, if $|\kappa_n| < 1$, the zero of this function lies inside the unit circle. From this result, it is not surprising that, as Jury has proved, the zeros of $\tilde{P}_N(z)$ are inside the unit circle if $|\kappa_n| < 1$ for all stages. That is, $\tilde{H}(z)$ in Eq. 9.53 is stable if and only if

$$|\kappa_n| < 1; \qquad n = 0, 1, \ldots, N - 1 \tag{9.66}$$

We turn now in our discussion of lattice structures to the formulation of a *nonrecursive lattice*, which has already been done in principle. Suppose, in Fig. 9.13, that we set $\nu_0 = 1$ and $\nu_n = 0$ for $n = 1, 2, \ldots, N$. Then the transfer function from f_m to g_m, which is now $\tilde{Y}_0(z)/\tilde{X}_N(z)$, is given by Eqs. 9.53 and 9.55 as

$$\tilde{H}(z) = \frac{\tilde{Y}_0(z)}{\tilde{X}_N(z)} = \frac{\tilde{Q}_0(z)}{\tilde{P}_N(z)} = \frac{1}{\tilde{P}_N(z)} \tag{9.67}$$

In other words, as we have noted previously, $\tilde{Y}_0(z) = \tilde{X}_0(z)$, therefore

$$[\tilde{H}(z)]^{-1} = \frac{\tilde{X}_N(z)}{\tilde{X}_0(z)} = \tilde{P}_N(z) \tag{9.68}$$

The function $\tilde{P}_N(z)$ is of course a polynomial, that is, a nonrecursive transfer function. This result indicates that we could obtain a nonrecursive lattice by inverting the lattice in Fig. 9.11 with the values of $[\nu_n]$ as above. Since Eqs. 9.44 and 9.46 are the same, such an inversion is obtained by using stage (a) in place of stage (b) in Fig. 9.12, so that x_m^N becomes the output signal. The result, a nonrecursive lattice, is shown in Fig. 9.14.

The algorithms for converting to and from the nonrecursive lattice in Fig. 9.14 have already been given in principle, since the two algorithms given previously convert between the lattice coefficients $[\kappa_n]$ and the coefficients of $\tilde{P}_N(z)$. In fact, the two routines described previously, SPLTOD and SPDTOL, could be used to convert between the direct form, Eq. 9.19, and the lattice in Fig. 9.14. However, there are two reasons for having routines especially tailored to nonrecursive systems. First, the values of ν_1 through ν_N as well as a_1 through a_N are all zero, so the extra storage is not needed. Second, values of κ_n greater than 1 in Fig. 9.14 do not indicate an unstable system — only a transfer function with zeros outside the unit circle.

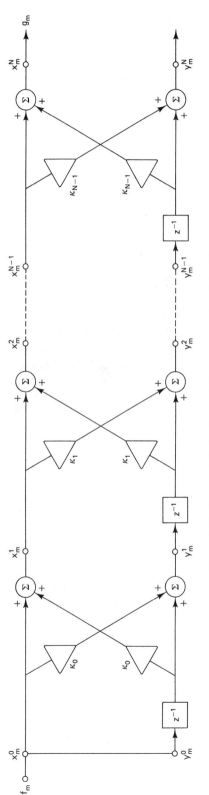

Figure 9.14 Symmetric lattice form of a linear nonrecursive digital filter.

The calling sequences for the nonrecursive lattice-to-direct and direct-to-lattice conversion routines are

```
CALL SPNLTD(KAPPA,N,B)
CALL SPNDTL(B,N,KAPPA,IERROR)
```

```
KAPPA(0:N - 1) = REAL lattice coefficients in Fig. 9.14
N              = Number of lattice stages (order of filter)
B(1:N)         = Nonrecursive filter coefficients in
                 H̃(z) = 1 + b₁z⁻¹ + · · · + b_Nz⁻ᴺ
IERROR         = 0 : no error
               = 1 : |κₙ| too close to 1.0−conversion not
                     possible
```

The leading coefficient of $\tilde{B}(z)$, $b_0 = 1$, is a natural consequence of the inversion described above, since $a_0 = 1$ in the recursive version. Any other value of b_0 is of course easily obtained if we write $\tilde{H}(z)$ in the form

$$\tilde{H}(z) = b_0\left(1 + \frac{b_1}{b_0}z^{-1} + \cdots + \frac{b_N}{b_0}z^{-N}\right) \tag{9.69}$$

We would then use the routines above with the coefficients $(b_n/b_0; n = 1, 2, \ldots, N)$ in Eq. 9.69, and attach the coefficient b_0 to the input or output of the lattice.

However, the approach in Eq. 9.69 does not allow symmetrical, linear-phase tranfers functions with $b_0 = b_N$ (see Section 8.4 as well as Section 9.9), because then p_N (and thus κ_{N-1}) would be 1, and the lattice would not be realizable (see Eqs. 9.63 and 9.64). An alternative approach which is therefore preferred is to subtract each input sample, f_m, from each output sample, g_m, thereby subtracting 1 from the transfer function and obtaining

$$\hat{H}(z) = b_1z^{-1} + b_2z^{-2} + \cdots + b_Nz^{-N} \tag{9.70}$$

Similarly, any *noncausal* transfer function, recursive or nonrecursive, could be transformed using any of the four routines described previously, simply by shifting the input sequence. In the numerator of Eq. 9.8, for example, we would write $\tilde{B}(z)$ in the form

$$\tilde{B}(z) = z^N(b_{-N} + b_{-N+1}z^{-1} + \cdots + b_Nz^{-2N}) \tag{9.71}$$

We would then use the routines with the $2N + 1$ coefficients in Eq. 9.71. (If the filter were nonrecursive, we would also alter the transfer function as in Eq. 9.69 or 9.70.) The following example serves to illustrate these procedures.

Example 9.9

Convert the filter given by Eq. 8.12, that is,

$$\tilde{H}(z) = \frac{1}{2} + \frac{2}{\pi}(z^{-1} + z)$$

from direct to lattice form, then back to direct form. Program DSA0909, shown here, shows both conversions. The direct coefficients are seen to be those above, that is, $2/\pi$, $\frac{1}{2}$, and $2/\pi$.

```
        PROGRAM DSA0909
C-DSA EXAMPLE 9.9 -- NONRECURSIVE LATTICE AND DIRECT COEFFICIENTS.
        REAL KAPPA(0:2),B(3)
        PI = 4.*ATAN(1.)
        B(1) = 2./PI
        B(2) = 1./2.
        B(3) = 2./PI
        CALL SPNDTL(B,3,KAPPA,IERROR)
        IF(IERROR.NE.0) STOP
        PRINT '(/'' LATTICE COEFFICIENTS:'',3F10.6)', KAPPA
        CALL SPNLTD(KAPPA,3,B)
        PRINT '(/'' DIRECT COEFFICIENTS: '',3F10.6/)', B
        STOP
        END
$ RUN DSA0909

LATTICE COEFFICIENTS: 0.461700 0.159262 0.636620

DIRECT COEFFICIENTS:  0.636620 0.500000 0.636620

FORTRAN STOP
$
```

Using the procedures described above, the lattice form is shown in Fig. 9.15. Note that the input, f_m, is shifted in accordance with Eq. 9.71 and subtracted from the output to obtain the form in Eq. 9.70. The shift in this case is z^2 to account for the unit shift (z^1) in $\tilde{H}(z)$ above plus the unit delay in the revised lattice transfer function, Eq. 9.70.

Other lattice structures are similar to the ones described previously are possible, and in general may be analyzed and synthesized along the lines described [see Mitra et al. (1977)]. For example, Fig. 9.16 shows a direct form along with two equivalent *asymmetric* lattice forms, whose coefficients are not equal in pairs as they were previously. These asymmetric forms are similar to Fig. 9.13 but do not require the ν coefficients.

Table 9.2 provides the essential analysis and synthesis formulas for the two asymmetric lattice forms in Fig. 9.16. The first form has stage 0 terminated with the coefficient 1 and requires $a_0 = b_N = 1$ in $\tilde{H}(z)$. The second form has stage 0 terminated with the coefficient γ, and requires $a_0 = 1$ and $a_N \neq 0$ in $\tilde{H}(z)$. In order to make Table 9.2 easier to type we have omitted the argument (z) from most of the functions. One should be able to compare the formulas with Eqs. 9.45, 9.46, 9.61, 9.63, and the matrix version of 9.64, and see that the development is similar.

Conversion routines for the asymmetric lattices in Table 9.2 are included in Appendix B. To convert from direct form to the first lattice form, the SPDTAL1 routine first sets GAIN equal to b_N and then divides $\tilde{B}(z)$ by b_N (which may not equal zero) so that the requirement $b_N = 1$ is met. The GAIN coefficient must then be used in series with the lattice. (Note that GAIN = 1 in Fig. 9.16.) Otherwise, the routines are similar to each other and easy to use.

Some direct forms will not convert to a particular lattice form. The direct-to-lattice conversion routines in Appendix B will indicate this property by returning IERROR = 1.

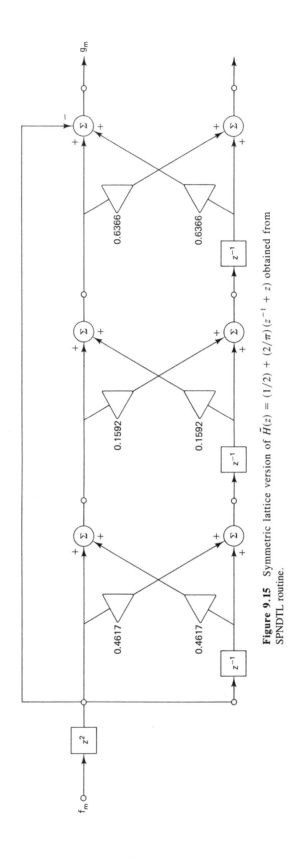

Figure 9.15 Symmetric lattice version of $\tilde{H}(z) = (1/2) + (2/\pi)(z^{-1} + z)$ obtained from SPNDTL routine.

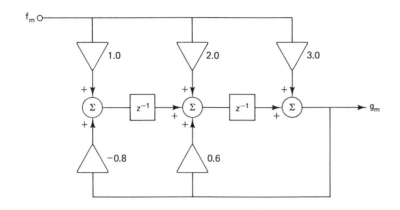

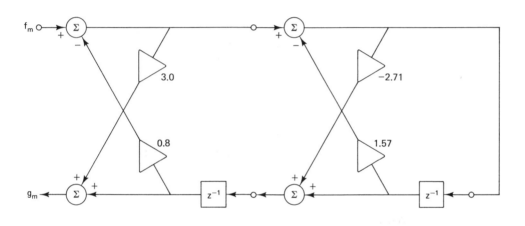

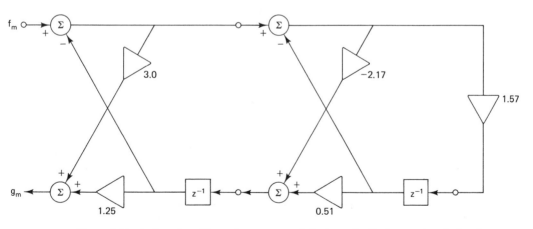

Figure 9.16 *A* direct-form filter and two asymmetric lattices, all with the same transfer function, $\tilde{H}(z) = (3 + 2z^{-1} + z^{-2})/(1 - 0.6z^{-1} + 0.8z^{-2})$.

TABLE 9.2. Asymmetric Recursive Lattices and Formulas

Stage
diagram

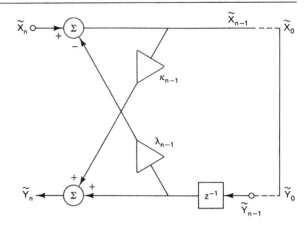

N-stage
formulas

$n = 1, \ldots, N$

$$\begin{bmatrix} \tilde{X}_n \\ \tilde{Y}_n \end{bmatrix} = \begin{bmatrix} 1 & \lambda_{n-1}z^{-1} \\ \kappa_{n-1} & z^{-1} \end{bmatrix} \begin{bmatrix} \tilde{X}_{n-1} \\ \tilde{Y}_{n-1} \end{bmatrix}$$

$$\tilde{P}_n = \frac{\tilde{X}_n}{\tilde{X}_0}; \quad \tilde{Q}_n = \frac{\tilde{Y}_n}{\tilde{X}_0}; \quad \tilde{X}_0 = \tilde{Y}_0$$

Lattice
to direct

$$\begin{bmatrix} \tilde{P}_n \\ \tilde{Q}_n \end{bmatrix} = \begin{bmatrix} 1 & \lambda_{n-1}z^{-1} \\ \kappa_{n-1} & z^{-1} \end{bmatrix} \cdots \begin{bmatrix} 1 & \lambda_0 z^{-1} \\ \kappa_0 & z^{-1} \end{bmatrix} \begin{bmatrix} 1 \\ 1 \end{bmatrix}$$

$$\tilde{H}(z) = \frac{\tilde{Y}_N}{\tilde{X}_N} = \frac{\tilde{Q}_N}{\tilde{P}_N} = \frac{q_0^N + \cdots + q_N^N z^{-N}}{1 + \cdots + p_N^N z^{-N}}$$

Polynomial
form

$$\begin{bmatrix} \tilde{P}_n \\ \tilde{Q}_n \end{bmatrix} = \begin{bmatrix} 1 + \cdots + \lambda_{n-1}z^{-n} \\ \kappa_{n-1} + \cdots + z^{-n} \end{bmatrix}; \quad n > 0$$

Direct to
lattice

$$\tilde{H}(z) = \frac{\tilde{B}(z)}{\tilde{A}(z)}; \quad a_0 = b_N = 1$$

$$\tilde{P}_N = \tilde{A}(z); \quad \tilde{Q}_N = \tilde{B}(z)$$

for $n = N, \ldots, 1$:

$$\kappa_{n-1} = q_0^n; \quad \lambda_{n-1} = p_n^n$$

$$D = 1 - \kappa_{n-1}\lambda_{n-1}$$

$$\begin{bmatrix} \tilde{P}_{n-1} \\ \tilde{Q}_{n-1} \end{bmatrix} = \frac{z}{D} \begin{bmatrix} z^{-1} & -\lambda_{n-1}z^{-1} \\ -\kappa_{n-1} & 1 \end{bmatrix} \begin{bmatrix} \tilde{P}_n \\ \tilde{Q}_n \end{bmatrix}$$

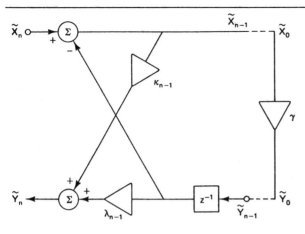

$$\begin{bmatrix} \tilde{X}_n \\ \tilde{Y}_n \end{bmatrix} = \begin{bmatrix} 1 & z^{-1} \\ \kappa_{n-1} & \lambda_{n-1}z^{-1} \end{bmatrix} \begin{bmatrix} \tilde{X}_{n-1} \\ \tilde{Y}_{n-1} \end{bmatrix}$$

$$\tilde{P}_n = \frac{\tilde{X}_n}{\tilde{X}_0}; \quad \tilde{Q}_n = \frac{\tilde{Y}_n}{\tilde{X}_0}; \quad \gamma\tilde{X}_0 = \tilde{Y}_0$$

$$\begin{bmatrix} \tilde{P}_n \\ \tilde{Q}_n \end{bmatrix} = \begin{bmatrix} 1 & z^{-1} \\ \kappa_{n-1} & \lambda_{n-1}z^{-1} \end{bmatrix} \cdots \begin{bmatrix} 1 & z^{-1} \\ \kappa_0 & \lambda_0 z^{-1} \end{bmatrix} \begin{bmatrix} 1 \\ \gamma \end{bmatrix}$$

$$\tilde{H}(z) = \frac{\tilde{Y}_N}{\tilde{X}_N} = \frac{\tilde{Q}_N}{\tilde{P}_N} = \frac{q_0^N + \cdots + q_N^N z^{-N}}{1 + \cdots + p_N^N z^{-N}}$$

$$\begin{bmatrix} \tilde{P}_n \\ \tilde{Q}_n \end{bmatrix} = \begin{bmatrix} 1 + \cdots + \gamma \prod\limits_{i=0}^{n-2} \lambda_i z^{-n} \\ \kappa_{n-1} + \cdots + \gamma \prod\limits_{i=0}^{n-1} \lambda_i z^{-n} \end{bmatrix}; \quad n > 1$$

$$\tilde{H}(z) = \frac{\tilde{B}(z)}{\tilde{A}(z)}; \quad a_0 = 1$$

$$\tilde{P}_N = \tilde{A}(z); \quad \tilde{Q}_N = \tilde{B}(z)$$

for $n = N, \ldots, 1$:

$$\kappa_{n-1} = q_0^n; \quad \lambda_{n-1} = q_n^n/p_n^n$$

$$D = \lambda_{n-1} - \kappa_{n-1}$$

$$\begin{bmatrix} \tilde{P}_{n-1} \\ \tilde{Q}_{n-1} \end{bmatrix} = \frac{z}{D} \begin{bmatrix} \lambda_{n-1}z^{-1} & -z^{-1} \\ -\kappa_{n-1} & 1 \end{bmatrix} \begin{bmatrix} \tilde{P}_n \\ \tilde{Q}_n \end{bmatrix}$$

for $n = 0$: $\gamma = q_0^0$

9.7 POLE–ZERO PLOTS

Besides the block diagrams discussed in the preceding section, another aid used in the analysis and design of digital filters is the pole–zero plot on the z-plane. Just as the block diagrams above are analogous to the schematic diagrams of continuous systems, the z-plane plot will be seen to be analogous to the s-plane plot for continuous filters. The s-plane plot, introduced in Chapter 3, is discussed later in connection with the synthesis of specific filters—here the emphasis is on relating the frequency response of a digital filter to its z-plane plot.

First, the term "z-plane" was introduced in Chapter 7. The substitution of $e^{j\omega T}$ for z, which was used to obtain the frequency transfer function $\overline{H}(j\omega)$, suggests that z be allowed to have complex values. Thus the z-plane is the complex plane, and a point on the plane at coordinates (x, y) represents the value $z = x + jy$. The real (x) axis is always horizontal.

Next, the poles and zeros of the function $\tilde{H}(z)$ can be defined. This function has already been shown to be a ratio of polynomials in z, as in Eq. 9.9:

$$\tilde{H}(z) = \frac{\tilde{B}(z)}{\tilde{A}(z)} = \frac{b_N z^{-N} + \cdots + b_1 z^{-1} + b_0}{a_N z^{-N} + \cdots + a_1 z^{-1} + 1} \tag{9.72}$$

$$= \frac{b_0 z^N + b_1 z^{N-1} + \cdots + b_N}{z^N + a_1 z^{N-1} + \cdots + a_N} \tag{9.73}$$

(Note that a_0 is one as above, so z^N in Eq. 9.73 now has a coefficient of 1.) To find the poles and zeros, the numerator and denominator of Eq. 9.73 must each be factored into N terms:

$$\tilde{H}(z) = b_0 \frac{(z - B_1)(z - B_2) \cdots (z - B_N)}{(z - A_1)(z - A_2) \cdots (z - A_N)} \tag{9.74}$$

where b_0 is assumed to be nonzero. This factoring may of course be a considerable problem in itself, sometimes requiring an iterative solution when N is greater than 3. On the other hand, if the A's and B's are given, the a's and b's—the filter coefficients—can be found by multiplying the factors together in Eq. 9.74. Each B_n is called a *zero* of $\tilde{H}(z)$ because $\tilde{H}(z)$ vanishes when z approaches this value. Similarly, A_n is a *pole* because $\tilde{H}(z)$ approaches infinity as z approaches A_n.

Example 9.10

Find and plot the poles and zeros of $\tilde{H}(z)$ if the filter algorithm is $g_m = f_{m-1} + 0.8 f_{m-2} - 0.6 g_{m-1} - 0.25 g_{m-2}$. The algorithm has $b_0 = 0$, $b_1 = 1$, $b_2 = 0.8$, $a_1 = 0.6$, and $a_2 = 0.25$. Therefore, using the form in Eq. 9.73, $\tilde{H}(z)$ is given by

$$\tilde{H}(z) = \frac{z + 0.8}{z^2 + 0.6z + 0.25} \tag{9.75}$$

or, in the pole–zero form of Eq. 9.74,

$$\tilde{H}(z) = \frac{z + 0.8}{(z + 0.3 + j0.4)(z + 0.3 - j0.4)} \tag{9.76}$$

The poles and zero are plotted in Fig. 9.17. The poles are seen to occur as a *conjugate pair* inside the *unit circle* on the z-plane. The significance of these properties will be seen shortly.

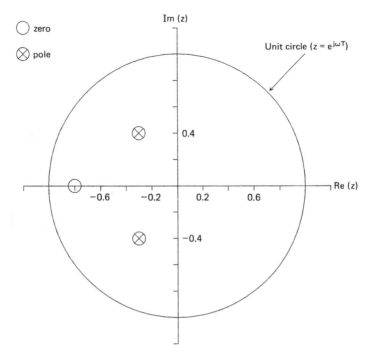

Im (z)

Unit circle (z = e^{jωT})

⊗   0.4

-0.6   -0.2   0.2   0.6   Re (z)

⊗   -0.4

Figure 9.17   Pole-zero plot for Example 9.10.

Having the poles and zeros of the filter function $\tilde{H}(z)$ on the z-plane, it is now easy to visualize the frequency response function. The latter is found by substituting $e^{j\omega T}$ for z in $\tilde{H}(z)$, and so Eq. 9.74 gives the frequency response as

$$\overline{H}(j\omega) = b_0 \frac{(e^{j\omega T} - B_1)(e^{j\omega T} - B_2) \cdots (e^{j\omega T} - B_N)}{(e^{j\omega T} - A_1)(e^{j\omega T} - A_2) \cdots (e^{j\omega T} - A_N)} \tag{9.77}$$

Each factor, $(e^{j\omega T} - A_n)$ or $(e^{j\omega T} - B_n)$, in Eq. 9.77 is represented on the z-plane by a *vector drawn from the pole or zero of $\tilde{H}(z)$ to the point $z = e^{j\omega T}$ on the unit circle.* The amplitude response of the filter is then found by multiplying and dividing the vector amplitudes, and the phase shift is found by adding and subtracting the vector angles.

Example 9.11

Given the filter in Example 9.10 with $g_m = f_{m-1} + 0.8f_{m-2} - 0.6g_{m-1} - 0.25g_{m-2}$, illustrate the amplitude gain and phase shift at $\omega = \pi/4T$. The pole–zero plot in Fig. 9.17 is repeated in Fig. 9.18, this time with vectors drawn to the point $z = e^{j\pi/4}$ on the unit circle, where $\omega = \pi/4T$. The vector magnitudes are V_1, U_1, and U_2 as shown in the figure, and so the amplitude gain is V_1/U_1U_2. Note how V_1, U_1, and U_2 in this example represent the magnitudes $|e^{j\omega T} - B_1|$, $|e^{j\omega T} - A_1|$, and $|e^{j\omega T} - A_2|$ in Eq. 9.77. The vector angles are β_1, α_1, and α_2, as illustrated, and so the filter phase shift is $\beta_1 - \alpha_1 - \alpha_2$.

The periodic form of $\overline{H}(j\omega)$ is immediately evident from the z-plane plot. As ωT increases from 0 to 2π, the point on the unit circle to which vectors are drawn starts at $z = 1$ and makes one complete revolution around the circle. The point again revolves around the

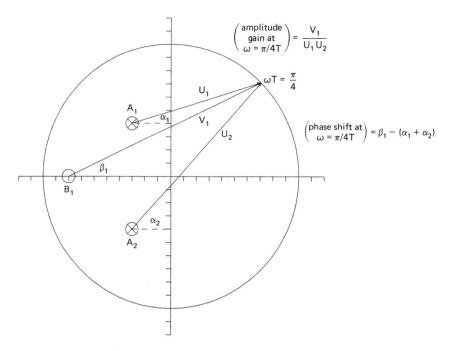

Figure 9.18 Gain and phase shift for Example 9.11.

circle as ωT goes from 2π to 4π, and so on. The transfer function, which changes according to the location of the point, must therefore be periodic with period $2\pi/T$.

Digital filters can be synthesized by placing poles and zeros on the z-plane to achieve given gain characteristics. This procedure is discussed in Chapter 12; however, some of the ground rules for placing the poles and zeros on the z-plane are already evident, and lend insight into the synthesis problem. These rules are:

1. Poles or zeros must either be real or occur in *conjugate pairs*.
2. *Adding a pole (or zero) at $z = 0$ causes $\overline{H}(j\omega)$ to be multiplied by $e^{-j\omega T}$ (or $e^{+j\omega T}$), and thus *affects only the phase shift* and not the amplitude gain of the filter.
3. *A pole (or zero) on the unit circle implies that $\overline{H}(j\omega)$ is infinite (or zero) at a specific frequency.*
4. *A pole outside the unit circle implies instability* in that the filter's response to a transient will increase rather than decay.
5. *Poles not on the positive real axis in general cause oscillation* in the output of the filter.

Each of these properties is shown quite easily. The first follows by multiplying the factors in Eq. 9.74 together to get the polynomials in Eq. 9.73, and noting that the A's and B's must occur in conjugate pairs for the a's and b's to be real. The second property follows from the shifting theorem, or by observing that the magnitude of the vector from the point $z = 0$ to the unit circle is always 1. Property 3 is evident because, if the pole or zero is on the unit circle, the corresponding vector length must decrease to zero at some specific frequency. Finally, properties 4 and 5 are demonstrated by solving partially for the impulse or step-function response as discussed above. In the solution, the term $z/(z - p)$, where p is a pole of $\tilde{H}(z)$, will occur in the partial-fraction expansion of $\tilde{F}(z)\tilde{H}(z)$. As in Eq. 9.24, the series for this term is

$$\frac{z}{z - p} = \sum_{m=0}^{\infty} p^m z^{-m} \tag{9.78}$$

If p is outside the unit circle so that $|p| > 1$, then p^m increases with m, and the filter output g_m increases without bound as m increases. If p is negative or complex, then p^m is of the form $r^m e^{jm\theta}$, so g_m has a component with phase $m\theta$ at time mT, that is, an oscillating component.

9.8 LINEAR AND ZERO PHASE SHIFT

The reader may recall that nonrecursive systems with zero phase shift were discussed in Chapter 8, and that they involved processing future as well as past samples of $f(t)$ as if the sample set $[f_m]$ were stored in a recording of some sort. Here the discussion is broadened to include recursive systems and linear as well as zero phase shift.

A transfer function with unit gain and *linear phase shift* is illustrated in Fig. 9.19. In the time domain the effect is simply to shift $f(t)$ by nT seconds, and in the frequency domain $F(s)$ is multiplied by e^{-nTs}, or $\tilde{F}(z)$ by z^{-n}. The shifting theorem guarantees the equivalence of these operations. Note that if $f(t)$ is a sinusoid at frequency ω, the transfer function $e^{-j\omega nT}$ depends on this frequency, but the effect is always to shift $f(t)$ by the constant amount, nT.

Zero phase shift is, of course, accomplished when $n = 0$ in the above example, and more generally when the transfer function is purely real. Filters with zero phase shift are sometimes very desirable in data processing instances where one desires not to shift $f(t)$ in the time domain, and particularly not to shift different frequency components of $f(t)$ by different amounts.

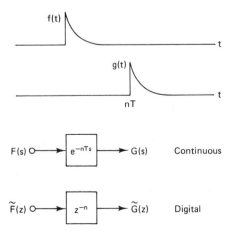

Figure 9.19 Unit gain and linear phase shift.

What properties of a system are necessary for linear (or zero) phase shift? This question is answered here in terms of the digital system, with the analog case being similar. For linear phase shift, the system frequency response must be of the form

$$\overline{H}(j\omega) = \overline{R}(j\omega)e^{-j\omega nT} \tag{9.79}$$

where $\overline{R}(j\omega)$ is a real function of ω. Thus $\tilde{H}(z)$ must be

$$\tilde{H}(z) = z^{-n}\tilde{R}(z) \tag{9.80}$$

in which $\tilde{R}(e^{j\omega T})$ is a real function of ω. But if $\tilde{R}(e^{j\omega T})$ is real, it must equal its own conjugate [i.e., $\tilde{R}(e^{-j\omega T})$], so $\tilde{R}(z)$ must equal $\tilde{R}(z^{-1})$. Thus the following theorem has been proved:

LINEAR PHASE SHIFT THEOREM

For a system to have linear phase shift, its transfer function $\tilde{H}(z)$ must be of the form

$$\tilde{H}(z) = z^{-n}\tilde{R}(z) \tag{9.81}$$

where $\tilde{R}(z)$ is a ratio of polynomials in z and

$$\tilde{R}(z) = \tilde{R}(z^{-1}) \tag{9.82}$$

The shift in time is nT seconds, with $n = 0$ for zero phase shift.

For nonrecursive systems, Eq. 9.81 can be satisfied by making the filter coefficients equal in pairs as described in Chapter 8. Since $\tilde{R}(z)$ is a ratio of nontrivial polynomials in the recursive case, however, the condition $\tilde{R}(z) = \tilde{R}(z^{-1})$ presents a special problem. Let $\tilde{R}(z)$ be expressed as

$$\tilde{R}(z) = \frac{b_0 z^N + b_1 z^{N-1} + \cdots + b_N}{z^N + a_1 z^{N-1} + \cdots + a_N} \tag{9.83}$$

$$b_n = b_{N-n} \quad \text{and} \quad a_n = a_{N-n} \quad \text{for all } n \tag{9.84}$$

with the conditions on the a's and b's being required for $\tilde{R}(z) = \tilde{R}(z^{-1})$. In particular, $a_N = 1$ in the denominator of Eq. 9.83; thus the product of all poles must be one, so *for any pole of $\tilde{R}(z)$ at z_0, there must also be a pole at $1/z_0$.* Now unless z_0 is on the unit circle in the z-plane, either z_0 or $1/z_0$ must be outside the unit circle, and so in general the attempt to obtain a recursive filter with linear phase shift appears to lead to an unstable design.

Fortunately there is a way out of this dilemma in nonreal-time situations, that is, when one's clock can be "reset" to agree with the position in the sample set $[f_m]$ after the entire set has been recorded. If $[f_m]$ is recorded, then of course it must be finite in time extent. For convenience let the nonzero samples of $f(t)$ be

$$[f_m] = [f_{-N}, f_{-N+1}, \ldots, f_0, \ldots, f_N] \tag{9.85}$$

so that there are $2N + 1$ samples in all. The *time reversal* of $[f_m]$ can now be defined as $[f_m]$ in reverse order, that is,

$$\begin{aligned}[f_m]^r &= [f_{-m}]\\ &= [f_N, f_{N-1}, \ldots, f_0, \ldots, f_{-N}]\end{aligned} \tag{9.86}$$

where the superscript r denotes time reversal. The z-transform of $[f_m]^r$ is also a reversal of $\tilde{F}(z)$ of sorts, that is,

$$\begin{aligned}\tilde{F}^r(z) &= \mathscr{Z}[f_m]^r\\ &= \sum_{m=-N}^{N} f_{-m} z^{-m} = \sum_{n=-N}^{N} f_n z^n\\ &= \tilde{F}(z^{-1})\end{aligned} \tag{9.87}$$

(In Eq. 9.87, the script $\mathscr{Z}$ stands for "z-transform of.")

Thus *time reversal implies substitution of z^{-1} for z*. Some fundamental relationships involving time reversal are as follows:

Reversal of sum: $[\tilde{A}(z) + \tilde{B}(z)]^r = \tilde{A}^r(z) + \tilde{B}^r(z)$ $\qquad$ (9.88)

Reversal of product: $[\tilde{A}(z)\tilde{B}(z)]^r = \tilde{A}^r(z)\tilde{B}^r(z)$ $\qquad$ (9.89)

Reversal of reversal: $[\tilde{F}^r(z)]^r = \tilde{F}(z)$ $\qquad$ (9.90)

Each of these is easily proved from the definition in Eq. 9.87.

Now, to achieve zero phase shift using any recursive $\tilde{H}(z)$, consider the two equivalent systems in Fig. 9.20. Each system involves two reversals of time series, and thus two new sample sets besides the original set $[f_m]$. Note, however, that only one set of $2N + 1$ storage cells is required, and the sample set can simply be processed in reverse (i.e., r implies the *concept* rather than the act of reversal). In either case, Eqs. 9.87 through 9.90 give the following result:

$$\begin{aligned}\tilde{G}(z) &= \{[\tilde{F}(z)\tilde{H}(z)]^r \tilde{H}(z)\}^r\\ &= \{[\tilde{F}(z)\tilde{H}(z)]^r\}^r \tilde{H}^r(z)\\ &= \tilde{F}(z)\tilde{H}(z)\tilde{H}(z^{-1})\end{aligned} \tag{9.91}$$

Thus the overall transfer function is

$$\tilde{H}_T(z) = \tilde{H}(z)\tilde{H}(z^{-1}) \tag{9.92}$$

$$\begin{aligned}\overline{H}_T(j\omega) &= \overline{H}(j\omega)\overline{H}(-j\omega)\\ &= |\overline{H}(j\omega)|^2\end{aligned} \tag{9.93}$$

So even though $\tilde{H}(z)$ may produce a nonlinear phase shift, $\tilde{H}_T(z)$ has zero phase shift and has the squared amplitude response of $\tilde{H}(z)$. Linear but nonzero phase shift simply involves placing a z^{-n} block in either diagram in Fig. 9.20, of course.

Example 9.12

Consider the effect of the system described in Fig. 9.21 on the waveform given by $f(t) = 10 \sin 2\pi t + 2 \sin 20\pi t$. The transfer function $\tilde{H}(z)$ is actually a lowpass filter sec-

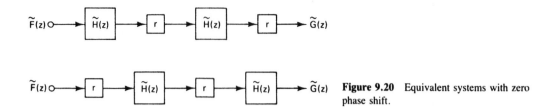

Figure 9.20 Equivalent systems with zero phase shift.

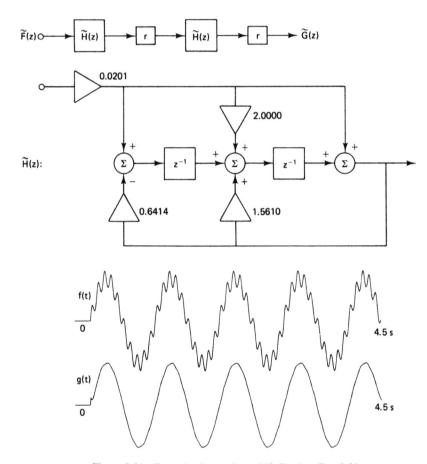

Figure 9.21 Example of zero-phase-shift filtering: $T = 0.01$ s.

tion from Chapter 12, and so the effect is to eliminate most of the 20% ripple on $f(t)$. The important feature here is the zero-phase-shift property exhibited in $g(t)$. The waveforms were constructed by drawing straight lines between sample points with $T = 0.01$ s.

9.9 FILTERING ROUTINES

In addition to the four routines described in Section 9.6 for converting between direct and lattice structures, we have included four filtering routines, SPCFIL, SPNFIL, SPLFIL, and SPNLFIL, for cascade, nonrecursive, lattice, and nonrecursive lattice filtering of a data sequence stored in a one-dimensional array. These routines do the filtering in place, replacing the original sequence with the filtered sequence. Initial conditions are set to zero internally.

The first routine, SPCFIL, will apply any linear filter in cascade form to the data. This routine is also applicable with a direct-form filter, which is just a cascade filter with one section. The second routine, SPNFIL, applies any nonrecursive filter in a similar manner.

The third routine, SPLFIL, will apply any recursive lattice in the form of Fig. 9.13 to a data sequence. The fourth routine, SPNLFIL, will apply any nonrecursive lattice in the form of Fig. 9.14 to a data sequence.

The calling sequences are as follows:

```
CALL SPCFIL(F,N,B,A,LI,NS,WORK,IERROR)
CALL SPNFIL(F,N,B,LI,WORK,IERROR)
CALL SPLFIL(F,N,KAPPA,NU,NS,WORK,IERROR)
CALL SPNLFIL(F,N,KAPPA,NS,WORK,IERROR)
```

```
F(0:N-1)        = Data sequence to be filtered and replaced
N               = Number of samples in data sequence
B(0:LI,NS)      = Numerator coefficients of cascade H(z)
A(1:LI,NS)      = Denominator coefficients of cascade H(z)
                  The transfer function of the nth cascade section is
```

$$H_n(z) = \frac{B(0,n) + B(1,n)z^{-1} + \cdots + B(LI,n)z^{-LI}}{1 + A(1,n)z^{-1} + \cdots + A(LI,n)z^{-LI}}$$

```
LI              = Last index in each cascade section
NS              = Number of cascade sections or lattice stages
KAPPA(0:NS-1)   = REAL lattice coefficients [κₙ] in Fig. 9.13 or 9.14
NU(0:NS)        = REAL lattice coefficients [νₙ] in Fig. 9.13
WORK            = Work array (need not be initialized), dimensioned
                  at least 2LI+2 for SPCFIL, LI+1 FOR SPNFIL, NS+1
                  for SPLFIL, or NS for SPNLFIL
IERROR          = 0; no errors
                  1; NS<1 or LI<1
                  2; Filter output exceeds 10^10 times maximum input.
```

All three filtering routines begin with zero initial conditions and work in the time domain without transforming the data. They are listed in Appendix B, and we note that all are simple and easy to modify for special applications. These routines should suffice for most engineering applications. Specific routines for lowpass, highpass, bandpass, and bandstop filtering are discussed in Chapter 12.

EXERCISES

The following filtering algorithms are used in the exercises below.

Filter A: $g_m = f_m + 0.2g_{m-1}$
Filter B: $g_m = 2f_m - 0.2g_{m-1}$
Filter C: $g_m = 2f_m + 0.7g_{m-1} - 0.1g_{m-2}$
Filter D: $g_m = f_m + 0.2g_{m-1} - 0.05g_{m-2}$
Filter E: $g_m = f_m + f_{m-1} + 0.2g_{m-1} - 0.05g_{m-2}$

1. Show the general form of the linear filtering algorithm when the output $g(t)$ is a function only of the past values of $f(t)$ as well as its own past values.
2. Find $\tilde{H}(z)$ for filters A and B.
3. Find $\tilde{H}(z)$ for filters C and D.
4. Find $\tilde{H}(z)$ for filter E.
5. What is the transfer function $\overline{H}(j\omega)$ of filter A when the sampling interval is 0.31416 s?
6. What is the effect on the transfer function if f_m is replaced by f_{m-2} in the algorithm of filter C?
7. Sketch the amplitude response of filter A.
8. Sketch the first period of the amplitude response of filter B.
9. Sketch the first period of the amplitude response of filter C.
10. What is the impulse response of filter B?
11. Give the impulse response of filter E.
12. Give the response of filter C to samples of the unit step function.
13. Plot the envelope of the step-function response of filter A.
14. Plot the envelope of the response of filter A to samples of the input shown.

15. Plot the envelope of the response of filter B to samples of the input shown.

16. Derive a general expression for the response of a filter described by

$$g_m + ag_{m-1} = f_m + bf_{m-1}$$

to the unit step function, u_m, in Eq. 9.28.

17. (Computer) Plot the responses of filters C and D to samples of the transient shown, with a sampling rate of 20,000 samples per second.

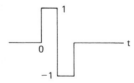

18. Draw filter C in the form of diagram 1.
19. Draw filter D in the form of diagram 2.
20. Give an algorithm similar to Eq. 9.38 for filter E.
21. Give an algorithm similar to Eq. 9.39 for filter E.

22. The transfer functions $\tilde{H}_1(z) = \tilde{B}_1(z)/\tilde{A}_1(z)$ and $\tilde{H}_2(z) = \tilde{B}_2(z)/\tilde{A}_2(z)$ are connected in a feedback loop as in Fig. 9.11. What must be true of the $\tilde{A}$'s and $\tilde{B}$'s in order for the overall transfer function to be stable?

23. Given $\tilde{H}(z) = (2 + 3z^{-1})/(1 - 0.8z^{-1})$:
    (a) Convert $\tilde{H}(z)$ to a lattice structure using Eqs. 9.60 through 9.65 (no computer).
    (b) Convert the lattice in part (a) back to direct form using Eqs. 9.53 through 9.55.
    (c) Draw a diagram of both forms.

24. Do Exercise 23 using filter C.

25. Do Exercise 23 using filter E.

26. Use the SPDTOL routine to convert filter E to a lattice structure. Draw a diagram of the lattice.

27. Use the SPLTOD routine to convert the result of Exercise 26 to direct form.

28. Use the SPLTOD routine to find the coefficients b_0 through b_N and a_1 through a_N for the direct form of a lattice with all κ values equal to 0.5 and all ν values equal to 1.0, for $N = 1$, 2, 3, and 4.

29. Do Exercise 28, this time with all κ values equal to 0.5, $\nu_0 = 1$, and ν_1 through ν_N equal to 0.

30. Use the SPNLTD routine to convert the nonrecursive lattice with all κ values equal to 0.5 to a direct form, and list the direct-form coefficients $b_0, \ldots, b_N$, for $N = 1, 2, 3$, and 4. Compare the results to those of Exercise 29 and explain.

31. Use the SPNDTL routine to find the lattice coefficients of $\tilde{H}(z) = 1 - 0.6z^{-1} + 0.25z^{-2}$. Draw a diagram of the lattice.

32. Use the SPNDTL routine to find the lattice equivalent of $\tilde{H}(z) = 0.2 - 0.4z^{-1} + 0.5z^{-2} - 0.4z^{-3} + 0.2z^{-4}$. Draw a diagram of the lattice.

33. Use the SPDTDL routine to convert the filter $\tilde{H}(z)$ in Fig. 9.20 to a lattice. Draw a diagram of the lattice.

34. Use the SPDTOL routine to convert

$$\tilde{H}(z) = \frac{z + 2 + z^{-1}}{1 - 0.6z^{-1} + 0.25z^{-2}}$$

to a lattice. Draw a diagram of the lattice.

35. Plot the pole and zero of filter B. Show the response vectors for 100 kHz when the sampling interval is 1 μs.

36. Plot the poles and zeros of filter E and show the response vectors as in Exercise 35.

37. Explain how any pole–zero plot represents not one but a set of digital filters. What parameter determines each element in the set? *Hint:* See Eq. 9.74.

38. Derive a general expression for the response of any digital filter at $\omega = 0$ in terms of the filter coefficients. *Hint:* See property 1 in Section 9.3.

39. What is the algorithm for a filter having a gain of 64 at $\omega = 0$ and poles and zeros as in Fig. 9.16?

40. Sketch diagrams for both cascade and parallel realizations of the following: $\tilde{H}(z) = (z^2 - 2z - 3)/(z^3 - z^2 + 1.04z - 0.32)$.

41. Plot the overall amplitude response in Fig. 9.17.

42. Compare (a) the impulse response of filter B with (b) the impulse response of filter B in a zero-phase-shift configuration.

43. Generate the following sequence:

$$f_m = \text{mod}[(m + 1)^3, 20] - 10; \quad m = 0, 1, \ldots, 9$$

Filter the sequence using SPCFIL with NS = 2 and the filter $\tilde{H}(z)$ given by

$$\tilde{H}(z) = \left(\frac{0.2 + 0.3z^{-1}}{1 - 0.5z^{-1}}\right)\left(\frac{0.1z^{-1} - 0.3z^{-2}}{1 - 0.4z^{-1} - 0.3z^{-2} + 0.5z^{-3}}\right)$$

List the input and output sequences.

44. Generate the sequence $[f_m]$ in Exercise 43. Using SPCFIL, filter $[f_m]$ with

$$\tilde{H}(z) = \frac{0.2 - 0.3z^{-1}}{1 - 0.4z^{-1} + 0.8z^{-2}}$$

List the input and output sequences.

45. Convert $\tilde{H}(z)$ in Exercise 44 to a recursive lattice and filter the sequence $[f_m]$ using SPLFIL. List the input and output sequences, and check your answer against the answer to Exercise 44.

46. Generate the sequence $[f_m]$ in Exercise 43. Using SPNFIL, filter $[f_m]$ using

$$\tilde{H}(z) = 1 + 0.5z^{-1} - 0.4z^{-2} + 0.3z^{-3}$$

List the input and output sequences.

47. Convert $\tilde{H}(z)$ in Exercise 46 to a nonrecursive lattice using SPNDTL and filter the sequence $[f_m]$ using SPLFIL. List the input and output sequences, and check your answer against the answer to Exercise 46.

48. Show that the recursive lattice in Fig. 9.13 with $\nu_N = 1$ and $\nu_N = 0$ for $n = N, 1, \ldots, 0$ is an "all-pass" filter with gain equal to 1 at all frequencies. *Hint:* Equations 9.52 and 9.53 will be useful. How might such a filter be used?

49. Given

$$\tilde{H}(z) = \frac{3 + 2z^{-1} + z^{-2}}{1 - 0.6z^{-1} + 0.8z^{-2}}$$

Derive asymmetric lattice diagrams of $\tilde{H}(z)$ using each of the two lattice formulas in Table 9.2. Check your answer against Fig. 9.16.

50. Develop explicit formulas to convert a nonrecursive filter given by $\tilde{H}(z) = 1 + b_1z^{-1} + b_2z^{-2}$ to the nonrecursive lattice in Fig. 9.14 with $N = 2$. Which filters cannot be converted?

51. Do the reverse of Exercise 50.

52. Develop explicit formulas to convert a recursive filter given by $\tilde{H}(z) = (b_0 + b_1z^{-1} + b_2z^{-2})/(1 + a_1z^{-1} + a_2z^{-2})$ to the following lattices with $N = 2$:
    (a) Fig. 9.13
    (b) Table 9.2 (left) after scaling by $1/b_2$
    (c) Table 9.2 (right)

53. Do the reverse of Exercise 52.

SOME ANSWERS

2. $\tilde{H}_A(z) = z/(z - 0.2)$, $\tilde{H}_B(z) = 2z/(z + 0.2)$ 4. $\tilde{H}_E(z) = (z^2 + z)/(z^2 - 0.2z + 0.05)$
5. $\overline{H}(j\omega) = (1 - 0.2e^{-0.31416j\omega})^{-1}$ 6. New $\overline{H}(j\omega) = e^{-2j\omega T} \times$ old $\overline{H}(j\omega)$
11. $g_m = (5(0.224)^m[\sin 1.11m + 0.224 \sin 1.11(m + 1)]$, $m \geq 0$
12. $g_m = 5 + (1/3)[0.2^m - 10(0.5)^m]$
14. $g_0 = 0$, $g_1 = (1/3)$, $g_2 = (2.2/3)$, $g_m = (1/3)[0.2^{m-1} + 2(0.2)^{m-2} + 3(0.2)^{m-3}]$, $m > 2$
16. $g_m = [1 + b + (a - b)(-a)^m]/(a + 1)$
20. $g_m = f_m + x_{m-1}^1$, $x_m^1 = f_m + x_{m-1}^2 + 0.2g_m$, $x_m^2 = -0.05g_m$
22. Zeros of $A_1A_2 + B_1B_2$ must be < 1.0 in magnitude.
23. $[\nu_n] = [4.4, 3.0]$; $\kappa_0 = -0.8$

24. $[\nu_n] = [2.0, 0.0, 0.0]; \quad [\kappa_n] = [0.778, -0.100]$

25. $[\nu_n] = [0.789, 1.0, 0.0]; \quad [\kappa_n] = [0.211, -0.050]$ **26.** See Exercise 25.

27. $[b_n] = [1.0, 1.0, 0.0]; \quad [a_n] = [0.20, -0.05]$

28. $N = 1: [b_n] = [1.5, 1.0]; \quad a_1 = 0.5$

     $N = 2: [b_n] = [2.00, 1.75, 1.00]; \quad [a_n] = [0.75, 0.50]$

     $N = 3: [b_n] = [2.500, 2.625, 2.000, 1.000]$

          $[a_n] = [1.000, 0.875, 0.500]$

29. $[b_n] = [1.0, 0, 0, \ldots]$ in all cases

     $N = 1: a_1 = 0.5$

     $N = 2: [a_n] = [0.75, 0.50]$

     $N = 3: [a_n] = [1.000, 0.875, 0.500]$

30. $N = 1: [b_n] = [1.0, 0.5]$

     $N = 2: [b_n] = [1.00, 0.75, 0.50]$

     $N = 3: [b_n] = [1.000, 1.000, 0.875, 0.500]$

31. (Fig. 9.14) $[\kappa_n] = [-0.48, 0.25]$

32. (Fig. 9.14 with input subtracted as in Fig. 9.15)

     $[\kappa_n] = [-1.074, -12.813, 0.934, -0.458, 0.200]$

33. $[\nu_n] = [0.075, 0.072, 0.020]$

     $[\kappa_n] = [-0.951, 0.641]$

43.

Input:	−9.00	−2.00	−3.00	−6.00	−5.00	6.00	−7.00	2.00	−1.00	−10.00
Output:	0.00	−0.18	0.07	0.85	1.04	1.28	1.53	1.17	0.35	0.30

44.

Input:	−9.00	−2.00	−3.00	−6.00	−5.00	6.00	−7.00	2.00	−1.00	−10.00
Output:	−1.80	1.58	2.07	−0.74	−1.15	2.83	−1.15	−0.22	0.03	−1.51

45.

Input:	−9.00	−2.00	−3.00	−6.00	−5.00	6.00	−7.00	2.00	−1.00	−10.00
Output:	−1.80	1.58	2.07	−0.74	−1.15	2.83	−1.15	−0.22	0.03	−1.51

46.

Input:	−9.00	−2.00	−3.00	−6.00	−5.00	6.00	−7.00	2.00	−1.00	−10.00
Output:	−9.00	−6.50	−0.40	−9.40	−7.40	5.00	−3.80	−5.40	4.60	−13.40

47.

Input:	−9.00	−2.00	−3.00	−6.00	−5.00	6.00	−7.00	2.00	−1.00	−10.00
Output:	−9.00	−6.50	−0.40	−9.40	−7.40	5.00	−3.80	−5.40	4.60	−13.40

50. Nonrecursive direct to lattice with $N = 2$:

$$\kappa_0 = b_1/(1 + b_2); \qquad \kappa_1 = b_2$$

     Eq. 9.64 fails if $b_2 = \pm 1$.

51. Nonrecursive lattice to direct with $N = 2$:

$$b_1 = \kappa_0(1 + \kappa_1); \qquad b_2 = \kappa_1$$

52. and **53.**

Fig. 9.13

Direct to lattice	$\kappa_0 = \dfrac{a_1}{1 + a_2}; \quad \kappa_1 = a_2$
	$\nu_0 = b_0 - b_2 a_2 + \dfrac{a_1(b_2 a_1 - b_1)}{1 + a_2}$
	$\nu_1 = b_1 - b_2 a_1; \quad \nu_2 = b_2$
Lattice to direct	$a_1 = \kappa_0(1 + \kappa_1); \quad a_2 = \kappa_1$
	$b_0 = \nu_0 + \kappa_0 \nu_1 + \kappa_1 \nu_2$
	$b_1 = \nu_1 + \kappa_0 \nu_2 (1 + \kappa_1)$
	$b_2 = \nu_2$

Table 9.2 (left)

Direct to lattice	$\kappa_0 = \dfrac{b_1' - b_0'a_1}{1 - b_0'a_2}; \quad b_0' = \dfrac{b_0}{b_2}$ $\kappa_1 = b_0'; \qquad\qquad b_1' = \dfrac{b_1}{b_2}$ $\lambda_0 = \dfrac{a_1 - a_2b_1'}{1 - b_0'a_2}$ $\lambda_1 = a_2$
Lattice to direct	$a_1 = \lambda_0 + \kappa_0\lambda_1; \quad a_2 = \lambda_1$ $b_0 = \kappa_1; \qquad\quad b_1 = \kappa_0 + \kappa_1\lambda_0; \quad b_2 = 1$

Table 9.2 (right)

Direct to lattice	$\kappa_0 = \dfrac{a_2(b_1 - b_0a_1)}{b_2 - b_0a_2}; \quad \kappa_1 = b_0$ $\lambda_0 = \dfrac{a_2(b_2 - b_0a_2)}{a_1b_2 - b_1a_2}; \quad \lambda_1 = \dfrac{b_2}{a_2}$ $\gamma = \dfrac{a_1b_2 - b_1a_2}{b_2 - b_0a_2}$
Lattice to direct	$a_1 = \gamma + \kappa_0; \quad a_2 = \gamma\lambda_0$ $b_0 = \kappa_1; \qquad\quad b_1 = \gamma\kappa_1 + \kappa_0\lambda_1$ $b_2 = \gamma\lambda_0\lambda_1$

REFERENCES

AHMED, N., and NATARAJAN, T., *Discrete-Time Signals and Systems,* Chap. 5. Reston, Va.: Reston, 1983.

GOLD, B., and RADER, C. M., *Digital Processing of Signals.* New York: McGraw-Hill, 1969.

GRAY, A. H., JR., and MARKEL, J. D., Digital Lattice and Ladder Filter Syntheses. *IEEE Trans. Audio Electroacoust.,* Vol. AU-21, December 1973, p. 491.

JURY, E. J., A Note on the Reciprocal Zeros of a Real Polynomial with Respect to the Unit Circle. *IEEE Trans. Commun. Technol.,* Vol. CT-11, June 1964, p. 292.

KAISER, J. F., Digital Filters, Chap. 7 in *System Analysis by Digital Computer,* ed. J. F. Kaiser and F. F. Kuo. New York: Wiley, 1966.

MITRA, S. K., KAMAT, P. S., and HUEY, D. C., Cascaded Lattice Realization of Digital Filters. *IEEE Trans. Circuit Theory Appl.,* Vol. 5, 1977, p. 3.

OPPENHEIM, A. V., and SCHAFER, R. W., *Digital Signal Processing,* Chaps. 1 and 4. Englewood Cliffs, N.J.: Prentice-Hall, 1975.

RABINER, L. R., and GOLD, B., *Theory and Application of Digital Signal Processing,* Chap. 2. Englewood Cliffs, N.J.: Prentice-Hall, 1975.

WAIT, J. V., Digital Filters, Chap. 5 in *Active Filters: Lumped, Distributed, Integrated, Digital, and Parametric,* ed. L. P. Huelsman. New York: McGraw-Hill, 1970.

10

Digital and Continuous Systems

10.1 INTRODUCTION

This chapter is a continuation of the digital filtering discussion in the last two chapters, with emphasis on comparing the performance of analogous digital and continuous systems. The linear digital system is viewed as an approximation to a continuous system with transfer function $H(s)$. Besides being an end in itself, this view lends further insight into the behavior of digital systems.

A simple linear approximation scheme, called the *zero-order approximation,* is introduced in this chapter. It is a generalization of procedure 2 in Section 8.8.

To begin the discussion here, suppose the digital filter is to be designed to approximate the linear system in Fig. 10.1 for an arbitrary input $f(t)$, that is, for any finite-valued input function whose sample set $[f_m]$ is provided in sequence. In a typical digital simulation of, say, a production process, a guidance system, or a vehicle, there are many linear subsystems like the one in Fig. 10.1, along with nonlinear subsystems, and many different input and output variables that cannot be predetermined. The digital filtering concepts generally are applicable when the linear subsystems can be described in terms of transfer functions, or when one is interested in spectral characteristics of the output variables.

The digital approximation or simulation of the system in Fig. 10.1 will be discussed for the case in which $H(s)$, expressed as a ratio of polynomials in s, is a *proper fraction,* that is, the degree of the numerator is less than that of the denominator. In this case the inverse transform $h(t)$ is defined, and the approximation scheme can include the use of $h(t)$. The following sections describe first the zero-order approximation to $H(s)$, and then the related zero-order approximation to the convolution integral.

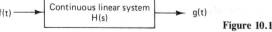

$$f(t) \longrightarrow \boxed{\begin{array}{c} \text{Continuous linear system} \\ H(s) \end{array}} \longrightarrow g(t)$$

Figure 10.1 System to be approximated.

219

10.2 ZERO-ORDER OR IMPULSE-INVARIANT APPROXIMATION

This type of approximation has been described by Kaiser (1966) and by Gold and Rader (1969). Let $H(j\omega)$ be the desired transfer function expressed as a Fourier transform. Since it is already known that the transfer function of a digital filter is a DFT, and since the DFT $\overline{H}(j\omega)$ is known to be the zero-order approximation to $H(j\omega)/T$ (as in Eq. 5.21), it is clearly possible to obtain a zero-order approximation to the system having the transfer function $H(j\omega)$ simply by using the digital filter having the transfer function $T\overline{H}(j\omega)$. Then, just as

$$G(s) = F(s)H(s) \tag{10.1}$$

for the continuous system, Eq. 9.5 of Chapter 9 gives the relationship for the digital simulation:

$$\overline{G}_0(j\omega) = \overline{F}(j\omega)\overline{H}_0(j\omega) \tag{10.2}$$

with the subscript 0 being used to indicate that $\overline{H}_0(j\omega)$ is a zero-order approximation to $H(j\omega)$. The digital output sample set is, of course, found from the inverse DFT of $\overline{G}_0(j\omega)$, and will be denoted $[g_{0m}]$, again to indicate the zero-order approximation. The continuous envelope $g_0(t)$ is, as before, limited to frequencies less than π/T, and if $F(j\omega)$ and $H(j\omega)$ are zero for $\omega \geq \pi/T$ then $g_0(t) = g(t)$ and the approximation is exact. [This is rarely the case in practical simulations, however—$f(t)$ is likely to contain discontinuities, etc.]

Summarizing and somewhat restating the above, the following steps would be taken to obtain a zero-order approximation to a proper, rational, continuous transfer function:

1. Begin with the desired transfer function, $H(s)$.
2. Determine $h(t)$ from a Laplace transform table.
3. Find $\tilde{H}_0(z) = \tilde{B}_0(z)/\tilde{A}_0(z)$, the z-transform of $Th(t)$, from a z-transform table, in order to make $\overline{H}_0(j\omega) = T\overline{H}(j\omega)$. [Alternatively, Appendix A eliminates step 2 by giving $\tilde{H}_0(z)/T$ directly in terms of $H(s)$.]
4. The digital filtering formula is then

$$g_{0m} = \sum_{n=0}^{N} b_{0n} f_{m-n} - \sum_{n=1}^{N} a_{0n} g_{0,\,m-n} \tag{10.3}$$

where the a_{0n}'s and b_{0n}'s are the coefficients of $\tilde{A}_0(z)$ and $\tilde{B}_0(z)$.

As mentioned before, the second step implies a finite $h(t)$, which in turn implies that $H(s)$ is proper. The initial value theorem (Truxal, 1955) states that

$$\lim_{t \to 0} h(t) = \lim_{s \to \infty} sH(s) \tag{10.4}$$

so that $h(0)$ would be infinite if $H(s)$ were not a proper ratio of polynomials. If $H(s)$ is not proper, a partial-fraction expansion can be used.

Clearly, because of step 3, the response of $\tilde{H}_0(z)$ to the impulse $1/T$ must consist of the sample set of $h(t)$. Thus the zero-order approximation is also *impulse-invariant* and is a generalization of this same method described in Section 8.8 for nonrecursive systems.

Example 10.1

Derive the zero-order algorithm to approximate $H(s) = 1/(s + 1)$. The following equations are the result of steps 1–4 above:

$$H(s) = \frac{1}{s + 1} \tag{10.5}$$

$$h(t) = e^{-t} \tag{10.6}$$

$$\tilde{H}_0(z) = \frac{\tilde{B}_0(z)}{\tilde{A}_0(z)} = \frac{T}{1 - e^{-T}z^{-1}} \tag{10.7}$$

$$g_{0m} = Tf_m + e^{-T}g_{0,m-1} \tag{10.8}$$

with Eq. 10.8 giving the desired result. The z-transform of e^{-t} is on line 151 in Appendix A. As with any rational function of $j\omega$, $H(j\omega)$ does not vanish beyond any finite frequency. Therefore, as suggested by Fig. 10.2, aliasing occurs in the DFT of $h(t)$. Figure 10.2 illustrates the case when $\pi/T = 10$ rad/s, with the solid line being a plot of $|H(j\omega)|$ and the dashed line a plot of $|\overline{H}_0(j\omega)|$ found by taking the magnitude of Eq. 10.7 with $z = e^{j\omega T}$.

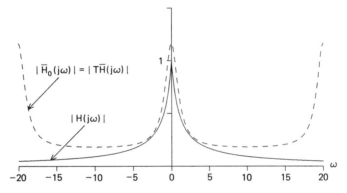

Figure 10.2 Continuous (solid) and impulse-invariant digital (dashed) amplitude responses, $H(s) = 1/(s + 1)$ and $T = \pi/10$.

Once the filtering formula has been derived as in Eq. 10.8 of Example 10.1, the digital response to an impulse or step function can be examined in closed form as in Chapter 9.

Example 10.2

Compare the impulse responses of the analog and digital filters in Example 10.1. The analog impulse response is given in Eq. 10.6, that is,

$$h(t) = \text{inverse transform of } \frac{1}{s + 1} \tag{10.9}$$

$$= e^{-t}$$

The digital impulse response (to the impulse $1/T$ at $m = 0$) can be found as in Section 9.4 of Chapter 9, that is,

$$g_{0m} = \text{inverse transform of } \frac{\tilde{H}_0(z)}{T} = \frac{z}{z - e^{-T}} \tag{10.10}$$

$$= e^{-mT}$$

and therefore *the impulse responses are identical at the sampling points.*

Example 10.3

Compare the step-function responses of the analog and digital filters in Example 10.1. The Laplace transform of the unit step function $u(t)$ is $1/s$, so the analog response in this case is

$$G(s) = F(s)H(s) = \frac{1}{s(s+1)} \tag{10.11}$$

Therefore,

$$g(t) = 1 - e^{-t} \tag{10.12}$$

Similarly, the digital response, from lines H and 155 in Appendix A, is

$$\tilde{G}_0(z) = \tilde{F}(z)\tilde{H}_0(z) = \frac{Tz^2}{(z-1)(z-e^{-T})} \tag{10.13}$$

Therefore,

$$g_{0m} = T\frac{1 - e^{-(m+1)T}}{1 - e^{-T}} \tag{10.14}$$

as in Example 9.4. In Fig. 10.3, the continuous and digital responses are compared for two values of T. The zero-order approximation, $[g_{0m}]$, of course approaches $g(t)$ as T diminishes. Also, it appears that most of the error could be eliminated simply by reducing the dc gain of $\overline{H}_0(j\omega)$. The reason for this observation is discussed in Chapter 11.

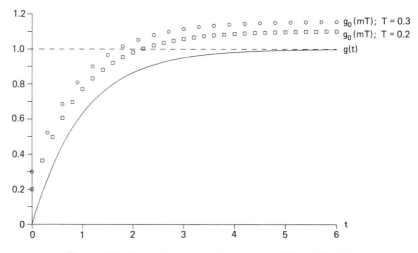

Figure 10.3 Zero-order step function response; $H(s) = 1/(s+1)$.

10.3 CONVOLUTION

In Chapter 7 the z-transform of the convolution of two sample sets was shown to be the product of two individual transforms. In this section the zero-order response, g_{0m} in Eq. 10.3, is analyzed as a zero-order approximation to the convolution integral:

$$g(t) = \int_0^t h(\tau)f(t-\tau)\,d\tau \tag{10.15}$$

To begin, consider the product of any pair of z-transforms, $\tilde{F}(z)$ and $\tilde{H}(z)$:

$$\tilde{G}(z) = \tilde{F}(z)\tilde{H}(z)$$

$$= \left(\sum_{m=0}^{\infty} f_m z^{-m} \right) \left(\sum_{n=0}^{\infty} h_n z^{-n} \right) \tag{10.16}$$

$$= \sum_{m=0}^{\infty} \sum_{n=0}^{\infty} f_m h_n z^{-(m+n)}$$

with the summations beginning at $m = 0$ and $n = 0$ because f_m and h_m are assumed to be zero for $m < 0$. The latter formulation of $\tilde{G}(z)$ has terms as follows:

$$\tilde{G}(z) = (f_0 h_0) z^0 + (f_0 h_1 + f_1 h_0) z^{-1} + (f_0 h_2 + f_1 h_1 + f_2 h_0) z^{-2} + \cdots \tag{10.17}$$

Reasoning inductively from Eq. 10.17, the coefficient of z^{-m} must be

$$f_0 h_m + f_1 h_{m-1} + \cdots + f_m h_0$$

but this coefficient is the inverse transform of $\tilde{G}(z)$, that is,

$$g_m = \sum_{n=0}^{m} f_n h_{m-n} = \sum_{n=0}^{m} h_n f_{m-n} \tag{10.18}$$

with the form on the right being obvious from the symmetry of Eq. 10.16. Thus, from Eqs. 10.16 and 10.18, the general convolution relationship for z-transforms is

$$\mathscr{Z} \left\{ \sum_{n=0}^{m} h_n f_{m-n} \right\} = \tilde{H}(z)\tilde{F}(z) \tag{10.19}$$

That is, the z-transform of a convolution is a product of z-transforms. This same result is seen in Section 7.3, property 5. In the zero-order approximation, $\tilde{H}_0(z)$ is the z-transform of $Th(t)$, so Eq. 10.19 gives

$$\tilde{G}_0(z) = \tilde{H}_0(z)\tilde{F}(z)$$

and

$$g_{0m} = T \sum_{n=0}^{m} h_n f_{m-n} \tag{10.20}$$

as a transform pair. Therefore, as stated above, the use of $\tilde{H}_0(z)$ produces a *zero-order approximation to the convolution integral* in Eq. 10.15.

This result of course applies to nonrecursive as well as recursive systems. In Chapter 8, the impulse response for nonrecursive systems was shown to be $[b_m]$, where $[b_m]$ represents the set of coefficients in

$$g_m = \sum_{n=0}^{N} b_n f_{m-n} \tag{10.21}$$

Since b_n/T is h_n in the impulse-invariant case, Eqs. 10.20 and 10.21 are seen to be equivalent.

Example 10.4

Show the zero-order convolution for the step-function response of $H(s) = 1/(s + 1)$. For this example $f(t)$ is one for $t \geq 0$ and $h(t)$ is e^{-t} as in the preceding examples. Therefore,

$$g_{0m} = T \sum_{n=0}^{m} h_n f_{m-n}$$

(10.22)

$$= T \sum_{n=0}^{m} e^{-nT}(1)$$

$$= T \frac{1 - e^{-(m+1)T}}{1 - e^{-T}}$$

(10.23)

just as in Eq. 10.14 of Example 10.3.

Figure 10.4, which is the basis for the following discussion, illustrates portions of the convolution at $t = 4T$, using the functions in Example 10.4. The continuous convolution is developed in the first three plots in the figure, so that the area under the lower left curve represents the integral in Eq. 10.15, that is, the "desired" area to be approximated by a digital computation. The zero-order approximation, Eq. 10.20, is illustrated at the lower right. In this latter illustration the cause of the error in Fig. 10.3 can be seen — the area of the discrete integrand extends both above and beyond that of the continuous integrand.

Thus the convolution analysis in Fig. 10.4 shows that the zero-order or impulse-invariant approximation leaves something to be desired in the way of accuracy. The question of how to achieve greater accuracy is left to the next chapter, which is on

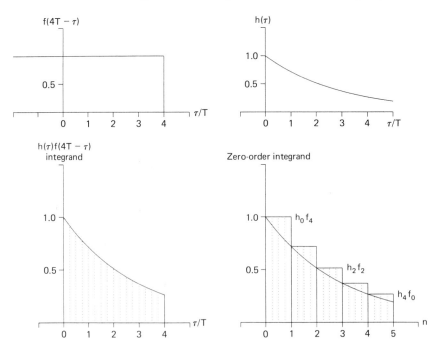

Figure 10.4 Convolution integrands for Example 10.4 at $t = 4T$.

Digital and Continuous Systems   Chap. 10

the subject of digital simulation. For now, the zero-order approximation offers at least the argument that there is a direct comparison between digital and continuous systems, and that digital systems can be designed with frequency response functions and convolutions similar to those of continuous systems.

10.4 FINAL VALUE THEOREMS

In the analysis of continuous linear systems, if the final value of a system response to a given input exists, it can be found by letting t approach infinity in the expression for the output $g(t)$. It can often be found more conveniently from the transform, $G(s)$, via the following argument from Truxal (1955), Chapter 1. If $g(t)$ reaches a constant final value, say K, then $g(t)$ can be expressed as the sum of a step function of amplitude K plus some other decaying function, that is,

$$g(t) = Ku(t) + g_1(t)e^{-at} \tag{10.24}$$

where $g_1(t)$ is finite-valued and $a > 0$. Using this representation of $g(t)$, the product of s times $G(s)$ becomes, from Appendix A, lines G and 101,

$$sG(s) = K + sG_1(s + a) \tag{10.25}$$

Thus, if $t \to \infty$ in Eq. 10.24 or if $s \to 0$ in Eq. 10.25 the result is K, the final value of $g(t)$. Therefore,

$$\boxed{\lim_{t \to \infty} g(t) = \lim_{s \to 0} sG(s)} \tag{10.26}$$

is the final value theorem for continuous systems.

Naturally there is a similar theorem for digital systems. Again, suppose g_m is always equal to K for m greater than some integer. Then, as in Eq. 10.24, one could write

$$g_m = K + g_{1m}e^{-maT} \tag{10.27}$$

with the same constraints on g_1 and a. Taking this time the product $(z - 1)\tilde{G}(z)$, the result is, from Appendix A, lines G and 101,

$$(z - 1)\tilde{G}(z) = zK + (z - 1)\tilde{G}_1(ze^{aT}) \tag{10.28}$$

Again, the final value K is obtained either by letting $m \to \infty$ in Eq. 10.27 or by letting $z \to 1$ in Eq. 10.28. Therefore,

$$\boxed{\lim_{m \to \infty} g_m = \lim_{z \to 1}(z - 1)\tilde{G}(z)} \tag{10.29}$$

is the final value theorem for digital systems.

Example 10.5

Determine the final value of the step-function response of $H(s) = 1/(s + 1)$. Equation 10.26 gives the result immediately:

$$\lim_{t \to \infty} g(t) = \lim_{s \to 0} sF(s)H(s)$$

$$= \lim_{s \to 0} s\left(\frac{1}{s} \cdot \frac{1}{s + 1}\right) \tag{10.30}$$

$$= 1$$

Example 10.6

Determine the final value of the step-function response of the zero-order approximation to $H(s) = 1/(s + 1)$. As in Eq. 10.7, the z-transfer function is

$$\tilde{H}_0(z) = \frac{Tz}{z - e^{-T}} \tag{10.31}$$

Hence Eq. 10.29 gives the final value:

$$\lim_{m \to \infty} g_m = \lim_{z \to 1}(z - 1)\tilde{F}(z)\tilde{H}_0(z)$$

$$= \lim_{z \to 1}(z - 1)\left(\frac{z}{z - 1}\frac{Tz}{z - e^{-T}}\right) \tag{10.32}$$

$$= \frac{T}{1 - e^{-T}}$$

Note that this final value approaches one as T approaches zero, as in Fig. 10.3, and could also have been obtained by letting $m \to \infty$ in Eq. 10.14.

10.5 POLE–ZERO COMPARISONS

Returning to the general case of Fig. 10.1, that is, the discrete approximation of any linear continuous system, it is often very useful to compare the poles and zeros of $H(s)$ with those of the approximation. In Chapter 9, characteristics of the poles and zeros of the z-transfer function, $\tilde{H}(z)$, were discussed. Frequency-domain characteristics were derived using the relationship

$$\overline{H}(j\omega) = \tilde{H}(e^{j\omega T}) \tag{10.33}$$

When s is substituted for $j\omega$, Eq. 10.33 becomes

$$\overline{H}(s) = \tilde{H}(e^{sT}); \qquad s = \sigma + j\omega \tag{10.34}$$

(Although $\overline{H}(s)$ is not a DFT in the strict sense, the bar will be retained as a reminder that $\overline{H}(s)$ is a periodic function of ω.) Thus, $\overline{H}(s)$ is the "continuous transfer function" of the digital filter, and can be compared with $H(s)$.

According to Eq. 10.34, the poles and zeros of $\tilde{H}(z)$ can be mapped from the z-plane into the s-plane under the transformation (Jury, 1964)

$$z = e^{sT} \tag{10.35}$$

This transformation, illustrated in Figs. 10.5 and 10.6, embodies some of the fundamental relationships between digital filters and other sampled-data systems, and should be examined carefully. Suppose a point on the z-plane is at polar coordinates (r, θ). Then, from Eq. 10.35, the corresponding point on the s-plane is at

$$s = \frac{1}{T} \log z = \frac{1}{T} \log r e^{j(\theta \pm 2n\pi)}; \qquad n = 0, 1, 2, \ldots \tag{10.36}$$

$$= \frac{1}{T} \log r + \frac{j}{T}(\theta \pm 2n\pi); \qquad n = 0, 1, 2, \ldots \tag{10.37}$$

That is, a single point on the z-plane transforms into an *infinite number of points* on the s-plane. In other words, letting $n = 0$, the infinite strip from $s = j\pi/T$ to

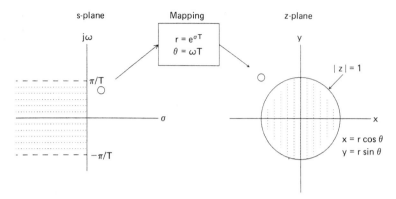

Figure 10.5 Mapping from s-plane to z-plane.

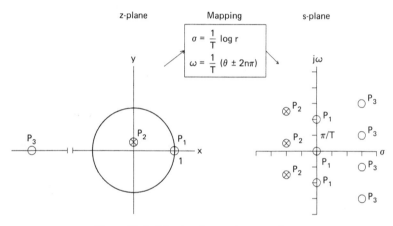

Figure 10.6 Mapping from z-plane to s-plane.

$s = -j\pi/T$ on the s-plane transforms under Eq. 10.35 into the entire z-plane. Note that the left and right halves of the s-plane map, respectively, into the interior and exterior of the circle $|z| = 1$, as indicated by the shading in Fig. 10.5. Some illustrations of Eq. 10.37 are given in Fig. 10.6. Note how each of the points P_1, P_2, and P_3 on the z-plane map into *sets* of points on the s-plane under Eq. 10.37.

Again, the periodicity of $\tilde{H}(s)$ is illustrated by the mapping in Fig. 10.6. The poles and zeros of $\tilde{H}(z)$ on the z-plane always map into a repeating pattern of poles and zeros on the s-plane. In the following examples, poles and zeros are mapped and compared for first and second-order continuous systems.

Example 10.7

The continuous transfer function is $H(s) = 1/(s + 1)$, the sampling interval is $T = 0.3$, and the zero-order simulation with $\tilde{H}_0(z) = Tz/(z - e^{-T})$ is used as in Examples 10.1 and 10.2. The pole–zero plots, consisting essentially of a single pole, are shown in Fig. 10.7. Note that the single pole at $e^{-0.3} = 0.74$ on the z-plane maps exactly into the "desired" pole at -1 on the s-plane, but also into the repeating poles of $\overline{H}_0(s)$ outside of the "primary strip" from $-\pi/0.3$ to $+\pi/0.3$ on the s-plane. Note also that the zero at $z = 0$ maps to the "point" at $s = -\infty$. As already discussed, this zero does not affect the amplitude response, but only the phase shift of the digital approximation.

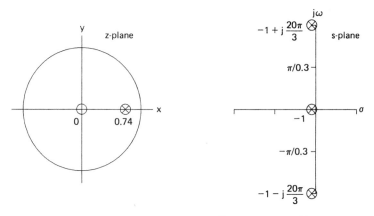

Figure 10.7 Pole–zero plots for $T\tilde{H}(z) = Tz/(z - e^{-T})$.

Example 10.8

Give a *root-locus plot* for the zero-order approximation to

$$H(s) = \frac{A(s + R)}{s^2 + Rs + 4} \tag{10.38}$$

The root-locus plot is, as its name implies, a plot of the loci of the roots (usually the poles) of the transfer functions as a coefficient changes. Such a plot is made in order to evaluate the effect of the coefficient (which is R in this case) on gain, oscillation, and so on.

As usual, the zero-order approximation to Eq. 10.38 is the z-transform of $Th(t)$, and line 201 of Appendix A thus gives $\tilde{H}_0(z)$ directly as

$$\tilde{H}_0(z) = T\left(\frac{A}{b - a}\right)\left[\frac{(R - a)z}{z - e^{-aT}} + \frac{(b - R)z}{z - e^{-bT}}\right] \tag{10.39}$$

$$a, b = \frac{1}{2}\left(R \pm \sqrt{R^2 - 16}\right)$$

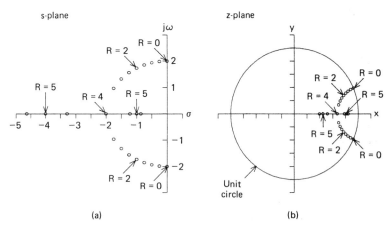

(a)   (b)

Figure 10.8 Root-locus plots for (a) $H(s) = A(s + R)/(s^2 + Rs + 4)$; (b) zero-order approximation to (a); $T = 0.2$.

Thus the poles of $H(s)$ on the s-plane and of $\tilde{H}_0(z)$ on the z-plane are as follows:

poles of $H(s)$
on the s-plane: $s_1, s_2 = \dfrac{1}{2}\left(-R \pm \sqrt{R^2 - 16}\right)$
(10.40)

poles of $\tilde{H}_0(z)$
on the z-plane: $z_1, z_2 = e^{-(RT/2)}e^{\pm(T/2)(\sqrt{R^2-16})}$
(10.41)

The root-locus plots for $T = 0.2$, made by varying R in Eqs. 10.40 and 10.41, are shown in Fig. 10.8. Note that the relative gain contributions of the digital and continuous poles vary similarly with R, and, in fact, that the root loci themselves are similar. Notice also that both systems become oscillatory in their responses to, say, an impulse input when R becomes less than 4, since at this point the poles of both systems leave the real axes.

10.6 CONCLUDING REMARKS

The emphasis in this chapter has been on comparing digital and continuous systems. Usually the starting point was the continuous transfer function, $H(s)$, as if a digital simulation of a continuous system were involved.

The reader has probably noticed that the zero-order digital transfer functions sometimes do not give very accurate approximations, particularly for larger values of T. Thus the zero-order approximation, although useful for conveying the fundamental relationships between analog and digital systems, does not appear to be accurate enough in most cases where digital simulation is the primary objective.

The next chapter emphasizes digital simulation and assumes that it is the main objective. Different schemes for deriving $\tilde{H}(z)$ are introduced. For simulation purposes, the new schemes will be seen to be improvements over the zero-order approximation, with greater accuracy being achieved for the same sampling interval T.

EXERCISES

1. Give the zero-order digital approximation to the circuit shown here. Provide a formula for g_{0m} as well as a filter diagram. *Hint:* See Chapter 3 for impedance functions.

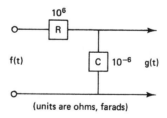

(units are ohms, farads)

2. Plot the response of the above circuit to the transient $f(t)$ shown here. Then, using $T = 0.25$ s, plot the zero-order digital response on the same graph.

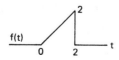

3. Give a formula for the zero-order digital approximation to the circuit shown here.

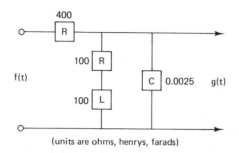

(units are ohms, henrys, farads)

4. Derive a zero-order algorithm and filter diagram if the desired transfer function is $H(s) = s/(s + a)^2$.

5. Show the zero-order convolution formula as applied to Exercise 2.

6. Give a formula in closed form for the zero-order response in Exercise 1 to a unit step function occurring *just after* $t = 0$.

7. Given the continuous system in Exercise 4 with $H(s) = s/(s + a)^2$, give a direct zero-order algorithm in the form of Eq. 9.38.

8. (Computer) Using $T = 0.25$, sketch and compare the amplitude responses $|H(j\omega)|$ and $|\bar{H}_0(j\omega)|$ in Exercise 3.

9. (Computer) Using $T = 0.25$, sketch and compare the continuous and zero-order digital responses for Exercise 3 to the ramp function used in Exercise 2.

10. Calculate the final value of the zero-order approximation in Exercise 1 when the input is a unit step function and $T = 0.25$.

11. Plot the digital and continuous poles and zeros for Exercise 1 on the s-plane using $T = 0.25$.

12. Using $T = 0.25$, derive a z-plane pole–zero plot for Exercise 3.

13. Make two root-locus plots as in Example 10.8 but using $H(s) = 1/(s^2 + Rs + 5)$.

14. Construct diagrams for the direct and parallel forms of the zero-order approximation to $H(s) = 2(s + 1)/(s^2 + 8s + 15)$.

15. What is the zero-order approximation to any $H(s)$ expanded into partial-fraction form with simple poles, that is,

$$H(s) = \sum_{n=1}^{N} \frac{A_n}{s + a_n}$$

if the terms are approximated one at a time?

SOME ANSWERS

1. $g_m = Tf_m + e^{-T}g_{m-1}$ **3.** $g_m = Tf_m - e^{-T}(\cos 2T)(Tf_{m-1} - 2g_{m-1}) - e^{-2T}g_{m-2}$

4. $g_m = T[f_m - (1 + aT)e^{-aT}f_{m-1}] + 2e^{-aT}g_{m-1} - e^{-2aT}g_{m-2}$

5. $g_m = T^2 e^{-mT} \sum_{n=0}^{\min(m,\,8)} n e^{nT}$

7. $g_m = Tf_m + x_{m-1}^1, \quad x_m^1 = -T(1 + aT)e^{-aT}f_m + x_{m-1}^2 + 2e^{-aT}g_m, \quad x_m^2 = -e^{-2aT}g_m$

10. 1.1302   **15.** $\tilde{H}_0(z) = \sum_{n=1}^{N} \frac{A_n Tz}{z - e^{-a_n T}}$

REFERENCES

AHMED, N., and NATARAJAN, T., *Discrete-Time Signals and Systems,* Chap. 6. Reston, Va.: Reston, 1983.

GOLD, B., and RADER, C. M., *Digital Processing of Signals,* Chap. 3. New York: McGraw-Hill, 1969.

JURY, E. I., *Theory and Application of the z-Transform Method.* New York: Wiley, 1964.

KAISER, J. F., Digital Filters, Chap. 7 in *System Analysis by Digital Computer,* ed. J. F. Kaiser and F. F. Kuo. New York: Wiley, 1966.

OPPENHEIM, A. V., and SCHAFER, R. W., *Digital Signal Processing,* Chap. 5. Englewood Cliffs, N.J.: Prentice-Hall, 1975.

STEIGLITZ, K., The Equivalence of Digital and Analog Signal Processing. *Inform. Control,* Vol. 8, 1965, p. 455.

TOU, J. T., *Digital and Sampled-Data Control Systems,* Chap. 5. New York: McGraw-Hill, 1959.

TRUXAL, J. G., *Control System Synthesis.* New York: McGraw-Hill, 1955.

11

Simulation

of Continuous Systems

11.1 INTRODUCTION

In Chapter 10 the emphasis was on comparing analog and digital systems, as if the digital processor were going to be used to approximate or simulate the continuous system. The subject is continued in this chapter, but here the emphasis is on the design and accuracy of different simulation methods. The discussion in this chapter is based on a deterministic approach to simulation in which the signals are essentially determined, that is, known in advance. A statistical approach, in which only signal statistics are known, is the subject of Chapter 14.

When the desired linear transfer function $H(s)$ is itself a rational function of s, as it must be for any realizable linear transfer function, the methods described here will provide simple and accurate digital simulations of $H(s)$. In general, by decreasing the sampling interval T, the error in the simulation can be made as small as desired in the interval $|\omega| < \pi/T$ without making the simulation unduly complex.

Digital simulation plays an important role in modern systems analysis. It is used to test complicated system designs on the computer, so that construction costs are limited to software until the design is completed. A few traditional examples are navigation, guidance, and control systems for aerospace vehicles; nuclear reactors, radar tracking, and communication systems. Simulation is also used to examine the response of existing systems to different inputs. Examples are the simulation of city traffic flow to study the impact of population increase, the simulation of ocean flow patterns to estimate future pollution dangers, simulations of living systems, etc. The list of references contains several texts on the general subject of digital simulation.

When the simulation involves inputs, outputs, and transfer functions, the signal-analysis or frequency-domain approach is applicable, and the designer of the digital simulation is concerned with programming an accurate approximation to an idealized "real world." The methods below apply particularly to this concern, but they are also useful in establishing a sort of general connection between linear continuous and digital systems.

11.2 CLASSIFICATION OF SIMULATION METHODS

A digital simulation of a continuous system, as defined here, involves constructing a transfer function $\tilde{H}(z)$ to represent each linear portion of the system given by an $H(s)$, plus accounting for any limits or other nonlinearities that may exist in the continuous system. (Nonlinearities are discussed in Section 11.6.)

The principal transfer functions, all of which are known to the reader by now, are pictured in Fig. 11.1. First there is the continuous function $H(s)$, with input $f(t)$ and output $g(t)$. The problem of digital simulation then consists of obtaining the z-transfer function $\tilde{H}(z)$, in a way such that the digital output g'_m is a good representation of $g(mT)$ at the sample points. The resultant transfer function, $\overline{H}(j\omega)$, is obtained by substituting $e^{j\omega T}$ for z in $\tilde{H}(z)$ and represents the frequency response of the digital simulation. Note that $\overline{H}(j\omega)$ here is *not* necessarily the transform of the sampled inverse transform of $H(s)$. The exact relationship between $\overline{H}(j\omega)$ and $H(s)$ depends on the simulation method, that is, on how $\tilde{H}(z)$ is derived from $H(s)$.

The basic error model used to judge the accuracy of the digital simulation is shown in Fig. 11.2. It is called a "discrete" error model because the error e_m is measured only at the sample points. It assumes that one is interested in the accuracy of the simulation only at these points rather than in a continuous approximation to $g(t)$. The discrete error model has been used by several authors [see Sage and Burt (1965), Greaves and Cadzow (1967), DeFigueiredo and Netravali (1971), and Rosko (1972)].

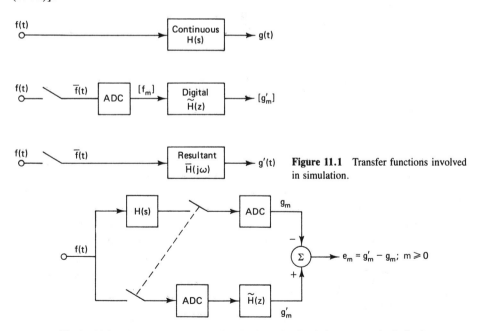

Figure 11.1 Transfer functions involved in simulation.

Figure 11.2 Discrete error model, showing how the simulation error e_m is obtained.

As in Fig. 11.2, the discrete error and its DFT are

$$e_m = g'_m - g_m$$
$$\overline{E}(j\omega) = \overline{G}'(j\omega) - \overline{G}(j\omega) \tag{11.1}$$
$$= \overline{F}(j\omega)\overline{H}(j\omega) - \overline{G}(j\omega) \tag{11.2}$$

and, in the discrete spectrum $\overline{E}(j\omega)$, the error is seen to depend on both $\overline{F}(j\omega)$ and $\overline{H}(j\omega)$, that is, on how well these functions represent their continuous counterparts. Since the error depends on the input $f(t)$ and yet the simulation design affects only $\overline{H}(j\omega)$, it seems reasonable to classify simulation methods according to the properties of $f(t)$. In particular, simulation methods are likely to differ according to whether the spectrum of $f(t)$ falls in one or the other of two broad classes:

Class 1: $|F(j\omega)|$ significantly greater than zero for $\omega \geq \pi/T$
Class 2: $|F(j\omega)|$ negligible for $\omega \geq \pi/T$.

These classes are suggested by the illustration in Fig. 11.3. The lowpass filter in the figure cuts off at $\omega = \pi/T$ so that no aliasing is possible when $f(t)$ is sampled at $1/T$ samples/s. Of course the lowpass filter need not exist — the figure simply emphasizes that $f(t)$ is frequency-limited in class 2 simulations.

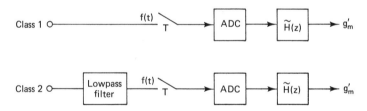

Figure 11.3 Two classes of digital simulation.

In class 1 simulations one cannot hope to obtain a completely satisfactory general-purpose simulation because aliasing is present. In effect, the sampling process does not convey enough information about $f(t)$ to allow the simulator to distinguish it from other functions, no matter how good the simulator. The approach taken in the next section, therefore, is to make the simulation accurate for one particular $f(t)$, and then to examine the accuracy of this simulation for other inputs.

For class 2 simulations, where the spectrum of $f(t)$ is limited to frequencies below π/T rad/s, one can attain simulations of arbitrarily good accuracy. The problem in this case is to make $\overline{H}(j\omega)$ approach $H(j\omega)$ in the interval $|\omega| < \pi/T$ so that the error in Eq. 11.1 will approach zero regardless of the exact form of $f(t)$. Methods for accomplishing this are discussed following the next section.

11.3 INPUT-INVARIANT SIMULATIONS

These simulations apply generally in either of the two classes described above. They are error-free for specific input functions, and are more or less accurate for others, depending on how much the others differ from the specified function. They are discussed first here because one of them, the impulse-invariant (zero-order) method, has already been discussed in Chapters 8 and 10.

The procedure for generalizing the impulse-invariant method and thus deriving simulations that are invariant for other $f(t)$'s can be summarized as follows. ($\mathcal{L}^{-1}$ signifies "inverse Laplace transform of," and $\mathcal{Z}$ stands for "z-transform of.")

1. Determine the desired transfer function, $H(s)$.
2. Let the invariant input function be $i(t)$.
3. Find the Laplace and z transforms of $i(t)$: $I(s)$ and $\tilde{I}(z)$.
4. The continuous output function is then $g(t) = \mathcal{L}^{-1}[H(s)I(s)]$.
5. Its z-transform is $\tilde{G}(z) = \mathcal{Z}\{\mathcal{L}^{-1}[H(s)I(s)]\}$.
6. The invariant digital simulation is then
   $$\tilde{H}(z) = \tilde{G}(z)/\tilde{I}(z) = \mathcal{Z}\{\mathcal{L}^{-1}[H(s)I(s)]\}/\tilde{I}(z).$$

Since $\tilde{G}(z)$ is by definition the z-transform of $g(t)$, one can see that g'_m and g_m in Fig. 11.1 are the same, $e_m = 0$ in Fig. 11.2, and the simulation is perfectly accurate when the input is $f(t) = i(t)$. These steps are illustrated in the following examples, which develop impulse- and step-invariant simulations for $H(s) = 1/(s + 1)$.

Example 11.1

Develop the impulse-invariant simulation of $H(s) = 1/(s + 1)$. The result should be the same as in Chapter 10. In this case $i(t) = \delta(t)$, $I(s) = 1$, and $\tilde{I}(z)$ is the z-transform of $\delta(t)$, or $1/T$ (see Appendix A, line 100). Step 4 gives $g(t) = e^{-t}$, and step 5 gives $\tilde{G}(z) = z/(z - e^{-T})$. Therefore, step 6 gives

$$\tilde{H}(z) = \frac{Tz}{z - e^{-T}} \tag{11.3}$$

just as in Chapter 10. Again, this simulation is impulse-invariant in the sense that when $f(t)$ is a unit impulse with sample $1/T$ at $m = 0$, the output

$$g'_m = Tf_m + e^{-T}g'_{m-1}$$
$$= e^{-mT} \tag{11.4}$$

is equal to g_m, the sample of $g(t)$ at $t = mT$.

Example 11.2

Develop the step-invariant simulation of $H(s) = 1/(s + 1)$. The six steps above yield the following result in this example.

$$H(s) = \frac{1}{s + 1}$$

$i(t) = u(t) = \text{unit step}^\dagger \text{ at } t = 0^-$

$$I(s) = \frac{1}{s}; \qquad \tilde{I}(z) = \frac{z}{z - 1}$$

(11.5)

$$g(t) = \mathcal{L}^{-1}\left[\frac{1}{s(s + 1)}\right] = 1 - e^{-t}; \qquad t \geq 0$$

$$\tilde{G}(z) = \frac{z}{z - 1} - \frac{z}{z - e^{-T}}$$

$$\tilde{H}(z) = \frac{1 - e^{-T}}{z - e^{-T}}$$

Here the computing formula is

$$g_m = (1 - e^{-T})f_{m-1} + e^{-T}g_{m-1}$$

(11.6)

which reduces to $1 - e^{-mT}$ when $f(t)$ is a unit step.

Thus the impulse-invariant simulation is perfectly accurate when the input $f(t)$ is an impulse function with sample $1/T$, the step-invariant simulation is perfect when $f(t)$ is a step function with samples $f_0, f_1, \ldots, = 1$, and so on. Now if these forms were accurate only for their specified inputs, they would be of rather limited practical interest. However, by linear superposition, an input-invariant simulation can also be shown to give zero error in response to any linear combination of specified input functions.

For example, consider the discrete error model in Fig. 11.4, in which the error e_m^* is always zero. Here the input to $H(s)$ is $f^*(t)$, the zero-order hold version of $f(t)$. The figure illustrates the following theorem. (As before, $\mathcal{Z}$ stands for "z-transform of" and $\mathcal{L}^{-1}$ stands for "inverse Laplace transform of.")

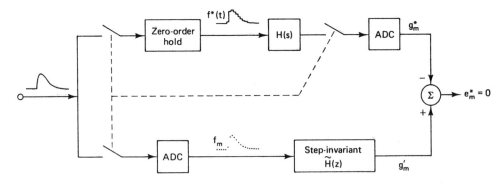

Figure 11.4 Discrete error model for the step-invariant simulation, showing that the error is zero when the input to $H(s)$ is the zero-order-hold version of $f(t)$.

---

†Throughout this chapter, the unit step function $u(t)$ is assumed to occur at $t = 0^-$, so that the sample at $t = 0$ is $u_0 = 1$.

ZERO-ORDER HOLD THEOREM

Let $H(s)$ be any linear transfer function and let

$$\tilde{H}(z) = \frac{z-1}{z}\; \mathscr{Z}\left\{\mathscr{L}^{-1}\left[\frac{H(s)}{s}\right]\right\} \tag{11.7}$$

Then $\tilde{H}(z)$ is an exact simulation of $H(s)$ for any input $f*(t)$ constructed of steps occurring at the sample points.

The proof of this theorem is just a generalization of the statement following steps 1–6 above. First, express $f*(t)$ as a superposition of step functions:

$$f*(t) = f_0 u(t) + \sum_{m=1}^{\infty} (f_m - f_{m-1}) u(t - mT) \tag{11.8}$$

Then the Laplace transform of $g*(t)$ must be

$$
\begin{aligned}
G*(s) &= H(s)F*(s) \\
&= \frac{H(s)}{s}\left(f_0 + \sum_{m=1}^{\infty} (f_m - f_{m-1})e^{-msT}\right)
\end{aligned} \tag{11.9}
$$

Here e^{-msT} implies a delay of mT seconds, and so the z-transform of $g*(t)$ must be

$$
\begin{aligned}
\tilde{G}*(z) &= \mathscr{Z}\left\{\mathscr{L}^{-1}\left[\frac{H(s)}{s}\right]\right\}\left(f_0 + \sum_{m=1}^{\infty} (f_m - f_{m-1})z^{-m}\right) \\
&= \mathscr{Z}\left\{\mathscr{L}^{-1}\left[\frac{H(s)}{s}\right]\right\}\left(\tilde{F}(z) - z^{-1}\tilde{F}(z)\right) \\
&= \tilde{H}(z)\tilde{F}(z) \\
&= \tilde{G}'(z)
\end{aligned} \tag{11.10}
$$

The second line in Eq. 11.10 follows from the definition of the z-transform and the third line follows from Eq. 11.7. The final result, $\tilde{G}*(z) = \tilde{G}'(z)$, proves the theorem (i.e., $e_m^* = 0$ in Fig. 11.4).

A similar result can be obtained for the first-order hold situation by replacing the step invariant $\tilde{H}(z)$ with the ramp-invariant $\tilde{H}(z)$. The discrete error model is shown in Fig. 11.5. Again the error is zero, this time provided that the input to $H(s)$ is the first-order hold version of $f(t)$ (i.e., a sequence of straight lines connecting the sample points). The first-order hold theorem corresponding with Fig. 11.5 is as follows:

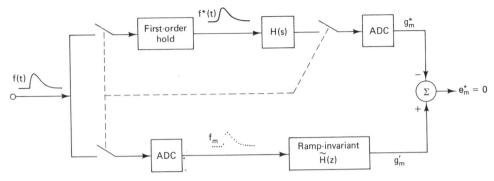

Figure 11.5 Discrete error model for the ramp-invariant simulation, showing that the error is zero when the input to $H(s)$ is the first-order-hold version of $f(t)$.

FIRST-ORDER HOLD THEOREM

Let $H(s)$ be any linear transfer function and let

$$\tilde{H}(z) = \frac{(z-1)^2}{Tz}\mathcal{Z}\left\{\mathcal{L}^{-1}\left[\frac{H(s)}{s^2}\right]\right\} \qquad (11.11)$$

Then $\tilde{H}(z)$ is an exact simulation of $H(s)$ for any input $f*(t)$ constructed of straight lines between sample values.

Here $\tilde{H}(z)$ is the ramp-invariant simulation of $H(s)$, constructed again as in steps 1–6 at the beginning of this section. The first-order hold theorem can be proved as in Eqs. 11.8 through 11.10 above, except that $f*(t)$ must now be written as a super-position of ramp functions instead of step functions. More instructively perhaps, the theorem can be proved by applying steps 1–6 above when $i(t)$ is a generalized ramp function:

$$i(t) = A(t - mT)u(t - mT) \qquad (11.12)$$

Thus, $i(t)$ is a ramp with slope A beginning at time $t = mT$. Here the Laplace transform is

$$I(s) = \frac{Ae^{-msT}}{s^2} \qquad (11.13)$$

and the z-transform is

$$\tilde{I}(z) = \frac{ATz}{z^m(z-1)^2} \qquad (11.14)$$

Simulation of Continuous Systems   Chap. 11

In $I(s)$, A is a constant factor and e^{-msT} represents a delay of m sample points, so the z-transform in step 5 above becomes

$$\tilde{G}(z) = \mathcal{Z}\left\{\mathcal{L}^{-1}\left[\frac{Ae^{-msT}H(s)}{s^2}\right]\right\}$$

$$= Az^{-m}\,\mathcal{Z}\left\{\mathcal{L}^{-1}\left[\frac{H(s)}{s^2}\right]\right\} \tag{11.15}$$

Therefore step 6 above gives

$$\tilde{H}(z) = \tilde{G}(z)/\tilde{I}(z)$$

$$= \frac{(z-1)^2}{Tz}\,\mathcal{Z}\left\{\mathcal{L}^{-1}\left[\frac{H(s)}{s^2}\right]\right\} \tag{11.16}$$

which is the formula for $\tilde{H}(z)$ in the theorem, Eq. 11.11. This result actually proves the theorem because it shows that the ramp-invariant simulation is independent of both A, the slope of the ramp function in Eq. 11.12, and m, its starting point. Therefore the simulation must be exact for any linear combination of these functions [i.e., for $f^*(t)$ in the theorem].

Example 11.3

Find the ramp-invariant simulation of $H(s) = 1/(s+1)$ and illustrate its response to $f(t) = 2te^{-t}$. According to the theorem above, $\tilde{H}(z)$ is found as follows:

$$\tilde{H}(z) = \frac{(z-1)^2}{Tz}\,\mathcal{Z}\left\{\mathcal{L}^{-1}\left[\frac{1}{s^2(s+1)}\right]\right\}$$

$$= \frac{(z-1)^2}{Tz}\,\mathcal{Z}\{t - 1 + e^{-t}\}$$

$$= \frac{(z-1)^2}{Tz}\left[\frac{Tz}{(z-1)^2} - \frac{(1-e^{-T})z}{(z-1)(z-e^{-T})}\right] \tag{11.17}$$

$$= \frac{(T-1+e^{-T})z - Te^{-T} + 1 - e^{-T}}{T(z-e^{-T})}$$

The computing formula corresponding to Eq. 11.17 is

$$g_m = \frac{1}{T}[(T-1+e^{-T})f_m - (Te^{-T}-1+e^{-T})f_{m-1}] + e^{-T}g_{m-1} \tag{11.18}$$

The effect of this formula is illustrated in Fig. 11.6, using $T = 0.6$. On the left is $f(t)$ and $f^*(t)$, the latter being composed of straight lines between sample points. On the right are $g(t)$ and the sample set $[g_m]$ from Eq. 11.18, the latter of course corresponding to $[g'_m]$ in Fig. 11.1. In this case $[g_m]$ is both (1) the output of $\tilde{H}(z)$ with input $[f_m]$, and (2) samples of the output of $H(s)$ with input $f^*(t)$.

The invariant simulations have been described by Jury (1964) and also by Rosko (1972), who provides examples of their application. They are useful when

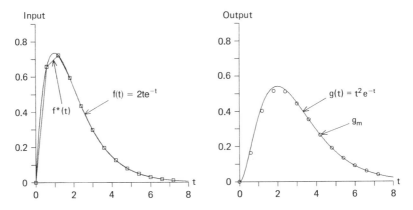

Figure 11.6 Inputs and outputs of $H(s) = 1/(s + 1)$ and its first-order hold simulation, using $f(t) = 2te^{-t}$ and $T = 0.6$.

$f(t)$ is not frequency-limited (class 1 simulations) because they provide an indication of the error in terms of how well $f(t)$ is represented by $f^*(t)$. If the representation is good, as it is in Fig. 11.6 for example, then one can usually expect satisfactory results from the simulation.

In fact, the exactness of the simulation for $f^*(t)$ can be used to establish a bound on the output error which is due to the difference between $f(t)$ and $f^*(t)$. The discrete error model for this exercise is shown in Fig. 11.7. Here the error is e_m as in Fig. 11.2 rather than e_m^* in Figs. 11.4 and 11.5. As noted above e_m is due entirely to the discrepancy between $f(t)$ and $f^*(t)$. From the figure, its envelope $e(t)$ is seen to be

$$e(t) = \mathscr{L}^{-1}\{H(s)[F^*(s) - F(s)]\}$$
$$= \mathscr{L}^{-1}\{H(s)E_i(s)\} \tag{11.19}$$

and the convolution theorem gives e_m, the sample of $e(t)$ at $t = mT$, as

$$e_m = \int_0^{mT} h(\tau)e_i(mT - \tau)\,d\tau \tag{11.20}$$

Now an upper bound on the magnitude of e_m can be established by letting $e_{i\,\text{max}}$ be the maximum magnitude of $e_i(t)$, that is, the maximum discrepancy (magnitude) between $f(t)$ and $f^*(t)$. From Eq. 11.20, this upper bound is

$$|e_n| \leq e_{i\,\text{max}} \int_0^{mT} |h(\tau)|\,d\tau; \qquad n \leq m \tag{11.21}$$

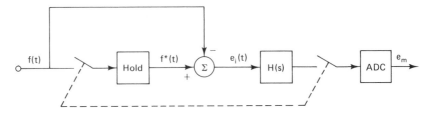

Figure 11.7 Overall discrete error model for invariant simulations, showing that the output error e_m is the result of passing the input error, $e_i(t)$, through $H(s)$.

Simulation of Continuous Systems   Chap. 11

where the subscript of e_m is changed to n to emphasize that the bound holds only in the time interval from $t = 0$ to $t = mT$. An absolute upper bound on the invariant simulation error is found by letting m increase without bound in Eq. 11.21. The result is

$$|e_m| \leq e_{i\max} \int_0^\infty |h(\tau)| \, d\tau; \qquad 0 \leq m < \infty \qquad (11.22)$$

This latter bound is useful provided the continuous impulse response, $h(t)$, has finite energy. Note that the bound is directly proportional to the maximum input error $e_{i\max}$, and can therefore usually be made as small as desired by reducing the time step T, and thus the discrepancy between $f(t)$ and $f^*(t)$. Error bounds for first and second-order systems are found in the exercises at the end of this chapter (see Exercises 23–25).

In conclusion, the invariant simulations described in this section are easy to obtain and useful in class 1 as well as class 2 simulations. They do not require that $f(t)$ be frequency-limited, and the simulation error can be assessed in terms of the discrepancy between $f(t)$ and its "held" version, $f^*(t)$.

11.4 OTHER SIMULATIONS

Two other simple simulation methods are described in this section. One is an adjusted version of the zero-order or impulse-invariant approximation and the other is a substitutional method, in which $\tilde{H}(z)$ is obtained by substituting a function of z for s in $H(s)$. Both methods are easiest to discuss and analyze if applied to class 2 simulations. Then, as described in Section 11.1, one of the objectives is to *match the continuous transfer function, $H(j\omega)$, with the resultant transfer function, $\overline{H}(j\omega)$, in the interval $|\omega| < \pi/T$*. This is a reasonable objective if $f(t)$ has negligible content at frequencies at or above $\omega = \pi/T$. If it can be achieved, then the simulation will be accurate in the sense that g'_m in Fig. 11.1 will be an accurate sample of $g(t)$.

The simplest approximation to $H(s)$ is the zero-order or impulse-invariant approximation, $\tilde{H}_0(z)$, already discussed in Chapter 10 and again in the preceding section. An example in Chapter 10 showed that, at least for $H(s) = 1/(s + 1)$, the difference between $\overline{H}_0(j\omega)$ and $H(j\omega)$ could be substantially reduced if $\tilde{H}_0(z)$ is simply multiplied by a constant. This approach, in which the *adjusted impulse-invariant approximation, $\tilde{H}_A(z)$,* is obtained by scaling $\tilde{H}_0(z)$, has been suggested by Fowler (1965).

The constant scaling factor is obtained by adjusting $\tilde{H}_0(z)$ so that its final value matches that of $H(s)$ for a step-function input. When the input is a step, the final value theorem (Chapter 10, Section 10.4) gives

$$\text{final value (continuous)} = \lim_{s \to 0} H(s) \qquad (11.23)$$

$$\text{final value (digital)} = \lim_{z \to 1} \tilde{H}_0(z) \qquad (11.24)$$

Assuming that these limits are finite and nonzero, the scaling factor is given by their ratio and the formula for the adjusted zero-order simulation is

$$\tilde{H}_A(z) = \frac{H(0)}{\tilde{H}_0(1)} \, \tilde{H}_0(z) \qquad (11.25)$$

With this scaled version of $\tilde{H}_0(z)$ the simulation will have the correct final value for a step input or any input that settles to a constant final value. In other words, the dc response of the simulation is exact, and $\overline{H}_A(j0) = H(j0)$. For this reason $\tilde{H}_A(z)$ is also called the "dc-adjusted" zero-order simulation.

If the limit in Eq. 11.23 or 11.24 is either zero or infinite, $\tilde{H}_0(z)$ can be adjusted to have the correct final-value response to some other input, such as an impulse or ramp. However, in these cases, it is often best to turn to one of the other simulation methods. The following example illustrates a case where $\tilde{H}_0(z)$ can be dc-adjusted.

Example 11.4

Find the dc-adjusted zero-order simulation of $H(s) = 1/(s + 1)$. The zero-order approximation (Eq. 11.3) is

$$\tilde{H}_0(z) = \frac{Tz}{z - e^{-T}} \tag{11.26}$$

and therefore, using Eq. 11.25,

$$\tilde{H}_A(z) = \frac{1}{T/(1 - e^T)} \left(\frac{Tz}{z - e^{-T}} \right) = \frac{(1 - e^{-T})z}{z - e^{-T}} \tag{11.27}$$

is the adjusted simulation.

Figure 11.8 illustrates the amplitude responses for this example, as well as the continuous and digital responses to $f(t) = 1 - e^{-t}$. Using $T = 0.4$ s, that is, an aliasing frequency of $\pi/T = 2.5\pi$, makes $F(j\omega) = 1/[j\omega(1 + j\omega)]$ small beyond this aliasing frequency, even though $H(j\omega)$ is not, and so the illustration is considered to be in class 2. Figure 11.8 should be compared with Fig. 10.2.

The second simple approximation method is the *substitutional* method, in which a function of z is substituted for s in $H(s)$ to obtain $\tilde{H}(z)$. The form having general use in digital approximations is the *bilinear substitution* (see Gibson (1963), Kaiser (1966), and Gold and Rader (1969)]. The bilinear substitution is

$$s \leftarrow A \frac{z - 1}{z + 1} \tag{11.28}$$

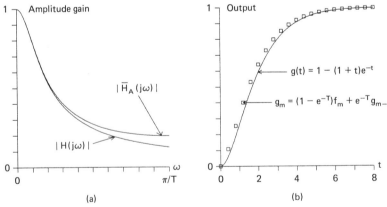

Figure 11.8 (a) Amplitude responses for adjusted zero-order simulation of $H(s) = 1/(s + 1)$; (b) analog and digital responses to $f(t) = 1 - e^{-t}$; $T = 0.4$.

Simulation of Continuous Systems   Chap. 11

in which A is a constant. This bilinear form has several properties of interest in digital simulation. It is rational so that $\tilde{H}(z)$ must be rational if $H(s)$ is a rational function. It also has the effect of mapping the interior of the unit circle in the z-plane onto the primary strip of the left half of the s-plane, a property common to $s = (1/T) \log z$ (see Chapter 10, Eqs. 10.35 and 10.36). In fact, with $A = 2/T$, Eq. 11.28 gives a crude approximation to $(1/T) \log z$, which accounts in a rough way for the validity of the bilinear simulation: The poles and zeros of $\tilde{H}(s)$ in the primary strip are roughly those of $H(s)$ in the primary strip. The bilinear substitution with $A = 2/T$ is called "Tustin's approximation," after A. Tustin.

Of greater interest, however, is the mapping of the $j\omega$ axis obtained by the bilinear substitution. In Eq. 11.28 a point $z = e^{j\omega T}$ on the unit circle of the z-plane transforms to a point $j\omega'$ on the s-plane as follows:

$$j\omega' = A \frac{e^{j\omega T} - 1}{e^{j\omega T} + 1}; \qquad = jA \tan\left(\frac{\omega T}{2}\right) \tag{11.29}$$

If the simulation of $H(s)$ is formed according to Eq. 11.28, then

$$\tilde{H}_B(z) = H\left(A \frac{z - 1}{z + 1}\right) \tag{11.30}$$

where B denotes the bilinear simulation. Then, according to Eq. 11.29, the simulation has the following frequency response:

$$\overline{H}_B(j\omega) = H(j\omega') = H\left(jA \tan\left(\frac{\omega T}{2}\right)\right) \tag{11.31}$$

With this result, one can see first of all that $\overline{H}_B(j\omega)$ and $H(j\omega)$ are equal at $\omega = 0$, that is, no dc adjustment is needed here. Second, $\overline{H}_B(j\omega)$ at $\omega = \pi/T$ is equal to $H(j\omega)$ at $\omega = \infty$ so that in effect $H(j\omega)$ for $0 \le \omega \le \infty$ is "compressed" into the interval $0 \le \omega \le \pi/T$ by the bilinear substitution. Finally, the constant A can be adjusted to make $\overline{H}_B(j\omega)$ equal to $H(j\omega)$ at some particular frequency less than π/T, say ω_0. That is, with $\omega' = \omega = \omega_0$ in Eq. 11.29,

$$A = \omega_0 \cot \frac{\omega_0 T}{2} \tag{11.32}$$

would make $\overline{H}_B(j\omega)$ equal to $H(j\omega)$ at $\omega = \omega_0$. These properties of the bilinear approximation are illustrated in the following example.

Example 11.5

Find the bilinear simulation of $H(s) = 1/(s + 1)$ such that $\overline{H}(j\omega) = H(j\omega)$ when ω is one-half the aliasing frequency. First, with $\omega_0 = \pi/2T$, Eq. 11.32 gives

$$A = \frac{\pi}{2T} \cot\left(\frac{\pi}{4}\right) = \frac{\pi}{2T} \tag{11.33}$$

Then the z-transfer function is found as in Eq. 11.30:

$$\tilde{H}_B(z) = H\left(A \frac{z - 1}{z + 1}\right) = \frac{1}{\dfrac{\pi}{2T}\left(\dfrac{z - 1}{z + 1}\right) + 1}$$

$$= \frac{2T(z + 1)}{(2T + \pi)z + 2T - \pi} \tag{11.34}$$

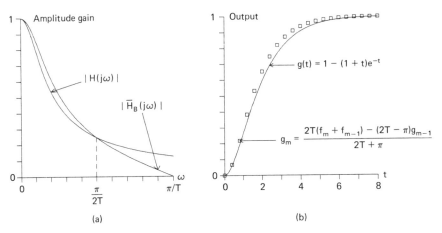

Figure 11.9 (a) Amplitude gain functions for bilinear simulation of $H(s) = 1/(s + 1)$; (b) continuous and digital responses to $f(t) = 1 - e^{-t}$; $T = 0.4$.

For comparison with Example 11.4, amplitude functions and responses to $f(t) = 1 - e^{-t}$ are given for $T = 0.4$ in Fig. 11.9. Compare Figs. 11.8 and 11.9.

The dc-adjusted zero order and bilinear methods discussed above, along with the step- and ramp-invariant method in Section 11.3, are probably the simplest methods that yield accurate approximations to $H(s)$, and this discussion of simple simulation methods will not consider further refinements, of which there are many. At present there is no general theory of simulation adequate to treat the classical numerical integration methods (Milne, Hamming, Runge–Kutta, Adams, etc.) [see Hamming (1973), Kelly (1967), and Rosko (1972), for example] together with the z-transform methods. The most accurate z-transform methods involve iterative adjustment of the poles and zeros of $\tilde{H}(z)$ to reduce the difference between $H(j\omega)$ and $\overline{H}(j\omega)$. DeFigueiredo and Netravali (1971) and Schroeder (1972) have provided examples of this approach, in which the derivation of $\tilde{H}(z)$ is treated as an optimization problem.

11.5 COMPARISON OF LINEAR SIMULATIONS

It is difficult to compare simulation methods in a general way, because accuracy depends on the ensemble of input functions as already discussed, and also on the system being simulated and the simulation method, as well as on the sampling interval T.

To give the reader an estimation of the quality of the simulation methods discussed above beyond that obtained from the preceding examples, two typical transfer functions are simulated in this section using three different methods:

1. Ramp-invariant
2. Adjusted zero order
3. Bilinear with $A = 2/T$

(Recall that the value $A = 2/T$ gives Tustin's approximation, or $s \approx (\log z)/T$.) The typical transfer functions are

$$H(s) = \frac{1}{s + 1} \tag{11.35}$$

Simulation of Continuous Systems   Chap. 11

and

$$H(s) = \frac{1}{s^2 + 2s + 5} \qquad (11.36)$$

The first is the simple first-degree system used above, and the second is a typical second-degree system with complex poles. In the simulations a relatively large value of T is chosen so that the errors of approximation can be compared easily. By reducing T, the errors can of course be reduced. The errors are illustrated in Figs. 11.10, 11.11, 11.13, and 11.14.

In the first case, the z-transfer functions are found in Eqs. 11.17, 11.27, and 11.30. The continuous and digital functions are

$$H(s) = \frac{1}{s + 1} \qquad (11.37)$$

$$\tilde{H}_R(z) = \frac{(T - 1 + e^{-T})z - (Te^{-T} - 1 + e^{-T})}{T(z - e^{-T})} \qquad (11.38)$$

$$\tilde{H}_A(z) = \frac{(1 - e^{-T})z}{z - e^{-T}} \qquad (11.39)$$

$$\tilde{H}_B(z) = \frac{T(z + 1)}{(T + 2)z + T - 2} \qquad (11.40)$$

The digital frequency responses, $\overline{H}_R(j\omega)$, $\overline{H}_A(j\omega)$, and $\overline{H}_B(j\omega)$ are obtained from the above by setting $z = e^{j\omega T}$, or from Eq. 11.31 in the bilinear case.

How should these functions be compared? If only an accurate simulation of *amplitude* response is desired, then comparisons such as those in Figs. 11.8 and 11.9 will illustrate the error. However, if accuracy in both amplitude and phase is required, then an error measure such as

$$E(\omega) = |H(j\omega) - \overline{H}(j\omega)| \qquad (11.41)$$

should be used. A normalized version of this measure is used by Rosko (1972). Obviously, $E(\omega)$ increases either with the magnitude error, the phase error, or both, and is thus a good overall indicator of error. It is the "distance" from $H(j\omega)$ to $\overline{H}(j\omega)$ at the frequency ω.

For example, for the adjusted zero-order approximation, the error function $E_A(\omega)$ is found as follows:

$$\begin{aligned} E_A(\omega) &= |H(j\omega) - \overline{H}_A(j\omega)| \\ &= |H(j\omega) - \tilde{H}_A(e^{j\omega T})| \qquad (11.42) \\ &= \left| \frac{1}{1 + j\omega} - \frac{1 - e^{-T}}{1 - e^{-T}e^{-j\omega T}} \right| \end{aligned}$$

For $H(s) = 1/(s + 1)$, the three error functions $E_A(\omega)$, $E_R(\omega)$ and $E_B(\omega)$, are plotted in Fig. 11.10 for $T = 0.4$ and in Fig. 11.11 for $T = 0.1$. Note that the vertical scales are logarithmic in these plots.

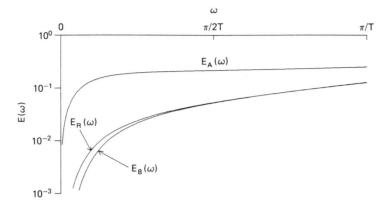

Figure 11.10 Plots of the simulation error, $E(\omega)$, for the adjusted zero-order, bilinear, and ramp-invariant simulations of $H(s) = 1/(s + 1)$; $T = 0.4$.

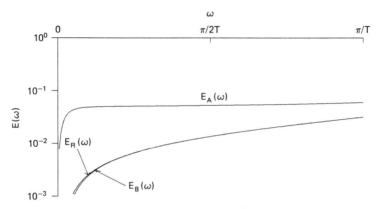

Figure 11.11 Plots of the simulation error, $E(\omega)$, for the adjusted zero-order, bilinear, and ramp-invariant simulations of $H(s) = 1/(s + 1)$; $T = 0.1$.

In the second representative case, Eq. 11.36 above, the continuous and digital transfer functions are

$$H(s) = \frac{1}{(s + 1)^2 + 2^2} \tag{11.43}$$

$$h(t) = \frac{1}{2} e^{-t} \sin 2t \tag{11.44}$$

$$\tilde{H}_A(z) = \frac{1}{5} \frac{(1 - 2e^{-T} \cos 2T + e^{-2T})z}{z^2 - 2ze^{-T} \cos 2T + e^{-2T}} \tag{11.45}$$

$$\begin{aligned}
\tilde{H}_R(z) &= \frac{(z - 1)^2}{25Tz} \left[\frac{pz - 2z^2}{(z - 1)^2} + \frac{2(z^2 - qz)}{z^2 - rz + u} \right] \\
&= \frac{(p - 2q + 2r - 4)z^2 + (2 + 4q - rp - 2u)z + (pu - 2q)}{25T(z^2 - rz + u)}
\end{aligned} \tag{11.46}$$

$$p = 5T + 2; \qquad q = e^{-T} \sec \theta \cos(2T - \theta); \qquad r = 2e^{-T} \cos 2T;$$

$$u = e^{-2T}; \qquad \theta = \tan^{-1}(3/4)$$

(11.47)

$$\tilde{H}_B(z) = \frac{T^2(z + 1)^2}{(5T^2 + 4T + 4)z^2 + 2(5T^2 - 4)z + (5T^2 - 4T + 4)}$$

The corresponding frequency response functions are found by substituting $e^{j\omega T}$ for z or, in the case of $\overline{H}_B(j\omega)$, substituting $A \tan(\omega T/2)$ for ω in $H(j\omega)$. The corresponding amplitude functions $|H(j\omega)|$, $|\overline{H}_A(j\omega)|$, $|\overline{H}_R(j\omega)|$, and $|\overline{H}_B(j\omega)|$ are plotted in Fig. 11.12 for $T = 0.4$. Note again that differences between these functions can be used only to compare amplitude errors; they do not include phase errors.

The total simulation errors, as before in the form $E(\omega) = |H(j\omega) - \overline{H}(j\omega)|$, are plotted in Figs. 11.13 and 11.14 for $T = 0.4$ and 0.1, respectively. For this particular example the ramp-invariant error, $E_R(\omega)$, is smallest over most of the range of ω for both values of T. Note however that in the bilinear simulation, the constant A

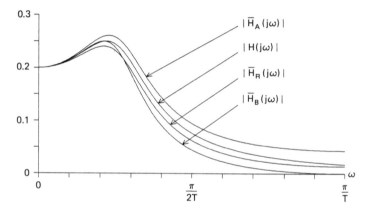

Figure 11.12 Amplitude functions for $H(j\omega) = 1/(5 - \omega^2 + 2j\omega)$ and the three simulations $\overline{H}_A(j\omega)$, $\overline{H}_B(j\omega)$, and $\overline{H}_R(j\omega)$; $T = 0.4$.

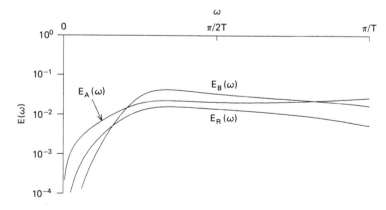

Figure 11.13 Plots of the simulation error, $E(\omega)$, for the adjusted zero-order, bilinear, and ramp-invariant simulations of $H(s) = 1/(s^2 + 2s + 5)$; $T = 0.4$.

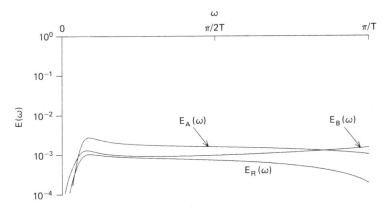

Figure 11.14 Plots of the simulation error, $E(\omega)$, for the adjusted zero-order, bilinear, and ramp-invariant simulations of $H(s) = 1/(s^2 + 2s + 5)$; $T = 0.1$.

could have been chosen as in Eq. 11.32 to force $E_B(\omega)$ to be zero at any single frequency. These examples are of course neither conclusive nor inclusive and only illustrate the use of $E(\omega)$ to compare digital simulations in class 2 situations.

Simulation errors using measures similar to $E(\omega)$ above or e_m in Section 11.2 have been documented by several authors. Jury (1964) and Rosko (1972) present plots of the frequency-domain errors for the integration operators $1/s$, $1/s^2$, and so on. Wait (1970) gives examples of frequency-domain errors in various simulations of first- and second-order systems, and Rosko (1972) provides a number of both time and frequency domain errors in simulations of various systems.

11.6 MULTIPLE AND NONLINEAR SYSTEMS

The simulation of a continuous system typically involves a number of different signals and a number of different transfer functions which are sometimes connected in feedback loops. Furthermore, nonlinearities may be present in the form of limits, thresholds, etc. The control system in Fig. 11.15 is an example. The closed-loop (overall) transfer function could not always be simulated all at once in this example for two reasons: First, no overall transfer function exists for large signals because of the nonlinear limiter in the system. Second, even for small signals, there are actually two overall transfer functions, one from f_1 to g and the other from f_2 to g.

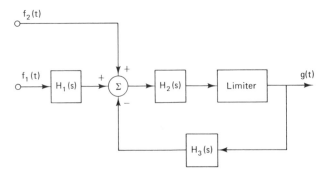

Figure 11.15 Example of a nonlinear continuous control system with multiple inputs.

There is no single approach to the simulation of such a complex structure that is best for all systems. Perhaps the most straightforward method would be to derive an $\tilde{H}(z)$ for $H_1(s)$, $H_2(s)$, and $H_3(s)$ as well as the limiter, that is, to simulate each block in the diagram individually. This would be usually less accurate than an overall simulation, however, because in the feedback loop (or in any closed loop for that matter), *not all of the signals can be computed as functions of present values of other signals.* The computations around the loop must start somewhere in terms of only previous values. Also, even if there were no closed loop, *the simulation of a product,* say $\tilde{H}(z)$ derived from $H(s) = H_1(s)H_2(s)$, is in general different from the product of simulations, that is, $\tilde{H}_1(z)\,\tilde{H}_2(z)$, depending on how the z-transfer functions are derived. An important exception to this general rule occurs with the bilinear approximation. Here, since $\tilde{H}(z)$ is found by replacing s with $A(z - 1)/(z + 1)$ in $H(s)$, $\tilde{H}(z) = \tilde{H}_1(z)\,\tilde{H}_2(z)$ follows from $H(s) = H_1(s)H_2(s)$. The following example is an illustration.

Example 11.6

For the system in Fig. 11.16 derive a simulation that gives an exact step-function response. Show how the simulation differs from the product of individual step-invariant simulations. The individual step-invariant simulations are essentially derived in Example 11.2, Section 11.3. From Appendix A, line 155, these are

$$\tilde{H}_1(z) = \frac{z-1}{z}\,\mathcal{Z}\left\{\mathcal{L}^{-1}\left[\frac{1}{s(s+1)}\right]\right\} = \frac{1-e^{-T}}{z-e^{-T}} \tag{11.48}$$

$$\tilde{H}_2(z) = \frac{z-1}{z}\,\mathcal{Z}\left\{\mathcal{L}^{-1}\left[\frac{2}{s(s+2)}\right]\right\} = \frac{1-e^{-2T}}{z-e^{-2T}} \tag{11.49}$$

Using the *product* of these two functions, the simulation of Fig. 11.16 would be

$$g_m = (1 - e^{-T})(1 - e^{-2T})f_{m-2} + (e^{-T} + e^{-2T})g_{m-1} - e^{-3T}g_{m-2} \tag{11.50}$$

The true step-invariant form for Fig. 11.16 is of course found (Appendix A, line 300) as follows:

$$\begin{aligned}
\tilde{H}(z) &= \frac{z-1}{z}\,\mathcal{Z}\left\{\mathcal{L}^{-1}\left[\frac{2}{s(s+1)(s+2)}\right]\right\} \\
&= \frac{z-1}{z}\left[\frac{z}{z-1} - \frac{2z}{z-e^{-T}} + \frac{z}{z-e^{-2T}}\right] \\
&= \frac{(1-e^{-T})^2(z+e^{-T})}{(z-e^{-T})(z-e^{-2T})}
\end{aligned} \tag{11.51}$$

From which the correct simulation, to be constrasted with Eq. 11.50 above, is

$$g_m = (1 - e^{-T})^2(f_{m-1} + e^{-T}f_{m-2}) + (e^{-T} + e^{-2T})g_{m-1} - e^{-3T}g_{m-2} \tag{11.52}$$

Note that the poles are the same in the two formulas, but that Eq. 11.51 adds a zero, causing an f_{m-1} term to appear in Eq. 11.52. To illustrate the error in Eq. 11.50, the two computed step function responses are shown in Fig. 11.17.

$H_1(s)$ $H_2(s)$

f(t) $\longrightarrow$ $\boxed{\dfrac{1}{s+1}}$ $\longrightarrow$ $\boxed{\dfrac{2}{s+2}}$ $\longrightarrow$ g(t)   **Figure 11.16** System to be simulated in Example 11.6.

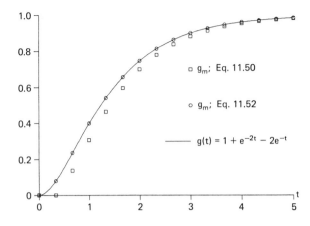

Figure 11.17 Step-invariant simulations for Fig. 11.16, showing that only Eq. 11.52 gives the accurate response to a step-function input. Sampling interval $T = \frac{1}{3}$.

Regarding nonlinearities, many are common in simulations. Six examples are shown in Table 11.1, which illustrates each type of nonlinear operation by giving its response to the sawtooth signal shown at the top of the table. A simulation formula is also given for each nonlinear operation, showing that the computing program for each is simple, amounting to only one or two Fortran statements.

Methods for simulating closed-loop systems with nonlinearities have been described by Hurt (1964), Sage and Burt (1965), Fowler (1965), and Rosko (1972). Generally, these methods involve simulating the individual blocks in the block dia-

TABLE 11.1 Nonlinearities

Nonlinearity	Response to $f(t)$	Digital Simulation
Limit	$g(t)$	$g_m = \min(f_m, L)$
Threshold		$g_m = 0;$ if $(f_m > M)$, $g_m = f_m$
Trigger		$g_m = 0;$ if $(f_m > M)$, $g_m = 1$
Logarithm		$g_m = \log f_m;$ $f_m > 0$
Reciprocal		$g_m = \dfrac{1}{f_m};$ $f_m \neq 0$
Square		$g_m = f_m^2$

gram, adding a unit delay in the feedback loop if necessary for realizability, and then adding blocks either inside or outside the loop to make the overall transfer function more accurate. Thus, the difficulties mentioned in connection with Fig. 11.15 and illustrated in Fig. 11.17 can be alleviated to some extent.

An approach to closed-loop simulation is illustrated here in a simple example. To follow the general plan in the preceding paragraph, some or all of the following steps can be taken to simulate a nonlinear, closed-loop system:

1. Replace the nonlinear elements temporarily with linear components.
2. After considering the known properties of $f(t)$, simulate each block in the block diagram using a method such as in Section 11.3 or 11.4.
3. Add a unit delay inside the closed loop if (and only if) necessary to make the closed-loop simulation realizable.
4. If possible, add a constant gain inside the closed loop to make the (overall) closed-loop poles agree with those of a simulation of the closed-loop $H(s)$.
5. Add a block at the input to make the overall $\tilde{H}(z)$ a desired (i.e., zero-order, step-invariant, ramp-invariant, etc.) model of $H(s)$.
6. Replace the nonlinear elements with suitable models as in Table 11.1.

The block diagram in Fig. 11.18 will now be used as an example. In the diagram, $H_1(s)$ represents a device whose output $y(t)$ is limited such that the magnitude of $g(t)$ cannot exceed one. The integrator, $H_2(s)$ in the feedback leg, causes $g(t)$ to oscillate in response to a transient input and to reach a final value of zero for a constant input, as seen in the form of the overall (linear) $H(s)$ for small signals:

$$H(s) = \frac{H_1(s)}{1 + H_1(s)H_2(s)} = \frac{s}{(s + 1)^2 + 9} \tag{11.53}$$

Step 1 above is accomplished in this example simply by letting the limiter have unit gain, that is, by modeling the system as in Eq. 11.53 for small signals. For step 2, suppose (for example) that the step-invariant form is chosen. Then the individual block models are

$$\tilde{H}_1(z) = \frac{z - 1}{z} \mathscr{Z}\left\{\mathscr{L}^{-1}\left[\frac{1}{s(s + 2)}\right]\right\} = \frac{1 - e^{-2T}}{2(z - e^{-2T})} \tag{11.54}$$

which is the same as Eq. 11.49 except for a constant, and

$$\tilde{H}_2(z) = \frac{z - 1}{z} \mathscr{Z}\left\{\mathscr{L}^{-1}\left[\frac{10}{s^2}\right]\right\} = \frac{10T}{z - 1} \tag{11.55}$$

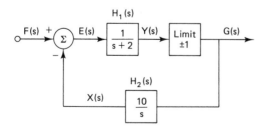

Figure 11.18 Nonlinear closed-loop system to be simulated.

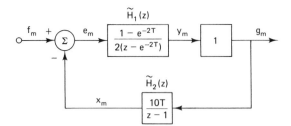

Figure 11.19 Partially completed simulation of Fig. 11.18 after step 2, using step-invariant blocks.

The result at the end of step 2 is shown in Fig. 11.19. The step-invariant model of the integrator has a property that is often very useful in closed-loop simulations, namely that there are more poles than zeros. Thus, in the feedback portion of Fig. 11.19,

$$x_m = 10Tg_{m-1} + x_{m-1} \qquad (11.56)$$

and the simulation is realizable without step 3 above, because the present value x_m (and therefore e_m) can be calculated in terms of past values of loop variables. (A counterexample is given below.)

Next, for step 4, the step-invariant model of the overall $H(s)$ is

$$
\begin{aligned}
\tilde{H}(z) &= \frac{z-1}{z} \, \mathscr{Z}\left\{ \mathscr{L}^{-1}\left[\frac{H(s)}{s}\right] \right\} \\
&= \frac{z-1}{z} \, \mathscr{Z}\left\{ \mathscr{L}^{-1}\left[\frac{1}{(s+1)^2 + 9}\right] \right\} \\
&= \frac{e^{-T}(z-1) \sin 3T}{3(z^2 - 2ze^{-T} \cos 3T + e^{-2T})}
\end{aligned}
\qquad (11.57)
$$

whereas the denominator of the overall transfer function of the partially completed model in Fig. 11.19 with a constant gain K anywhere in the closed loop is

$$\text{denominator} = z^2 - (1 + e^{-2T})z + 5KT(1 - e^{-2T}) + e^{-2T} \qquad (11.58)$$

Obviously no value of K will make the roots of Eq. 11.58 the same as those of the denominator of $\tilde{H}(z)$ in Eq. 11.57, and step 4 is not possible in this case. (Again, a counter example is given below.)

To accomplish step 5, an input block, $\tilde{H}_3(z)$, is added to Fig. 11.19 to make the overall transfer function step-invariant, that is, equal to $\tilde{H}(z)$ in Eq. 11.57. That is, as in Fig. 11.20,

$$\frac{\tilde{H}_3(z)\tilde{H}_1(z)}{1 + \tilde{H}_1(z)\tilde{H}_2(z)} = \tilde{H}(z) \text{ in Eq. 11.57} \qquad (11.59)$$

and solving for $\tilde{H}_3(z)$,

$$
\begin{aligned}
\tilde{H}_3(z) &= \left(\frac{1}{\tilde{H}_1(z)} + \tilde{H}_2(z)\right)\tilde{H}(z) \\
&= \frac{2e^{-T} \sin 3T}{3(1 - e^{-2T})} \frac{z^2 - (1 + e^{-2T})z + e^{-2T} + 5T(1 - e^{-2T})}{z^2 - 2e^{-T}z \cos 3T + e^{-2T}}
\end{aligned}
\qquad (11.60)
$$

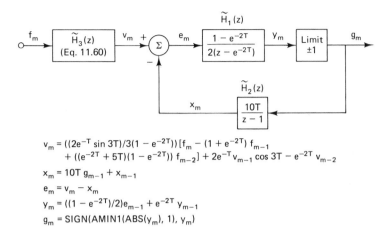

$$v_m = ((2e^{-T} \sin 3T/3(1 - e^{-2T}))[f_m - (1 + e^{-2T}) f_{m-1}$$
$$+ ((e^{-2T} + 5T)(1 - e^{-2T})) f_{m-2}] + 2e^{-T} v_{m-1} \cos 3T - e^{-2T} v_{m-2}$$
$$x_m = 10T g_{m-1} + x_{m-1}$$
$$e_m = v_m - x_m$$
$$y_m = ((1 - e^{-2T})/2)e_{m-1} + e^{-2T} y_{m-1}$$
$$g_m = \text{SIGN(AMIN1(ABS}(y_m), 1), y_m)$$

Figure 11.20 Completed step-invariant simulation of Fig. 11.18 using step-invariant blocks.

Finally, step 6 is accomplished by placing the limiter in the digital model. The completed step-invariant simulation is shown along with the computing algorithm in Fig. 11.20. The simulation in Fig. 11.20 obviously can be realized if the computations are made in the order given. The formula for g_m in the figure is the Fortran expression for g_m as the limited version of y_m.

Before giving a sample simulation using Fig. 11.20, it is instructive to reconstruct the step-invariant simulation with another choice for the blocks in step 2. Using the adjusted zero-order form for each block, the construction at the end of step 2 is shown in Fig. 11.21. Each block is modeled as in Section 11.4 (see Example 11.4). The forward block is dc-adjusted while the feedback block has the correct impulse response, and requires no adjustment.

The important difference between Figs. 11.19 and 11.21 is that the latter is not realizable because the present value of each variable in the loop must be computed in terms of other present loop values, for example,

$$x_m = 10T g_m + x_{m-1} \tag{11.61}$$

As mentioned previously, the computation must start somewhere in terms of only past loop values. Therefore, step 3 in this case requires placing a unit delay somewhere in the loop to make the simulation possible. Choosing (arbitrarily) to place the delay in the feedback portion, the result is shown in Fig. 11.22. Note that x_m is now found as in Eq. 11.56 instead of Eq. 11.61, and so the closed-loop simulation is realizable.

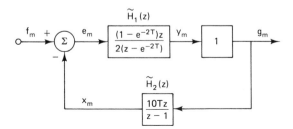

Figure 11.21 Partially completed simulation of Fig. 11.18 after step 2, using adjusted zero-order blocks.

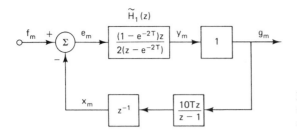

$\tilde{H}_1(z)$

Figure 11.22 Partially completed simulation of Fig. 11.18 after step 3, using adjusted zero-order blocks.

Proceeding to step 4 for this case, one can again attempt to make the closed-loop poles of Fig. 11.22 agree with those of a closed-loop model of $H(s)$. The denominator of the zero-order model of $H(s)$ is the same as in Eq. 11.57, that is,

$$\begin{matrix} \text{denominator of zero-order} \\ \text{model of } H(s) \end{matrix} = z^2 - 2e^{-T}z \cos 3T + e^{-2T} \qquad (11.62)$$

but the denominator of the overall transfer function of Fig. 11.22 with a constant gain K in the loop is now

$$\text{denominator} = z^2 + [5KT(1 - e^{-2T}) - 1 - e^{-2T}]z + e^{-2T} \qquad (11.63)$$

In this case one can adjust K to make the roots of Eq. 11.63 agree with those in Eq. 11.62. The solution for K is

$$5KT(1 - e^{-2T}) - 1 - e^{-2T} = -2e^{-T} \cos 3T,$$

or

$$K = \frac{1 - 2e^{-T} \cos 3T + e^{-2T}}{5T(1 - e^{-2T})} \qquad (11.64)$$

If K is inserted (arbitrarily) into the feedback portion, the result is as illustrated in Fig. 11.23, with the closed-loop poles now agreeing with those of Eq. 11.62. Finally, to accomplish step 5 above and also make the present model comparable to the simulation in Fig. 11.20, an input transfer function $\tilde{H}_3(z)$ can be derived to make the present overall simulation step-invariant. The equation for $\tilde{H}_3(z)$ is similar to Eq. 11.60:

$$\tilde{H}_3(z) = \left(\frac{1}{\tilde{H}_1(z)} + Kz^{-1}\tilde{H}_2(z)\right)\tilde{H}(z) \qquad (11.65)$$

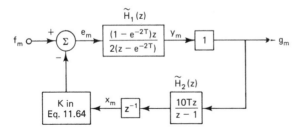

$\tilde{H}_1(z)$

$\tilde{H}_2(z)$

Figure 11.23 Partially completed simulation of Fig. 11.18 after step 4, using adjusted zero-order blocks.

Simulation of Continuous Systems   Chap. 11

where $\tilde{H}_1(z)$ and $\tilde{H}_2(z)$ are given in Fig. 11.23, K in Eq. 11.64, and $\tilde{H}(z)$ in Eq. 11.57. When these functions are placed in Eq. 11.65, the result is

$$\tilde{H}_3(z) = \frac{2e^{-T}\sin 3T}{3(1 - e^{-2T})z} \tag{11.66}$$

After performing step 6, that is, placing the limiter, the completed step-invariant simulation using zero-order blocks is as shown in Fig. 11.24. Again, as in Fig. 11.20, the formula for g_m is a Fortran statement describing the limiter.

It is instructive to compare the step-invariant simulations in Figs. 11.20 and 11.24. Although the two models are different, they are guaranteed via the construction of $\tilde{H}_3(z)$ to give identical, exact results when the input $f(t)$ is composed of steps and is small so that the operation is linear. This behavior is illustrated in Fig. 11.25,

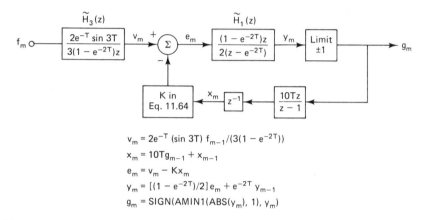

$$v_m = 2e^{-T}(\sin 3T)f_{m-1}/(3(1 - e^{-2T}))$$
$$x_m = 10Tg_{m-1} + x_{m-1}$$
$$e_m = v_m - Kx_m$$
$$y_m = [(1 - e^{-2T})/2]e_m + e^{-2T}y_{m-1}$$
$$g_m = \text{SIGN(AMIN1(ABS}(y_m), 1), y_m)$$

Figure 11.24 Completed step-invariant simulation of Fig. 11.18 using adjusted zero-order blocks.

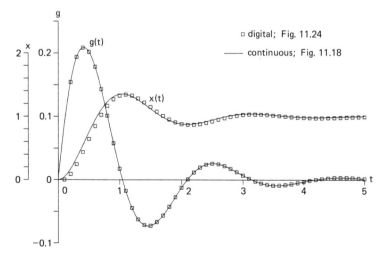

Figure 11.25 Continuous and simulated responses to unit step function, step size $T = 0.1$.

which shows the integrator output, $x(t)$, as well as the normal output, $g(t)$, when the input is a unit step function at $t = 0^-$. Both simulations (Figs. 11.20 and 11.24) of $g(t)$ are exact, but there is a slight error in the simulation of $x(t)$. This error is almost the same for both simulations when the step size is $T = 0.1$, and of course increases when T is made larger and decreases when T is decreased.

Figure 11.26 illustrates nonlinear operation in response to a step input with amplitude 6 at $t = 0^-$. Again, both simulations are exact until $g(t)$ reaches 1.0 and the operation becomes nonlinear. After the limit is reached the two simulations of $g(t)$ differ, the simulation in Fig. 11.24 being the more accurate of the two. The improved accuracy with Fig. 11.24 is due mainly to the closed-loop pole adjustment possible with the zero-order blocks but not with the step-invariant blocks. The same is true for $x(t)$, that is, the simulation with zero-order blocks is more accurate. Notice also, in Fig. 11.25 as well as Fig. 11.26, that simulations of $x(t)$ would be more accurate if they were shifted one time step to the left, thus eliminating the artificial delay in the feedback loop.

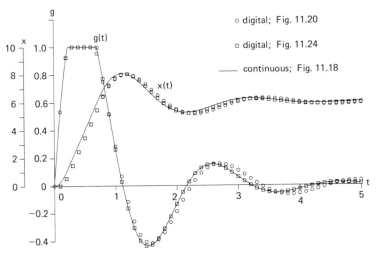

Figure 11.26 Continuous and simulated responses to step function with amplitude = 6; step size $T = 0.1$.

11.7 CONCLUDING REMARKS

In this chapter some of the easiest methods for simulating continuous transfer functions have been introduced. Simulation per se has really been only a part of the objective; the remainder has been to develop some of the interesting relationships between digital and continuous systems. The usefulness of these relationships is not limited to simulation; for example, the bilinear substitution is used in Chapter 12 to derive digital filters.

The question of the overall accuracy of the different simulation methods is difficult to address in any general way. As shown in this chapter, errors can be measured in the time domain, in the frequency domain, or under specified input conditions with nonlinear operation.

In the final analysis, simulation always involves a mixture of art, science, luck, and varying degrees of honesty. It is generally possible to model a complex process so that the outcome is as the scientist wishes to see it rather than a faithful reproduction of reality, and one must be careful not to let misleading results emerge from the complexity of the simulation.

EXERCISES

1. Show that g_m in Eq. 11.6 is correct when the input function is $f(t) = u(t)$.

2. Derive a simulation of $H(s) = 1/(s + a)$ that will be precise if the input is $f(t) = Ae^{-at}$. Give the computing formula.

3. Do Exercise 2 for $H(s) = 1/(s + b)$.

4. Using the answer to Exercise 3, compute the output for $f_m = Ae^{-maT}$ and confirm that it is correct.

5. Give $\tilde{H}(z)$ for the ramp-invariant simulation of $H(s) = 1/(s + a)$.

6. Provide a computing formula for the step-invariant simulation of $H(s) = s/(s^2 + 2s + 5)$.

7. Plot $|\tilde{H}(j\omega)|$ together with $|H(j\omega)|$, where $\overline{H}(j\omega)$ is the response of the step-invariant simulation of $H(s) = 1/(s + 1)$. Use $T = 0.3$ s.

8. Prove that the final value of the dc-adjusted zero order simulation of $H(s) = 1/(s + a)$ is correct when the input is a step function.

9. Derive $\tilde{H}(z)$ for the adjusted zero-order approximation to $H(s) = 1/s$.

10. Do Exercise 9 for $H(s) = 1/(s + 1)^2$. Give a computing formula.

11. Determine $\tilde{H}(z)$ for the bilinear approximation to $H(s) = 1/(s + 1)^2$ that is accurate at $\omega = 0$ and at one-fourth the Nyquist frequency.

12. (Computer.) Plot the logarithmic amplitude response of $H(s) = 1/[s(s + 1)]$ together with its bilinear simulation, with the latter accurate at 1 Hz. Use $T = 0.25$.

13. Simulate the system in Fig. 11.16 using the bilinear approximation, accurate at one-half the Nyquist frequency.

14. Simulate the system below. Use the step-invariant approximation.

15. Give an impulse-invariant simulation algorithm for the system shown. Adjust the dc gain appropriately.

16. What is the response of $H(s)$ in Section 11.6, Eq. 11.53 to a unit step function at $t = 0$?

17. In Fig. 11.22, give the algorithm for y_m when the delay is inserted in the forward part of the loop.

18. In Fig. 11.23, what is K if the delay is moved to the forward portion of the loop?

19. Modify Fig. 11.24 by placing both K and z^{-1} in the forward part of the loop. What is $\tilde{H}_3(z)$ in the modified diagram?

20. Give the bilinear blocks needed to simulate Fig. 11.18. Use Tustin's approximation.

21. Using the bilinear blocks in the preceding problem along with a feedback delay, construct a complete bilinear simulation of Fig. 11.18. What is $\tilde{H}_3(z)$ in this case?

22. Develop ramp-invariant blocks for the nonlinear system below.

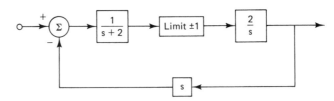

23. Given that $e_{i\max}$ is the maximum difference between $f(t)$ and its held version, $f^*(t)$, as in Eq. 11.22, what is the upper bound on the simulation error when $H(s) = A/(s + a)$?

24. Solve Exercise 23 for $H(s) = A/[(s + a)^2 + b^2]$.

25. Solve Exercise 23 for $H(s) = As/[(s + a)^2 + b^2]$.

SOME ANSWERS

2. $g_m = e^{-aT}[Tf_{m-1} + g_{m-1}]$ **3.** $g_m = (e^{-aT} - e^{-bT})f_{m-1}/(b - a) + e^{-bT}g_{m-1}$

5. $\tilde{H}(z) = \dfrac{(aT + e^{-aT} - 1)z - (aTe^{-aT} + e^{-aT} - 1)}{a^2 T(z - e^{-aT})}$

6. $g_m = e^{-T}\sin 2T(f_{m-1} - f_{m-2})/2 + 2e^{-T}\cos 2Tg_{m-1} - e^{-2T}g_{m-2}$ **9.** $\tilde{H}(z) = Tz/(z - 1)$

10. $g_m = (1 - e^{-T})^2 f_{m-1} + 2e^{-T}g_{m-1} - e^{-2T}g_{m-2}$

11. $\tilde{H}(z) = \dfrac{(z + 1)^2}{[(A + 1)z - (A - 1)]^2}; \quad A = \dfrac{\pi}{4T}\cot\left(\dfrac{\pi}{8}\right)$

13. $\tilde{H}(z) = \dfrac{8T^2(z + 1)^2}{[(\pi + 2T)z - (\pi - 2T)][(\pi + 4T)z - (\pi - 4T)]}$ **16.** $g(t) = (1/3)e^{-t}\sin 3t$

18. $K = \dfrac{1 - 2e^{-T}\cos 3T + e^{-2T}}{5T(1 - e^{-2T})}$ **19.** $\tilde{H}_3(z) = \dfrac{10Te^{-T}\sin 3T}{3(1 - 2e^{-T}\cos 3T + e^{-2T})}$

21. $\tilde{H}_3(z) = \dfrac{2(T + 1)z^3 + (5T^2 - 4)z^2 + 2(5T^2 - T + 1)z + 5T^2}{(5T^2 + 2T + 2)z^3 + 2(5T^2 - 2)z^2 + (5T^2 - 2T + 2)z}$

23. $|e_m| \leq (A/a)e_{i\max}$ **24.** $|e_m| \leq \dfrac{A\coth(a\pi/2b)}{a^2 + b^2}e_{i\max}$

25. $|e_m| \leq \dfrac{Ae^V\operatorname{csch}(a\pi/2b)}{\sqrt{a^2 + b^2}}e_{i\max}; \quad V = (a/b)\tan^{-1}(a/b)$

REFERENCES

CESCHINO, F., and KUNTZMANN, J., *Numerical Solution of Initial Value Problems*. Englewood Cliffs, N.J.: Prentice-Hall, 1966.

DEFIGUEIREDO, R. J. P., and NETRAVALI, A. N., Optimal Spline Digital Simulators of Analog Filters. *IEEE Trans. Commun. Technol.*, Vol. CT-18, No. 6, November 1971, p. 711.

FOWLER, M. E., A New Numerical Method for Simulation. *Simulation*, May 1965, p. 324.

GIBSON, J. E., *Nonlinear Automatic Control*, Chap. 4. New York: McGraw-Hill, 1963.

GOLD, B., and RADER, C. M., *Digital Processing of Signals*. New York: McGraw-Hill, 1969.

GREAVES, C. J., and CADZOW, J. A., The Optimal Discrete Filter Corresponding to a Given Analog Filter. *IEEE Trans. Automatic Control*, Vol. AC-13, June 1967, p. 304.

HAMMING, R. W., *Numerical Methods for Scientists and Engineers,* 2nd ed. New York: McGraw-Hill, 1973.

HILDEBRAND, F. B., *Finite-Difference Equations and Simulations.* Englewood Cliffs, N.J.: Prentice-Hall, 1968.

HURT, J. M., New Difference-Equation Technique for Solving Nonlinear Differential Equations. *AFIPS Conf. Proc.,* Vol. 25, 1964, p. 169.

JURY, E. I., *Theory and Application of the z-Transform Method.* New York: Wiley, 1964.

KAISER, J. F., Digital Filters, Chap. 7 in *System Analysis by Digital Computer,* ed. J. F. Kaiser and F. F. Kuo. New York: Wiley, 1966.

KELLY, L. G., *Handbook of Numerical Methods and Applications.* Reading, Mass.: Addison-Wesley, 1967, Chap. 19.

LAPIDUS, L., and SEINFELD, J. H., *Numerical Solution of Ordinary Differential Equations.* New York: Academic Press, 1971.

Numerical Techniques for Real-Time Flight Simulation, IBM Corporation Manual E20-0029-1, 1964.

RAGAZZINI, J. R., and FRANKLIN, G. F., *Sampled-Data Control Systems,* Chap. 4. New York: McGraw-Hill, 1958.

REA, J. L., *z-Transformation Techniques in Digital Realization of Coaxial Equalizers.* Sandia Laboratories SC-RR-72 0524, September 1972.

REITMAN, J., *Computer Simulation Applications.* New York: Wiley, 1971.

ROSKO, J. S., *Digital Simulation of Physical Systems.* Reading, Mass.: Addison-Wesley, 1972.

SAGE, A. P., and BURT, R. W., Optimum Design and Error Analysis of Digital Integrators for Discrete System Simulation. *AFIPS Conf. Proc.,* Vol. 27, Pt. 1, 1965, p. 903.

SCHROEDER, D. H., *A New Optimization Procedure for Digital Simulation,* Ph.D. dissertation. University of New Mexico, Albuquerque, 1972.

WAIT, J. V., Digital Filters, Chap. 5 in ed. L. P. Huelsman. *Active Filters: Lumped, Distributed, Integrated, Digital, and Parametric,* New York: McGraw-Hill, 1970.

12

Analog and Digital Filter Design

12.1 INTRODUCTION

Some practical filter designs, both analog and digital, are introduced in this chapter. Here, the word "filter" is used in the restricted sense: a filter is assumed to be a system for passing the spectral content of an input signal in a certain specified band of frequencies. In other words, the filter transfer function forms a "window" in the frequency domain through which a portion of the input spectrum is allowed to pass.

The idealized amplitude characteristics of four basic filter types are illustrated (for $\omega \geq 0$) in Figure 12.1. Of these, the lowpass characteristic is, in a sense, the most fundamental. Sections 11.2 through 11.4 are concerned only with the design of lowpass analog and digital filters. In Section 11.5 a systematic method for transforming lowpass filters into highpass, bandpass, and bandstop filters is described. Section 11.6 describes some digital filtering routines, and Section 11.7 then describes a method applicable, in general, to the synthesis of digital filters and in particular to the synthesis of multichannel digital bandpass filters. Finally, Section 11.8 describes the errors caused by finite word lengths.

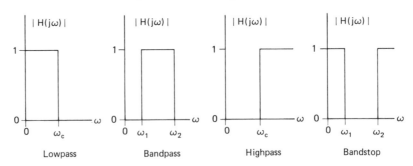

Figure 12.1 Idealized amplitude characteristics.

At the outset, something should be said comparing the design of analog and digital filters. The design of analog filters is an extensively explored subject. Storer (1957), Guillemin (1957), and Kuo (1962), for example, provide detailed discussions of the Butterworth and Chebyshev filters, described below, as a part of the broader subject of linear network design. In general, the complete design of an analog filter involves two steps:

1. Deriving $H(s)$, usually by locating poles and zeros at appropriate points on the s-plane.
2. Realizing $H(s)$ using obtainable linear components.

The first step, discussed in this chapter, is of more general interest in that it lends insight into analog pole–zero synthesis, and also in that it is used further in the design of digital filters. The second step, not discussed here, involves special problems in analog realization such as isolation between stages, power loss in components, and so on—problems that do not exist in digital design.

Further, concerning the design of digital filters, two approaches are discussed here. The first involves designing an analog filter (placing poles and zeros on the s-plane) and then converting it to digital form (mapping the poles and zeros to the z-plane) to achieve the desired characteristic. The second involves placing poles and zeros directly on the z-plane to achieve a given characteristic. Both approaches, particularly the first, again emphasize the close relationship between analog and digital systems discussed in Chapters 10 and 11.

12.2 BUTTERWORTH FILTERS

The design of analog lowpass filters that approximate the ideal characteristic in Fig. 12.1 has been, as mentioned, a much-explored subject. This design problem, called the "approximation problem" by Guillemin (1957), has led to some more-or-less standard analog lowpass designs, one of which is the Butterworth filter. The Butterworth filter can be described in simple terms after defining a few preliminary items summarized in Fig. 12.2.

Figure 12.2 is first of all a plot of the power gain, that is, the squared amplitude response, versus frequency. It is of course an even function of ω, only the right half of the plot being shown. The power gain function, $|H(j\omega)|^2$, is preferred over the amplitude response, $|H(j\omega)|$, in describing the characteristics of filters. (Note that the two functions are, moreover, identical for the idealized cases in Fig. 12.1.) In general the power gain of a filter is expressed in one of two ways:

$$\text{power gain} = |H(j\omega)|^2$$
$$\text{power gain (dB)} = 10 \log_{10}|H(j\omega)|^2 \tag{12.1}$$

(The abbreviation "dB" stands for "decibels.")

Furthermore, anticipating the fact that the ideal lowpass characteristic cannot be realized in finite form, Fig. 12.2 defines some basic regions and parameters generally useful in filter design. There is a *cutoff frequency*, ω_c, that marks the upper limit of a *passband* from zero to ω_c on the frequency axis and a *rejection frequency*, ω_r, which is greater than ω_c and marks the beginning of a *stopband* from ω_r to infinity.

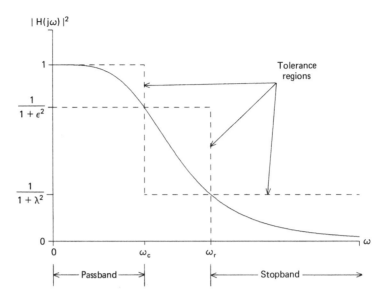

Figure 12.2 Lowpass filter power gain characteristic.

Between the passband and the stopband, that is, in the interval $\omega_c < |\omega| < \omega_r$, there is a transition band in which the power gain rapidly decreases. The gain parameters, λ and ϵ, determine the tolerances the designer is willing to accept, as follows:

$$\left.\begin{array}{ll} \text{passband:} \quad |\omega| < \omega_c; & |H(j\omega)|^2 > \dfrac{1}{1 + \epsilon^2} \\[4mm] \text{stopband:} \quad |\omega| > \omega_r; & |H(j\omega)|^2 < \dfrac{1}{1 + \lambda^2} \end{array}\right\} \tag{12.2}$$

A typical lowpass characteristic within the tolerances is shown in the figure. One can see that the ideal rectangular characteristic is approached if ϵ approaches zero, ω_r approaches ω_c, and λ approaches infinity.

The *Butterworth* analog filter has a power gain of the following general form:

$$|H_B(j\omega)|^2 = \frac{1}{1 + \epsilon^2(\omega/\omega_c)^{2N}} \tag{12.3}$$

in which the subscript B designates the Butterworth transfer function and N is called the *order* of the filter and ϵ and ω_c are as defined above. Such a filter is said to be *maximally flat* near $\omega = 0$ and $\omega = \infty$. It has a maximum number of vanishing derivatives in these neighborhoods for the degree of polynomial involved. The order N is determined as follows from the remaining design parameter, λ: When ω equals ω_r, Eq. 12.3 gives

$$|H_B(j\omega_r)|^2 = \frac{1}{1 + \lambda^2} = \frac{1}{1 + \epsilon^2(\omega_r/\omega_c)^{2N}}$$

Analog and Digital Filter Design   Chap. 12

Therefore,

$$N \geq \frac{\log(\lambda/\epsilon)}{\log(\omega_r/\omega_c)} \qquad (12.4)$$

(Since the logarithms in Eq. 12.4 appear in a ratio, any base may be used.) Thus the designer must make a choice between the squareness of the power gain characteristic and the smallness of N. The relative improvement in this characteristic as N is increased is illustrated in Fig. 12.3 for the case where $\epsilon = 1$. Typically, the designer chooses values for λ and ω_r, uses Eq. 12.4 to determine N, possibly adjusts λ and/or ω_r if N can be made larger or if N must be made smaller, and in this manner eventually arrives at a choice for N. (The transient response time of the filter, which increases more or less in proportion to N, also affects the choice of N in some cases.)

Once N is determined, Eq. 12.3 gives the power gain of the Butterworth filter. The s-plane poles of $|H_B(s)|^2$ are found by setting the denominator of Eq. 12.3 equal to zero, with $s = j\omega$. That is, there are poles where

$$\epsilon^2 \left(-\frac{s^2}{\omega_c^2} \right)^N + 1 = 0$$

Therefore,

$$\text{poles at } s_n = \omega_c \epsilon^{-1/N} e^{j\pi(2n+N-1)/2N}; \qquad n = 1, 2, 3, \ldots, 2N \qquad (12.5)$$

Thus, as illustrated in Fig. 12.4 for $N = 3$, the poles of the power gain function lie on a circle with radius $\omega_c \epsilon^{-1/N}$ in the s-plane. The poles of the transfer function itself, $H_B(s)$, are those in the left-half plane, since

$$|H_B(s)|^2 = H_B(s)H_B(-s) \qquad (12.6)$$

and therefore Eq. 12.5 gives the poles of the transfer function when n equals $1, 2, \ldots$, and N. The following is an example of the derivation of $H_B(s)$ from typical requirements.

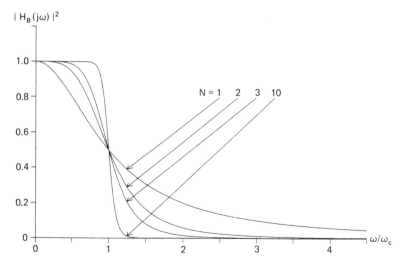

Figure 12.3 Butterworth filter power gain for $\epsilon = 1$.

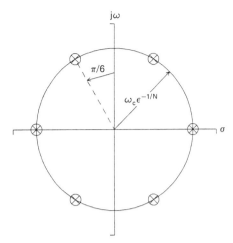

Figure 12.4 Poles of Butterworth filter on the s-plane; $N = 3$.

Example 12.1

Design a maximally flat (Butterworth) analog filter to meet the following specifications:

passband (0-ω_c)	0–100 krad/s
minimum power gain at ω_c	0.5(-3 dB)
start of stopband (ω_r)	150 krad/s
maximum power gain at ω_r	0.1(-10 dB)

From the definitions in Eqs. 12.1 and 12.2, ϵ and λ are determined to be 1 and 3, respectively. Next, the order N is found from Eq. 12.4 to be

$$N \geq \frac{\log(\lambda/\epsilon)}{\log(\omega_r/\omega_c)} \geq \frac{\log 3}{\log 1.5}; \quad N = 3 \tag{12.7}$$

Finally, using Eq. 12.5 to obtain the poles in the left-half plane, the transfer function of the filter is obtained:

$$\left. \begin{aligned} H_B(s) &= \frac{-s_1 s_2 s_3}{(s - s_1)(s - s_2)(s - s_3)} \\ s_n &= \omega_c \epsilon^{-1/N} e^{j\pi(n+1)/3}; \quad n = 1, 2, 3 \\ &= 10^5 e^{j\pi(n+1)/3}; \quad n = 1, 2, 3 \end{aligned} \right\} \tag{12.8}$$

The power gain characteristic for this example is illustrated in Fig. 12.3, being the case where $N = 3$ and $\epsilon = 1$. Note that the magnitude of $H_B(s)$ in Eq. 12.8 is in fact the square root of the power gain, Eq. 12.3, in this case. The product $s_1 s_2 s_3$ in the numerator would be obtained if, in Eq. 12.3, the numerator and denominator were multiplied by $\epsilon^{-2}\omega_c^{2N}$ before taking the square root. Equation 12.8 suggests, in fact, a general form for $H_B(s)$ that must be used to make the dc power gain, found by letting $s = 0$ in $|H_B(j\omega)|^2$, equal to 1:

$$H_B(s) = \frac{(-1)^N s_1 s_2 \cdots s_N}{(s - s_1)(s - s_2) \cdots (s - s_N)} \tag{12.9}$$

Here $s_1 \cdots s_N$ are the left-plane poles of those in Eq. 12.5, and clearly the dc power gain is $|H_B(0)|^2 = 1$. Again, note that the numerator equals $\epsilon^{-1}\omega_c^N$.

Example 12.2

Plot the poles, the power gain in dB, and the phase shift for the following Butterworth filter:

$$N = 2$$

$$\epsilon = 1$$

$$\omega_c = 100 \text{ rad/s}$$

The plots are shown in Fig. 12.5. With N equal to 2, there are two poles in each half-plane on the circle with radius 100, as specified by Eq. 12.5. the power gain in dB is found from Eqs. 12.1 and 12.3:

$$\text{dB} = 10 \log_{10} |H_B(j\omega)|^2$$

$$= -10 \log_{10}\left[1 + \left(\frac{\omega}{100}\right)^4\right] \tag{12.10}$$

When ω is large, the *slope* of the dB function in Eq. 12.10 is seen to be about 12 dB per octave. That is, when ω doubles, the power gain drops about 12 dB. This slope is called the *rolloff rate* of the power gain function and is seen in general to be *6N dB per octave* for the Butterworth filter.

On the s-plane plot, the power gain may also be considered as the square of $\omega_c^2 / P_1 P_2$, where P_1 and P_2 are the respective distances from the poles to the operating frequency ω. A final point concerning the power gain here is that it is also illustrated in Fig. 12.3, being the curve for $N = 2$, so that the different shape of the dB-plot in Fig. 12.5 can be compared.

The phase contributions α_1 and α_2 are also shown in the pole–zero plot in Fig. 12.5, and, from the geometry of the plot, these are seen to be

$$\left. \begin{aligned} \alpha_1 &= \tan^{-1} \frac{\omega + 100/\sqrt{2}}{100/\sqrt{2}} \\[2mm] \alpha_2 &= \tan^{-1} \frac{\omega - 100/\sqrt{2}}{100/\sqrt{2}} \end{aligned} \right\} \tag{12.11}$$

and, of course, the total phase shift of the filter is

$$\phi(\omega) = -(\alpha_1 + \alpha_2) \tag{12.12}$$

which is also plotted in Fig. 12.5.

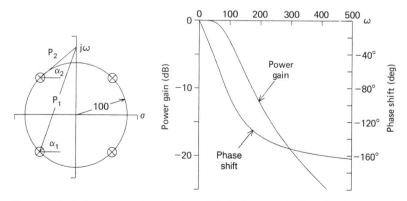

Figure 12.5   Poles, power gain, and phase shift for Butterworth filter, $N = 2$, $\omega_c = 100$.

In summary, Butterworth filters are characterized by a smooth power gain characteristic having maximum flatness in both the passband and the stopband, along with reasonably sharp cutoff. By sacrificing some of the flatness and allowing a ripple in either the passband or the stopband, a sharper cutoff can be obtained with the same number of poles. The Chebyshev filters, described next, follow this approach.

12.3 CHEBYSHEV FILTERS

If the Butterworth filter is to be altered so as to allow ripple to occur in, say, the passband, then it is natural to desire that, for a fixed number N of poles on the s-plane, the maximum gain excursions caused by the ripple be held to a minimum. This in turn leads to the use of the *Chebyshev polynomial* of degree N, $V_N(x)$, defined as follows:

$$V_0(x) = 1$$
$$V_1(x) = x$$
$$V_2(x) = 2x^2 - 1$$
$$V_3(x) = 4x^3 - 3x$$
$$\vdots$$
$$V_N(x) = 2xV_{N-1}(x) - V_{N-2}(x); \qquad N > 1$$

These polynomials have the following property, discovered by P. L. Chebyshev:

Of all polynomials of degree N with leading coefficient equal to 1, the polynomial

$$\frac{V_N(x)}{2^{N-1}}$$

has the smallest maximum magnitude in the interval $|x| \leq 1$. This smallest maximum magnitude is, in fact, 2^{1-N}.

For N equal 1, the property is obvious. Examples for N equal 2 and 3 are illustrated in Fig. 12.6.

Thus the Chebyshev polynomial produces equal ripple, that is, ripple of constant magnitude, in the interval $|x| \leq 1$. The problem now is to translate this equal ripple into the passband or stopband of a power gain characteristic. This is done with the two following types of functions of $V_N(x)$:

$$\text{Type 1:} \quad |H_C(j\omega)|^2 = \frac{1}{1 + \epsilon^2 V_N^2(\omega/\omega_c)} \qquad (12.14)$$

$$\text{Type 2:} \quad |H_C(j\omega)|^2 = \frac{1}{1 + \epsilon^2 V_N^2(\omega_r/\omega_c)/V_N^2(\omega_r/\omega)} \qquad (12.15)$$

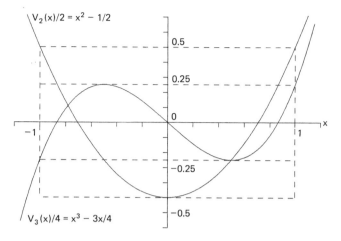

Figure 12.6 Chebyshev polynomials in the interval $|x| < 1$.

Note the similarity of Eq. 12.14 to the Butterworth characteristic in Eq. 12.3. Here, the function $V_N(\omega/\omega_c)$ is used in place of just $(\omega/\omega_c)^N$ in Eq. 12.3 to obtain the type 1 form in Eq. 12.14. The type 1 characteristic translates the equal ripple to the *passband* of the power gain curve, whereas the type 2 characteristic translates it into the *stopband*.

These properties are illustrated in Fig. 12.7, which compares the type 1 and type 2 Chebyshev power gains with the Butterworth power gain. The illustrations are for the specific case with $N = 3$, $\epsilon = 0.2$, and $\omega_r = 2\omega_c$. The sharper cutoff of the Chebyshev characteristics is illustrated by the lower stopband gain for the Chebyshev filter, which, in this example, is $1/(1 + \lambda^2) \approx 0.04$, compared with about 0.28 for the Butterworth filter. In general, a comparison of the three power gain curves can be tabulated as follows:

Filter	$|H(j\omega)|^2$ in Passband	$|H(j\omega)|^2$ in Stopband
Butterworth	Maximally flat	Maximally flat
Chebyshev type 1	Equal ripple between 1 and $1/(1 + \epsilon^2)$	Maximally flat
Chebyshev type 2	Maximally flat	Equal ripple between 0 and $1/(1 + \lambda^2)$

Just as for the Butterworth filter, the order N of the Chebyshev filters can be found in terms of ϵ, λ, ω_c, and ω_r. Furthermore, as suggested by Fig. 12.7, N is found in the same manner for both the type 1 and type 2 filters. To demonstrate the latter and to derive an expression for N, a closed expression for $V_N(x)$ in Eq. 12.13 is needed. Such an expression (Storer, 1957) is (see Table 1.2)

$$V_N(x) = \cosh(N \cosh^{-1} x)$$
$$= \cos(N \cos^{-1} x) \qquad (12.16)$$

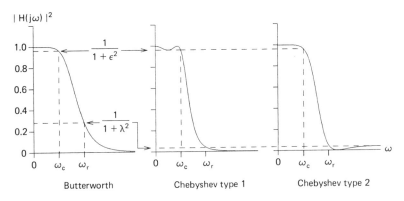

$|H(j\omega)|^2$

1.0

0.8

0.6

0.4

0.2

0

$\dfrac{1}{1+\epsilon^2}$

$\dfrac{1}{1+\lambda^2}$

0   ω_c ω_r 0 ω_c ω_r 0 ω_c ω_r ω

Butterworth   Chebyshev type 1   Chebyshev type 2

Figure 12.7 Comparison of power gains; $N = 3$, $\epsilon = 0.2$, and $\omega_r = 2\omega_c$.

To derive an expression for N, let $\omega = \omega_r$ and $|H_C(j\omega)|^2 = 1/(1 + \lambda^2)$ in either Eq. 12.14 or 12.15. The result is

$$\frac{1}{1 + \lambda^2} = \frac{1}{1 + \epsilon^2 V_N^2(\omega_r/\omega_c)} \tag{12.17}$$

Note that either equation gives this result, because $V_N^2(1) = 1$ in Eq. 12.15. Equations 12.16 and 12.17 now provide the solution for N, which must of course be an integer:

$$\frac{\lambda}{\epsilon} = V_N\!\left(\frac{\omega_r}{\omega_c}\right) = \cosh\!\left[N \cosh^{-1}\!\left(\frac{\omega_r}{\omega_c}\right)\right]$$

Therefore,

$$N \geq \frac{\cosh^{-1}(\lambda/\epsilon)}{\cosh^{-1}(\omega_r/\omega_c)} \tag{12.18}$$

This result for either of the Chebyshev filters is analogous to Eq. 12.4 for the Butterworth filter, but gives a smaller N for the same parameter values.

The poles of the two Chebyshev filters are of course different and will therefore be considered separately [see Guillemin (1957) for a similar treatment]. For the type 1 filter, the poles are found by setting the denominator of Eq. 12.14 equal to zero, with s in place of $j\omega$ as the argument:

$$1 + \epsilon^2 V_N^2\!\left(\frac{s}{j\omega_c}\right) = 0$$

Therefore,

$$V_N\!\left(\frac{s}{j\omega_c}\right) = \pm\frac{j}{\epsilon} \tag{12.19}$$

Recalling from Eq. 12.16 that $V_N(x)$ equals $\cos(N \cos^{-1} x)$ and noting that the inner term, $\cos^{-1}(s/j\omega_c)$ in this case, is complex, let this term be represented by $\gamma + j\alpha$, so that

$$\cos^{-1}\left(\frac{s}{j\omega_c}\right) = \gamma + j\alpha$$

Therefore,

$$s = \omega_c[j \cos(\gamma + j\alpha)]$$
$$= \omega_c(\sinh \alpha \sin \gamma + j \cosh \alpha \cos \gamma) \tag{12.20}$$

The function V_N, given in terms of α and γ, now becomes

$$V_N\left(\frac{s}{j\omega_c}\right) = \cos[N(\gamma + j\alpha)]$$
$$= \cos N\gamma \cosh N\alpha - j \sin N\gamma \sinh N\alpha \tag{12.21}$$

where α and γ are real variables. Equating this latter form of V_N to $\pm j/\epsilon$ in Eq. 12.19 gives the following results: First, since $\cosh N\alpha$ has to be nonzero for α real,

$$\cos N\gamma = 0$$

Therefore,

$$\gamma_n = \frac{2n - 1}{2N}\pi; \qquad n = 1, 2, \ldots, 2N \tag{12.22}$$

(Note that these are closely related to the angles of the Butterworth poles in Eq. 12.5.) Secondly, since $\sin N\gamma = \pm 1$ as a consequence of Eq. 12.22,

$$\sinh N\alpha = \pm\frac{1}{\epsilon}$$

Therefore,

$$\alpha = \pm\frac{1}{N} \sinh^{-1}\frac{1}{\epsilon} \tag{12.23}$$

Thus the Chebyshev type 1 poles are given by Eqs. 12.20, 12.22, and 12.23. Although this derivation is somewhat complicated, the result is quite simple. It is even more appealing if $\beta_n - \pi/2$, where β_n is the Butterworth pole angle in Eq. 12.5, is substituted for γ_n. (Note that Eq. 12.22 is thus related to Eq. 12.5.) Then the *type 1 Chebyshev poles* are as follows:

$$s_n = \omega_c(\sinh \alpha \cos \beta_n + j \cosh \alpha \sin \beta_n);$$
$$\alpha = \frac{1}{N} \sinh^{-1}\frac{1}{\epsilon}; \qquad \beta_n = \frac{2n + N - 1}{2N}\pi; \qquad n = 1, 2, \ldots, 2N \tag{12.24}$$

As suggested by the form of Eq. 12.24, the poles of $|H_C(s)|^2$ lie on an ellipse in the s-plane. An example is given in Fig. 12.8 for $N = 3$. The figure shows the geometry

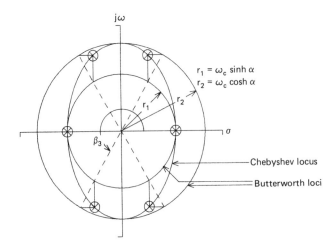

$r_1 = \omega_c \sinh \alpha$
$r_2 = \omega_c \cosh \alpha$

——— Chebyshev locus

=== Butterworth loci

Figure 12.8 Poles of the Chebyshev type 1 filter with $N = 3$.

associated with Eq. 12.24, and in fact suggests a systematic construction of the Chebyshev poles, given Butterworth poles for the same N. As mentioned above, the Butterworth poles (on the circles in the figure) are at the angles $[\beta_n]$. The construction is then as shown in the figure. The poles of $H_C(s)$ are taken to be those on the left-half plane, that is, those for n going from 1 to N in Eq. 12.24. The following example shows the design of a type 1 filter from typical requirements.

Example 12.3

Design a type 1 Chebyshev filter to meet the following specifications:

passband	0–100 krad/s
minimum power gain at ω_c	0.9615(-0.17 dB)
start of stop band	200 krad/s
maximum power gain at ω_r	0.0385(-14.1 dB)

From the definitions of ϵ and λ (e.g., see Fig. 12.7), these parameters are determined from the specifications to be $\epsilon = 0.2$ and $\lambda = 5.0$. Next, N can be found using Eq. 12.18:

$$N \geq \frac{\cosh^{-1}(\lambda/\epsilon)}{\cosh^{-1}(\omega_r/\omega_c)} = 2.97; \qquad N = 3 \tag{12.25}$$

Equation 12.24 now gives the following poles for this example:

$$\alpha = \tfrac{1}{3} \sinh^{-1}(5) = 0.771$$

$$\beta_1 = \frac{2\pi}{3}, \; \beta_2 = \pi, \; \beta_3 = \frac{4\pi}{3} \text{ in left-half plane}$$

$$s_2 = 10^5(\sinh 0.771)(-1) = -0.850 \times 10^5$$

$$s_1, s_3 = 10^5[(0.850)(-0.500) \pm j(1.31)(0.866)]$$

$$= 10^5(-0.425 \pm j1.13)$$

The poles s_1, s_2, and s_3 are located as in the left half of Fig. 12.8, except that here the ellipse is more circular. Thus this type 1 filter has the transfer function

$$H_C(s) = \frac{-s_1 s_2 s_3}{(s - s_1)(s - s_2)(s - s_3)} \tag{12.26}$$

with the poles given as above. The power gain, $|H_C(j\omega)|^2$, is shown in Fig. 12.7.

For this example the factor $(-s_1 s_2 s_3)$ in the numerator of Eq. 12.26 is the correct choice, assuming that the ripple is to be between 1 and $1/(1 + \epsilon^2)$ as in Fig. 12.7, because then $H_C(0) = 1$. Note carefully, however, that this procedure is only correct when N is odd. When N is even, the $V_N^2(0)$ in Eq. 12.14 is one instead of zero, and the correct value of $H_C(0)$ is

$$\sqrt{\frac{1}{1 + \epsilon^2}}$$

The transfer function $H_C(s)$ must therefore be scaled by this amount when N is even. Specifically, when N is even, $H_C(s)$ has the form

$$H_C(s) = \frac{s_1 s_2 \cdots s_N}{(s - s_1)(s - s_2) \cdots (s - s_N)\sqrt{1 + \epsilon^2}} \tag{12.27}$$

Turning now to the *type 2 Chebyshev* filter, with equal ripple in the stopband, Eq. 12.15 suggests that the transfer function has zeros as well as poles. The poles and zeros can be found by an approach similar to the above if the substitutions

$$\nu = \frac{\omega_r \omega_c}{\omega}; \qquad \hat{\epsilon} = \frac{1}{\epsilon V_N(\omega_r/\omega_c)} \tag{12.28}$$

are made in Eq. 12.15 to obtain the power gain in the following form:

$$|H_C(j\nu)|^2 = \frac{1}{1 + \hat{\epsilon}^{-2}/V_N^2(\nu/\omega_c)} = \frac{\hat{\epsilon}^2 V_N^2(\nu/\omega_c)}{1 + \hat{\epsilon}^2 V_N^2(\nu/\omega_c)} \tag{12.29}$$

The similarity of Eqs. 12.29 and 12.14 suggests that the poles of, say, $|H_C(\tau)|^2$, where $\tau = u + j\nu$, can be found by steps similar to the steps above for the type 1 filter. One can then map the poles and zeros of $H_C(\tau)$ from the τ-plane onto the s-plane using the transformation $s = \omega_r \omega_c/\tau$, as suggested by Eq. 12.28. This procedure for the type 2 filter is illustrated in the following example:

Example 12.4

Design a type 2 Chebyshev filter (flat in the passband, equal ripple in the stopband) having the same properties as the filter in Example 12.3. First, recall that the procedures for finding ϵ, λ, and N are the same for either type 1 or type 2 filter. Therefore,

$$\epsilon = 0.2; \qquad \lambda = 5.0; \qquad N = 3$$

as in the preceding example. Next, as in Eq. 12.28,

$$\nu = \frac{\omega_r \omega_c}{\omega} = \frac{2 \times 10^{10}}{\omega}$$

$$\hat{\epsilon} = \frac{1}{\epsilon V_N(\omega_r/\omega_c)} = \frac{1}{\epsilon V_3(2)} = 0.1923 \tag{12.30}$$

Next, following the procedure given above, the left-half-plane poles of Eq. 12.29 are found as follows:

$$\alpha = \frac{1}{3} \sinh^{-1} \frac{1}{\hat{\epsilon}} = 0.784$$

$$\beta_1 = \frac{2\pi}{3}, \quad \beta_2 = \pi, \quad \beta_3 = \frac{4\pi}{3}$$

$$\tau_2 = -\omega_c \sinh \alpha = -0.866 \times 10^5$$

$$\tau_1, \tau_3 = \omega_c(0.500 \sinh \alpha \pm j0.866 \cosh \alpha)$$

$$= (-0.433 \pm j1.15)10^5$$

These poles are mapped to the s-plane under $s = \omega_r\omega_c/\tau$, giving

$$s_2 = 2 \times 10^{10}/\tau_2 = -2.31 \times 10^5$$

$$s_1, s_3 = (-0.574 \pm j1.52)10^5$$

Finally, Eq. 12.29 has zeros where $V_3(\tau/\omega_c) = 0$, or at $\tau = 0$ and $\pm j0.866\omega_c$. The zero at $\tau = 0$ becomes a zero at $s = \infty$, and the other two zeros are at

$$s_4, s_5 = \frac{\omega_r\omega_c}{\tau} = \pm j2.31 \times 10^5 \tag{12.31}$$

The transfer function for this example is, therefore,

$$H_C(s) = \frac{-s_1s_2s_3(s - s_4)(s - s_5)}{s_4s_5(s - s_1)(s - s_2)(s - s_3)} \tag{12.32}$$

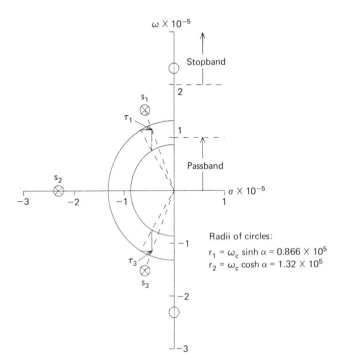

Radii of circles:
$r_1 = \omega_c \sinh \alpha = 0.866 \times 10^5$
$r_2 = \omega_c \cosh \alpha = 1.32 \times 10^5$

Figure 12.9 Poles and zeros of the Chebyshev type 2 filter in Example 12.4.

with the poles and zeros given as above. Since this type 2 filter has $\epsilon = 0.2$, $N = 3$, and $\omega_r = 2\omega_c$, the power gain function, $|H_C(j\omega)|^2$, is shown on the right in Fig. 12.7. Figure 12.9 illustrates the pole and zero locations. Note the location of the poles and zeros relative to the passband and stopband in Fig. 12.9 and also note the construction of s_1 and s_3 from τ_1 and τ_3, which are in turn constructed as in Fig. 12.8. Finally, in this example, note that $H_C(s)$ in Eq. 12.32 is constructed so that $H_C(0) = 1$. Unlike the type 1 Chebyshev filter, the type 2 filter has $H_C(0) = 1$ for both even and odd values of N.

Besides the Butterworth and Chebyshev filters described above, the *elliptic* filter (Storer, 1957; Gillemin, 1957; Gold and Rader, 1969) is another standard analog design. The elliptic filter has a power gain function with equal ripple in both passband and stopband, and an even sharper cutoff. However, instead of continuing with the subject of analog filters, the discussion in this chapter turns now to the design of digital filters.

12.4 DIGITAL FILTERS VIA THE BILINEAR TRANSFORMATION

This approach to digital filter design enables one to take advantage of known analog designs, such as the Butterworth and Chebyshev filters. The bilinear transformation is viewed as a transformation from the s-plane to the z-plane, which allows the conversion of analog poles and zeros into digital poles and zeros. Appendix B contains Fortran routines that embody the procedure outlined in this section for designing Butterworth digital filters.

The reader will recall that the bilinear transformation is among the simulation methods described in Chapter 11. Indeed, having already studied digital simulation, it would be natural to arrive at a digital filter design by attempting to simulate the corresponding analog filter as accurately as possible.

However, *accurate simulation is not the primary objective* here. If the analog filter characteristic can be improved rather than simulated, that is, made more rectangular in the transformation, then the resulting digital filter will be preferred over an accurate simulation. An example of such an improvement is illustrated in Fig. 12.10. For the figure, assume that an analog filter with the transfer function

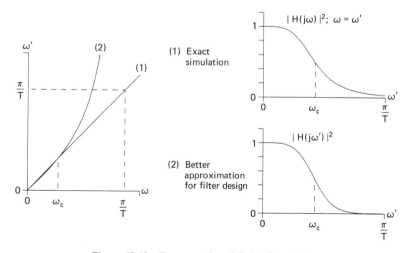

Figure 12.10 Two examples of digital filter design.

$H(j\omega)$ has been designed. Case 1 in the figure illustrates a perfect simulation of $H(j\omega)$. The plot on the left has $\omega' = \omega$, where ω' is the frequency variable for the digital simulation as in Chapter 11, Section 11.4. Therefore the power gain of the resulting digital filter, shown at the upper right in the figure, is exactly $H(j\omega)$ for $|\omega| < \pi/T$. Case 2, on the other hand, shows a bilinear transformation of ω into ω' that results in the improved power gain characteristic, $|H(j\omega')|^2$, shown at the lower right. The improved gain characteristic cuts off faster because $\omega' > \omega$ for $\omega > \omega_c$, and has a flatter top because $\omega' < \omega$ for $\omega < \omega_c$.

Figure 12.10, then, suggests that a bilinear transformation used for producing a digital filter from an analog filter should produce an improved form similar to curve 2 on the left. The bilinear transformation with $A = 1$ is the simplest mapping of the $j\omega$ axis to the unit circle on the z-plane:

$$s \leftarrow \frac{z - 1}{z + 1} \tag{12.33}$$

Being only an approximation to $s = \log z$, it produces a frequency transformation similar to curve 2 in Fig. 12.10, as shown by letting $z = e^{j\omega T}$:

$$j\omega' = \frac{e^{j\omega T} - 1}{e^{j\omega T} + 1}$$

$$= \frac{e^{j\omega T/2} - e^{-j\omega T/2}}{e^{j\omega T/2} + e^{-j\omega T/2}}$$

Therefore,

$$\omega' = \tan \frac{\omega T}{2} \tag{12.34}$$

A plot of ω' versus ωT is given in Fig. 12.11. It is similar to curve 2 in Fig. 12.10, except that Fig. 12.11 is of course independent of ω_c and other properties of the continuous filter.

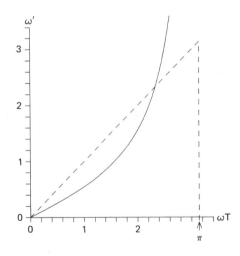

Figure 12.11 Bilinear frequency transformation, $\omega' = \tan (\omega T/2)$.

Analog and Digital Filter Design   Chap. 12

To adjust for the fact that ω_c and ω_r are also transformed, the following design procedure has been devised [see Gold and Rader (1969), Kaiser (1966), and Golden and Kaiser (1964)]:

PROCEDURE FOR USING THE BILINEAR TRANSFORMATION

1. Start with the desired values of ω_c and ω_r or similar critical frequencies. (For a bandpass filter, for example, the endpoints of the passband would be used in place of ω_c and ω_r (see Example 12.8). Find the transformed values ω_c' and ω_r' using Eq. 12.34, that is,

$$\omega' = \tan\left(\frac{\omega T}{2}\right)$$

2. Design an analog filter with transfer function $H_A(s)$ just as in the preceding sections, but using ω_c' and ω_r' in place of ω_c and ω_r.

3. Transform $H_A(s)$ into $\tilde{H}(z)$ using

$$s \leftarrow \frac{z-1}{z+1}$$

and $\tilde{H}(z)$ is then the z-transfer function of the desired digital filter.

The correctness of the "adjustment" of ω_c and ω_r in this procedure is shown by the following argument: According to step 3,

$$\tilde{H}(z) = H_A\left(\frac{z-1}{z+1}\right) \tag{12.35}$$

But then, as shown previously, the transfer function of the digital filter is

$$\overline{H}(j\omega) = \tilde{H}(e^{j\omega T}) \tag{12.36}$$

Therefore, from Eqs. 12.34 through 12.36,

$$\overline{H}(j\omega) = H_A(j\omega') \tag{12.37}$$

and so step 2 of the procedure produces the desired gain values $\overline{H}(j\omega_c)$ and $\overline{H}(j\omega_r)$ at the frequencies ω_c and ω_r.

The following typical examples, in which digital Butterworth and Chebyshev filters are designed, serve to illustrate this procedure.

Example 12.5

Design a digital Butterworth filter to satisfy the following specifications:

Sampling interval $T = 100\ \mu s$

Power gain between 0 and -0.7 dB from 0 to 1000 Hz

Power gain down to at least -10 dB at 1200 Hz

Following the above procedure, the first step is to transform the frequencies ω_c and ω_r, which in this example are 2000π and 2400π rad/s, respectively. The transformed values are

$$\left.\begin{aligned} \omega_c' &= \tan\frac{\omega_c T}{2} = \tan\frac{2000\pi \times 10^{-4}}{2} = 0.32492 \\[2mm] \omega_c' &= \tan\frac{\omega_r T}{2} = \tan\frac{2400\pi \times 10^{-4}}{2} = 0.39593 \end{aligned}\right\} \qquad (12.38)$$

The next step is to design an analog filter using ω_c' and ω_r'. If the minimum passband gain is -0.7 dB at ω_c', then

$$10\log_{10}|H_A(j\omega_c')|^2 = 10\log_{10}\frac{1}{1+\epsilon^2} = -0.7$$

Therefore,

$$\log_{10}(1+\epsilon^2) = 0.070 \quad \text{and} \quad \epsilon = 0.41821 \qquad (12.39)$$

and if the maximum stopband gain is -10 dB at ω_r', then

$$10\log_{10}|H_A(j\omega_r')|^2 = 10\log_{10}\frac{1}{1+\lambda^2} = -10$$

Therefore,

$$\lambda = 3 \qquad (12.40)$$

Having found ω_c', ω_r', ϵ, and λ, the order N of the filter can be specified using Eq. 12.4:

$$N \geq \frac{\log(\lambda/\epsilon)}{\log(\omega_r'/\omega_c')} = 9.97$$

Therefore,

$$N = 10 \qquad (12.41)$$

Therefore, as in Eqs. 12.9 and 12.5, the transfer function of the analog filter is

$$H_A(s) = \frac{s_1 s_2 \cdots s_{10}}{(s - s_1)(s - s_2)\cdots(s - s_{10})};$$
$$s_n = \omega_c' \epsilon^{-1/N} e^{j\pi(2n+N-1)/2N}; \qquad n = 1, 2, \ldots, N \qquad (12.42)$$
$$= 0.35452 e^{j\pi(2n+9)/20}; \qquad n = 1, 2, \ldots, 10$$

The third and final step is now to translate $H_A(s)$ into $\tilde{H}(z)$ using the bilinear substitution:

$$\tilde{H}(z) = H_A\left(\frac{z-1}{z+1}\right)$$

$$= \frac{s_1 s_2 \cdots s_{10}}{\left(\dfrac{z-1}{z+1} - s_1\right)\left(\dfrac{z-1}{z+1} - s_2\right)\cdots\left(\dfrac{z-1}{z+1} - s_{10}\right)} \qquad (12.43)$$

$$= \frac{s_1 s_2 \cdots s_{10}(z+1)^{10}}{[(1 - s_1)z - (1 + s_1)]\cdots[(1 - s_{10})z - (1 + s_{10})]}$$

with s_1 through s_{10} as in Eq. 12.42. $\tilde{H}(z)$ is the desired z-transfer function. The power gain of this digital filter could be found by substituting $e^{j\omega T}$ for z and finding the squared magnitude of Eq. 12.43, but it can be found much more easily, as shown by Golden and Kaiser (1964), by using Eqs. 12.37 and 12.34:

$$|\overline{H}(j\omega)|^2 = |H_A(j\omega')|^2$$
$$= \left| H_A\left(j\,\tan\frac{\omega T}{2} \right) \right|^2 \tag{12.44}$$

Thus it is only necessary to substitute ω'_c for ω_c and ω' for ω in the Butterworth power gain, Eq. 12.3, to obtain

$$|\overline{H}(j\omega)|^2 = \frac{1}{1 + \epsilon^2 \left(\dfrac{\tan(\omega T/2)}{\tan(\omega_c T/2)} \right)^{2N}} \tag{12.45}$$

The power gain function for this example is plotted in Fig. 12.12. The figure shows the power gain of the digital filter in dB, as well as the power gain of an analog Butterworth filter with the same N and ϵ, thus illustrating the design improvement effected by the bilinear transformation and predicted above in Fig. 12.10.

The digital transfer function $\tilde{H}(z)$ in Eq. 12.43 can be realized conveniently in the cascade form described in Chapter 9, Section 9.5. To accomplish this, note that s_n and s_{11-n} in Eq. 12.42 are complex conjugates. That is,

$$\left. \begin{array}{l} s_n = Re^{j\theta_n}; \quad R = \omega'_c\epsilon^{-1/10},\, \theta_n = \dfrac{2n + 9}{20}\pi \\[2ex] s_{11-n} = Re^{-j\theta_n} \end{array} \right\} \tag{12.46}$$

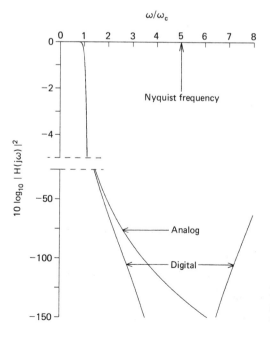

Figure 12.12 Butterworth filter power gain for $N = 10$, $\epsilon = 0.41821$, and $\omega_c T = 0.2\pi$.

Next, factor Eq. 12.43 into five parts as follows:

$$\tilde{H}(z) = \tilde{H}_1(z)\tilde{H}_2(z)\tilde{H}_3(z)\tilde{H}_4(z)\tilde{H}_5(z);$$ (12.47)

$$\tilde{H}_n(z) = \frac{s_n s_{11-n}(z+1)^2}{[(1-s_n)z - (1+s_n)][(1-s_{11-n})z - (1+s_{11-n})]}$$

$$= \frac{R^2(z+1)^2}{[(1-Re^{j\theta_n})z - (1+Re^{j\theta_n})][(1-Re^{-j\theta_n})z - (1+Re^{-j\theta_n})]}$$ (12.48)

$$= \frac{R^2(z^2 + 2z + 1)}{(1+R^2-2R\cos\theta_n)z^2 - 2(1-R^2)z + (1+R^2+2R\cos\theta_n)}$$

with n going from 1 to 5 in the latter. The realization given by Eqs. 12.47 and 12.48 is diagrammed in Fig. 12.13. The parameters are left in general form so that the figure applies to any digital Butterworth filter of order 10. In the lower diagram, note the multiplication of the input by R^2 in the numerator of Eq. 12.48 and the division of the output by the coefficient of z^2 in the denominator of Eq. 12.48. By altering the number of blocks in the upper diagram, the figure can obviously be applied to any digital Butterworth filter with N even. The slight modification for N odd is left as an exercise.

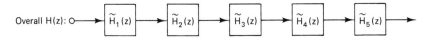

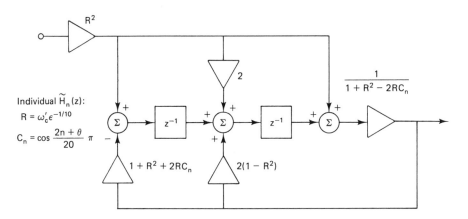

Figure 12.13 Cascade realization of the digital Butterworth filter; $N = 10$.

Example 12.6

To compare the Butterworth filter above with a Chebyshev filter of the same order, suppose a digital Chebyshev filter is to be designed according to the following:

Sampling interval $T = 100\ \mu s$

Power gain between 0 and -0.7 dB from 0 to 1000 Hz

Power gain maximally flat in stopband and down to at least -10 dB at 1040 Hz

Analog and Digital Filter Design   Chap. 12

Here, λ, ϵ, and ω_c are the same as in Example 12.5, but ω_r is much smaller—the transition band between ω_c and ω_r has only one-fifth of its former width. Still N turns out to be as before:

$$N \geq \frac{\cosh^{-1}(\lambda/\epsilon)}{\cosh^{-1}(\omega_r'/\omega_c')} = \frac{\cosh^{-1}(3/0.41821)}{\cosh^{-1}(0.33887/0.32492)}$$

$$= 9.105$$

Therefore,

$$N = 10 \qquad\qquad (12.49)$$

Thus a much sharper cutoff is achieved with the Chebyshev filter of the same order. Following step 2 of the design procedure outlined above, and using Eqs. 12.27 and 12.24 the analog filter is designed as follows (the type 1 filter is chosen to provide flat power gain in the stopband):

$$H_A(s) = \frac{s_1 s_2 \cdots s_{10}}{(s - s_1)(s - s_2) \cdots (s - s_{10})\sqrt{1 + \epsilon^2}};$$

$$s_n = \omega_c'(\sinh \alpha \cos \beta_n + j \cosh \alpha \sin \beta_n); \qquad n = 1, 2, \ldots, 10 \qquad (12.50)$$

$$= 0.05241 \cos \frac{2n + 9}{20}\pi + j0.32912 \sin \frac{2n + 9}{20}\pi$$

Step 3 of the procedure then gives the z-transfer function of the desired digital filter:

$$\tilde{H}(z) = H_A\left(\frac{z - 1}{z + 1}\right) \qquad\qquad (12.51)$$

$$= \frac{s_1 s_2 \cdots s_{10}(z + 1)^{10}}{[(1 - s_1)z - (1 + s_1)] \cdots [(1 - s_{10})z - (1 + s_{10})]\sqrt{1 + \epsilon^2}}$$

with s_n as in Eq. 12.50. A cascade realization of $\tilde{H}(z)$ has the same form as Fig. 12.13 (but different parameter values of course), and can be obtained in the same manner.

The digital power gain for this example, similar to Eq. 12.45 for the preceding example, is

$$|\overline{H}(j\omega)|^2 = |H_A(j\omega')|^2$$

$$= \frac{1}{1 + \epsilon^2 V_N^2\left(\dfrac{\tan(\omega T/2)}{\tan(\omega_c T/2)}\right)} \qquad\qquad (12.52)$$

The power gain in dB is plotted in Fig. 12.14 for comparison with the Butterworth characteristic in Fig. 12.12. The poles and zeros for this example are plotted on the z-plane in Fig. 12.15. As shown by Eq. 12.51, there is a zero of order 10 at $z = -1$, and there are 10 poles at

$$z_n = \frac{1 + s_n}{1 - s_n}; \qquad n = 1, 2, \ldots, 10 \qquad (12.53)$$

with s_n as in Eq. 12.50. Note the location of the poles relative to the passband. Since the cutoff frequency is 1000 Hz and one-half the sampling frequency is 5000 Hz, the passband extends over one-fifth of the unit circle, or out to 36 degrees in this case.

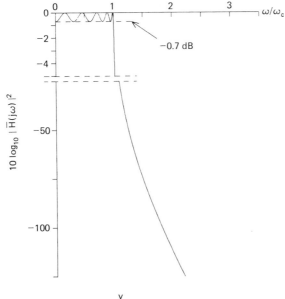

Figure 12.14 Digital type 1 Chebyshev power gain for $N = 10$, $\epsilon = 0.41821$, and $\omega_c T = 0.2\pi$.

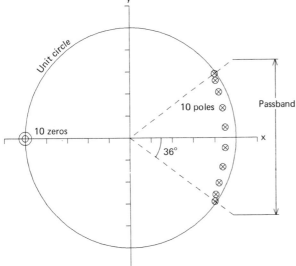

Figure 12.15 Poles and zeros of the digital type 1 Chebyshev filter; $N = 10$, $\epsilon = 0.41821$, and $\omega_c T = 0.2\pi = 36°$.

In summary, the bilinear transformation provides a convenient and useful means for employing analog design methods in the construction of digital filters. The analog power gain curve is not preserved in the transformation but is in fact improved, so that the digital filter has an even more desirable characteristic than its analog counterpart.

12.5 FREQUENCY TRANSFORMATIONS

The conversion of lowpass filters into highpass, bandpass, and bandstop filters is described quite clearly by Guillemin (1957) and by Kuo (1962). The procedures are relatively simple and apply to both analog and digital filters.

Each conversion procedure involves the substitution of a function of s, say $s'(s)$, into a lowpass analog transfer function, $H(s)$, to provide a new transfer function, $H(s')$. Consequently, the conversions have in common the following important properties:

1. The converted transfer function, $H(j\omega')$, has *precisely the same values* as the lowpass function, $H(j\omega)$, but at different values of ω, that is, at different frequencies.

2. Digital filters designed using the bilinear transformation can be converted simply by converting the analog versions before applying the substitution $s \leftarrow (z - 1)/(z + 1)$.

The conversion formulas are given in Table 12.1. As stated above, each conversion involves the replacement of s with a new function of s in the lowpass transfer function, $H(s)$. The lowpass case is shown first as a trivial case, in which s is replaced with itself. On the right in the table are the resulting transformations of the frequency domain, obtained by letting $s = j\omega$. In the highpass case, the passband of the original lowpass filter, that is, $|\omega| < \omega_c$, is mapped to the domain where $|\omega| > \omega_c$ as shown, thus creating the desired new passband at frequencies greater than ω_c (see Example 12.7).

The bandpass transformation is a bit more complicated, and note that it doubles the order of the filter, since s^2 is involved in the transformation. As shown in the table, the original lowpass filter must be designed with a cutoff frequency

$$\omega_c = \omega_2 - \omega_1 \tag{12.54}$$

that is, a *cutoff frequency equal to the bandwidth of the bandpass filter*. The correctness of this, as well as line 3 of Table 12.1, can be seen by the following argument: Call the lowpass and bandpass transfer functions respectively $H_L(j\omega)$ and $H_B(j\omega)$. This substitution in line 3 of the table then produces

$$j\omega \leftarrow \frac{-\omega^2 + \omega_1\omega_2}{j\omega}$$

TABLE 12.1 Conversion of Lowpass Filters

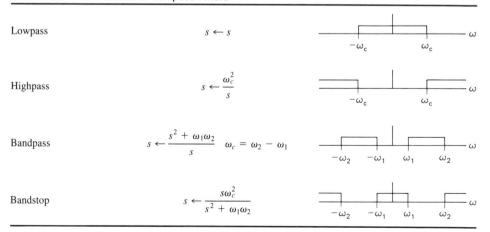

Therefore,

$$H_B(j\omega) = H_L\left(j\,\frac{\omega^2 - \omega_1\omega_2}{\omega}\right)$$

and

$$H_B(\pm j\omega_1) = H_L[\pm j(\omega_1 - \omega_2)] = H_L(\mp j\omega_c)$$
$$H_B(\pm j\omega_2) = H_L[\pm j(\omega_2 - \omega_1)] = H_L(\pm j\omega_c)$$

(12.55)

The two latter equations demonstrate the mapping of the endpoints of the passband shown on line 3 of the table. More generally, the second equation produces the desired result: The lowpass gain for $|\omega| < \omega_c$ equals the bandpass gain for $\omega_1 < |\omega| < \omega_2$, given the relationship in Eq. 12.54.

The final (bandstop) transformation in Table 12.1 follows from the preceding argument. By simply inverting the bandpass substitution function and multiplying by ω_c^2, the "pass" regions in the bandpass frequency domain are mapped to the "stop" regions in the bandstop frequency domain, and vice versa, as shown.

One point about all of the transformations should be noted in passing. All of them will create a perfect gain characteristic from a perfect lowpass characteristic. The reader can verify this by applying each transformation to Fig. 12.1 and obtaining the three functions on the right from the lowpass function on the left. However, when the lowpass characteristic is *not* rectangular (and of course it never is in practice), then the skirts of the gain function are warped in the transformation. The warping does not usually cause problems, as suggested below.

Two examples are presented to illustrate the transformations in Table 12.1. The first produces an analog highpass filter, and the second a digital bandpass filter. In both cases, the Butterworth lowpass filter is used as a starting point.

Example 12.7

Convert the lowpass Butterworth analog filter with $N = 10$ and $\epsilon = 1$ into a highpass filter. The lowpass transfer function, given by Eqs. 12.9 and 12.5, is

$$\left.\begin{aligned} H_L(s) &= \frac{s_1 s_2 \cdots s_{10}}{(s - s_1)(s - s_2) \cdots (s - s_{10})} \\ s_n &= \omega_c e^{j\pi(2n+9)/20}; \qquad n = 1, 2, \ldots, 10 \end{aligned}\right\}$$

(12.56)

Using line 2 of Table 12.1, s is now replaced with ω_c^2/s in $H_L(s)$ to obtain the highpass transfer function, $H_H(s)$:

$$H_H(s) = H_L\left(\frac{\omega_c^2}{s}\right)$$

$$= \frac{s_1 s_2 \cdots s_{10}}{(\omega_c^2/s - s_1)(\omega_c^2/s - s_2) \cdots (\omega_c^2/s - s_{10})}$$

(12.57)

$$= \frac{s^{10}}{(s - \omega_c^2/s_1)(s - \omega_c^2/s_2) \cdots (s - \omega_c^2/s_{10})}$$

Thus the original poles in Eq. 12.56 are mapped to themselves on the s-plane, and a multiple zero is introduced at $s = 0$. The highpass power gain function from Eq. 12.3 is

$$|H_H(j\omega)|^2 = \left| H_L\left(\frac{\omega_c^2}{j\omega}\right) \right|^2$$

$$= \frac{1}{1 + (\omega_c/\omega)^{20}}$$

(12.58)

A dB plot of this function is given in Fig. 12.16. The lowpass power gain, $|H_L(j\omega)|^2$, is also plotted for comparison. The curves are normalized by plotting power gain in dB versus ω/ω_c.

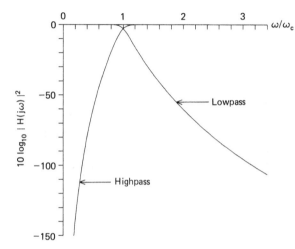

Figure 12.16 High- and lowpass Butterworth analog power gains; $N = 10$, $\epsilon = 1$.

Example 12.8

Convert the lowpass Butterworth filter with $N = 10$ and $\epsilon = 1$ into a digital bandpass filter having a passband from 0.2 to 0.3 times the sampling frequency. As mentioned previously, the procedure is first to convert the lowpass filter into an appropriate bandpass analog filter, then to use the bilinear transformation on the latter to obtain the digital filter. For the first step, the statement of the problem gives

$$\omega_1 = 2\pi\left(\frac{0.2}{T}\right) = \frac{2\pi}{5T}; \qquad \omega_2 = 2\pi\left(\frac{0.3}{T}\right) = \frac{3\pi}{5T}$$

(12.59)

These frequencies are now transformed according to the procedure in the preceding section:

$$\omega_1' = \tan\frac{\omega_1 T}{2} = \tan\left(\frac{\pi}{5}\right) = 0.7265$$

$$\omega_2' = \tan\frac{\omega_2 T}{2} = \tan\left(\frac{3\pi}{10}\right) = 1.3764$$

(12.60)

Next, ω_c' is found as in Table 12.1 or Eq. 12.54:

$$\omega_c' = \omega_2' - \omega_1' = 0.6499$$

(12.61)

The appropriate lowpass power gain is, therefore,

$$|H_L(j\omega)|^2 = \frac{1}{1 + (\omega/\omega_c')^{2N}} \tag{12.62}$$

with $N = 10$ and ω_c' as above, and using $\epsilon = 1$ in this case, from Table 12.1 or Eq. 12.55, the analog bandpass power gain is

$$|H_B(j\omega)|^2 = \left| H_L\left(j\frac{\omega^2 - \omega_1'\omega_2'}{\omega} \right) \right|^2$$

$$= \frac{1}{1 + \left(\dfrac{\omega^2 - \omega_1'\omega_2'}{\omega\omega_c'} \right)^{2N}} \tag{12.63}$$

From this, the power gain of the digital bandpass filter is found as in Eq. 12.44:

$$|\overline{H}(j\omega)|^2 = |H_B(j\,\tan(\omega T/2))|^2$$

$$= \frac{1}{1 + \left(\dfrac{\tan^2(\omega T/2) - \omega_1'\omega_2'}{\omega_c'\,\tan(\omega T/2)} \right)^{2N}} \tag{12.64}$$

This power gain is plotted in Fig. 12.17. The comparable analog gain, Eq. 12.63 with unprimed parameters, is also plotted for comparison. The development of the z-transfer function proceeds along these same lines, starting with Eq. 12.56 in place of Eq. 12.62, and substituting the bilinear form into the analog bandpass transfer function.

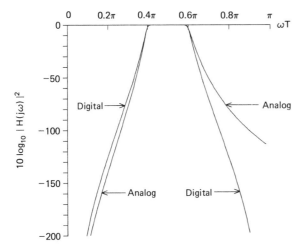

Figure 12.17 Power gain of Butterworth bandpass filters; $N = 10$.

It is instructive to find the poles of $\tilde{H}(z)$ by transforming the poles of the analog lowpass filter (Eq. 12.56), first into analog bandpass poles and then into digital bandpass poles on the z-plane. This is illustrated in Fig. 12.18. To obtain the analog bandpass poles, line 3 of Table 12.1 gives

$$s_L = \frac{s_B^2 + \omega_1'\omega_2'}{s_B} \tag{12.65}$$

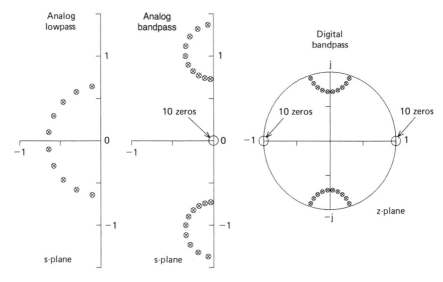

Figure 12.18 Transformation of poles (in Example 12.8) to construct a digital bandpass filter from an analog lowpass function. $\omega_c' = 0.6499$, $\omega_1' = 0.7265$, and $\omega_2' = 1.3764$.

where s_L is a lowpass pole and s_B a bandpass pole. Therefore

$$s_B = \tfrac{1}{2}(s_L \pm \sqrt{s_L^2 - 4\omega_1'\omega_2'}) \tag{12.66}$$

and so there are the two sets of analog bandpass poles shown in the center plot. Next, the bilinear transformation,

$$s = \frac{z - 1}{z + 1} \tag{12.67}$$

gives the digital bandpass poles at

$$z_B = \frac{1 + s_B}{1 - s_B} \tag{12.68}$$

and these poles are shown on the right-hand plot. Note how they span the digital passband between the angles 0.4π and 0.6π, giving the result in Fig. 12.17. Note that the transformations also introduce 10 zeros into the analog bandpass function and 20 zeros into the digital bandpass function. The proof of this fact is left as an exercise for the reader.

12.6 DIGITAL FILTER ROUTINES

The DSA library in Appendix B contains two routines that may be used to design four types of Butterworth digital filters. The SPLHBW routine is used for lowpass and highpass Butterworth filters and the SPBBBW routine is used for bandpass and bandstop Butterworth filters. The calling sequences are, respectively,

```
CALL SPLHBW (ITYPE,FC,NS,B,A,IERROR)
CALL SPBBBW (ITYPE,F1,F2,NS,B,A,IERROR)
```

```
ITYPE = Filter type = 1 (lowpass,bandpass), 2 (highpass,
        bandstop)
FC,F1,F2 = Cutoff (−3dB) frequencies in Hz−s. (Sampling frequency
        = 1.0)
    NS = Number of two−pole sections in cascade. Must be even
        for bandpass and bandstop filters.
B(0:2,NS) = Numerator coefficients.
A(2,NS) = Denominator coefficients. The nth section has
```

$$H_n(z) = \frac{B(0,n) + B(1,n)z^{-1} + B(2,n)z^{-2}}{1 + A(1,n)z^{-1} + A(2,n)z^{-2}} \tag{12.69}$$

```
IERROR = 0: no errors
        1: ITYPE not valid
        2: FC,F1, or F2 not in (0.0,0.5) or F1≥F2
        3: NS not >0 or not even with SPBBBW.
```

In these calling sequences we note that each filter consists of NS two-pole sections in cascade. The routines return all of the coefficients in the real arrays B(0:2, NS) and A(2, NS).

For lowpass filter design, the construction by SPLHBW with ITYPE = 1 is similar to that in Example 12.5. With $\epsilon = 1$ so that the power gain is down 3 dB at the cutoff frequency, the nth section has analog poles at

$$s_n, s_n^* = \omega_c' e^{\pm j\pi(2n+2NS-1)/(4NS)} \tag{12.70}$$

Note that the number of poles, N, is twice the number of sections, NS. These poles are mapped to the z-plane in accordance with the bilinear transformation, Eq. 12.33; that is, as in Eq. 12.53,

$$z_n = \frac{1 + s_n}{1 - s_n} \tag{12.71}$$

Let $H_{Ln}(z)$ represent the transfer function of the nth digital lowpass section. As shown in Eq. 12.43, the factor $(1 - s_n)(1 - s_n^*)$ becomes the leading denominator coefficient of $\tilde{H}_{Ln}(z)$. So the transfer function is

$$\begin{aligned}\tilde{H}_{Ln}(z) &= \frac{\omega_c'^2(z + 1)^2/[(1 - s_n)(1 - s_n^*)]}{(z - z_n)(z - z_n^*)}\\ &= \frac{\omega_c'^2(1 + 2z^{-1} + z^{-2})/[(1 - s_n)(1 - s_n^*)]}{1 - 2\,\text{Re}[z_n]\,z^{-1} + |z_n|^2 z^{-2}}\end{aligned} \tag{12.72}$$

For highpass filter design, the construction by SPLHBW with ITYPE = 2 proceeds nearly as just described. Since the substitution of $\omega_c'^2/s$ for s maps each analog pole into its complex conjugate, the denominator of a two-pole transfer function, $\tilde{H}_{Hn}(z)$, is the same as the denominator of $\tilde{H}_{Ln}(z)$. As seen in Eq. 12.57, the factor s^2 replaces $s_n s_n^*$ in the numerator of each analog section, causing $(z + 1)^2$ to be replaced by $(z - 1)^2$ in each digital section. The result, similar to Eq. 12.72, is

$$\tilde{H}_{Hn}(z) = \frac{(1 - 2z^{-1} + z^{-2})/[(1 - s_n)(1 - s_n^*)]}{1 - 2\,\text{Re}[z_n]\,z^{-1} + |z_n|^2 z^{-2}} \tag{12.73}$$

In summary, SPLHBW is essentially an implementation of Eqs. 12.70 through 12.73.

For bandpass and bandstop design, the transformation is a little more complicated. First, the number of sections (NS) must be an even integer, because each lowpass section leads to two bandpass sections under the transformations in Table 12.1. As seen in the discussion with Eq. 12.10, since $N = 2$ times the number of lowpass sections, the power gain rolloff rate is

$$
\begin{aligned}
\text{dB/octave} &= 12 \text{ (NS) (lowpass, highpass)} \\
&= 6 \text{ (NS) (bandpass, bandstop)}
\end{aligned}
\tag{12.74}
$$

For each transformation we begin with a lowpass section of the form

$$
H_{Ln}(s) = \frac{\omega_c'^2}{(s - s_{Ln})(s - s_{Ln}^*)}; \qquad n = 1, \ldots, \frac{NS}{2}
$$

$$
\omega_c' = \omega_2' - \omega_1'
\tag{12.75}
$$

$$
s_{Ln} = \omega_c' e^{j\pi(2n + NS - 1)/2NS}
$$

We first transform $H_{Ln}(s)$ into two bandpass or bandstop sections in cascade, given by

$$
H_{Bn1}(s)H_{Bn2}(s) = \frac{P^2(s)}{(s - s_{Bn1})(s - s_{Bn1}^*)(s - s_{Bn2})(s - s_{Bn2}^*)}
$$

$$
\begin{aligned}
P(s) &= \omega_c' s && \text{(bandpass)} \\
&= s^2 + \omega_1'\omega_2' && \text{(bandstop)}
\end{aligned}
\tag{12.76}
$$

$$
s_{Bn1}, s_{Bn2} = \frac{1}{2}\left(s_{Ln} \pm \sqrt{s_{Ln}^2 - 4\omega_1'\omega_2'}\right); \qquad n = 1, \ldots, \frac{NS}{2}
$$

The reader can verify these results by substituting the bandpass and bandstop functions in Table 12.1 into Eq. 12.75. We note that the bandpass and bandstop poles are identical, as were the lowpass and highpass poles.

Next we transform each section in Eq. 12.76 using the bilinear transformation, Eq. 12.33. The z-transfer function for each section is given by

$$
\tilde{H}_{Bni}(z) = \frac{\tilde{P}(z)}{(z - z_{Bni})(z - z_{Bni}^*)}; \qquad i = 1, 2
$$

$$
\begin{aligned}
\tilde{P}(z) &= \frac{\omega_c'(1 - z^{-2})}{(1 - s_{Bni})(1 - s_{Bni}^*)} && \text{(bandpass)} \\
&= \frac{(z - 1)^2 + \omega_1'\omega_2'(z + 1)^2}{(1 - s_{Bni})(1 - s_{Bni}^*)} && \text{(bandstop)}
\end{aligned}
\tag{12.77}
$$

$$
z_{Bni} = \frac{1 + s_{Bni}}{1 - s_{Bni}}; \qquad n = 1, \ldots, \frac{NS}{2}
$$

Again, these results can be verified by making the bilinear substitution into each section in Eq. 12.76. The routine SPBBBW, whose calling sequence has been described, implements these transformations to obtain the digital bandpass or bandstop coefficients.

Besides the Butterworth design routines just described, there are also two simple filtering routines. Each of these calls the corresponding design routine to design the appropriate filter and then filters a one-dimensional data sequence, replacing the sequence with its filtered version. The calling sequences are

```
CALL SPFIL1(X,N,ITYPE,FC,NS,WRK,IERROR)
CALL SPFIL2(Y,N,ITYPE,F1,F2,NS,WRK,IERROR)
```

```
     X(0:N-1) = Data sequence to be filtered and replaced
            N = Number of samples to be processed
        ITYPE = 1 (lowpass) or 2 (highpass) in SPFIL1
              = 1 (bandpass) or 2 (bandstop) in SPFIL2
     FC,F1,F2 = Cutoff (3-dB) frequencies; 0.0<F<0.5.
   WRK(8*NS+3) = Work array, dimensioned at least 8*NS+3.
                 Need not be initialized.
       IERROR = 0: No errors
                1: ITYPE not 1 or 2
                2: FC, F1, or F2 not valid
                3: NS not greater than 0.
```

The SPFIL1 routine is for lowpass and highpass filtering, and SPFIL2 is for bandpass and bandstop filtering. Both routines filter in the forward direction but, by modifying a single statement, may be made to filter in reverse. In the listing in Appendix B, the reader can see that this statement and its modifications are

$$\text{forward:} \quad \text{DO 3 K} = 0, \text{N} - 1$$
$$\text{reverse:} \quad \text{DO 3 K} = \text{N} - 1, 0, -1 \tag{12.78}$$

To conclude this section on digital filter routines, we look at two examples. The first is an example of filter design.

Example 12.9

Use the SPBBBW routine to design the 10-section bandpass filter in Example 12.8. The calling sequence in this case is

```
CALL SPBBBW(1,0.2,0.3,10,B,A,IERROR)
```

The B and A arrays in this case are dimensioned B(0:2, 10) and A(2, 10). The coefficients produced by the routine are as follows:

N	B(0,N)	B(1,N)	B(2,N)	A(1,N)	A(2,N)
1	0.21560	0.00000	−0.21560	0.58484	0.91191
2	0.40562	0.00000	−0.40562	−0.58484	0.91191
3	0.20576	0.00000	−0.20576	0.49423	0.76074
4	0.36634	0.00000	−0.36634	−0.49423	0.76074
5	0.20562	0.00000	−0.20562	0.37685	0.64250
6	0.32806	0.00000	−0.32806	−0.37685	0.64250
7	0.21475	0.00000	−0.21475	0.23722	0.55909
8	0.29183	0.00000	−0.29183	−0.23722	0.55909
9	0.23296	0.00000	−0.23296	0.08118	0.51515
10	0.25934	0.00000	−0.25934	−0.08118	0.51515

The A coefficients in Example 12.9 of course correspond with the z-plane poles in Fig. 12.18. For example, for the first section, the poles are given by

$$z_1^2 + 0.58484z_1 + 0.91191 = 0$$

$$z_1, z_1^* = -0.2924 \pm j0.9091$$

(12.79)

The next example illustrates the use of both of the filtering routines.

Example 12.10

Generate the following data sequence:

$$f_k = \sin[2\pi k(0.04)] + 5\sin[2\pi k(0.1)] + 2\sin[2\pi k(0.3)]; \qquad k = 0, \ldots, 250$$

The sequence is seen to have components at 0.04, 0.1, and 0.3 Hz-s. First, use a band-stop filter to remove the component at 0.1 Hz-s. Then use a lowpass filter to remove the component at 0.3 Hz-s. The calling sequences and results are shown in Fig. 12.19. A four-section bandstop filter with a notch from 0.08 to 0.12 Hz-s is used to reduce the large component at 0.1 Hz-s, and then a two-section lowpass filter reduces the high-frequency component at 0.3 Hz-s. The result, at the bottom of Fig. 12.19, is approximately the unit sine wave at 0.04 Hz-s. Note the startup transients, which distort the initial parts of the filtered waveforms.

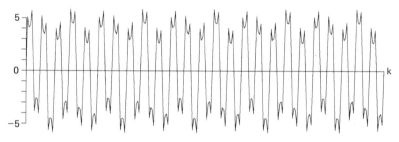

CALL SPFIL2(F,251,2,0.08,0.12,4,WRK,IE)

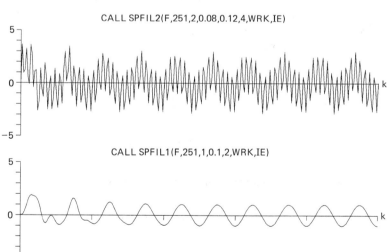

CALL SPFIL1(F,251,1,0.1,2,WRK,IE)

Figure 12.19 Example of bandstop filtering followed by lowpass filtering.

12.7 FREQUENCY-SAMPLING DIGITAL FILTERS

The preceding sections have shown how to obtain a digital filter from an analog design via the bilinear transformation. Here a direct method for designing digital filters is introduced. The method is based on sampling a desired amplitude spectrum $|H(j\omega)|$, and obtaining filter coefficients that are functions of the sample set $[|H_k|]$. As shown in this section, this approach is particularly useful when only a small percentage of the H_k's are nonzero, or when several bandpass functions are desired simultaneously.

The frequency-sampling filter is, in the first place, a *finite-impulse-response* filter. That is, $\tilde{H}(z)$ can be expressed in polynomial rather than rational form. The nonrecursive formulas, Eqs. 8.1, 8.5, and 8.34 can therefore be used here as a starting point:

$$g_m = \sum_{n=0}^{N-1} b_n f_{m-n} \tag{12.80}$$

$$\tilde{H}(z) = \sum_{n=0}^{N-1} b_n z^{-n} \tag{12.81}$$

$$\overline{H}(j\omega) = \sum_{n=0}^{N-1} b_n e^{-jn\omega T} \tag{12.82}$$

If the reader has any doubts on the meaning of these formulas, he should review their derivation in Chapter 8. Actually, they are slightly altered above, so that g_m is a function only of the N past and present samples of $f(t)$. The filter coefficients are the set $[b_n]$ and b_n is also the response at time nT to an impulse with amplitude one at time zero. (The impulse response is thus finite and lasts for NT seconds.)

In Chapter 8, the coefficients $[b_n]$ were chosen in a natural way to make $\overline{H}(j\omega)$ a least-squares approximation to a desired $H(j\omega)$ within the interval $|\omega| < \pi/T$. Here these coefficients are to be derived from *samples* of $H(j\omega)$ instead. To begin the discussion of how this derivation is to take place, let $[B_k]$ be the DFT of $[b_n]$ so that

$$B_k = \sum_{n=0}^{N-1} b_n e^{-j(2\pi nk/N)}$$

$$b_n = \frac{1}{N} \sum_{k=0}^{N-1} B_k e^{j(2\pi nk/N)} \tag{12.83}$$

Then the z-transfer function, Eq. 12.81, becomes

$$\tilde{H}(z) = \sum_{n=0}^{N-1} \left(\frac{1}{N} \sum_{k=0}^{N-1} B_k e^{j(2\pi nk/N)} \right) z^{-n}$$

$$= \frac{1}{N} (1 - z^{-N}) \sum_{k=0}^{N-1} \frac{B_k}{1 - e^{j(2\pi k/N)} z^{-1}} \tag{12.84}$$

with the latter form being the result of summing over n. This changes the transfer function into a recursive form, and the problem is now to relate $[B_k]$ to the "desired" sample set $[|H_k|]$ mentioned above, in a manner such that $\tilde{H}(z)$ is realizable.

To solve this problem, suppose first that the desired amplitude function, $|H(j\omega)|$, is sampled in the interval $0 \leq \omega \leq \pi/T$ as in Fig. 12.20 to produce the sample set $[|H_k|]$. Suppose further that B_k and H_k are equal in amplitude, so that

$$|B_k| = H_k = |H_k| \qquad \text{for } k \leq \frac{N}{2} \tag{12.85}$$

(From this point on, as indicated in Eq. 12.85, the amplitude bars will be dropped from H_k. The latter will always be a sample of $|H(j\omega)|$, as in Fig. 12.20.) Consider now the value of $|\overline{H}(j\omega)|$, the amplitude response of the digital filter, at $\omega_m = 2\pi m/NT$, that is, at any one of the sample points on the frequency axis, with $m \leq N/2$. This value is

$$
\begin{aligned}
|\overline{H}(j\omega_m)| &= |\tilde{H}(e^{j\omega_m T})| = |\tilde{H}(e^{j(2\pi m/N)})| \\
&= \left| \frac{1}{N} \sum_{n=0}^{N-1} \sum_{k=0}^{N-1} B_k e^{j(2\pi nk/N)} e^{-j(2\pi nm/N)} \right| \\
&= \left| \frac{1}{N} \sum_{k=0}^{N-1} B_k \sum_{n=0}^{N-1} e^{j[2\pi n(k-m/N)]} \right| \\
&= |B_m| \\
&= H_m
\end{aligned}
\tag{12.86}
$$

The second line in Eq. 12.86 follows from the first by Eq. 12.84, the fourth follows by summing over n—then only the term with $k = m$ is nonzero, and the final line follows from Eq. 12.85. Thus, *given $|B_k| = H_k$, the amplitude response of the digital filter equals the desired values at the sample points.*

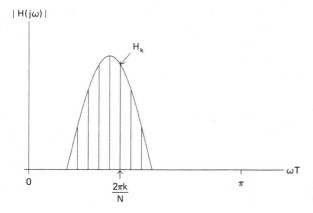

Figure 12.20 Sampled amplitude spectrum.

The remaining problem is to adjust the relative *phases* of the B_k's so that $|\overline{H}(j\omega)|$ will provide a smooth approximation to $|H(j\omega)|$ *between* the sample points. To solve this problem, write $\overline{H}(j\omega)$ in the following form, again using Eq. 12.84:

$$\overline{H}(j\omega) = \tilde{H}(e^{j\omega T})$$

$$= \frac{1}{N}(1 - e^{-jN\omega T})\sum_{k=0}^{N-1}\frac{B_k}{1 - e^{j(2\pi k/N)}e^{-j\omega T}} \tag{12.87}$$

$$= \frac{1}{N}e^{-j[(N-1)\omega T/2]}\sin\left(\frac{N\omega T}{2}\right)\sum_{k=0}^{N-1}\frac{B_k e^{-j(k\pi/N)}}{\sin\left(\frac{\omega T}{2} - \frac{\pi k}{N}\right)}$$

This gives the amplitude and phase portions of $\overline{H}(j\omega)$ in somewhat more explicit form. Now suppose there is only *one* nonzero B_k in Eq. 12.87. Without loss of generality, let this coefficient be B_0. The single-sample amplitude response is then

$$|\overline{H}_0(j\omega)| = \frac{|B_0|}{N}\left|\frac{\sin(N\omega T/2)}{\sin(\omega T/2)}\right|$$

$$\approx |B_0|\left|\frac{\sin(N\omega T/2)}{N\omega T/2}\right| \tag{12.88}$$

with the approximation being good when ωT is small. Thus, a single B_k produces a $(\sin x)/x$-like contribution to $|\overline{H}(j\omega)|$, so if the B_k's can be added in phase, $|\overline{H}(j\omega)|$, should be a smooth reconstruction of $|H(j\omega)|$, reminiscent of Whittaker's reconstruction (Section 5.6).

To examine the phases of two adjacent contributions to $|H(j\omega)|$ from, say, the B_k and B_{k+1} terms, Fig. 12.21 is provided. In the figure it is assumed that $N = 51$, $k = 9$, and $|B_9| = |B_{10}| = 1$. The two curves overlayed in the center show the amplitudes of the B_9 and B_{10} terms, similar to Eq. 12.88. The two upper curves show the phases (ϕ_9 and ϕ_{10}) of these terms. The terms are seen to be *in phase except between the sample points*, that is, except between the twin peaks in the figure, where they are 180 degrees out of phase. Thus the two terms, $\overline{H}_9(j\omega)$ and $H_{10}(j\omega)$, should be *subtracted* to provide a smooth reconstruction between sample points. This difference is shown at the bottom of Fig. 12.21, which shows a smooth construction near the sample points as well as supression of the sidelobes of the individual contributions shown above.

From the above discussion, as well as the prior conclusion in Eq. 12.85 that $|B_k| = H_k$, one would conclude that B_k should be equal to H_k with alternating sign, that is, $B_k = (-1)^k H_k$, so that $|\overline{H}(j\omega)|$ would be smooth and equal to $|H(j\omega)|$ at the sample points. This is a valid conclusion except that B_k and B_{N-k} must be complex conjugates (since B_k is real, this of course implies $B_k = B_{N-k}$) for the digital filter to be realizable, that is, for the b_n's to be real (see Eq. 12.83). Only the first half of the B_k's are independent, which is also seen from the fact that the samples corresponding

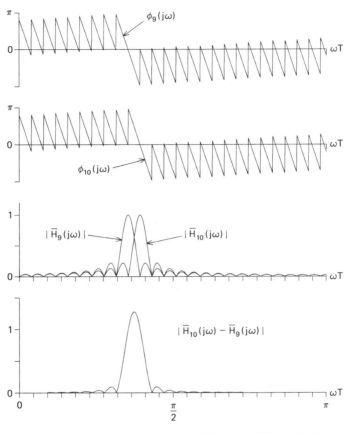

Figure 12.21 Phases (ϕ_9 and ϕ_{10}) and amplitudes of $\overline{H}_9(j\omega)$ and $\overline{H}_{10}(j\omega)$ for $|B_9| = |B_{10}| = 1$, and amplitude $|\overline{H}_{10}(j\omega) - \overline{H}_9(j\omega)|$; $N = 51$.

to these B_k's span the interval from $\omega = 0$ to $\omega = \pi/T$. Therefore, a correct specification of the B_k's is

$$\left.\begin{array}{c} B_k = (-1)^k H_k \\[2mm] B_{N-k} = B_k \end{array}\right\}; \qquad k \le \frac{N}{2} \tag{12.89}$$

This is valid for N odd or even, although from here on it is convenient to assume that B_0 is zero, and $B_{N/2}$ is zero if N is even.

With this assumption and with the B_k's assigned as in Eq. 12.89, the digital filter transfer functions follow from Eq. 12.84:

$$\tilde{H}(z) = \frac{1}{N}(1 - z^{-N}) \sum_{k<N/2} (-1)^k H_k \left[\frac{1}{1 - e^{j(2\pi k/N)}z^{-1}} + \frac{1}{1 - e^{j[2\pi(N-k)/N]}z^{-1}} \right]$$

$$= \frac{2}{N}(1 - z^{-N}) \sum_{k<N/2} (-1)^k H_k \frac{1 - z^{-1}\cos(2\pi k/N)}{1 - 2z^{-1}\cos(2\pi k/N) + z^{-2}} \tag{12.90}$$

Therefore,

$$\overline{H}(j\omega) = \tilde{H}(e^{j\omega T})$$

$$= \frac{2}{N} je^{-j(N\omega T/2)} \sin\left(\frac{N\omega T}{2}\right) \tag{12.91}$$

$$\times \sum_{k<N/2} (-1)^k H_k \frac{\cos \omega T - \cos(2\pi k/N) + j \sin \omega T}{\cos \omega T - \cos(2\pi k/N)}$$

Figure 12.22 illustrates the digital amplitude response, $|\overline{H}(j\omega)|$, for a case with $N = 31$ and 6 nonzero samples of the desired amplitude response. As shown, Eq. 12.91 gives a smooth reconstruction of $|H(j\omega)|$, which is exact at the sample points.

Some thought must now be given to the nature and implementation of the digital filter represented by Eq. 12.90. For one thing, how can the filter be recursive and still have a finite impulse response? The answer is that $\tilde{H}(z)$ does not really have poles in an essential sense. In Eq. 12.90, the poles appear at $e^{\pm j(2\pi k/N)}$ on the unit circle in the z-plane. However, the term $(1 - z^{-N})$ provides N zeros at exactly these same points, equally spaced around the unit circle. Each nonzero sample H_k, then, brings to the digital filter a conjugate pair of poles which cancel a corresponding pair of zeros at $e^{\pm j(2\pi k/N)}$ on the unit circle.

This cancellation of zeros by poles brings up a practical consideration (Gold and Rader, 1969): Although the cancellation is exact in theory, it will not be in practice due to finite word length in the computer. The $\cos 2\pi k/N$ terms in Eq. 12.90 will in general be accurate only to some fixed number of significant digits. An effective solution to this problem is to move the poles and zeros *slightly inside* the unit circle. Let

$$r = 1 - 2^{-M} \tag{12.92}$$

where M is no larger than the number of bits used to represent the significant digits of the filter coefficients. A value $12 \leq M \leq 27$ is suggested by Gold and Rader

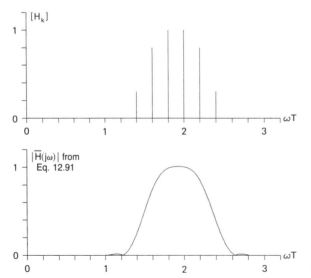

Figure 12.22 Sampled and digital (reconstructed) amplitude functions; $N = 31$.

(1969). To move the poles and zeros in to radius r, Eq. 12.90 is modified simply as follows:

$$\tilde{H}(z) = \frac{2}{N}(1 - r^N z^{-N}) \sum_{k<N/2} (-1)^k H_k \frac{1 - rz^{-1}\cos(2\pi k/N)}{1 - 2rz^{-1}\cos(2\pi k/N) + r^2 z^{-2}} \qquad (12.93)$$

The overall implementation of Eq. 12.93 is shown in Fig. 12.23. The term $(1 - r^N z^{-N})$ is represented on the left in the figure, and the E's on the right represent the nonzero terms in the sum over k.

Each E in Fig. 12.23 is called an *elemental filter,* and there must be an E_k for each nonzero frequency sample, H_k. A diagram of an E_k is represented in Fig. 12.24. Note, however, that the multiplication by $r \cos 2\pi k/N$ actually need only take place once if desired, since the terms $2e_{km}$ and $(-1)^k H_k x_m$ could be added before instead of after the multiplication takes place.

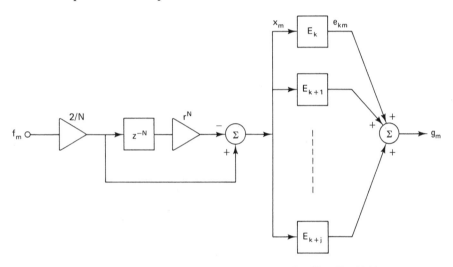

Figure 12.23 Overall diagram of frequency-sampling filter; Eq. 12.93.

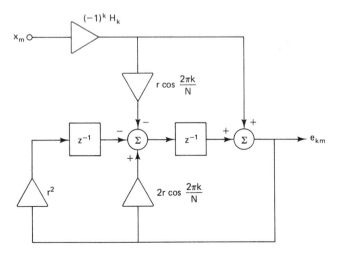

Figure 12.24 The kth elemental filter, E_k, in Fig. 12.23.

The forms illustrated in Figs. 12.23 and 12.24 suggest that the frequency-sampling filter should be particularly useful where (1) only a few H_k's are nonzero in a single bandpass function, or (2) several bandpass functions are to be implemented at once. In cases such as these one can take advantage of the N zeros of the "$1 - r^N z^{-N}$" part of the filter, while using only a few elemental filters at a time. An example of case (2) is illustrated in Fig. 12.25. Here there are three bandpass functions, each with its own set of elemental filters, but connected in common through the $(1 - r^N z^{-N})$ terms (see also Example 12.11).

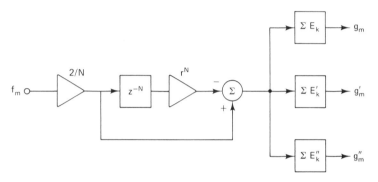

Figure 12.25 Three bandpass filters.

The following steps serve to summarize the design procedure for frequency-sampling digital filters:

1. Ascertain that T is small enough that $|H(j\omega)|$ is negligible for $\omega > \pi/T$.
2. Choose the frequency-sampling interval, $2\pi/N$ (on the ωT-axis as in Fig. 12.20), remembering that N also specifies the filter memory size (z^{-N} in Fig. 12.25) as well as the impulse duration (NT seconds).
3. Obtain the sample set, $[H_k] = [|H(j2\pi k/N)|]$.
4. Implement the filter as in Eq. 12.93 or Figs. 12.23 and 12.24.
5. Repeat for other filters as in Fig. 12.25.

In practice, step 3 can be something of an art, and may require a few trials before a satisfactory sample set is attained. The following example illustrates this final point, as well as the entire procedure above.

Example 12.11

Two bandpass digital filters, F_1 and F_2, are required to meet the following specifications:

$$\text{sampling interval } T = 25 \ \mu s$$

$$\text{passband} = 4\text{–}5 \text{ kHz for } F_1; \quad 6\text{–}7 \text{ kHz for } F_2$$

$$\text{power gain at ends of passbands} = 0.5 = -3 \text{ dB}$$

$$\text{duration of impulse response} = 5 \text{ ms}$$

Following the steps in the procedure above, one proceeds as follows:

Step 1. The folding frequency here is $1/2T = 20$ kHz, which is well beyond both passbands.

Step 2. If the impulse response duration is 5 ms, then

$$N = \frac{0.005}{T} = 200 \qquad (12.94)$$

Step 3. With one revolution around the z-plane representing 40 kHz in this example, and with $N = 200$, the zeros on the unit circle must be 200 Hz apart. Thus, in the F_1 passband for example, there are sample points at 4, 4.2, 4.4, 4.6, 4.8, and 5 kHz, that is, at angles $2\pi k/N$ with $k = 20$–25. With -3 dB specified at the passband ends, one might therefore choose $[H_k]$ for F_1 and F_2 as follows:

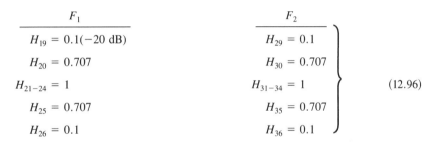

F_1	F_2	
$H_{20} = 0.707\ (-3\ \text{dB})$	$H_{30} = 0.707$	
$H_{21-24} = 1$	$H_{31-34} = 1$	(12.95)
$H_{25} = 0.707$	$H_{35} = 0.707$	

Using these values for F_1 and F_2, the resulting power gain curves are shown in Fig. 12.26. If the sidelobes are objectionable, the designer may choose to suppress them with additional nonzero H_k's on either side of the passband. For example, suppose that $[H_k]$ is modified as follows for F_1 and F_2:

F_1	F_2	
$H_{19} = 0.1(-20\ \text{dB})$	$H_{29} = 0.1$	
$H_{20} = 0.707$	$H_{30} = 0.707$	
$H_{21-24} = 1$	$H_{31-34} = 1$	(12.96)
$H_{25} = 0.707$	$H_{35} = 0.707$	
$H_{26} = 0.1$	$H_{36} = 0.1$	

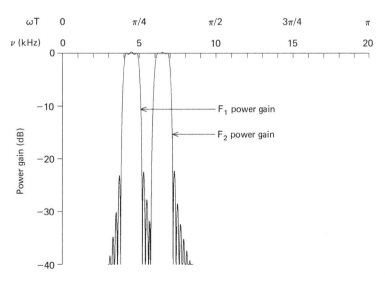

Figure 12.26 Gain curves for Example 12.11, using Eq. 12.91 with sample sets in Eq. 12.95.

The result is shown in Fig. 12.27. The bandpass characteristics are a bit fatter than in Fig. 12.26, but the sidelobes are noticably suppressed.

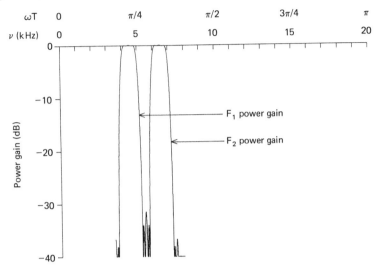

Figure 12.27   Improved gain curves for Example 12.11 using Eq. 12.91 with Eq. 12.96.

Steps 4 and 5. These final steps are partially completed in Fig. 12.28. The overall design is shown in the figure, with an output for each of the two passbands. A typical elemental filter section is shown in Fig. 12.24.

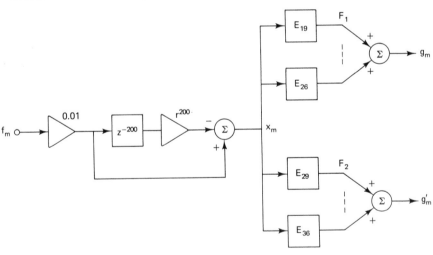

Figure 12.28   Filter diagram for Example 12.11.

A filtering algorithm for F_1 in Example 12.11, obtained from Eq. 12.93, is

$$\tilde{H}(z) = 0.01(1 - r^{200}z^{-200}) \sum_{k=19}^{26} (-1)^k H_k \frac{1 - rz^{-1}\cos(0.01\pi k)}{1 - 2rz^{-1}\cos(0.01\pi k) + r^2 z^{-2}}$$

$$(12.97)$$

Therefore,

$$x_m = 0.01(f_m - r^{200}f_{m-200})$$

$$e_{km} = (-1)^k H_k[x_m - rx_{m-1}\cos(0.01\pi k)] + 2re_{k,m-1}\cos(0.01\pi k) - r^2 e_{k,m-2}$$

$$\text{(12.98)}$$

$$g_m = e_{19,m} + e_{20,m} + \cdots + e_{26,m}$$

Using a value of $r = 1 - 2^{-19}$ in these computing formulas, the response to a single sample $f_0 = 1/T = 4 \times 10^4$ at $t = 0$ is found to be as in Fig. 12.29. In the figure, $g(t)$ was constructed by drawing straight lines between the sample points $[g_m]$. As expected, the impulse response dies out at the end of 5 ms. In Fig. 12.30, the response to a sine wave with a frequency of 5 kHz starting at $t = 0$ is plotted. Since

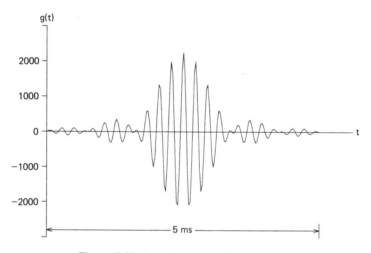

Figure 12.29 Impulse response; Example 12.11.

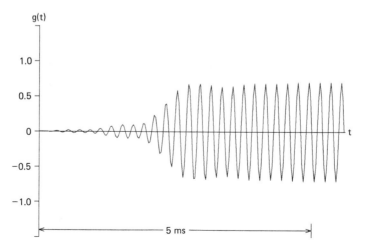

Figure 12.30 Response to $f(t) = \sin(10,000\pi t)$ for $t > 0$; Example 12.11.

Sec. 12.7 Frequency-Sampling Digital Filters

the gain of F_1 is -3 dB at 5 kHz, the amplitude of $g(t)$ is 0.707 after stability is reached.

12.8 ERRORS DUE TO FINITE WORD LENGTHS

Round-off or truncation errors due to finite word lengths cause problems in many computing situations, and digital filtering is no exception. This section is a brief discussion of these errors as they occur generally in digital filters, that is, not just for the filters discussed in the present chapter. There is a considerable amount of literature on finite-word-length effects, and only some fundamental points are presented here along with a few examples. The latter are constructed by making the word length unusually short so that the errors become more noticeable. More extensive literature on finite-word-length effects can be found in Chapter 4 of Gold and Rader (1969), Section 5.11 of Wait (1970), Part 3 of Rabiner and Rader (1972), and Chapter 9 of Oppenheim and Schafer (1975).

An error model appropriate to the discussion of truncation errors is shown in Fig. 12.31. the model is similar to those in Chapter 11 in that an error, e_m, is produced as the difference between the output sample value, g_m, and an ideal output g_m^*. The ideal output is produced theoretically by digitizing $f(t)$ using an infinite (for all practical purposes) ananlog-to-digital converter, ADC*, and filtering the samples through $\tilde{H}(z)$ using a computer with infinite word length. The actual output, g_m, emerges from a finite ADC followed by a finite signal processor.

The sources of e_m may be considered in three separate categories:

1. *Signal quantization.* Because of the finite ADC, f_m is a truncated (or rounded) version of $f(mT)$ as discussed in Chapter 4, Section 4.4.
2. *Coefficient quantization.* The coefficients of $\tilde{H}(z)$ are generally derived from the continuum, often using e^{-aT}, cos bT, and so on, as in the examples in this chapter. The finite-length words in the signal processor thus require small adjustments of these coefficient values in order to realize the filter.
3. *Product round-off.* In the algorithm for g_m, each product (of a coefficient times a sample value) requires rounding or truncation in order to fit the result into the original word size. Even if there were no errors in the first two categories, product round-off could cause g_m to differ from g_m^*.

As depicted in Fig. 12.31, g_m is different from the ideal, g_m^*, first because $[f_m]$ is not the true sample set of $f(t)$, and second, because $\tilde{H}(z)$ is not an exact realization of $\tilde{H}^*(z)$. The first of these reasons is reflected in category 1 above, and the second is reflected in categories 2 and 3.

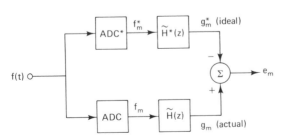

Figure 12.31 Truncation error model. Error e_m is the difference between the actual output and the ideal, infinite-word-length output.

Analog and Digital Filter Design   Chap. 12

An example of signal quantization (category 1) and its effect on the output of a (perfectly realized) second-order filter is shown in Fig. 12.32. The block diagram shows a continuous signal, $f(t) = te^{-t}$, being digitized by a 4-bit ADC whose overall range is ± 0.4. The amplitude units are of course those of $f(t)$ (e.g., volts, amperes, degrees, meters, etc.). Thus one bit is equivalent to $0.8/2^4$, or 0.05. Assuming that the ADC always rounds to the nearest quantum level, the maximum input quantizing error is then $\pm 0.05/2 = \pm 0.025$. The input sample set and the associated input quantizing error, $f_m - f_m^*$, are shown on the left in Fig. 12.32.

On the right in Fig. 12.32, the filter output, g_m, is plotted along with e_m, the output error. Note that e_m is *not* the error obtained by quantizing $g(t)$, but rather the result of passing the input error through the filter, as illustrated in Fig. 12.33. Thus in this example the output error is a smoothed version of the input quantizing error. In general, the effect of filtering on the input quantizing error is best treated in the

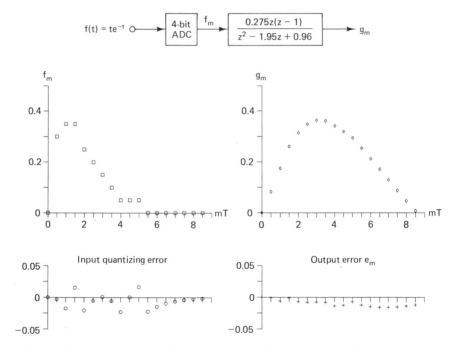

Figure 12.32 Illustration of the input and output quantizing error. Range of the 4-bit ADC is ± 0.4; step size is $T = 0.5$.

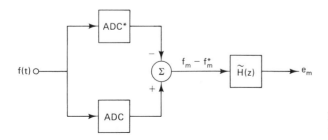

Figure 12.33 Output error due solely to input quantizing error, as in Fig. 12.32.

frequency domain. This is usually done by multiplying the power spectrum of the input error by $|\overline{H}(j\omega)|^2$, a subject discussed later.

However, it is interesting to note here that a system like the one shown in Fig. 12.32 can be used, under certain conditions, to improve the accuracy of an ADC. Let $f(t)$ be sampled at, say, $2n$ times its highest frequency, and suppose that the filter in Fig. 12.32 is a lowpass filter cutting off at $1/n$ times the sampling frequency. Then the filter will pass the samples of the signal, $f(t)$, correctly, but will tend to reduce or smooth out the quantizing error, whose spectrum tends to be more or less flat over the range from zero to the sampling frequency. This idea is closely associated with the subject of digital interpolation discussed in Chapter 5, Section 5.7 [see also Claasen et al. (1980)].

Consider now categories 2 and 3, which concern the imperfect realization of an ideal $\tilde{H}^*(z)$. The first type of imperfection is due to finite-length coefficients in $\tilde{H}(z)$ when the coefficients of $\tilde{H}^*(z)$ are taken from the continuum. The second type is due to product round-off during the filtering process. The derivation of the error from both imperfections is shown in Fig. 12.34, which suggests that the error could be examined by comparing the output of $\tilde{H}(z)$ with that of a "perfect" version of $\tilde{H}(z)$, that is, the output of a processor with a very large word length.

The finite-length coefficient (category 2) errors can be viewed as errors caused by the slight displacement of the poles and zeros of $\tilde{H}^*(z)$, which are in turn caused by truncating (or rounding) the coefficients. For example, let

$$\tilde{H}(z) = \frac{\tilde{B}(z)}{z^2 - 2a_1 z + a_2} \tag{12.99}$$

and suppose a_1 and a_2 are implemented as 4-bit binary fractions (from 0.0001 to 0.1111) or as 5-bit binary fractions (from 0.00001 to 0.11111). All of the possible complex pole locations inside the unit circle on the z-plane are illustrated in Fig. 12.35 for these two cases. Since the pattern is the same in each quadrant, only the first quadrant is illustrated. Thus, in this illustration, no error would result if the poles of $\tilde{H}^*(z)$ coincided with a set of dots, but otherwise each pole would have to move to a nearby dot to become a pole of $\tilde{H}(z)$. Note that the dots in Fig. 12.35 are evenly spaced over the real axis but not over the imaginary axis because the real part of each pole in Eq. 12.99 is simply a_1, while the imaginary part is $\pm\sqrt{a_2 - a_1^2}$. It is also possible for category 2 errors to produce instability by causing one or more poles to move outside the circle, as described by Kaiser (1965).

A specific example of the effect of finite-length coefficients on filter gain is shown in Fig. 12.36. The filter is the five-section Butterworth filter derived in

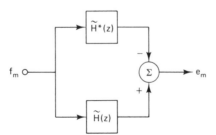

Figure 12.34 Output error due solely to imperfect realization of $\tilde{H}^*(z)$.

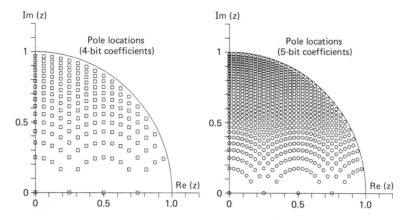

Figure 12.35 Possible pole locations inside the unit circle for $\tilde{H}(z)$ in Eq. 12.99 using 4- and 5-bit, fixed-point coefficients.

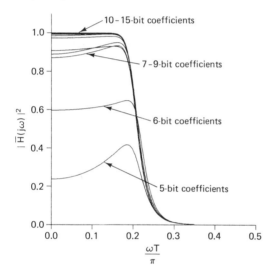

Figure 12.36 Coefficient rounding effects on the power gain of the digital Butterworth filter in Fig. 12.13.

Section 4, Example 12.5. An important point to note in this example is that the theoretical power gain is given by Eq. 12.45 but, with coefficient rounding, the actual power gain, assuming the filter is realized in cascade form as in Fig. 12.13, is found by taking the squared magnitude in Eq. 12.47, that is,

$$|\overline{H}(j\omega)|^2 = \prod_{n=1}^{5} |\tilde{H}_n(e^{j\omega T})|^2 \qquad (12.100)$$

where each factor in the product is found by substituting $z = e^{j\omega T}$ into Eq. 12.47 with rounded coefficients. (The answer to Exercise 37 gives each factor in Eq. 12.100 explicitly.) Fixed-point coefficients are used in this illustration, that is, each n-bit coefficient has one bit to the left and $n - 1$ bits to the right of the binary point. The important observation in Fig. 12.36 is that, due to the rather arbitrary movement of the poles as suggested by Fig. 12.35, the gain function is distorted in arbitrary ways as the coefficient word size changes.

The filter in the example of Fig. 12.36 was assumed to be in cascade form. In general the *cascade and parallel forms of implementation should be preferred over the direct form* (Section 9.5) because the coefficient accuracy requirements increase with the order of the difference equation (Gold and Rader, 1969).

Regarding product round-off (category 3) errors, it is difficult to provide any simple guidelines, particularly in floating-point systems. These errors depend not only on the filter characteristics but also on signal characteristics, since both are involved in the rounded products. Gold and Rader (1969) provide some round-off noise formulas for different filters. However, as suggested by Wait (1970), it is often best to measure both category 2 and category 3 errors by programming the filter on a large general-purpose computer and varying the word length for a specified set of inputs. The filter "word length" can be adjusted simply by rounding and masking in the general-purpose computer. This empirical approach is relatively safe and easy, and it allows one to study other trade-offs in the filter design (e.g., number of sections, step size, etc.), as well.

EXERCISES

1. Derive a formula for the filter parameter ϵ in terms of dB_c, the power gain in dB at the end of the passband.

   The following filter characteristics are used in the exercises below.

	Filter A	Filter B	Filter C
Passband	0–10 kHz	0–10 kHz	5–15 kHz
Minimum power gain at ω_c	0.5	0.5	—
Start of stopband	15 kHz	11.75 kHz	—
Maximum power gain at ω_r	0.1	0.1	—

2. What is the order of filter A if the Butterworth design is used?
3. What order is required for filter A if the Chebyshev design is chosen?
4. What is the order of Butterworth filter B?
5. What is the order of Chebyshev filter B?
6. What are the coordinates of pole s_3 of Butterworth filter A?
7. Determine the complete transfer function of Butterworth filter A.
8. Express the power gain of Butterworth filter A in dB.
9. Derive the phase shift, $\phi(\omega)$ for Butterworth filter A.
10. Find and plot the poles of Butterworth filter B.
11. Plot the power gain and phase shift for Butterworth filter B.
12. Find the coordinates of pole s_2 of Chebyshev type 1 filter A.
13. Plot the poles of Chebyshev type 1 filter A on the s-plane.
14. Plot the power gain of Chebyshev type 1 filter A.
15. Plot the poles and zeros of Chebyshev type 2 filter A.
16. Plot the power gain in dB of Chebyshev type 2 filter B.
17. Given a sampling rate of 50,000 samples per second, design digital Butterworth filter A. Give the computing formula for g_m.
18. Plot the power gains of the analog and digital versions of Butterworth filter A in dB, and explain the difference between the two curves. Use $T = 10 \ \mu s$, and limit to -60 dB.

19. Draw a schematic realization of digital Butterworth filter A with $T = 10$ μs.

20. Plot the power gain in dB of digital Chebyshev type 1 filter A using $T = 20$ μs.

21. Give a set of computing formulas for digital Butterworth filter B using $T = 20$ μs.

22. Convert analog Butterworth filter A to a high-pass analog filter. Where are the poles on the s-plane?

23. Convert the analog filter in exercise 22 to a digital filter and give a formula for the power gain. Use $T = 20$ μs.

24. What are the frequencies ω_1 and ω_2 if analog filter A is to be converted into analog filter C?

25. What are the frequencies ω_1', ω_2', and ω_c' if an analog filter is converted to digital filter C and $T = 20$ μs?

26. Give a formula for the power gain of digital filter C if the latter is derived from a Chebyshev type 2 filter similar to filter B. Use $T = 20$ μs.

27. Construct an elemental filter diagram similar to the one in Fig. 12.24, but having four coefficients (triangles), one of which is just "2".

28. Design digital filter C using frequency sampling. Use $T = 20$ μs and a memory size of 20. Choose the set $[H_k]$ so that each element is either one or zero. Give an equation for $\bar{H}(z)$.

29. Construct a complete diagram of the preceding filter C.

30. Plot the power gain in dB of the preceding filter C.

31. Adjust the set $[H_k]$ to suppress the sidelobes in the preceding power gain, and plot the improved power gain in dB. *Hint:* See Example 12.11.

32. Use the SPLHBW routine to design a lowpass Butterworth filter with two poles and cutoff frequency at 0.1 times the sampling frequency. Print the coefficients.

33. Use a routine from this chapter to design a highpass Butterworth filter with two sections, cutoff at 20 Hz, and time step $T = 1$ ms. Print all coefficients, section by section.

34. Use a routine from this chapter to design digital Butterworth filter C, assuming four sections and time step $T = 10$ ms. Print the coefficients.

35. Use the SPFIL1 routine to remove the high-frequency component from

$$f(t) = \sin(2\pi t) + \sin(4\pi t)$$

Use a four-section lowpass filter on a sequence of 200 samples at 20 samples per second. Plot the waveform before and after filtering.

36. Use the SPFIL2 routine to obtain the impulse response of a 10-section Butterworth version of Filter C with time step $T = 10$ μs.

37. Express each term in the formula for the power gain of a digital Butterworth filter using Eqs. 12.100 and 12.48.

SOME ANSWERS

1. $\epsilon = \sqrt{10^{-dB_c/10} - 1}$ **2.** $N = 3$ **3.** $N = 2$ **4.** $N = 7$ **5.** $N = 4$

6. $2\pi \cdot 10^4(-0.5 - j0.866)$

7. $\dfrac{8\pi^3 \cdot 10^{12}}{(s - 2\pi \cdot 10^4 e^{j2\pi/3})(s - 2\pi \cdot 10^4 e^{j4\pi/3})(s + 2\pi \cdot 10^4)}$

8. $-10 \log_{10}\left[1 + \left(\dfrac{\omega}{2\pi \cdot 10^4}\right)^6\right]$ **10.** $s_n = 2\pi \cdot 10^4 e^{j\pi(n+3)/7}; \quad n = 1, 2, \ldots, 7$

12. $-20219 - j48813$ **14.** $|H_C(j\omega)|^2 = \dfrac{1}{0.0004\nu^4 - 0.04\nu^2 + 2}; \quad \nu = $ frequency in kHz.

18. $|H_B(j\omega)|^2 = \dfrac{10^6}{\nu^6 + 10^6};\quad |\overline{H}_B(j\omega)|^2 = \dfrac{0.00118}{0.00118 + \tan^6(0.01\pi\nu)};\quad \nu = \text{frequency in kHz}$

20. $|\overline{H}_C(j\omega)|^2 = \dfrac{1}{14.36\,\tan^4(10^{-5}\omega) - 7.58\,\tan^2(10^{-5}\omega) + 2}$

22. $s_n = 2\pi \times 10^4 e^{\,j[(n+1)\pi/3]};\quad n = 1, 2, \ldots, 6$ **24.** $\omega_1 = 10^4\pi;\quad \omega_2 = 3 \times 10^4\pi$

28. $\tilde{H}(z) = \dfrac{1 - r^{20}z^{-20}}{10} \displaystyle\sum_{k=2}^{6} (-1)^k \dfrac{z^2 - rz\,\cos(k\pi/10)}{z^2 - 2rz\,\cos(k\pi/10) + r^2}$

32. B: 0.067455   0.134911   0.067455
     A: −1.142981 0.412802

34. Sect. 1 B: 0.242721   0.000000   −0.242721
              A: −1.044314 0.720630
     Sect. 2 B: 0.321176   0.000000   −0.321176
              A: −1.780623 0.878036
     Sect. 3 B: 0.225977   0.000000   −0.225977
              A: −1.105472 0.468727
     Sect. 4 B: 0.273857   0.000000   −0.273857
              A: −1.487822 0.631797

REFERENCES

BUTTERWORTH, S., On the Theory of Filter Amplifiers. *Wireless Engr.,* Vol. 1, 1930, p. 536.

CHAN, D. S. K., and RABINER, L. R., Analysis of Quantization Errors in the Direct Form for Finite Impulse Response Digital Filters. *IEEE Trans. Audio Electroacoust.,* Vol. AU-21, No. 4, August 1973, p. 354.

CLAASEN, T. A. C. M., ET AL., Signal Processing Method for Improving the Dynamic Range of A/D and D/A Converters. *IEEE Trans. Acoust. Speech Signal Process.,* October 1980, p. 529.

GOLD, B., and RADER, C. M., *Digital Processing of Signals,* Chaps. 2, 3, and 4. New York: McGraw-Hill, 1969.

GOLDEN, R. M., and KAISER, J. F., Design of Wideband Sampled-Data Filters. *Bell System Tech. J.,* July 1964, p. 1533.

GUILLEMIN, E. A., *Synthesis of Passive Networks,* Chap. 14. New York: Wiley, 1957.

KAISER, J. F., Some Practical Considerations in the Realization of Linear Digital Filters. *Proc. 3rd Annual Conf. Circuit Systems Theory,* 1965, p. 621.

KAISER, J. F., Digital Filters, Chap. 7 in *System Analysis by Digital Computer,* ed. J. F. Kaiser and F. F. Kuo. New York: Wiley, 1966.

KUO, F. F., *Network Analysis and Synthesis,* Chap. 12. New York: Wiley, 1962.

OPPENHEIM, A. V. (ed.), *Papers on Digital Signal Processing,* p. 43. Cambridge, Mass.: MIT Press, 1969.

OPPENHEIM, A. V., and SCHAFER, R. W., *Discrete-Time Signal Processing.* Englewood Cliffs, N.J.: Prentice-Hall, 1989.

RABINER, L. R., and RADER, C. M. (eds.), *Discrete-Time Signal Processing.* New York: IEEE Press, 1972.

STORER, J. E., *Passive Network Synthesis,* Chap. 30. New York: McGraw-Hill, 1957.

VAN VALKENBURG, M. E., *Introduction to Modern Network Synthesis,* Chap. 13. New York: Wiley, 1960.

WAIT, J. V., Digital Filters, Chap. 5 in *Active Filters: Lumped, Distributed, Integrated, Digital, and Parametric,* ed. L. P. Huelsman. New York: McGraw-Hill, 1970.

13

Review of Random Functions, Correlation, and Power Spectra

13.1 RANDOM FUNCTIONS

Situations arise throughout the applications of signal analysis in which a function $f(t)$ cannot be specified exactly or is not completely predictable. Sometimes such a function is called noise to indicate that it is undesirable as well as unpredictable, but an unpredictable signal can also occur when the cause or generator of the signal is not specified completely. Such functions are treated as *random functions,* so that their statistical properties are used instead of their unspecified actual values.

Some of the fundamental aspects of these statistical properties applicable in the present discussion are reviewed in this and the following sections. The fields of statistics and probability are very broad as well as fundamental to many branches of science, and of course no attempt is made here to provide a comprehensive coverage. The reference list contains some excellent texts in these areas.

The most important statistical property of any random function $f(t)$ is its *probability function,* which is supposed to be a measure of the likelihood of occurrence of each value of $f(t)$. The probability function of $f(t)$ is defined on the *one-dimensional sample space* of $f(t)$. This is simply a line containing points corresponding to all of the possible values of $f(t)$.

For example, if the values of $f(t)$ are determined by tossing a coin at each value of t, then the sample space consists of two points representing heads and tails. Or, if $f(t)$ is the pointing angle of a weather vane, then the sample space is a continuous line from 0 to 2π radians.

If the sample space is discrete (contains a countable number of points), then the probability function is defined as $P(f_n)$, such that

$$P(f_n) \geq 0 \quad \text{and} \quad \sum_{n=1}^{N} P(f_n) = 1 \tag{13.1}$$

where N, the number of points in the sample space, can be any integer and the values of n enumerate the points (f_n) in the sample space. In the above example of the coin, letting $n = 1$ and 2 for heads and tails, $P(f_1) = 1/2$ and $P(f_2) = 1/2$ are the values of the probability function with $N = 2$.

In the case of the *continuous* sample space, which is of greater interest here, the probability function is written $p(f)$, is called the probability *density* function (pdf), and is defined such that

$$p(f) \geq 0 \quad \text{and} \quad \int_{-\infty}^{\infty} p(f)\, df = 1 \tag{13.2}$$

in a manner analogous to Eq. 13.1. In this notation the argument f is not $f(t)$ but rather a possible value of $f(t)$, that is, a point in the sample space of $f(t)$. Correspondingly, the probability of some value of $f(t)$ between $f = a$ and $f = b$ is

$$P\{a \leq f \leq b\} = \int_a^b p(f)\, df \tag{13.3}$$

Before examining important specific examples of the pdf, a few of its fundamental properties can be defined in general terms. Examples of these properties are given below along with the pdf examples.

First, the *mean value* (expected value, average value) of any function of f, say $y(f)$, is defined

$$E(y) = \int_{-\infty}^{\infty} y(f)p(f)\, df \tag{13.4}$$

so that $E(y)$ is essentially the sum of all values of $y(f)$, each value being weighted by a corresponding value of $p(f)$. If the integral in Eq. 13.4 does not converge, then $y(f)$ has no mean value.

Two particular mean value functions are so important that they are given special symbols. The first is the mean value of f itself, given the symbol μ_f. From Eq. 13.4 with $y(f) = f$, the mean value of f is

$$\mu_f \equiv E(f) = \int_{-\infty}^{\infty} fp(f)\, df \tag{13.5}$$

Thus μ_f is essentially the sum of pdf-weighted values of f, that is, the "center of mass" coordinate of $p(f)$.

The second important expected value function is the *variance* of f, which is given the symbol σ_f^2, The variance is the most common measure of the variability of $f(t)$ about its mean value, μ_f. It is defined as the *expected squared deviation of f from its mean value*, i.e., σ_f^2 is the expected value of $y(f) = (f - \mu_f)^2$. Accordingly,

$$\sigma_f^2 \equiv E[(f - \mu_f)^2];$$

$$\sigma_f^2 = \int_{-\infty}^{\infty} (f - \mu_f)^2 p(f)\, df$$

$$= \int_{-\infty}^{\infty} f^2 p(f)\, df - 2\mu_f \int_{-\infty}^{\infty} fp(f)\, df + \mu_f^2 \int_{-\infty}^{\infty} p(f)\, df$$

$$= E(f^2) - 2\mu_f \mu_f + \mu_f^2$$

Thus

$$\sigma_f^2 = E(f^2) - \mu_f^2 \tag{13.6}$$

so that the variance turns out to be the difference between the mean squared value of f and the square of μ_f. The square root of σ_f^2, σ_f, is called the *standard deviation* of f, and is the standard measure of the deviation of f from its mean value, the measure having the same units as f.

An important fundamental problem arising often in signal analysis is that of determining pdf of a function $y(f)$, given $p(f)$. This problem is solved in general by viewing the product $p(f)\,df$ as the probability that $f(t)$ lies between f and $f + df$, that is, as a vanishingly small increment of area above the sample space of f. In the simplest case, illustrated in Fig. 13.1, $y(f)$ is a unique, one-to-one mapping between the f and y sample spaces, so that $p(f)\,df$ and $q(y)\,dy$ are the same probabilities. That is,

$$\text{areas} \quad q(y)|dy| = p(f)|df| \tag{13.7}$$

or, assuming that $y(f)$ has a derivative,

$$q[y(f)] = \frac{p(f)}{|dy/df|} \tag{13.8}$$

where $|dy/df|$ is the absolute value of the derivative of $y(f)$.

Example 13.1

Given the function $p(f)$ along with the linear relationship $y = af + b$, what is the pdf of y? The answer is that $|dy/df| = |a|$ in this case and therefore

$$q(y) = \frac{1}{|a|}\, p(f) \tag{13.9}$$

Note that if $a = 1$, the two pdf's are the same, showing that adding or subtracting a constant from $f(t)$ changes the argument but not the form of the pdf.

Two useful corollaries follow from the result in Eq. 13.8 and Example 13.1. First,

$$\mu_{af+b} = a\mu_f + b \tag{13.10}$$

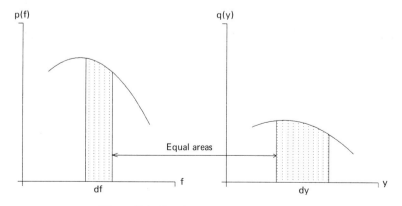

Figure 13.1 Equal probabilities as equal areas.

That is, if $y = af + b$, then the mean value of y is $a\mu_f + b$. This follows by using Eqs. 13.8 and 13.9 in the definition of the mean, Eq. 13.5, assuming that in the definition of μ_y, dy is a positive increment, and is thus equal to $|a|df$ in this case:

$$\mu_{af+b} = \int_{-\infty}^{\infty} (af + b)\frac{p(f)}{|a|}\,dy$$

$$= a\int_{-\infty}^{\infty} fp(f)\,df + b\int_{-\infty}^{\infty} p(f)\,df \qquad (13.11)$$

$$= a\mu_f + b$$

Second, the variance of $af + b$ is

$$\sigma_{af+b}^2 = a^2\sigma_f^2 \qquad (13.12)$$

which is shown by using Eqs. 13.8 through 13.10 in the variance definition. Again letting $y = af + b$, the proof is as follows:

$$\sigma_{af+b}^2 = \int_{-\infty}^{\infty} (y - \mu_{af+b})^2 q(y)\,dy$$

$$= \int_{-\infty}^{\infty} [af + b - (a\mu_f + b)]^2\frac{p(f)}{|a|}(|a|\,df) \qquad (13.13)$$

$$= a^2\int_{-\infty}^{\infty} (f - \mu_f)^2 p(f)\,df$$

$$= a^2\sigma_f^2$$

The methods used in the above demonstrations are generally applicable to finding the mean or variance of other functions of f [see Dwass (1970), Chap. 7].

13.2 UNIFORM AND NORMAL DENSITY FUNCTIONS

Turning now to specific examples of the pdf, two forms are of particular interest in digital signal analysis. The first is the *uniform* pdf, shown in Fig. 13.2. The uniform pdf expresses the assumption that all values of $f(t)$ between the values a and b are "equally likely." ("Equally likely" has real meaning only in the discrete case; nevertheless, the notion has heuristic appeal here in the continuous case.) Note again in Fig. 13.2 that the sample space is the horizontal (f) axis (the interval from a to b), and that the integral of $p(f)$ satisfies Eq. 13.2. When $p(f)$ is uniform, $f(t)$ is called a uniform variate.

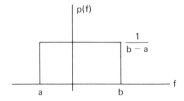

Figure 13.2   Uniform probability density function.

The *mean value* of a uniform variate distributed as in Fig. 13.2 is, according to Eq. 13.5,

$$\mu_f = \int_{-\infty}^{\infty} fp(f)\,df$$

$$= \frac{1}{b-a}\int_a^b f\,df \tag{13.14}$$

$$= \frac{b+a}{2}$$

As expected, the mean value is halfway between a and b in Fig. 13.2. The *variance* of the uniform variate is, from Eq. 13.6,

$$\sigma_f^2 = E(f^2) - \mu_f^2$$

$$= \frac{1}{b-a}\int_a^b f^2\,df - \frac{(b+a)^2}{4} \tag{13.15}$$

$$= \frac{(b-a)^2}{12}$$

Thus the *standard deviation* of a uniform variate is just $1/\sqrt{12}$ times the total range of values, $b - a$.

The second pdf of interest is the Gaussian, or *normal* probability density function, illustrated in Fig. 13.3. This function is of interest here primarily because in many applications of signal analysis, *Gaussian noise* is assumed, which means that the pdf of the noise function is as shown in Fig. 13.3. The general form of the normal pdf is

$$N(\mu, \sigma) \equiv p(f) = \frac{1}{\sigma\sqrt{2\pi}} e^{-(f-\mu)^2/2\sigma^2} \tag{13.16}$$

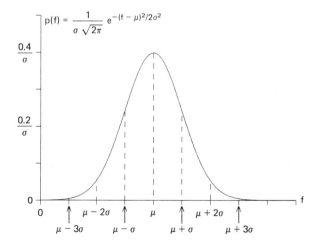

Figure 13.3 Normal probability density function, $N(\mu, \sigma)$.

in which μ and σ are μ_f and σ_f, the mean and standard deviation of f, as shown below. First, to see that $N(\mu, \sigma)$ has a unit integral as in Eq. 13.2, the substitution $x = (f - \mu)/(\sigma\sqrt{2})$ can be made so that the integral is

$$\int_{-\infty}^{\infty} N(\mu, \sigma)\, df = \frac{1}{\sigma\sqrt{2\pi}} \int_{-\infty}^{\infty} e^{-(f-\mu)^2/2\sigma^2}\, df$$

$$= \frac{1}{\sqrt{\pi}} \int_{-\infty}^{\infty} e^{-x^2}\, dx \qquad (13.17)$$

The latter integral can be evaluated simply by evaluating its square:

$$\left[\int_{-\infty}^{\infty} N(\mu, \sigma)\, df \right]^2 = \frac{1}{\pi} \iint_{-\infty}^{\infty} e^{-x^2} e^{-y^2}\, dx\, dy$$

$$= \frac{1}{\pi} \int_{0}^{2\pi} \int_{0}^{\infty} e^{-r^2} r\, dr\, d\theta \qquad (13.18)$$

$$= 2 \int_{0}^{\infty} r e^{-r^2}\, dr$$

$$= 1$$

thus proving that $N(\mu, \sigma)$ has a unit integral. (The result in the second line of Eq. 13.18 is obtained by switching from Cartesian to polar coordinates, that is, $x^2 + y^2 = r^2$ and $dx\, dy = r\, dr\, d\theta$.)

The demonstration that μ is the mean value of f is made by placing the normal pdf into Eq. 13.5 and again making the substitution, $x = (f - \mu)/(\sigma\sqrt{2})$:

$$\mu_f = \int_{-\infty}^{\infty} f N(\mu, \sigma)\, df$$

$$= \frac{1}{\sigma\sqrt{2\pi}} \int_{-\infty}^{\infty} f e^{-(f-\mu)^2/2\sigma^2}\, df \qquad (13.19)$$

$$= \sigma\sqrt{\frac{2}{\pi}} \int_{-\infty}^{\infty} x e^{-x^2}\, dx + \mu \int_{-\infty}^{\infty} \frac{1}{\sqrt{\pi}} e^{-x^2}\, dx$$

$$= \mu$$

In the third line of Eq. 13.19, the first integral is zero and the second, according to Eq. 13.18, is one, and so $\mu_f = \mu$ follows. Similarly, $\sigma_f = \sigma$ can be demonstrated. Let $\mu = 0$ so that Eq. 13.6 gives $\sigma_f^2 = E(f^2)$, and Eq. 13.4 then gives

$$\sigma_f^2 = \int_{-\infty}^{\infty} f^2 N(0, \sigma)\, df$$

$$= \frac{1}{\sigma\sqrt{2\pi}} \int_{-\infty}^{\infty} f^2 e^{-f^2/2\sigma^2}\, df \qquad (13.20)$$

$$= \sigma^2$$

with the integration being done by parts here.

Unfortunately, when $N(\mu, \sigma)$ is the pdf of f, the probability that f lies between some pair of values (as defined in Eq. 13.3) cannot be easily expressed as it obviously can be, for example, in the case of the uniform pdf. Therefore, normalized versions of this probability are tabulated in handbooks. [See, for example, the handbook by Burington (1965). Burington tabulates the integral ϕ in Eq. 13.21 as well as related integrals of $N(\mu, \sigma)$.] Typically, values of the integral of $N(0, 1)$ are tabulated; that is, if x is distributed according to $N(0, 1)$, values of the probability that x is less than some value T,

$$
\begin{aligned}
\phi(T) &= P\{x < T\} \\
&= \int_{-\infty}^{T} N(0, 1)\, dx
\end{aligned}
\tag{13.21}
$$

are tabulated versus values of T. Given these tabulated values, the more general form of $P\{a \le f \le b\}$ in Eq. 13.3, with $p(f) = N(\mu, \sigma)$, can be found by using Eqs. 13.10 and 13.12 with the substitution

$$
x = (f - \mu)/\sigma
\tag{13.22}
$$

All of this becomes more clear in the following examples.

Example 13.2

If the values of a random function $f(t)$ are distributed so that $p(f) = N(1, 2)$, what is the probability that $f(t)$ is positive? Letting

$$
x = \frac{f - \mu_f}{\sigma_f} = \frac{f}{2} - \frac{1}{2}
\tag{13.23}
$$

Eq. 13.10 gives $\mu_x = \frac{1}{2} - \frac{1}{2} = 0$, and Eq. 13.12 gives $\sigma_x = (\frac{1}{2})(2) = 1$, and thus the pdf of x is $N(0, 1)$. Then, according to Eq. 13.23, x is greater than $-\frac{1}{2}$ when f is positive, and so

$$
\begin{aligned}
P\{f > 0\} &= P\left\{x > -\frac{1}{2}\right\} \\
&= 1 - \phi\left(-\frac{1}{2}\right) = \phi\left(\frac{1}{2}\right)
\end{aligned}
\tag{13.24}
$$

which, if found in a handbook, is about 0.69.

Example 13.3

If the pdf of a function $f(t)$ is $N(2, 3)$, what is the probability that $f(t)$ is in the interval $(-4, +4)$? Letting

$$
x = \frac{f - \mu_f}{\sigma_f} = \frac{f - 2}{3}
\tag{13.25}
$$

Eqs. 13.10 and 13.12 again show that the pdf of x is $N(0, 1)$. Next, if f is between -4 and $+4$, then x must be between -2 and $+\frac{2}{3}$. The situation is illustrated in Fig. 13.4, and the required probability is given in the figure.

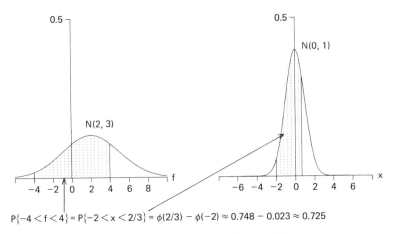

$$P\{-4 < f < 4\} = P\{-2 < x < 2/3\} = \phi(2/3) - \phi(-2) \approx 0.748 - 0.023 \approx 0.725$$

Figure 13.4 Probabilities in Example 13.3.

13.3 MULTIVARIATE DENSITY FUNCTIONS

The discussion so far has concerned a single random variate whose values are taken from a one-dimensional sample space, that is, a line. Generalization to an n-dimensional sample space for n random variates, $x_1, x_2, \ldots, x_n$, follows in a natural way. For continuous variates, for example, the probability density function can be denoted $p(x_1, x_2, \ldots, x_n)$, and, as for $n = 1$ in Eq. 13.2,

$$p(x_1, x_2, \ldots, x_n) \geq 0,$$

and

$$\int \int_{-\infty}^{\infty} \cdots \int p(x_1, x_2, \ldots, x_n) \, dx_1 \, dx_2 \cdots dx_n = 1 \qquad (13.26)$$

An illustration with $n = 2$ is given in Fig. 13.5. In this case the sample space is the $x_1 x_2$ plane, and $p(x_1, x_2)$ is called the *joint* pdf of x_1 and x_2 and in this illustration is the same for all pairs (x_1, x_2) inside the shaded rectangle in the $x_1 x_2$ plane. Corresponding to Eq. 13.3, the joint probability that x_1 is between α_1 and β_1 *and* x_2 is between α_2 and β_2 is

$$P\{\alpha_1 \leq x_1 \leq \beta_1, \alpha_2 \leq x_2 \leq \beta_2\} = \int_{\alpha_1}^{\beta_1} \int_{\alpha_2}^{\beta_2} p(x_1, x_2) \, dx_1 \, dx_2 \qquad (13.27)$$

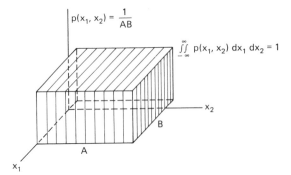

Figure 13.5 Joint probability density function.

Several other properties of the joint pdf are of general interest. To simplify the notation, let the two variates be x and y and denote the pdf by $p(x, y)$. Then, as in Eq. 13.27, the probability that x is between x and $x + dx$ while y has any value is

$$P = \int_x^{x+dx} \int_{-\infty}^{\infty} p(x, y) \, dy \, dx \qquad (13.28)$$

Since y can have any value, the probability on the left is really just the unconditional probability $q(x) \, dx$, $q(x)$ being the pdf of x, while the integral on the right becomes just the integral over y times dx. Hence, dividing by dx, the pdf of x is

$$q(x) = \int_{-\infty}^{\infty} p(x, y) \, dy \qquad (13.29)$$

Note that x and y can be exchanged in this formula and $q(y)$ is then the pdf of y.

It may be that the variates x and y do not depend on each other, so that knowledge of x does not help to determine the value of y, and vice versa. The variates are called *independent* [or "mutually independent" (see Feller, 1957)] if, for all x and y,

$$p(x, y) = q(x)r(y) \qquad (13.30)$$

where $q(x)$ is the pdf of x and $r(y)$ is the pdf of y.

The definition of the *expected value* of any function $F(x, y)$ is analogous to Eq. 13.4. Using $E(F)$ to represent this expected value,

$$E(F) = \iint\limits_{-\infty}^{\infty} F(x, y)p(x, y) \, dx \, dy \qquad (13.31)$$

Also, there is a *covariance function* similar to the variance defined above. The covariance is the *expected product of the deviations of x and y from their mean values:*

$$\begin{aligned} \sigma_{xy}^2 &= \iint\limits_{-\infty}^{\infty} (x - \mu_x)(y - \mu_y)p(x, y) \, dx \, dy \\ &= \iint\limits_{-\infty}^{\infty} xyp(x, y) \, dx \, dy - \mu_x\mu_y \\ &= E(xy) - \mu_x\mu_y \end{aligned} \qquad (13.32)$$

with the latter form resulting from the use of Eqs. 13.29 and 13.31. Thus the covariance is the difference between the expected product and the product of the expected values. Note that if x and y are *independent*, then Eqs. 13.30 and 13.31 give

$$E(xy) = \mu_x\mu_y \qquad (13.33)$$

and so the covariance function is zero if the variates are independent.

These properties of the joint pdf are applied in discussions to follow and are generally applicable when more than one random function is being considered or where correlation properties are concerned.

13.4 STATIONARY AND ERGODIC PROPERTIES

In applications of probability theory to signal analysis, a fundamental and interesting question arises as to the "real" nature of the sample space on which a probability function or a pdf is defined. Assuming that samples of a random function $f(t)$ are to be taken, are they to be taken at different values of t, or are different values to be selected from an ensemble of functions at the same value of t? These choices are illustrated in Fig. 13.6. The first possibility is that sampling could proceed *horizontally* from $t = 0$ over *any one* of the $f(t)$'s, with samples being taken at *different values of t*. The second possibility is that sampling could proceed *vertically* in the figure over *the population* (ensemble) of $f(t)$'s, for *the same value of t*. Situations analogous to both of these possibilities do in fact exist in practice. Examples of ensembles of random functions are blood pressure versus age in a population of humans [i.e., for each person in the population, $f(t)$ = blood pressure and t = age], temperature versus time in a population of reactor fuel elements, and so on.

The stationary and ergodic properties, both of which are assumed in many signal analysis procedures, are defined here in terms of the above ensemble of functions. Suppose that the pdf of $f(t)$, $p(f)$, is defined over the ensemble of functions (vertically in Fig. 13.6). Then, as suggested in Fig. 13.6, $p(f)$ itself can be viewed as a function of t. In a like manner the joint pdf, $p_\tau(f, f_\tau)$, in which f is a value of $f(t)$ taken from the ensemble at time t and f_τ is a similar value taken τ seconds later, can also be viewed as a function of t. (The subscript τ serves as a reminder that p_τ also depends on τ, which is assumed to be a fixed delay.)

The random function $f(t)$ is defined to be *stationary* if (and only if), for all τ, $p_\tau(f, f_\tau)$ is constant over t. This in turn implies that $p(f)$ is also constant over t because, as in Eq. 13.29,

$$p(f) = \int_{-\infty}^{\infty} p_\tau(f, f_\tau)\, df \qquad (13.34)$$

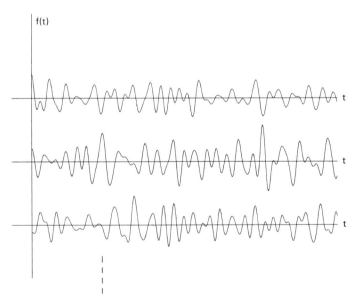

Figure 13.6 Ensemble or population of random functions of t.

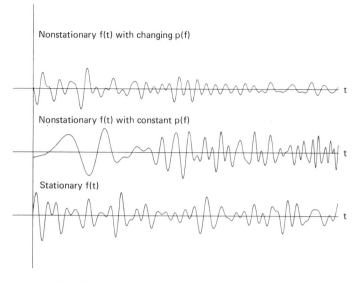

Nonstationary f(t) with changing p(f)

Nonstationary f(t) with constant p(f)

Stationary f(t)

Figure 13.7 Examples of nonstationary and stationary random functions.

and $p(f)$ is thus constant over t if the integrand is constant over t. The implication does not hold in reverse, however, as illustrated in Fig. 13.7, which shows first a nonstationary $f(t)$ with $p(f)$ changing; second, a nonstationary $f(t)$ with $p(f)$ constant; and finally, a stationary $f(t)$. The impression in the second case is that the frequency content of $f(t)$ is changing even though $p(f)$ is constant, and therefore that $p_\tau(f, f_\tau)$ must be related to the spectrum of $f(t)$. This important relationship is discussed in the next two sections.

If $f(t)$ is *ergodic* in addition to being stationary, then any one of the ensemble in Fig. 13.6 has statistical properties that are the *same as those of the entire ensemble*. For example, the time average of a stationary, ergodic $f(t)$ is the same as the average taken over the set of functions at any value of t. More thorough and precise definitions of the stationary and ergodic properties are presented with examples by Laning and Battin (1956). In the theory to follow, random functions are usually assumed to have the stationary and ergodic properties over some specified time interval.

13.5 CORRELATION FUNCTIONS

As in the preceding section there is often a need to measure or specify the interdependence or association between signal values — to answer questions such as, "Is the present value of $f(t)$ correlated with or dependent upon its own past values, or upon values of some other function $g(t)$?" Thus a quantitative indicator of correlation between signals is needed.

The correlation function defined here is meant to fill this need and to be applicable to stationary random functions as well as deterministic functions. For deterministic functions the idea of statistical dependence or independence as defined in Eq. 13.30 does not apply, however, and the correlation function becomes more a measure of similarity between functions.

The correlation function is defined in the same manner for either random or determined functions: It is the average product of two functions, $f(t)$ and $g(t)$, shifted relative to each other by the amount τ. Assume first that $f(t)$ and $g(t)$ are *deterministic* functions. (Random functions are treated below.) The formula for the correlation function of $f(t)$ and $g(t)$ is

$$\phi_{fg}(\tau) = \lim_{T \to \infty} \frac{1}{T} \int_{-T/2}^{T/2} f(t)g(t + \tau)\,dt \qquad (13.35)$$

and $\phi_{fg}(\tau)$ is obviously the average product just mentioned.

Note first that Eq. 13.35 resembles the test for orthogonality in Chapter 2, Section 2.3—in fact, if $f(t)$ and $g(t + \tau)$ are orthogonal in the interval of length T about $t = 0$, then $\phi_{fg}(\tau)$ is zero. Thus $\phi_{fg}(\tau)$ can be viewed as a *measure of orthogonality* of $f(t)$ and $g(t + \tau)$ over the entire t domain. The correlation function is thus determined from all values of $f(t)$ and $g(t)$ rather than a particular set of values.

The fundamental nature of $\phi_{fg}(\tau)$ is demonstrated clearly by Truxal (1955) who shows that *if a linear system is to be designed using the least-squares criterion, then the input and output signals are adequately described by specifying only their correlation functions.* [Actually these concepts date back to Wiener (1930).] Truxal's demonstration proceeds as follows: Suppose the linear system given by $H(s)$ in Fig. 13.8 is to be designed and that the desired output $d(t)$ cannot be attained exactly,

Figure 13.8   Linear system.

so that the design process involves minimizing the squared error between $d(t)$ and $g(t)$, the latter being the actual output. The average squared error in this case is

$$E_{av}^2 = \lim_{T \to \infty} \frac{1}{T} \int_{-T/2}^{T/2} [d(t) - g(t)]^2\,dt \qquad (13.36)$$

But $g(t)$, the actual output, can be given in terms of the convolution integral:

$$g(t) = \int_{-\infty}^{\infty} h(\tau)f(t - \tau)\,d\tau \qquad (13.37)$$

When this integral is substituted into Eq. 13.36 and the squaring operation is carried out, the result is

$$E_{av}^2 = \lim_{T \to \infty} \frac{1}{T} \int_{-T/2}^{T/2} \left[\left(\int_{-\infty}^{\infty} h(\tau)f(t - \tau)\,d\tau \cdot \int_{-\infty}^{\infty} h(x)f(t - x)\,dx \right) + d^2(t) \right.$$
$$\left. - 2d(t) \int_{-\infty}^{\infty} h(\tau)f(t - \tau)\,d\tau \right] dt \qquad (13.38)$$

where x is used as a dummy variable to avoid confusion. Next, the terms in Eq. 13.38 are rearranged as follows:

$$
E_{av}^2 = \iint\limits_{-\infty}^{\infty} \left(h(\tau)h(x) \lim_{T\to\infty} \frac{1}{T} \int_{-T/2}^{T/2} f(t-\tau)f(t-x)\,dt \right) d\tau\,dx + \lim_{T\to\infty} \frac{1}{T} \int_{-T/2}^{T/2} d^2(t)\,dt
$$

$$
- 2 \int_{-\infty}^{\infty} h(\tau) \lim_{T\to\infty} \frac{1}{T} \int_{-T/2}^{T/2} d(t)f(t-\tau)\,dt\,d\tau \tag{13.39}
$$

$$
= \iint\limits_{-\infty}^{\infty} h(\tau)h(x)\phi_{ff}(\tau-x)\,d\tau\,dx + \phi_{dd}(0) - 2 \int_{-\infty}^{\infty} h(\tau)\phi_{fd}(\tau)\,d\tau
$$

with the last line being obtained by using the definition of the correlation function in Eq. 13.35. Thus $f(t)$ and $d(t)$ appear in the formula for E_{av}^2 only through their correlation functions and the above statement is proved. (The subject of least-squares design is continued in the next chapter.)

Note that two of the three correlation functions in Eq. 13.39 have repeated subscripts. These functions, of the form

$$
\phi_{ff}(\tau) = \lim_{T\to\infty} \frac{1}{T} \int_{-T/2}^{T/2} f(t)f(t+\tau)\,dt \tag{13.40}
$$

are called *autocorrelation functions* because they are measures of the correlation of a function with its own past, present, or future values. The autocorrelation function has several distinguishing properties:

1. $\phi_{ff}(\tau)$ is an even function; that is, $\phi_{ff}(\tau) = \phi_{ff}(-\tau)$. If $-\tau$ is used in place of τ in Eq. 13.40, the integral does not change — effectively, the two $f(t)$'s are still shifted relative to each other by the amount τ.

2. $|\phi_{ff}(\tau)| \leq \phi_{ff}(0)$; that is, the average product is maximal when $f(t)$ is multiplied by itself without shifting. For a proof of this property, see Truxal (1955), or Wiener (1930).

3. $\phi_{ff}(0)$ is nonzero only for functions whose integrals over t do not converge absolutely, because obviously the integral in Eq. 13.40 must increase with T in order for $\phi_{ff}(\tau)$ to be nonzero in the limit. Thus Eq. 13.40 is not useful if $f(t)$ is a single transient that decays to zero with increasing t. An alternative autocorrelation function,

$$
\rho_{ff}(\tau) = \int_{-\infty}^{\infty} f(t)f(t+\tau)\,dt \tag{13.41}
$$

which, when compared with Eq. 13.40, is seen to be the total rather than the average product of $f(t)$ and $f(t+\tau)$, is useful in these cases.

4. $\phi_{ff}(\tau)$ carries no phase information about $f(t)$. This is seen by realizing that Eq. 13.40 gives the same result for $f(t+\alpha)$ as for $f(t)$, that is, that a shift in $f(t)$ does not change $\phi_{ff}(\tau)$.

5. $\phi_{ff}(0)$ is the average value of $f^2(t)$, as seen by setting $\tau = 0$ in Eq. 13.40. This value is defined to be the *average power* in $f(t)$. The use of "power" for $f^2(t)$ is a generalization of the physical meaning of the term. If $f(t)$ is an electrical potential or current measured in volts or amperes, then $f^2(t)$ is the instantaneous power in watts dissipated in a 1-Ω resistor "fed" by $f(t)$.

Although Eq. 13.40 is valid for $\phi_{ff}(\tau)$ when $f(t)$ is deterministic and has nonzero average power, there are two other forms of the autocorrelation function that are also important. First, when $f(t)$ is *periodic*, the product $f(t)f(t + \tau)$ can be averaged over a single period of length $2\pi/\omega_0$ where ω_0 is the fundamental frequency, as follows:

$$f(t) \text{ periodic:} \quad \phi_{ff}(\tau) = \frac{\omega_0}{2\pi} \int_{-\pi/\omega_0}^{\pi/\omega_0} f(t)f(t + \tau)\, dt$$

$$= c_0^2 + 2 \sum_{n=1}^{\infty} |c_n|^2 \cos n\omega_0\tau \qquad (13.42)$$

In this result the c_n are the complex Fourier coefficients and the second line is obtained from the first by substituting the complex Fourier series (Chapter 2, Section 2.4) for $f(t)$ and $f(t + \tau)$.

Second, suppose that $f(t)$ is a *stationary random* function. Since by definition $f(t)$ is not known exactly in this case, Eq. 13.40 is not formally applicable. Instead the average product $f(t)f(t + \tau)$ can be viewed as an expected product in the statistical sense. As in Section 13.4, let f represent a value of $f(t)$, f_τ a value of $f(t + \tau)$, and $p_\tau(f, f_\tau)$ the joint pdf of f and f_τ. Then, as in Eq. 13.31, the autocorrelation function is

$$f(t) \text{ random:} \quad \phi_{ff}(\tau) = E(f\!f_\tau) = \iint_{-\infty}^{\infty} f\!f_\tau\, p_\tau(f, f_\tau)\, df\, df_\tau \qquad (13.43)$$

This is the formula for the autocorrelation function of a random variate. If f and f_τ are independent, then $E(f\!f_\tau)$ becomes $E(f)E(f_\tau)$, or μ_f^2, as shown in Eq. 13.33. Therefore the behavior of $\phi_{ff}(\tau)$ for large τ becomes a possible test for periodic content in a function, that is,

$$\lim_{\tau \to \infty} \phi_{ff}(\tau) = \begin{cases} \text{nonzero if } f(t) \text{ has dc or periodic content} \\ \text{zero if } f(t) \text{ is not periodic and } \mu_f = 0. \end{cases} \qquad (13.44)$$

(The test becomes ambiguous for "almost periodic" functions — see Wiener, 1930.)

Some examples of cases to which these two forms of $\phi_{ff}(\tau)$ apply are illustrated in Fig. 13.9. In the first case Eq. 13.40 applies directly, and Eq. 13.42 can be used in the second. In the third example, the application of Eqs. 13.43 and 13.44 may be made as follows: Consider any unit interval of τ, for example, $(0, 1), (1, 2)$, and so on. First, when τ is less than a, identical pulses of f and f_τ in this interval coincide over a smaller interval, $a - \tau$. Within this smaller interval, the expected product $E(f\!f_\tau)$ is just $E(A^2)$, or b^2, the variance of A. Thus,

$$\text{for } \tau < a, \quad E(f\!f_\tau) = \frac{a - \tau}{1}(\sigma_A^2) = b^2(a - \tau) \qquad (13.45)$$

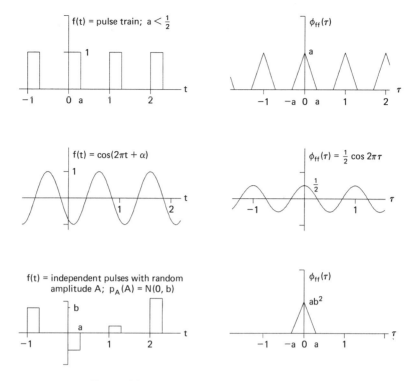

Figure 13.9 Examples of the autocorrelation function.

Next, when τ is not less than a, a given pulse of f cannot overlap the *same* pulse in f_τ and so, since the pulse amplitudes are independent and have zero mean,

$$\text{for } \tau \geq a, \quad E(f f_\tau) = E(f)E(f_\tau) = 0 \tag{13.46}$$

The result in Eqs. 13.45 and 13.46 is $\phi_{ff}(\tau)$ on the lower right in Fig. 13.9.

13.6 POWER AND ENERGY SPECTRA

The usual measures of energy and power are closely related to the autocorrelation function. Property 5 above gives $\phi_{ff}(0)$ as the average value of $f^2(t)$ or *average power* in $f(t)$.

On the other hand, it follows that $\rho_{ff}(0)$ in Eq. 13.41 must be the *total energy* in $f(t)$; because, from Eq. 13.41,

$$\rho_{ff}(0) = \int_{-\infty}^{\infty} f^2(t)\, dt \tag{13.47}$$

Obviously, if $\rho_{ff}(0)$ is finite the average power is zero and, conversely, if $\phi_{ff}(0)$ is nonzero, the total energy must be infinite. A *transient* is defined here as a function $f(t)$ having *finite energy* so that the integral in Eq. 13.47 converges.

For transient waveforms the Fourier transform $F(j\omega)$ exists, as does the transform, $R_{ff}(j\omega)$, of the correlation function $\rho_{ff}(\tau)$. In fact, the latter transform is

$$R_{ff}(j\omega) = \int_{-\infty}^{\infty} \rho_{ff}(\tau)e^{-j\omega\tau}\,d\tau$$

$$= \int_{-\infty}^{\infty} f(t)\int_{-\infty}^{\infty} f(t+\tau)e^{-j\omega\tau}\,d\tau\,dt \tag{13.48}$$

$$= \int_{-\infty}^{\infty} f(t)e^{j\omega t}\,dt\int_{-\infty}^{\infty} f(x)e^{-j\omega x}\,dx; \qquad x = \tau + t$$

$$= F(-j\omega)F(j\omega) = |F(j\omega)|^2$$

with the third line being obtained by substituting $x = \tau + t$ into the inner integral in the second line. Thus $R_{ff}(j\omega)$ turns out to be the square of the amplitude spectrum of $f(t)$. On the other hand, $R_{ff}(j\omega)$ is also related to the total energy, $\rho_{ff}(0)$ in Eq. 13.47, since

$$\rho_{ff}(\tau) = \frac{1}{2\pi}\int_{-\infty}^{\infty} R_{ff}(j\omega)e^{j\omega\tau}\,d\omega, \tag{13.49}$$

and hence

$$\rho_{ff}(0) = \frac{1}{2\pi}\int_{-\infty}^{\infty} R_{ff}(j\omega)\,d\omega = \frac{1}{2\pi}\int_{-\infty}^{\infty} |F(j\omega)|^2\,d\omega \tag{13.50}$$

Since the total energy $\rho_{ff}(0)$ in $f(t)$ is $1/2\pi$ times the integral of $R_{ff}(j\omega)$, $R_{ff}(j\omega) = |F(j\omega)|^2$ must be the *energy spectrum*, or energy spectral density function of $f(t)$, giving the energy density in units of energy per hertz. (The units are "per hertz" because $d\nu = d\omega/2\pi$, and $d\nu$ is in hertz if $d\omega$ is in radians per second.) Since $|F(j\omega)|$ is the amplitude spectrum of $f(t)$, it is not unreasonable to find that $|F(j\omega)|^2$ is the energy spectrum. (The term "power gain" is also used for $|F(j\omega)|^2$ when $F(j\omega)$ is a transfer function as in Chap. 12. Thus, the power gain of a filter is the energy spectrum of its impulse response.)

Example 13.4

Describe the correlation function and the energy content of the transient $f(t) = e^{-at}$ for $t \geq 0$, illustrated in Fig. 13.10. First, the transient autocorrelation function is

$$\rho_{ff}(\tau) = \int_{-\infty}^{\infty} f(t)f(t+\tau)\,dt$$

$$= \int_{max(0,\,-\tau)}^{\infty} e^{-a(2t+\tau)}\,dt = \frac{1}{2a}e^{-a|\tau|}$$

and is also illustrated in Fig. 13.10. Next, the energy spectral function, computed either as $R_{ff}(j\omega)$ or $|F(j\omega)|^2$, is $1/(\omega^2 + a^2)$, also as illustrated. Finally, the total energy, $\rho_{ff}(0)$, is $1/2a$.

The situation is analogous for periodic or random waveforms, except that here the energy is infinite and so power must be used instead of energy. Just as the energy spectrum of a transient function is $R(j\omega)$, the *power spectrum of a random function* is

$$\Phi_{ff}(j\omega) = \int_{-\infty}^{\infty} \phi_{ff}(\tau)e^{-j\omega\tau}\,d\tau \tag{13.51}$$

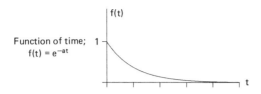
Function of time; 1
$f(t) = e^{-at}$

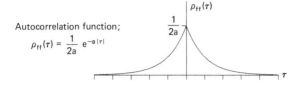
Autocorrelation function;

$\rho_{ff}(\tau) = \dfrac{1}{2a} e^{-a|\tau|}$

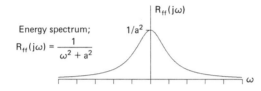
Energy spectrum;

$R_{ff}(j\omega) = \dfrac{1}{\omega^2 + a^2}$

Figure 13.10 Functions in Example 13.4.

That is, the power spectrum is the Fourier transform of the autocorrelation function. The total power in $f(t)$ is

$$\text{average or total power} = [f^2(t)]_{\text{avg}} = \phi_{ff}(0)$$
$$= \frac{1}{2\pi} \int_{-\infty}^{\infty} \Phi_{ff}(j\omega)\, d\omega \qquad (13.52)$$

which follows from the definition of the inverse transform as Eq. 13.50 follows from Eq. 13.49. Again, the units of $\Phi_{ff}(j\omega)$ are units of power per hertz, analogous to the units of energy per hertz for $R_{ff}(j\omega)$. The result in Eq. 13.52 is known as *Parseval's Theorem*.

Note that $\Phi_{ff}(j\omega)$, the Fourier transform of $\phi_{ff}(\tau)$, does not converge if $f(t)$ has periodic content, because if so $\phi_{ff}(\tau)$ does not vanish with increasing τ (see Eq. 13.44). If $f(t)$ is random with nonzero mean it is considered to have both random and periodic content, the periodic content in this case being just a dc component. In cases where $f(t)$ has both random and periodic content, it is useful to define the power spectrum of $f(t)$ as the continuous function $\Phi_{ff}(j\omega)$ combined with one or more impulse functions with areas $2\pi|c_n|^2$ at frequencies $\pm n\omega_0$ so that the total power in $f(t)$ can still be found as $1/2\pi$ times the integral of the power spectrum as in Eq. 13.52. On the other hand, if $f(t)$ is random with zero mean, then Eq. 13.52 gives the total power as just the variance, σ_f^2, of $f(t)$, since σ_f^2 is the average of $f^2(t)$ in this case. Thus, the power spectrum $\Phi_{ff}(j\omega)$ gives the distribution of the variance of $f(t)$ over frequency.

Example 13.5

Find the power spectrum of the random waveform in Fig. 13.9 with $b = 1$. The autocorrelation function $\phi_{ff}(\tau)$, given in Fig. 13.9, is repeated with $b = 1$ in Fig. 13.11. The power spectrum is the Fourier transform of $\phi_{ff}(\tau)$:

$$\Phi_{ff}(j\omega) = \int_{-a}^{a} (a - |\tau|)e^{-j\omega\tau}\,d\tau$$

$$= \frac{4}{\omega^2}\sin^2\!\left(\frac{\omega a}{2}\right)$$

and is also plotted. The total power is $\phi_{ff}(0) = a$, being also $1/2\pi$ times the integral of $\Phi_{ff}(j\omega)$.

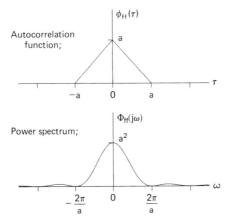

Autocorrelation function;

Power spectrum;

Figure 13.11 Functions in Example 13.5.

A summary of the principal formulas for autocorrelation functions, power spectra, and energy spectra is given in Table 13.1. Note that the power density spectrum of the periodic function is given as a series of impulse functions as just described, so that its dimensions are those of $\Phi_{ff}(j\omega)$, namely units of power per hertz.

TABLE 13.1. Summary of Continuous Formulas

Function	Type of $f(t)$	Formula
Autocorrelation	Transient	$\rho_{ff}(\tau) = \displaystyle\int_{-\infty}^{\infty} f(t)f(t + \tau)\,dt$
Autocorrelation	Random	$\phi_{ff}(\tau) = E(ff_\tau) = \displaystyle\lim_{T\to\infty}\frac{1}{T}\int_{-T/2}^{T/2} f(t)f(t + \tau)\,dt$
Autocorrelation	Periodic	$\phi_{ff}(\tau) = c_0^2 + 2\displaystyle\sum_{n=1}^{\infty} |c_n|^2 \cos n\omega_0\tau$
Energy density	Transient	$R_{ff}(j\omega) = \displaystyle\int_{-\infty}^{\infty} \rho_{ff}(\tau)e^{-j\omega\tau}\,d\tau = |F(j\omega)|^2$
Power density	Random	$\Phi_{ff}(j\omega) = \displaystyle\int_{-\infty}^{\infty} \phi_{ff}(\tau)e^{-j\omega\tau}\,d\tau$
Power density	Periodic	$\Phi_{ff}(j\omega) = 2\pi\displaystyle\sum_{n=-\infty}^{\infty} |c_n|^2\delta(\omega - n\omega_0)$
Total energy	Transient	$\rho_{ff}(0) = \displaystyle\int_{-\infty}^{\infty} f^2(t)\,dt = \frac{1}{2\pi}\int_{-\infty}^{\infty} R_{ff}(j\omega)\,d\omega$
Total power	Random	$\phi_{ff}(0) = \sigma_f^2 = \dfrac{1}{2\pi}\displaystyle\int_{-\infty}^{\infty} \Phi_{ff}(j\omega)\,d\omega$
Total power	Periodic	$\phi_{ff}(0) = \displaystyle\sum_{n=-\infty}^{\infty} |c_n|^2$

13.7 CORRELATION FUNCTIONS AND POWER SPECTRA OF SAMPLED SIGNALS

The correlation function and the power spectrum have different but analogous forms when the signals are sampled. As stated previously, the correlation function is a function of the time shift between two signals. When the signals are sampled the time shifts become discrete, and are integer multiples of the sample period T. Thus, the correlation function becomes discrete or "sampled." Consequently the power (or energy) spectrum, which is the Fourier transform of the correlation function, is described by the Fourier summation in Eq. 5.3 rather than the Fourier integral. As in previous sections the definition of correlation differs slightly depending on the nature of the signal.

For a deterministic sample sequence, $[f_m]$, with infinite energy, the autocorrelation function is defined as

$$\phi_{ff}(n) = \lim_{N \to \infty} \frac{1}{2N + 1} \sum_{m=-N}^{N} f_m f_{m+n} \qquad (13.53)$$

This definition also applies to stationary random processes which are *ergodic* since in such cases time averages are equivalent to ensemble averages. The following properties of the autocorrelation function here parallel those in Section 13.5.

1. The autocorrelation is an even function, that is,

$$\phi_{ff}(-n) = \phi_{ff}(n) \qquad (13.54)$$

2. The autocorrelation function has a maximum at $n = 0$, so that

$$|\phi_{ff}(n)| \leq \phi_{ff}(0) \qquad (13.55)$$

3. $\phi_{ff}(0)$ is nonzero only for functions whose summations in Eq. 5.3 do not converge. For functions with finite energy an alternate autocorrelation function similar to Eq. 13.41 is used, namely

$$\rho_{ff}(n) = \sum_{m=-\infty}^{\infty} f_m f_{m+n} \qquad (13.56)$$

4. For $[\phi_{ff}(n)]$ with no periodic component,

$$\lim_{n \to \infty} \phi_{ff}(n) = \mu_f^2 \qquad (13.57)$$

5. $[\phi_{ff}(n)]$ carries no phase information about $[f_m]$; that is, the autocorrelation function of $[f_{m+k}]$ is identical to that of $[f_m]$.
6. $\phi_{ff}(0)$ represents the average power of $[f_m]$.
7. Unlike $\phi_{ff}(\tau)$, $[\phi_{ff}(n)]$ is discrete; that is, it is defined only for integer values of the argument n.

For a periodic sequence, $[f_m]$, the autocorrelation function is defined as follows:

$$\phi_{ff}(n) = \frac{1}{N} \sum_{m=0}^{N-1} f_m f_{m+n} \qquad (13.58)$$

where N is the number of samples per period. In this case the autocorrelation function is also periodic with the same period N. The periodicity is easily verified by letting n equal $n + N$ in the definition above and using the fact that $[f_m]$ is periodic.

Finally, for stationary random processes which are not ergodic the autocorrelation function is given by Eq. 13.43 with $\tau = nT$, that is,

$$\phi_{ff}(n) = E(f_m f_{m+n}) = \iint_{-\infty}^{\infty} ff_{nT}\, p_{nT}(f, f_{nT})\, df_{nT} \tag{13.59}$$

where $p_{nT}(f, f_{nT})$ is the joint pdf of f and f_{nT}.

The definitions of cross-correlation follow in a straightforward manner from the definitions of autocorrelation above. For example, the cross-correlation between two stationary random processes $[f_m]$ and $[g_m]$ is defined

$$\phi_{fg}(n) = \lim_{N \to \infty} \frac{1}{2N + 1} \sum_{m=-N}^{N} f_m g_{m+n} \tag{13.60}$$

Unlike the autocorrelation, the cross-correlation is not, in general, an even function of n. That is, $\phi_{fg}(n) \neq \phi_{fg}(-n)$ in general. However, if we exchange f and g, we change the direction of the shift, and therefore

$$\phi_{fg}(n) = \phi_{gf}(-n) \tag{13.61}$$

Thus, for cross-correlation, the order of the subscripts (which determines which function is shifted ahead of the other) is important.

The *discrete power spectrum* is found by taking the z-transform of the autocorrelation function.

$$\tilde{\Phi}_{ff}(z) = \sum_{n=-\infty}^{\infty} \phi_{ff}(n) z^{-n} \tag{13.62}$$

By letting $z = e^{j\omega T}$ we obtain the power spectrum as a function of frequency,

$$\tilde{\Phi}_{ff}(e^{j\omega T}) = \overline{\Phi}_{ff}(j\omega) = \sum_{n=-\infty}^{\infty} \phi_{ff}(n) e^{-jn\omega T} \tag{13.63}$$

which is seen to be the Fourier transform of the autocorrelation sequence, $[\phi_{ff}(n)]$. It is common to refer to both $\tilde{\Phi}_{ff}(z)$ and $\overline{\Phi}_{ff}(j\omega)$ as the "power spectrum" of $[f_m]$, letting the argument, z or $j\omega$, distinguish between the two within the context of the discussion. The autocorrelation can be retrieved from either by using the appropriate inverse transform (see Eq. 7.13, for example):

$$\phi_{ff}(n) = \frac{1}{2\pi j} \oint \tilde{\Phi}_{ff}(z) z^{n-1}\, dz \tag{13.64}$$

or

$$\phi_{ff}(n) = \frac{T}{2\pi} \int_{-\pi/T}^{\pi/T} \overline{\Phi}_{ff}(j\omega) e^{jn\omega T}\, d\omega \tag{13.65}$$

Note that when these expressions are evaluated at $n = 0$, the result is the total power of the signal, that is,

$$\text{total power} = \phi_{ff}(0) = \frac{1}{2\pi j} \oint \tilde{\Phi}_{ff}(z) \, dz/z$$

$$= \frac{T}{2\pi} \int_{-\pi/T}^{\pi/T} \overline{\Phi}_{ff}(j\omega) \, d\omega \tag{13.66}$$

Using the fact that $\phi_{ff}(n)$ is an even function, one can easily verify the following property of the power spectrum:

$$\tilde{\Phi}_{ff}(z) = \tilde{\Phi}_{ff}(z^{-1}) \tag{13.67}$$

This in turn implies that

$$\overline{\Phi}_{ff}(j\omega) = \overline{\Phi}_{ff}^*(j\omega) = \overline{\Phi}_{ff}(-j\omega) \tag{13.68}$$

Therefore, $\overline{\Phi}_{ff}(j\omega)$ is both *real* and *even*. In addition, because $\overline{\Phi}_{ff}(j\omega)$ is the Fourier transform of a sample sequence, this function is periodic with period equal to ω_s, the sampling frequency of $[f_m]$ and hence of $[\phi_{ff}(n)]$. A final important property of $\overline{\Phi}_{ff}(j\omega)$ is that it is nonnegative. Rather than prove this property formally, we simply note that any power spectrum is a distribution of power (i.e., mean-squared value) over frequency, and that negative power in any band of frequencies would not be physically possible.

The discrete power spectrum must be given special consideration when the sampled signal has a nonzero mean or a periodic component. In such cases the auto-correlation function does not vanish for increasing n and the summation in Eq. 13.63 does not converge. As in Table 13.1, it is again useful in such cases to define the discrete power spectrum of these components to be $2\pi|\overline{c}_m|^2$ at frequencies $\pm m\omega_0$ ($m = 0$ for the dc component), where $\overline{c}_m$ is the Fourier series coefficient of the sampled periodic signal given in Eq. 5.29. With this definition for the power spectrum, it is possible to use the inverse transforms in Eq. 13.65 to recover $\phi_{ff}(n)$.

The *discrete cross-power spectrum* is found by transforming the cross-correlation function. Thus, in terms of both z and ω, we have

$$\tilde{\Phi}_{fg}(z) = \sum_{n=-\infty}^{\infty} \phi_{fg}(n) z^{-n} \tag{13.69}$$

$$\tilde{\Phi}_{fg}(e^{j\omega T}) = \overline{\Phi}_{fg}(j\omega) = \sum_{n=-\infty}^{\infty} \phi_{fg}(n) e^{-jn\omega T} \tag{13.70}$$

As with autocorrelation the cross-correlation function can be retrieved from the cross-power spectrum by using the inverse transform relations:

$$\phi_{fg}(n) = \frac{1}{2\pi j} \oint \tilde{\Phi}_{fg}(z) z^{n-1} \, dz \tag{13.71}$$

$$\phi_{fg}(n) = \frac{T}{2\pi} \int_{-\pi/T}^{\pi/T} \overline{\Phi}_{fg}(j\omega) e^{jn\omega T} \, d\omega \tag{13.72}$$

Since the cross-correlation function is not in general an even function, the cross-power spectrum exhibits properties somewhat different from the auto spectrum defined above. Using the relationship in Eq. 13.61, one can show that

$$\tilde{\Phi}_{fg}(z) = \tilde{\Phi}_{gf}(z^{-1}) \tag{13.73}$$

Correspondingly, in terms of frequency we have

$$\overline{\Phi}_{fg}(j\omega) = \overline{\Phi}_{gf}(-j\omega) = \overline{\Phi}_{gf}^{*}(j\omega) \tag{13.74}$$

We note here that the cross-power spectrum will in general have a nonzero phase and can in fact take on negative values.

The definitions of power spectra discussed in this section can be applied to functions with finite energy by using $\rho(n)$ (see Eq. 13.56) in place of $\phi(n)$. In this case the spectrum is an *energy spectrum* rather than a power spectrum. Its properties are the same as those of the power spectrum discussed above. The power spectrum provides the distribution of signal power over frequency, and the energy spectrum provides the distribution of signal energy over frequency.

13.8 DISCRETE-TIME RANDOM PROCESSES AND LINEAR FILTERING

In this section we explore the effect of linear filtering on the correlation functions and power spectra of discrete-time random processes. To simplify the discussion we will assume that all signals are stationary random signals so that the correlation functions can be expressed using the expected value operator. The results derived under this assumption can be applied to any type of sampled signal as long as the appropriate definition of correlation is used.

Consider the block diagram in Fig. 13.12, which shows two signals $[x_m]$ and $[f_m]$ which are passed through linear time-invariant digital filters $\tilde{D}(z)$ and $\tilde{H}(z)$ to produce outputs $[y_m]$ and $[g_m]$. As we have observed previously, the filter outputs are related to the inputs via the convolution equations,

$$y_m = \sum_{l=-\infty}^{\infty} d_l x_{m-l} \tag{13.75}$$

$$g_m = \sum_{l=-\infty}^{\infty} h_l f_{m-l} \tag{13.76}$$

Let us assume that the auto- and cross-correlation functions or, equivalently, the auto- and cross-power spectra of $[x_m]$ and $[f_m]$, are known. We seek expressions relating the correlation functions and power spectra of the output signals in Fig. 13.12.

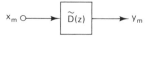

Figure 13.12 Signals used to explore the effect of linear filtering on correlation functions and their spectra.

First, we develop an expression for the cross-correlation $\phi_{xg}(n)$. By definition,

$$\phi_{xg}(n) = E[x_m g_{m+n}]$$

Substituting for g from Eq. 13.76 and simplifying gives

$$\phi_{xg}(n) = E\left[x_m \sum_{l=-\infty}^{\infty} h_l f_{m+n-l}\right]$$

$$= \sum_{l=-\infty}^{\infty} h_l E[x_m f_{m+n-l}] \qquad (13.77)$$

$$= \sum_{l=-\infty}^{\infty} h_l \phi_{xf}(n-l)$$

Thus, $\phi_{xg}(n)$ is the convolution of $\phi_{xf}(n)$ with the impulse response of the filter. Convolution in the time domain implies multiplication in the transform domain, and therefore the power spectrum is $\tilde{\Phi}_{xg}(z)$ given by

$$\tilde{\Phi}_{xg}(z) = \tilde{H}(z)\tilde{\Phi}_{xf}(z)$$
$$\text{with } \tilde{H}(z) = \tilde{G}(z)/\tilde{F}(z) \qquad (13.78)$$

If we let $x_m = f_m$ in Eqs. 13.77 and 13.78, we obtain expressions for the cross-correlation and cross-power spectrum between the input and output of a discrete-time linear filter:

$$\phi_{fg}(n) = \sum_{l=-\infty}^{\infty} h_l \phi_{ff}(n - l) \qquad (13.79)$$

$$\tilde{\Phi}_{fg}(z) = \tilde{H}(z)\tilde{\Phi}_{ff}(z)$$
$$\text{with } \tilde{H}(z) = \tilde{G}(z)/\tilde{F}(z) \qquad (13.80)$$

Next we develop an expression for the cross-correlation between $[y_m]$ and $[g_m]$. Again, by definition,

$$\phi_{yg}(n) = E[y_m g_{m+n}]$$

Substituting for y_m from Eq. 13.75 and simplifying, we find that

$$\phi_{yg}(n) = E\left[\sum_{l=-\infty}^{\infty} d_l x_{m-l} g_{m+n}\right]$$

$$= \sum_{l=-\infty}^{\infty} d_l E[x_{m-l} g_{m+n}] \qquad (13.81)$$

$$= \sum_{l=-\infty}^{\infty} d_l \phi_{xg}(n + l)$$

In this result, using $-l = m$, we have the convolution of $[d_{-m}]$ with the correlation function $\phi_{xg}(m)$. Thus, the cross-power spectrum is given by

$$\tilde{\Phi}_{yg}(z) = \tilde{D}(z^{-1})\tilde{\Phi}_{xg}(z)$$
$$\text{with } \tilde{D}(z) = \tilde{Y}(z)/\tilde{X}(z) \qquad (13.82)$$

where the property in Eq. 7.8 has been used to transform $[d_{-l}]$. Substituting Eq. 13.78 for $\tilde{\Phi}_{xg}(z)$ gives

$$\tilde{\Phi}_{yg}(z) = \tilde{D}(z^{-1})\tilde{H}(z)\tilde{\Phi}_{xf}(z) \tag{13.83}$$

Further, with $x_m = f_m$, $\tilde{D}(z) = \tilde{H}(z)$, and therefore $y_m = g_m$, we obtain the following expression for the power spectrum of $[g_m]$:

$$\tilde{\Phi}_{gg}(z) = \tilde{H}(z^{-1})\tilde{H}(z)\tilde{\Phi}_{ff}(z)$$

$$\text{with } \tilde{H}(z) = \tilde{G}(z)/\tilde{F}(z) \tag{13.84}$$

With $z = e^{j\omega T}$ this expression becomes $\overline{\Phi}_{gg}(j\omega) = |\overline{H}(j\omega)|^2 \overline{\Phi}_{ff}(j\omega)$, and thus the power spectrum of a filter output is equal to the power spectrum of the input weighted by the power gain function of the filter.

The *power gain* of any linear digital filter was given in Chapter 12 as $|\overline{H}(j\omega)|^2$, and thus we have

$$\begin{aligned}
\text{power gain} &= \frac{\text{output power}}{\text{input power}} \\[2mm]
&= \frac{\overline{\Phi}_{gg}(j\omega)}{\overline{\Phi}_{ff}(j\omega)} \\[2mm]
&= |\overline{H}(j\omega)|^2 \\[2mm]
&= \tilde{H}(z)\tilde{H}(z^{-1}) \qquad \text{with } z = e^{j\omega T} \\[2mm]
&= |\tilde{H}(z)|^2
\end{aligned} \tag{13.85}$$

TABLE 13.2. Summary of Discrete Formulas

Function	Formula
Power spectrum of $[f_m]$	$\tilde{\Phi}_{ff}(z) = \displaystyle\sum_{n=-\infty}^{\infty} \phi_{ff}(n)z^{-n}$
Cross-power spectrum	$\tilde{\Phi}_{fg}(z) = \displaystyle\sum_{n=-\infty}^{\infty} \phi_{fg}(n)z^{-n}$ $= \tilde{\Phi}_{gf}(z^{-1})$
Autocorrelation	$\phi_{ff}(n) = \dfrac{1}{2\pi j} \displaystyle\oint \tilde{\Phi}_{ff}(z)z^{n-1}\, dz$
Cross-correlation	$\phi_{fg}(n) = \dfrac{1}{2\pi j} \displaystyle\oint \tilde{\Phi}_{fg}(z)z^{n-1}\, dz$
Total power	$E[f_m^2] = \phi_{ff}(0)$ $= \dfrac{1}{2\pi j} \displaystyle\oint \tilde{\Phi}_{ff}(z)z^{-1}\, dz$
Linear filter	$\tilde{G}(z) = \tilde{H}(z)\tilde{F}(z)$ $\tilde{\Phi}_{gg}(z) = \tilde{H}(z)\tilde{H}(z^{-1})\tilde{\Phi}_{ff}(z)$ $\tilde{\Phi}_{fg}(z) = \tilde{H}(z)\tilde{\Phi}_{ff}(z)$ Power gain $= \tilde{H}(z)\tilde{H}(z^{-1}) = |\tilde{H}(e^{j\omega T})|^2$

Using these results and the properties developed in the preceding section, one can derive other relationships involving signals and transfer functions. For example, the cross spectrum $\tilde{\Phi}_{gx}(z)$ can be found using Eq. 13.78 and the properties in Eqs. 7.8 and 13.73. The result is

$$\tilde{\Phi}_{gx}(z) = \tilde{H}(z^{-1})\tilde{\Phi}_{xf}(z^{-1}) = \tilde{H}(z^{-1})\tilde{\Phi}_{fx}(z) \qquad (13.86)$$

Table 13.2 contains a summary of some of the most useful relationships and other formulas.

13.9 COMPUTING ROUTINES

Three routines relating to this chapter are included in Appendix B and on the floppy disk. The first two generate samples of approximately uniform and Gaussian variates, and the third is useful for computing certain expected-value functions. All three routines are functions rather than subroutines.

The first two routines are random number generators:

```
FUNCTION SPUNIF(ISEED)
FUNCTION SPGAUS(ISEED)


SPUNIF = Sample of a uniform random variate distributed over the
         interval (0,1)

SPGAUS = Sample of a Gaussian variate with probability density
         N(0,1)

ISEED  = Integer set initially to any value, then left alone.
```

These random number functions are typical of those found in computer libraries, except that they are meant to be portable and not computer-dependent. However, to achieve a reasonable nonrepeating sequence length, we have assumed that the computer allows integers up to 2^{24} in magnitude and has double-precision words with at least 14 decimal digits of accuracy.

The sequence length of SPUNIF is 2^{24} samples. SPGAUS generates an approximately normal variate by summing 12 uniform samples, so its effective usable sequence length is $2^{24}/12$, or 1,398,101 samples. Various tests have been run to test these functions, two of which are shown in the following examples.

Example 13.6

Generate uniform and Gaussian sequences of length 10^5 and plot histograms with 20 intervals. Start with ISEED = 123. The following program would generate the histogram in the array H(0:19) for the uniform sequence:

```
        REAL H(0:19)
        DATA H/20*0./
        ISEED=123
        DO 1 K=1,100 000
          N=20.*SPUNIF(ISEED)
          H(N)=H(N)+1.E-5
      1 CONTINUE
```

In this program we see that each uniform random sample is multiplied by 20 to obtain the histogram interval. A histogram value is then incremented by 0.00001, so that, with 10^5 samples, the sum of the histogram values will equal 1. The two resulting histograms are shown in Fig. 13.13 and, with 10^5 samples, are seen to be close to the theoretical pdf's.

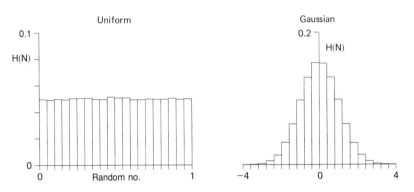

Figure 13.13 Histograms in Example 13.6 illustrating uniform and Gaussian density functions.

Example 13.7

Estimate the autocorrelation functions of several Gaussian sequences of length 25000. After setting ISEED initially, each of the sequences is generated as follows:

```
          DO 1 K=0,24999
            F(K)=SPGAUS(ISEED)
        1 CONTINUE
```

Then, to estimate the autocorrelation function, a finite version of Eq. 13.53 is used:

$$\hat{\phi}_{ff}(k) = \frac{1}{N-k} \sum_{m=0}^{N-k-1} f_m f_{m+k}; \qquad k = 0, \ldots, k_{\max} \qquad (13.87)$$

In this example we have $N = 25,000$ and we let $k_{\max} = 5000$. Each value of $\hat{\phi}_{ff}(k)$ for $k = 0$ through $k_{\max}$ is, then, in accordance with Eq. 13.87, the average of products $f_m f_{m+k}$, and thus the best unbiased estimate of $\hat{\phi}_{ff}(k)$. Plots of $\hat{\phi}_{ff}(k)$ for three different sequences are shown in Fig. 13.14. We note that $\hat{\phi}_{ff}(0)$ in each case is near 1.0 because $\sigma_f^2 = E[f^2] = 1.0$, and also that $\hat{\phi}_{ff}(k)$ is near zero for all $n > 0$. This latter property indicates that the samples in $[f_m]$ are statistically *independent* and that $[f_m]$ is a *white* Gaussian sequence as described in Chapter 15, Section 15.2.

The third routine, SPEXV, is a function that is used to obtain any moment of any real data sequence. The ith moment of the sequence $[x_m]$ is found as the expected value, $E[x_m^i]$. Thus, the first moment is the mean, the second is the mean-squared value, and so on. These expected values are computed as

$$\text{SPEXV} = \frac{1}{N} \sum_{m=0}^{N-1} x_m^i \qquad (13.88)$$

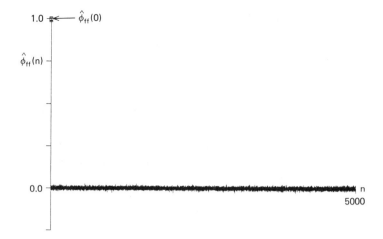

Figure 13.14 Plots of the estimated autocorrelation functions of three different sequences generated using SPGAUS.

The function SPEXV is defined as follows:

```
FUNCTION SPEXV(X,N,I)
```

$$
\begin{aligned}
\text{SPEXV} &= \text{Expected value defined in Eq. 13.88} \\
\text{X(0:N-1)} &= \text{Data sequence} \\
\text{I} &= \text{Moment in Eq. 13.88-1 for mean value, 2 for mean-} \\
& \quad\;\; \text{squared value, etc.}
\end{aligned}
$$

EXERCISES

1. Suppose the samples of $f(t)$ at $t = 0$, ± 1, ± 2, and so on, are determined by throwing two dice, with each sample value f_n being the sum thrown for time t_n:
   (a) Sketch the probability function $P(f_n)$.
   (b) Discuss whether or not $f(t)$ is a stationary function.
2. Given the exponential pdf below, calculate **(a)** μ_x, and **(b)** σ_x^2.

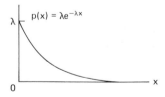

3. If the values of $f(t)$ are distributed uniformly from -1 to $+1$, what is the average value of $g(t) = 3f(t) + 4$? How does σ_f^2 compare with σ_g^2?
4. If the values of $f(t)$ are distributed as in problem 3, what is the pdf of $y(t) = f^2(t)$? *Hint:* A one-to-one mapping between sample spaces does not exist; therefore a revised version of Eq. 13.8 must be found.

5. If $f(t)$ is normally distributed, what is the probability that $f(t)$ is within one standard deviation of its mean value? Give your answer in terms of the normalized function $\phi(T)$ described in Section 2, and, if possible, look up the numerical value.

6. Sketch the uniform pdf's with
   (a) $\mu = 0$ and $\sigma^2 = \frac{1}{12}$.
   (b) $\mu = \frac{1}{2}$ and $\sigma^2 = \frac{1}{12}$.

7. If the pdf of $f(t)$ is $N(100, 1)$, what is the probability that $f(t)$ is between 98 and 102? Answer as in Exercise 5.

8. If the pdf of $f(t)$ is $N(0, 4)$ what are the chances of measuring $f(t) = 3$, assuming that the measuring instrument rounds off to the nearest integer? Answer as in Exercise 5.

9. Derive the covariance function of the joint pdf in Fig. 13.5.

10. Give a physical example of a stationary function that is not ergodic.

11. Given $f(t)$ with pdf uniform from -1 to 1, show how to generate $g(t)$ with pdf uniform from 0 to 2.

12. Prove that the expected value of the product of two independent variates x and y is $\mu_x \mu_y$, and therefore that the covariance is zero as in Eq. 13.33.

13. Compute the autocorrelation function, the energy spectrum, and the total energy for $f(t)$ shown here.

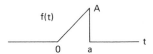

14. Do Exercise 13 for the sinusoidal transient given here.

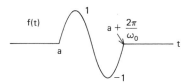

15. Derive the first autocorrelation function in Fig. 13.9.

16. Derive the second autocorrelation function in Fig. 13.9.

17. Derive the third autocorrelation function in Fig. 13.9.

18. Describe the correlation function $\phi_{fg}(\tau)$ in the case where one of the functions $f(t)$ or $g(t)$ is a unit impulse at $t = 0$.

19. **(a)** Compute the autocorrelation function of $f(t)$ as shown, where $f(t)$ consists of a series of rectangular pulses whose amplitudes are independent and distributed uniformly in the interval $(-1, +1)$.

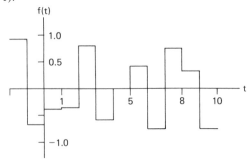

(b) Compute the power spectrum $\Phi_{ff}(j\omega)$.

20. Do Exercise 19 for the case where $f(t)$ is distributed uniformly in the interval $(0, 1)$.

21. Two random variates, $x(t)$ and $y(t)$, have identical power spectra given by $\Phi_{xx}(j\omega) = \Phi_{yy}(j\omega)$, but $p(x) = N(0, 2)$ and $q(y)$ is uniform. What is σ_y^2?

22. Given the autocorrelation function $\phi_{ff}(\tau)$ of a random variate $f(t)$ with mean μ_f equal to zero,
    (a) Derive a formula for $\phi_{gg}(\tau)$ for the function $g(t) = Af(t) + B$.
    (b) Derive $\Phi_{gg}(j\omega)$ in terms of $\Phi_{ff}(j\omega)$.

23. Find the autocorrelation function and energy spectrum of a single rectangular pulse of width a and amplitude b.

24. An FIR filter has the transfer function

$$\tilde{H}(z) = \sum_{n=0}^{N} b_n z^{-n}$$

Express the power gain of this filter as a function of z and $[b_n]$.

25. A white Gaussian sequence, $[f_n]$, with parameters $\mu_f = 0$ and $\sigma_f = 3$ is processed by an integration algorithm,

$$g_m = f_m + g_{m-1}$$

Derive an expression for the discrete output power spectrum, $\tilde{\Phi}_{gg}(z)$.

26. Derive an expression for the autocorrelation function, $\phi_{ff}(n)$, of a general sinusoidal waveform given by $f_m = A\,\cos(m\omega_0 T + \theta)$.

SOME ANSWERS

2. $1/\lambda$, $1/\lambda^2$ **3.** $\mu_g = 4$, $\sigma_f^2 = 1/3$, $\sigma_g^2 = 3$ **4.** $q(y) = 1/(2\sqrt{y})$, $0 \le y \le 1$
5. $\phi(1) - \phi(-1) \approx 0.68$ **7.** $\phi(2) - \phi(-2) \approx 0.95$ **8.** $\phi(0.875) - \phi(0.625) \approx 0.08$

9. 0   **11.** $g(t) = 1 + f(t)$ **13.** $\rho_{ff}(\tau) = \dfrac{A^2(a - |\tau|)^2(2a + |\tau|)}{6a^2}$ for $-a \le \tau \le a$

14. $\rho_{ff}(\tau) = \left[\left(\dfrac{2\pi}{\omega_0} - |\tau| \right) \cos \omega_0 \tau + \left(\dfrac{1}{\omega_0} \right) \sin \omega_0 |\tau| \right] \Big/ 2$ for $-\dfrac{2\pi}{\omega_0} \le \tau \le \dfrac{2\pi}{\omega_0}$

19. (a) $\phi_{ff}(\tau) = \dfrac{1 - |\tau|}{3}$ for $|\tau| \le 1$, **(b)** $\Phi_{ff}(j\omega) = \left(\dfrac{4}{3\omega^2} \right) \sin^2 \left(\dfrac{\omega}{2} \right)$

20. (See answers to Exercises 19 and 22)   **21.** 4
22. (a) $\phi_{gg}(\tau) = A^2 \phi_{ff}(\tau) + B^2$, **(b)** $\Phi_{gg}(j\omega) = A^2 \Phi_{ff}(j\omega) + B^2 \delta(\omega)$

25. $\Phi_{gg}(z) = 3/(2 - z - z^{-1})$ **26.** $\phi_{ff}(n) = \dfrac{A^2}{2} \cos n\omega_0 T$

REFERENCES

Blanc-Lapierre, A., and Fortet, R., *Theory of Random Functions*, Vol. 1. New York: Gordon and Breach, 1965.

Burington, R. S., *Handbook of Mathematical Tables and Formulas*, 4th ed. New York: McGraw-Hill, 1965.

Derman, C., Gleser, L. J., and Olkin, I., *A Guide to Probability Theory and Application*. New York: Holt, Rinehart and Winston, 1973.

Dwass, M., *Probability Theory and Applications*. New York: W. A. Benjamin, 1970.

Feller, W., *An Introduction to Probability Theory and Its Applications*, 2nd ed., Vol. 1. New York: Wiley, 1957.

GOODE, HARRY H., and MACHOL, R. E., *System Engineering,* Chaps. 5 and 6. New York: McGraw-Hill, 1957.

LANING, J. H., and BATTIN, R. H., *Random Processes in Automatic Control.* New York: McGraw-Hill, 1956.

MIDDLETON, D., *Statistical Communication Theory.* New York: McGraw-Hill, 1960.

SCHWARTZ, M., *Information Transmission, Modulation and Noise,* 2nd ed., Chap. 6. New York: McGraw-Hill, 1970.

TRUXAL, J. G., *Control System Synthesis*, Chap. 7. New York: McGraw-Hill, 1955.

WIENER, N., Generalized Harmonic Analysis. *Acta Math.,* Vol. 54, 1930, p. 117.

14

Least-Squares
System Design

14.1 INTRODUCTION

The least-squares principle is widely applicable to the design of digital signal processing systems. In Chapter 2 we saw how to fit a general linear combination of functions to a desired function or sequence of samples, and how the finite Fourier series, as an example, provides a least-squares fit to a sequence of data. In Chapter 8 we used the results of Chapter 2 to obtain a least-squares approximation to a desired FIR filter function. In this chapter we wish to extend the least-squares concept to the design of other kinds of signal processing systems.

First, we describe briefly a variety of tasks such as prediction, modeling, equalization, and interference canceling, where least-squares design is useful. We show that all of these tasks lead to the same least-squares design problem, which is to match a particular signal to a desired signal so that the difference between the two signals is minimal in the least-squares sense.

Next, we discuss the solution to the least-squares system design problem, which amounts for nonrecursive systems to the inversion of a matrix of correlation coefficients, similar to the symmetric coefficient matrix in Chapter 2, Eq. 2.9. We show that finding this solution is equivalent to finding the minimum point on a quadratic *mean-squared-error performance surface,* and we introduce software to perform this task. We include various examples of least-squares design. The examples illustrate a variety of system configurations used in different applications, and also some of the different ways to estimate correlation coefficients.

14.2 APPLICATIONS OF LEAST-SQUARES DESIGN

In this section we describe some system configurations where least-squares design is applicable. The first configuration, illustrated in Fig. 14.1, is the *linear predictor.* The prediction concept is illustrated in its simplest form in Fig. 14.1(a). The coefficients of a causal linear system, $\tilde{H}(z)$, are adjusted (as suggested by the slanting

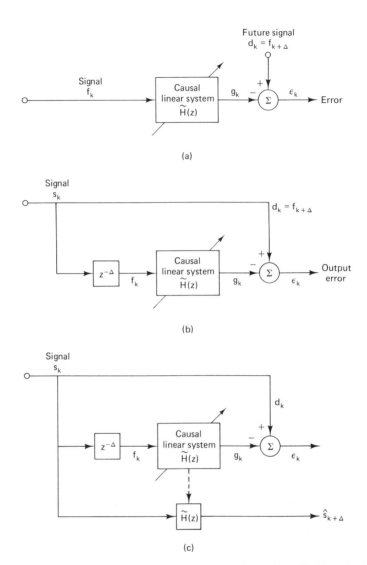

(a)

(b)

(c)

Figure 14.1 Linear predictor configurations. (a) Simplest form; (b) realizable form; (c) augmented form.

arrow in the figure) to minimize the mean-squared error, $E[\epsilon_k^2]$, thereby making the system output, g_k, the best least-squares approximation to a desired future signal, $d_k = f_{k+\Delta}$. The adjustment process is not realizable in this form, however, unless each future signal value is available at time k. Therefore, we use the realizable form in Fig. 14.1(b), which is equivalent to the form in Fig. 14.1(a), but (with s_k delayed to produce the system input, f_k) has the future value, $f_{k+\Delta}$, equal to the present signal value, s_k. Figure 14.1(b) is the most common form of the linear predictor, even though it does not actually predict a future value of the signal, s_k. In most signal processing applications the error ϵ_k, rather than a future value of s_k, is needed. If an actual prediction of a stationary signal s_k is needed, the augmented form in Fig. 14.1(c) may be implemented. Here s_k is sent to a copy of the predictor, $\tilde{H}(z)$, which produces the predicted future value, $\hat{s}_{k+\Delta}$.

The linear predictor is useful in waveform encoding and data compression (Bordley, 1983), spectral estimation (Chapter 15), spectral line enhancement (Zeidler et al., 1978), event detection (Clark and Rodgers, 1981); Stearns and Vortman, 1981), and other areas.

The second configuration where least-squares design is applicable is called *modeling* or *system identification*. The concept is illustrated in Fig. 14.2. Here a linear system, $\tilde{H}(z)$, models or identifies an unknown plant consisting of an unknown system with internal noise. The least-squares design forces the linear system output, g_k, to be a least-squares approximation to the desired plant output, d_k, for a particular input signal, f_k. When f_k has spectral content at all frequencies and when the plant noise contributes at most a small part of the power in d_k, we expect $\tilde{H}(z)$ to be similar to the transfer function of the plant's unknown system. Note, however, that $\tilde{H}(z)$ itself is not necessarily a least-squares approximation, as it was, for example, in Chapter 8. Thus, the modeling concept is applicable where the best approximation to a signal, rather than to a transfer function, is the objective.

The type of modeling illustrated in Fig. 14.2 has a wide range of applications, including modeling in the biological, social, and economic sciences (Kailath, 1974), in adaptive control systems (Landau, 1979; Franklin and Powell, 1980; Widrow and Stearns, 1985), in digital filter design (Widrow et al., 1981), and in geophysics (Widrow and Stearns, 1985).

Another application of least-squares system design, known as *inverse modeling* or *equalization*, is illustrated in Fig. 14.3. Here the desired output of $\tilde{H}(z)$, again labeled d_k, is $s_{k-\Delta}$, a delayed version of the input signal. To cause the output g_k to approximate d_k, $\tilde{H}(z)$ is adjusted to model the inverse of, or equalize, the unknown system with internal noise. When the unknown system and the linear system are both causal the delay, $z^{-\Delta}$, is used to compensate for the propagation delay through the two systems in cascade. When the internal noise of the unknown system is only a small part of f_k, the output of the equalizer, g_k, is a good approximation to the delayed input, $s_{k-\Delta}$. Hence the linear system is used here to invert, or equalize, the effect of the unknown system on the input signal.

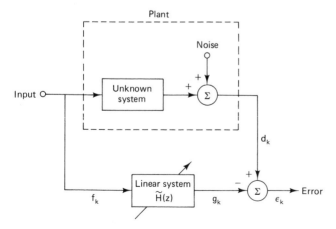

Figure 14.2 Least-squares modeling of an unknown plant.

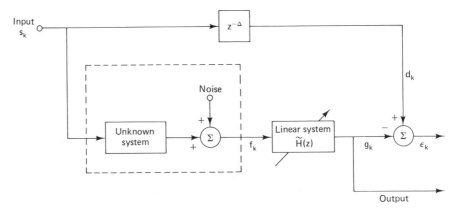

Figure 14.3 Least-squares inverse modeling, or equalization.

Equalization is used in communication systems to remove distortion by compensating for nonuniform channel gain and multipath effects, and to improve the signal-to-noise ratio if the channel introduces bandlimited noise [see Lucky (1966) and Gersho (1969) for examples]. Equalization and inverse modeling are also used in adaptive control (Widrow and Stearns, 1985), speech analysis (Itakura and Saito, 1971), deconvolution (Griffiths et al., 1977), digital filter design (Widrow and Stearns, 1985), and other areas.

Our final example of a configuration where least-squares design is used, *interference canceling,* is illustrated in Fig. 14.4. The interference-canceling principle is applicable in cases where there is a signal, s_k, with additive noise, n_k, and also a source of correlated noise, n_k'. Ideally, n_k and n_k' are correlated with each other but not with s_k, although the principle is applicable even if the signal and noise are correlated. The least-squares design objective is to adjust $\tilde{H}(z)$ so that its output, g_k, is a least-squares approximation to n_k, thereby canceling, by subtraction, the noise from the incoming waveform. When the signal and noise are independent, this is equivalent to minimizing the mean-squared value of the error, ϵ_k, because the independent noise cannot be made to cancel the signal, s_k. The delay is placed in the configuration to compensate for propagation through the causal linear system and allow the noise sequences, n_k and n_k', to be aligned in time.

Interference canceling is an interesting and sometimes preferable alternative to bandpass filtering for improving the signal-to-noise ratio. For example, suppose we have an underground seismic sensor that receives, in addition to a seismic signal component s_k, an acoustic noise component, n_k, coupled into the ground from the at-

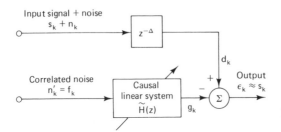

Figure 14.4 Basic configuration for interference cancelling.

Least-Squares System Design   Chap. 14

mosphere above, and the two components have similar spectra so that n_k cannot be removed from $(s_k + n_k)$ using bandpass filtering. We could then add an aboveground microphone to the system to receive n_k', an acoustic component correlated with (but not exactly the same as) n_k, and process the signals as in Fig. 14.4 to reduce the acoustic noise and increase the signal-to-noise ratio. The number of examples of this type is limited only by one's imagination.

Although the applications illustrated in Figs. 14.1 through 14.4 are distinctly different from each other, they all involve the same least-squares design problem. The important feature common to all applications is the following. There is a linear system, $\tilde{H}(z)$, with adjustable coefficients. These coefficients are adjusted to cause the output, g_k, of the linear system to be a least-squares approximation to the desired signal, d_k, and thereby minimize the mean-squared error, $E[\epsilon_k^2]$. We now proceed to examine the nature of the resulting least-squares design problem, using a geometrical interpretation.

14.3 THE MEAN-SQUARED-ERROR PERFORMANCE FUNCTION

If we compare Figs. 14.1 through 14.4 we can see a common least-squares design problem, which is illustrated in Fig. 14.5. The parameters of a causal linear system, $\tilde{H}(z)$, are to be adjusted or selected to minimize a mean-squared error, $E[\epsilon_k^2]$. [Actually, there is no need in what follows to assume that $\tilde{H}(z)$ is causal, but we assume causality for convenience, and to include real-time applications.] For example, if $\tilde{H}(z)$ is a linear system of the form $\tilde{B}(z)/\tilde{A}(z)$, the parameters to be selected are $[b_n]$ and $[a_n]$, the coefficients of $\tilde{B}(z)$ and $\tilde{A}(z)$.

The error, ϵ_k, is the difference between the desired signal, d_k, and the linear system output, g_k:

$$\epsilon_k = d_k - g_k \tag{14.1}$$

Let us now assume that the signals in Fig. 14.5 are stationary so that expected values and correlation functions are defined. Then the mean-squared error (MSE) is

$$
\begin{aligned}
\text{MSE} &= E[\epsilon_k^2] \\
&= E[(d_k - g_k)^2] \\
&= E[d_k^2] + E[g_k^2] - 2E[d_k g_k] \\
&= \phi_{dd}(0) + \phi_{gg}(0) - 2\phi_{dg}(0)
\end{aligned}
\tag{14.2}
$$

In the last line we have substituted the correlation functions discussed in Sections 13.7 and 13.8. Using the relationships in Table 13.2, we convert Eq. 14.2 to a relationship of power spectra,

$$\text{MSE} = \phi_{dd}(0) + \frac{1}{2\pi j} \oint [\tilde{\Phi}_{gg}(z) - 2\tilde{\Phi}_{dg}(z)]\frac{dz}{z} \tag{14.3}$$

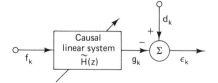

Figure 14.5 Basic elements of least-squares design common to Figs. 14.1 through 14.4.

Next, with $z = e^{j\omega T}$, a point on the unit circle, we have again from Table 13.2 the transfer relationships

$$\tilde{\Phi}_{gg}(z) = \tilde{H}(z)\tilde{H}(z^{-1})\tilde{\Phi}_{ff}(z) \qquad (14.4)$$

and

$$\tilde{\Phi}_{dg}(z) = \tilde{H}(z)\tilde{\Phi}_{df}(z) \qquad (14.5)$$

Substituting these into Eq. 14.3, we have

$$\boxed{\text{MSE} = \phi_{dd}(0) + \frac{1}{2\pi j} \oint [\tilde{H}(z^{-1})\tilde{\Phi}_{ff}(z) - 2\tilde{\Phi}_{df}(z)]\tilde{H}(z)\frac{dz}{z}} \qquad (14.6)$$

In this expression the MSE function is given in terms of the linear transfer function, $\tilde{H}(z)$. This is a general expression for the stochastic-signal case in which we have not specified the form of $\tilde{H}(z)$, but have specified only that $\tilde{H}(z)$ is linear and has adjustable parameters. Nevertheless, suppose that

$$M = \text{number of adjustable parameters in } \tilde{H}(z) \qquad (14.7)$$

Then Eq. 14.6 describes the MSE as a function of M variables, that is, as a surface in $(M + 1)$-dimensional space. The least-squares design problem is to find the parameters that specify a global minimum on this surface, which by definition must be everywhere positive.

In the next section we restrict the discussion to the case where $\tilde{H}(z)$ is a nonrecursive filter with M coefficients, or parameters, to adjust. In this case we will see that Eq. 14.6 can be simplified significantly and that the MSE becomes a quadratic function of the M coefficients, describing a bowl-shaped surface in $(M + 1)$-dimensional space, and resulting in a set of M linear least-squares equations just like those discussed at the beginning of Chapter 2.

14.4 NONRECURSIVE LEAST-SQUARES DESIGN: STATIONARY CASE

Having the results from the preceding section in which stationary signals were assumed, we now assume that the least-squares system, $\tilde{H}(z)$, is nonrecursive with coefficients $b_0, b_1, \ldots, b_{M-1}$ as in Fig. 14.6. We have therefore

$$\tilde{H}(z) = \sum_{n=0}^{M-1} b_n z^{-n} \qquad (14.8)$$

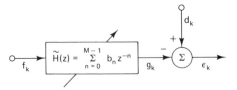

Figure 14.6 Elements of nonrecursive least-squares design.

If we substitute this representation for $\tilde{H}(z)$ into Eq. 14.6 and exchange the order of summation and integration, we obtain

$$\text{MSE} = \phi_{dd}(0) + \sum_{m=0}^{M-1} \sum_{n=0}^{M-1} \frac{b_m b_n}{2\pi j} \oint \tilde{\Phi}_{ff}(z) z^{m-n} \frac{dz}{z} - 2 \sum_{n=0}^{M-1} \frac{b_n}{2\pi j} \oint \tilde{\Phi}_{df}(z) z^{-n} \frac{dz}{z}$$

(14.9)

To this result we apply the correlation relationships in Table 13.2 and we obtain

$$\text{MSE} = \phi_{dd}(0) + \sum_{m=0}^{M-1} \sum_{n=0}^{M-1} b_m b_n \phi_{ff}(m-n) - 2 \sum_{n=0}^{M-1} b_n \phi_{fd}(n) \qquad (14.10)$$

This is the formula for the MSE with a nonrecursive system. Its principal feature is that the MSE is a *quadratic* function of the coefficients, $[b_n]$. We can see that the MSE is quadratic because the coefficients appear only to first and second degree in the formula.

The optimal coefficients, that is, those coefficients that minimize the MSE in Eq. 14.10, are denoted with an overbar as follows:

$$\text{optimal coefficients} = \overline{b}_0, \overline{b}_1, \ldots, \overline{b}_{M-1} \qquad (14.11)$$

To determine these optimal coefficients we note that the MSE in Eq. 14.10 is, by its nature, always positive and therefore, since it is a quadratic function, it describes a bowl-shaped surface in $(M + 1)$-dimensional Cartesian space. Thus the MSE must have a single global minimum, which may be found by setting its derivatives with respect to the b's in Eq. 14.10 equal to zero. This minimum might be distributed; that is, the bowl might have a flat bottom, but local minima cannot exist.

Taking derivatives of the MSE in Eq. 14.10 with respect to each of the b's gives

$$\frac{\partial(\text{MSE})}{\partial b_n} = 2 \sum_{m=0}^{M-1} b_m \phi_{ff}(m-n) - 2\phi_{fd}(n); \qquad n = 0, 1, \ldots, M-1 \qquad (14.12)$$

When these M derivatives are set equal to zero we have M simultaneous equations for the optimal coefficients:

$$\sum_{m=0}^{M-1} \overline{b}_m \phi_{ff}(m-n) = \phi_{fd}(n); \qquad n = 0, 1, \ldots, M-1 \qquad (14.13)$$

As with any set of simultaneous linear equations, these are solved for the $\overline{b}$'s by inverting a matrix, in this case the matrix of ϕ_{ff}'s. To express the solution in simple form, let

$$\mathbf{R} = \begin{bmatrix} \phi_{ff}(0) & \phi_{ff}(1) & \phi_{ff}(2) & \cdots & \phi_{ff}(M-1) \\ \phi_{ff}(1) & \phi_{ff}(0) & \phi_{ff}(1) & \cdots & \phi_{ff}(M-2) \\ \phi_{ff}(2) & \phi_{ff}(1) & \phi_{ff}(0) & \cdots & \phi_{ff}(M-3) \\ \vdots & \vdots & \vdots & & \vdots \\ \phi_{ff}(M-1) & \phi_{ff}(M-2) & \phi_{ff}(M-3) & \cdots & \phi_{ff}(0) \end{bmatrix} \qquad (14.14)$$

$$\mathbf{B} = [b_0 \quad b_1 \quad b_2 \quad \cdots \quad b_{M-1}]^{\mathrm{T}} \tag{14.15}$$

$$\mathbf{P} = [\phi_{fd}(0) \quad \phi_{fd}(1) \quad \phi_{fd}(2) \quad \cdots \quad \phi_{fd}(M-1)]^{\mathrm{T}} \tag{14.16}$$

In this notation, $\mathbf{R}$ is called the *correlation matrix* or *R-matrix* of the signal f, $\mathbf{B}$ is called the *coefficient vector* or *weight vector,* and $\mathbf{P}$ is called the *cross-correlation vector* or *P-vector*. We also use $\overline{\mathbf{B}}$ to denote the vector of optimal coefficients.

With the notation introduced in Eqs. 14.14 through 14.16, the formula for optimal coefficients in Eq. 14.13 may be expressed as

$$\mathbf{R}\overline{\mathbf{B}} = \mathbf{P} \quad \text{or} \quad \overline{\mathbf{B}} = \mathbf{R}^{-1}\mathbf{P} \tag{14.17}$$

We can also use this notation in the mean-squared-error formula, Eq. 14.10. If we premultiply the $\mathbf{R}$-matrix by the transpose of the weight vector and then postmultiply by the weight vector, we obtain the double sum in Eq. 14.10. The single sum is just the inner product of the $\mathbf{P}$-vector and the weight vector. Thus, Eq. 14.10 becomes

$$\mathrm{MSE} = \phi_{dd}(0) + \mathbf{B}^{\mathrm{T}}\mathbf{R}\mathbf{B} - 2\mathbf{P}^{\mathrm{T}}\mathbf{B} \tag{14.18}$$

The minimum MSE formula then follows by substituting Eq. 14.17 into Eq. 14.18:

$$(\mathrm{MSE})_{\min} = \phi_{dd}(0) + (\mathbf{R}^{-1}\mathbf{P})^{\mathrm{T}}\mathbf{R}\overline{\mathbf{B}} - 2\mathbf{P}^{\mathrm{T}}\overline{\mathbf{B}} \tag{14.19}$$

Using the identity $(\mathbf{AB})^{\mathrm{T}} = \mathbf{B}^{\mathrm{T}}\mathbf{A}^{\mathrm{T}}$, which is true for any matrices $\mathbf{A}$ and $\mathbf{B}$, and also recalling that $\mathbf{R}$ in Eq. 14.14, and therefore $\mathbf{R}^{-1}$, is symmetric, we simplify Eq. 14.19 and obtain

$$\begin{aligned} (\mathrm{MSE})_{\min} &= \phi_{dd}(0) - \mathbf{P}^{\mathrm{T}}\overline{\mathbf{B}} \\ &= \phi_{dd}(0) - \mathbf{P}^{\mathrm{T}}\mathbf{R}^{-1}\mathbf{P} \end{aligned} \tag{14.20}$$

Thus we have a simple expression for the minimum mean-squared error for a non-recursive least-squares system. An equivalent expression without vector and matrix notation could just as easily have been obtained by substituting Eq. 14.13 into Eq. 14.10:

$$(\mathrm{MSE})_{\min} = \phi_{dd}(0) - \sum_{n=0}^{M-1} \overline{b}_n \phi_{fd}(n) \tag{14.21}$$

The matrix notation has been introduced here because it is used commonly in the literature on least-squares and adaptive systems. For stochastic signals, we can summarize the important least-squares design formulas obtained thus far, with and without matrix notation, from Eqs. 14.10, 14.13, 14.17, 14.18, 14.20, and 14.21, as follows:

QUADRATIC ERROR SURFACE

$$\text{MSE} = \phi_{dd}(0) + \sum_{m=0}^{M-1}\sum_{n=0}^{M-1} b_m b_n \phi_{ff}(m-n) - 2\sum_{n=0}^{M-1} b_n \phi_{fd}(n)$$

$$= \phi_{dd}(0) + \mathbf{B}^T\mathbf{R}\mathbf{B} - 2\mathbf{P}^T\mathbf{B} \qquad (14.22)$$

OPTIMAL COEFFICIENTS

$$\sum_{m=0}^{M-1} \overline{b}_m \phi_{ff}(m-n) = \phi_{fd}(n); \qquad n = 0, 1, \ldots, M-1$$

$$\mathbf{R}\overline{\mathbf{B}} = \mathbf{P} \qquad (14.23)$$

MINIMUM MEAN-SQUARED ERROR

$$(\text{MSE})_{\min} = \phi_{dd}(0) - \sum_{n=0}^{M-1} \overline{b}_n \phi_{fd}(n)$$

$$= \phi_{dd}(0) - \mathbf{P}^T\overline{\mathbf{B}} \qquad (14.24)$$

As stated previously, the MSE in Eq. 14.22 is represented by an M-dimensional, quadratic performance function in an $(M+1)$-dimensional space whose dimensions correspond with the b's. The solution for the optimal coefficients in Eq. 14.23 designates a global minimum on this surface, and the value of the MSE at this minimum is given by Eq. 14.24. A simple example of this concept with one parameter ($M = 1$) is shown in Fig. 14.7. In this case there is a single global minimum of the MSE function at $b_0 = \overline{b}_0$.

We have seen that the least-squares design of a nonrecursive system is accomplished by solving Eq. 14.23 for the optimal parameters. Because of the special nature of the **R**-matrix, a special type of algorithm is used to solve Eq. 14.23. The **R**-matrix is called a *Toeplitz* matrix, and its special nature is seen in Eq. 14.14. The matrix is symmetrical and, in addition, each row or column is a rearrangement of the elements of any other row or column.

The special algorithm used (with variations) to solve Eq. 14.23 in the stochastic-signal case is called *Levinson's algorithm*. Descriptions of Levinson's algorithm, including computer codes, can be found in the literature [see Blahut (1985), or Stearns and David (1987)]. Later in this chapter we will again use SPSOLE, the general-purpose algorithm introduced in Chapter 2, to obtain the solution of Eq. 14.23.

A final point about stochastic nonrecursive least-squares design concerns the elements of Eq. 14.23, which are the correlation coefficients, $\phi_{ff}(n)$ and $\phi_{fd}(n)$, for $0 \le n < M$. For some design problems these coefficients are known or assumed exactly. In other applications they must be estimated. In still other applications they may drift slowly with time, which leads to the design of *adaptive* signal-processing systems (Honig and Messerschmitt, 1984; Widrow and Stearns, 1985), that is, systems that adjust **B** continually as they continuously seek the solution to Eq. 14.23.

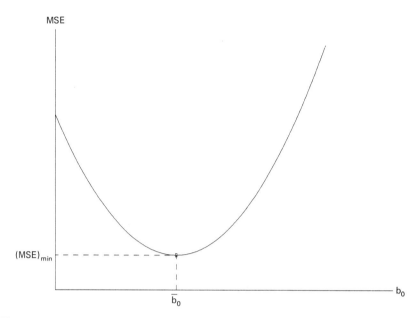

MSE

$(MSE)_{min}$

$\overline{b}_0$

b_0

Figure 14.7 Example of a one-parameter MSE performance function. The minimum MSE is at $b_0 = \overline{b}_0$.

14.5 A DESIGN EXAMPLE

We now consider a specific example to illustrate the use of the formulas derived in Section 14.4. Perhaps the easiest type of system to understand and to solve is the least-squares *predictor* with a simple periodic input, an example of which is shown in Fig. 14.8. The predictor in Fig. 14.8 is an example of Fig. 14.1(b) with the delay (Δ) set to one time step. It is called a *one-step predictor*.

We can see in Fig. 14.8 that the one-step predictor will exactly cancel the signal, s_k, if the coefficients can be adjusted to shift the phase of f_k, which equals s_{k-1}, by just the right amount. Let us see whether or not this can be done. First, the correlation coefficients of s_k are found by averaging over one period of s_k:

$$
\begin{aligned}
\phi_{ss}(n) &= E[s_k s_{k+n}] \\
&= \frac{1}{12} \sum_{k=0}^{11} \left[\sqrt{2} \sin\left(\frac{2\pi k}{12}\right) \right] \left[\sqrt{2} \sin\left(\frac{2\pi(k+n)}{12}\right) \right] \\
&= \frac{1}{12} \left[\sum_{k=0}^{11} \cos\left(\frac{2\pi n}{12}\right) - \sum_{k=0}^{11} \cos\left(\frac{2\pi(2k+n)}{12}\right) \right] \\
&= \cos\left(\frac{2\pi n}{12}\right); \qquad n = 0, 1, 2, \dots
\end{aligned}
\tag{14.25}
$$

In this result, we obtained line 3 from line 2 by using Eq. 1.3 and then we obtained line 4 by noting that the second sum in line 3 is zero because the cosine function is summed over exactly two cycles (see Chapter 13, Exercise 26).

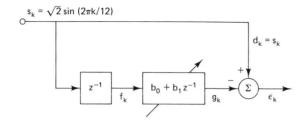

$s_k = \sqrt{2}\sin(2\pi k/12)$

$d_k = s_k$

z^{-1} f_k $b_0 + b_1 z^{-1}$ g_k Σ ϵ_k

Figure 14.8 One-step predictor with two coefficients and sinusoidal input.

The correlation coefficients in Eq. 14.25 are the elements of the **R**-matrix, and also of the **P**-vector in this example. For the **P**-vector, we note in Fig. 14.8 that $d_k = s_k$ and $f_k = s_{k-1}$, so the **P**-vector elements in Eq. 14.16 are $\phi_{fd}(0) = \phi_{ss}(1)$, $\phi_{fd}(1) = \phi_{ss}(2)$, and so on. Thus, for this example, we have from Eq. 14.25

$$\mathbf{R} = \begin{bmatrix} \phi_{ss}(0) & \phi_{ss}(1) \\ \phi_{ss}(1) & \phi_{ss}(0) \end{bmatrix} = \begin{bmatrix} 1 & \dfrac{\sqrt{3}}{2} \\ \dfrac{\sqrt{3}}{2} & 1 \end{bmatrix} \tag{14.26}$$

$$\mathbf{P} = \begin{bmatrix} \phi_{ss}(1) \\ \phi_{ss}(2) \end{bmatrix} = \begin{bmatrix} \dfrac{\sqrt{3}}{2} \\ \dfrac{1}{2} \end{bmatrix} \tag{14.27}$$

With these results we can illustrate the functions in Eq. 14.22 through Eq. 14.24 of the preceding section. The mean-squared error given by Eq. 14.22 is

$$\mathrm{MSE} = \phi_{ss}(0) + \mathbf{B}^\mathrm{T}\mathbf{R}\mathbf{B} - 2\mathbf{P}^\mathrm{T}\mathbf{B}$$

$$= 1 + \begin{bmatrix} b_0 & b_1 \end{bmatrix} \begin{bmatrix} 1 & \dfrac{\sqrt{3}}{2} \\ \dfrac{\sqrt{3}}{2} & 1 \end{bmatrix} \cdot \begin{bmatrix} b_0 \\ b_1 \end{bmatrix} - 2 \begin{bmatrix} \dfrac{\sqrt{3}}{2} & \dfrac{1}{2} \end{bmatrix} \cdot \begin{bmatrix} b_0 \\ b_1 \end{bmatrix} \tag{14.28}$$

$$= b_0^2 + b_1^2 + \sqrt{3}\,b_0 b_1 - \sqrt{3}\,b_0 - b_1 + 1$$

The MSE in this case is a three-dimensional bowl-shaped surface, which can be illustrated in two dimensions by plotting *error contours*, that is, graphs of b_1 versus b_0 with the MSE held constant. Error contours of Eq. 14.25 are shown in Fig. 14.9.

In Eq. 14.23 we have the optimal coefficient formula, which in this example is

$$\mathbf{R}\overline{\mathbf{B}} = \mathbf{P};$$

$$\begin{bmatrix} 1 & \dfrac{\sqrt{3}}{2} \\ \dfrac{\sqrt{3}}{2} & 1 \end{bmatrix} \cdot \begin{bmatrix} \overline{b_0} \\ \overline{b_1} \end{bmatrix} = \begin{bmatrix} \dfrac{\sqrt{3}}{2} \\ \dfrac{1}{2} \end{bmatrix} \tag{14.29}$$

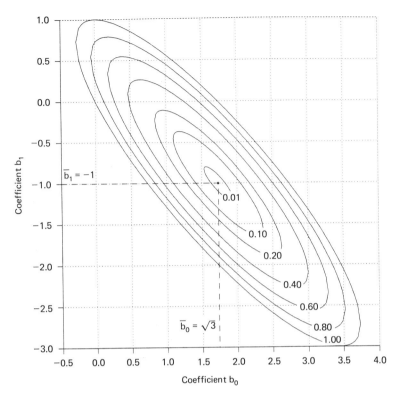

Figure 14.9 Error contours of the predictor in Fig. 14.8. MSE values are shown near the contours. Optimal coefficient values $\bar{b}_0 = \sqrt{3}$ and $\bar{b}_1 = -1$ are at the center.

Therefore, the solution for the optimal coefficients is

$$
\begin{bmatrix} \bar{b}_0 \\ \bar{b}_1 \end{bmatrix} = \begin{bmatrix} 1 & \dfrac{\sqrt{3}}{2} \\ \dfrac{\sqrt{3}}{2} & 1 \end{bmatrix}^{-1} \cdot \begin{bmatrix} \dfrac{\sqrt{3}}{2} \\ \dfrac{1}{2} \end{bmatrix}
\tag{14.30}
$$

We note from Chapter 1, Eq. 1.30 that the inverse of any nonsingular two-by-two matrix is

$$
\begin{bmatrix} a & b \\ c & d \end{bmatrix}^{-1} = \frac{1}{ad - bc} \begin{bmatrix} d & -b \\ -c & a \end{bmatrix}
\tag{14.31}
$$

For larger matrices the inverse is more complicated, and in general a computing algorithm is needed as described in the preceding section. Applying this formula to Eq. 14.30, we have

$$
\begin{bmatrix} \bar{b}_0 \\ \bar{b}_1 \end{bmatrix} = \frac{1}{1 - 3/4} \begin{bmatrix} 1 & \dfrac{-\sqrt{3}}{2} \\ \dfrac{-\sqrt{3}}{2} & 1 \end{bmatrix} \cdot \begin{bmatrix} \dfrac{\sqrt{3}}{2} \\ \dfrac{1}{2} \end{bmatrix} = \begin{bmatrix} \sqrt{3} \\ -1 \end{bmatrix}
\tag{14.32}
$$

These optimal weight values are seen at the center of the error contours in Fig. 14.9, where the MSE is at its minimum. From Eq. 14.24, we see that the minimum MSE is

$$(\text{MSE})_{\min} = \phi_{dd}(0) = \mathbf{P}^T \overline{\mathbf{B}}$$

$$= 1 - \begin{bmatrix} \dfrac{\sqrt{3}}{2} & \dfrac{1}{2} \end{bmatrix} \cdot \begin{bmatrix} \sqrt{3} \\ -1 \end{bmatrix} = 0 \tag{14.33}$$

Thus, in Fig. 14.8, the optimal predictor is able to shift the phase of the delayed sine wave and achieve exact cancellation, resulting in $\epsilon_k = 0$. This example is artificial because the signal s_k is simple and is known exactly, and also because the filter size is only $M = 2$. It is meant to illustrate the quadratic error surface in Eq. 14.22 and to show a derivation of the optimal coefficients in Eq. 14.23.

As a final step in this simple example, let us observe the effect of adding a third coefficient, b_2, to the filter in Fig. 14.8. Using Eqs. 14.25 through 14.27 again, the optimal coefficient relation in Eq. 14.23 is now

$$\mathbf{R}\overline{\mathbf{B}} = \begin{bmatrix} 1 & \dfrac{\sqrt{3}}{2} & \dfrac{1}{2} \\[2mm] \dfrac{\sqrt{3}}{2} & 1 & \dfrac{\sqrt{3}}{2} \\[2mm] \dfrac{1}{2} & \dfrac{\sqrt{3}}{2} & 1 \end{bmatrix} \cdot \begin{bmatrix} \overline{b}_0 \\[1mm] \overline{b}_1 \\[1mm] \overline{b}_2 \end{bmatrix} = \begin{bmatrix} \dfrac{\sqrt{3}}{2} \\[2mm] \dfrac{1}{2} \\[2mm] 0 \end{bmatrix} = \mathbf{P} \tag{14.34}$$

In this case the $\mathbf{R}$-matrix is singular and has no inverse. That is, Eq. 14.34 imposes constraints on $\overline{b}_0$, $\overline{b}_1$, and $\overline{b}_2$ but does not provide specific values. These constraints may be expressed as

$$
\begin{aligned}
b_0 - b_2 &= \sqrt{3} \\
b_0 + \sqrt{3}\,b_1 + 2b_2 &= 0
\end{aligned}
\tag{14.35}
$$

They imply a *distributed minimum* of the error surface as described in the preceding section, where the MSE is everywhere equal to zero. The solution with $b_2 = 0$ is seen to agree with the two-weight solution in Eq. 14.32.

14.6 NONRECURSIVE LEAST-SQUARES DESIGN: GENERAL CASE

In the three preceding sections we have discussed least-squares design with stationary signals of infinite extent and we have seen that the optimal coefficients are obtained from the correlation functions of these signals. The design method is just the same with nonstationary signals, except that *covariance functions* (Rabiner and Schafer, 1978) are used in place of correlation functions.

We begin again with the nonrecursive system and the signals shown in Fig. 14.6. However, we now assume that the signals may be stochastic and stationary but are not necessarily so and, in any case, a *total squared error* (TSE), given by

$$\text{TSE} = \sum_{k=0}^{K-1} \epsilon_k^2 \tag{14.36}$$

is to be minimized. That is, *we are designing the system for a particular time frame* with $k = 0, 1, \ldots, K - 1$ in which K samples of the error sequence $[\epsilon_k]$, are given and the total squared error is to be minimized. Using the variables in Fig. 14.6 we can expand Eq. 14.36 in terms of the signal values and filter coefficients to obtain

$$\text{TSE} = \sum_{k=0}^{K-1} \epsilon_k^2$$

$$= \sum_{k=0}^{K-1} \left[d_k - \sum_{n=0}^{M-1} b_n f_{k-n} \right]^2 \tag{14.37}$$

$$= \sum_{k=0}^{K-1} d_k^2 + \sum_{n=0}^{M-1} \sum_{m=0}^{M-1} b_n b_m \sum_{k=0}^{K-1} f_{k-n} f_{k-m} - 2 \sum_{n=0}^{M-1} b_n \sum_{k=0}^{K-1} d_k f_{k-n}$$

We now define the following covariance functions, which are similar to the correlation functions in Eq. 14.10:

$$r_{xx}(m, n) = \sum_{k=0}^{K-1} x_{k-m} x_{k-n} \tag{14.38}$$

$$r_{xy}(n) = \sum_{k=-n}^{K-n-1} x_k y_{k+n}$$

With these definitions, Eq. 14.37 becomes

$$\text{TSE} = r_{dd}(0, 0) + \sum_{n=0}^{M-1} \sum_{m=0}^{M-1} b_n b_m r_{ff}(m, n) - 2 \sum_{n=0}^{M-1} b_n r_{fd}(n) \tag{14.39}$$

This expression for the TSE is similar to that for the MSE in Eq. 14.10. We note that the covariance functions used here require knowledge of the signal sequences beyond the range $(0, K - 1)$. Specifically, from the way the functions in Eq. 14.38 are used in Eq. 14.39, we see that the following sequences are needed:

$$[f_k] = [f_{-M+1} \quad f_{-M+2} \quad \cdots \quad f_0 \quad \cdots \quad f_{K-1}]$$
$$[d_k] = [d_0 \quad \cdots \quad d_{K-1}] \tag{14.40}$$

That is, these sequences are required to produce the error sequence, $[\epsilon_0, \ldots, \epsilon_{k-1}]$, so that the TSE may be minimized.

With the TSE expression Eq. 14.39, the principal results for the error surface and the least-squares coefficients in Eqs. 14.22 through 14.24 are applicable, provided that the covariance functions in Eq. 14.38 are substituted for the correlation functions used previously. That is, the **R**-matrix with covariance elements is

$$\mathbf{R} = \begin{bmatrix} r_{ff}(0, 0) & r_{ff}(0, 1) & r_{ff}(0, 2) & \cdots & r_{ff}(0, M - 1) \\ r_{ff}(1, 0) & r_{ff}(1, 1) & r_{ff}(1, 2) & \cdots & r_{ff}(1, M - 1) \\ r_{ff}(2, 0) & r_{ff}(2, 1) & r_{ff}(2, 2) & \cdots & r_{ff}(2, M - 1) \\ \vdots & \vdots & \vdots & & \vdots \\ r_{ff}(M - 1, 0) & r_{ff}(M - 1, 1) & r_{ff}(M - 1, 2) & \cdots & r_{ff}(M - 1, M - 1) \end{bmatrix}$$

$$\tag{14.41}$$

Similarly, the *P*-vector with covariance elements is

$$\mathbf{P} = [r_{fd}(0) \quad r_{fd}(1) \quad r_{fd}(2) \quad \cdots \quad r_{fd}(M-1)]^{\mathrm{T}} \tag{14.42}$$

the **R**-matrix in Eq. 14.41 is symmetrical because, from the definition in Eq. 14.38,

$$r_{ff}(i,j) = r_{ff}(j,i) \tag{14.43}$$

However, the **R**-matrix is no longer a Toeplitz matrix as it was for stationary signals in Eq. 14.14, and it cannot be inverted with special algorithms that are applicable only to Toeplitz matrices. It must be inverted using the standard methods of matrix algebra.

In summary, in the general case where the signals are not necessarily stationary, we use the total squared error defined in Eq. 14.36 to optimize the design for a particular sampling period from $k = 0$ through $k = K - 1$. This approach leads to the revised versions of **R** and **P** in Eqs. 14.41 and 14.42, that is, to the use of covariance functions in place of correlation functions in Eqs. 14.22 through 14.24. The revised version of Eq. 14.22 was given in Eq. 14.39 and the revised version of Eqs. 14.23 and 14.24 are

---

OPTIMAL COEFFICIENTS

$$\mathbf{R}\overline{\mathbf{B}} = \mathbf{P} \tag{14.44}$$

MINIMUM TOTAL SQUARED ERROR

$$(\text{TSE})_{\min} = r_{dd}(0,0) - \mathbf{P}^{\mathrm{T}}\overline{\mathbf{B}} \tag{14.45}$$

As a final point, we note that when the signals f_k and d_k are periodic (and hence stationary) and K is an integral number of periods, the **R**-matrix in Eq. 14.41 is Toeplitz and the minimum TSE in Eq. 14.45 is just K times the minimum MSE in Eq. 14.24. Similarly, when f_k and d_k are random stationary signals, the minimum-TSE solution approaches the minimum-MSE solution as K increases.

14.7 LEAST-SQUARES DESIGN ROUTINES

We now describe two subroutines, SPSOLE and SPLESQ, for solving $\mathbf{RB} = \mathbf{P}$ in Eq. 14.23 for **B** and designing least-squares systems when **R** and **P** are *either* the correlation arrays in Eqs. 14.14 and 14.16 or the corresponding covariance arrays in Eqs. 14.41 and 14.42. The routines, which are listed in Appendix B, are simpler and less reliable than some of the more complex routines found in scientific libraries. In the case where **R** is a Toeplitz matrix, the equation-solving routine (SPSOLE) is also less efficient than the Levinson algorithms mentioned previously. Nevertheless, SPSOLE and SPLESQ are easy to use and will suffice for most nonrecursive least-squares designs.

The first routine, SPSOLE, which was introduced in Chapter 2, is specifically for solving the linear equations given by $\mathbf{RB} = \mathbf{P}$, where $\mathbf{R}$ is a correlation or covariance matrix. Its calling sequence is

```
CALL SPSOLE (DD,LR,LC,M,IERROR)
```

```
DD(0:LR,0:LC) = double-precision augmented matrix [R,P]
           LR = Last row index of DD
           LC = Last column index of DD
            M = Number of linear equations to solve
       IERROR = 0: no errors
                1: LR < M-1 or LC < M; DD not large enough
                2: No solution due to singularity or near-
                   singularity of R.
```

When SPSOLE is called, the $\mathbf{R}$-matrix should be contained in the DD array such that $DD(I,J) = r_{ff}(I,J)$ for I and J in the range $[0, M - 1]$. The $\mathbf{P}$-vector should also be stored such that $DD(I,M) = r_{fd}(I)$. When SPSOLE is executed, the solution replaces the $\mathbf{P}$-vector so that

$$\overline{\mathbf{B}} = \mathbf{R}^{-1}\mathbf{P} = [DD(I,M); \quad I = 0, 1, \ldots, M - 1]^{\mathrm{T}} \tag{14.46}$$

The matrix DD is declared double precision in order to reduce roundoff errors within SOLE, which uses the Gauss–Jordan elimination method [Kelly, 1967] to solve the system of linear equations. Thus, DD must be declared double precision in the user's program, although it is normally filled with single-precision elements from $\mathbf{R}$ and $\mathbf{P}$. In any case, the solution vector in Eq. 14.46 should be considered to be of single-precision accuracy, even though it is a double-precision vector.

As a final point concerning SPSOLE, we note that although $\mathbf{R}$ is by definition a symmetric (and sometimes Toeplitz) matrix, symmetry of the coefficient matrix is not required by SPSOLE, which therefore runs correctly for a broader class of linear equations that includes the class $\mathbf{RB} = \mathbf{P}$.

The second subroutine is SPLESQ, which begins with the two input signal vectors, $[f_k]$ and $[d_k]$ in Eq. 14.40, generates $\mathbf{R}$ and $\mathbf{P}$, and then uses SPSOLE internally to solve for the least-squares coefficient vector, $\overline{\mathbf{B}}$, and the minimum total squared error, $(\mathrm{TSE})_{\min}$. The calling sequence variables for SPLESQ are illustrated in Fig. 14.10. The calling sequence is

```
CALL SPLESQ(F(-M+1),D(0),K,B,M,TSEMIN,IERROR)
```

```
F(-M+1:K-1) = Input data sequence [f_k] - see Eq. 14.40
   D(0:K-1) = Desired input sequence [d_k] - see Eq. 14.40
          K = Number of samples in [d_k]
   B(0:M-1) = Optimal coefficient vector (output)
          M = Number of filter coefficients, up to 50
     TSEMIN = Minimum TSE - see Eq. 14.45
     IERROR = 0: no errors
            = 1-2: error in SPSOLE execution
            = 3: M out of range [1,50]
```

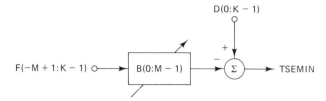

Figure 14.10 Variables used in the SPLESQ calling sequence.

When SPLESQ is executed, covariance coefficients are computed and stored in an internal double-precision matrix, which becomes the augmented matrix (*DD*) used in SPSOLE. (The maximum value of *M*, MMAX = 50, corresponds with the dimension of *DD*.) We emphasize that the first index of the *F* array must be $(-M + 1)$, where *M* is the filter length. The samples with negative indices are needed for the covariance computations, as seen in Eq. 14.40. As discussed previously, if f_k and d_k are periodic signals and *K* spans an integral number of periods, each covariance coefficient becomes just *K* times the corresponding correlation coefficient.

After the covariance coefficients are computed and SPSOLE is called, the optimal coefficient vector is stored in B(0:M-1). Finally, the minimum total squared error, TSEMIN, is computed using Eq. 14.45.

In summary, two routines for use in least-squares design have been described in this section. Examples of their use are presented in the next section.

14.8 MORE DESIGN EXAMPLES

In this section we consider two more examples of nonrecursive least-squares design in order to illustrate the use of the routines just described, and also to design systems other than the predictor designed previously. In the first example, illustrated in Fig. 14.11, we are to design a least-squares equalizer for an "unknown channel," which in this case consists of an all-pole transfer function given by

$$\tilde{U}(z) = \frac{e^{-0.1}\sin(0.1\pi)z^{-1}}{1 - 2e^{-0.1}\cos(0.1\pi)z^{-1} + e^{-0.2}z^{-2}} \tag{14.47}$$

This description of the unknown channel assures us of a simple design for the all-zero equalizer, $\tilde{B}(z)$, whose zeros must cancel the poles of $\tilde{U}(z)$. To obtain the equalizer, we assume that $\tilde{U}(z)$ is unknown but that we have sequences of the signals s_k and f_k. These are shown in Fig. 14.12, where s_k is seen to be an impulse at $k = 0$ and f_k is the impulse response of $\tilde{U}(z)$ given in Appendix A, line 206. Thus, we are designing an equalizer that corrects the impulse response of the unknown channel

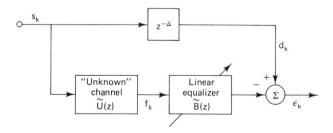

Figure 14.11 Equalizer for a noise-free channel.

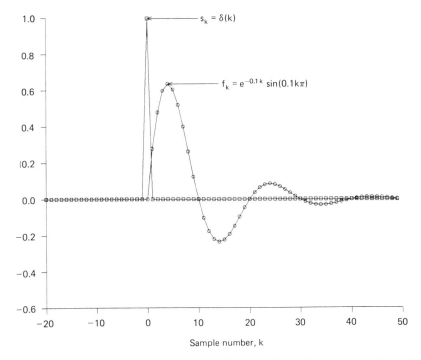

$s_k = \delta(k)$

$f_k = e^{-0.1k} \sin(0.1k\pi)$

Figure 14.12 Sample sequences $[s_k]$ and $[f_k]$ for $-20 \leq k < 50$, equalizer example in Fig. 14.11.

accurately. Since the impulse function has a broad spectrum, our equalizer should be reasonably good for other inputs as well.

Suppose that the sequences s_k and f_k in Fig. 14.11 are stored in two arrays, $S(-20:49)$ and $F(-20:49)$, and that the delay, Δ, is specified by the integer variable IDELT, which is positive. Then a least-squares equalizer with M coefficients is obtained via the following calling sequence:

CALL
SPLESQ(F(-M+1),S(-IDELT),50,B,M,TSEMIN(M),IE) (14.48)

Comparing this calling sequence with the description in the preceding section, we note that F(-M + 1) is the first sample of the input sequence $[f_k]$ and S(-IDELT) is the first sample of $[d_k]$. The next parameter in the calling sequence is $K = 50$, which defines the TSE time frame—see Eq. 14.36. The next item is B, the coefficient vector, which we assume has been dimensioned B(0:M − 1) or larger. The next parameters are M, the number of coefficients, and TSEMIN(M), the minimum total squared error. The latter was not an array in the preceding section, but here we wish to observe the minimum TSE as a function of M. The final parameter, IE, is the error indicator, which should be checked following the execution of SPLESQ.

The results of the execution of Eq. 14.48 for different values of M and IDELT are plotted in Fig. 14.13. Here we have plotted TSEMIN(M) versus M for different values of IDELT. The plots are seen to agree with our knowledge of the "unknown" system, $\tilde{U}(z)$ in Eq. 14.47. The linear equalizer, $\tilde{B}(z)$, becomes an exact inverse

Least-Squares System Design   Chap. 14

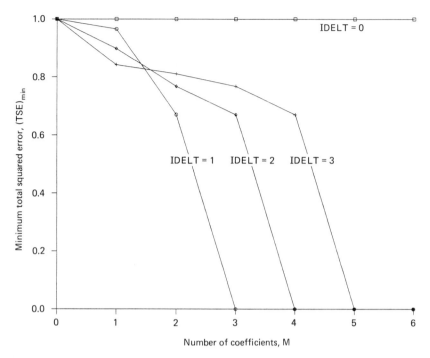

Figure 14.13 Results from the least-squares equalizer example in Fig. 14.11. TSEMIN(M) is plotted versus M for four values of the delay, $\Delta = $ IDELT.

of $\tilde{U}(z)$ when $\Delta = 1$ and M reaches 3, when $\Delta = 2$ and M reaches 4, and so on. When $\Delta = 0$, an equalizer cannot be obtained because $\tilde{B}(z)$ is causal and $\tilde{U}(z)$ is seen to have a unit delay, that is, the numerator of Eq. 14.47 contains z^{-1}.

Each point on each plot in Fig. 14.13, of course, implies a vector $[\bar{b}_l]$ of optimal coefficients that minimize the mean-squared impulse-response error. The $\mathbf{\bar{B}}$-vector solutions for all points in Fig. 14.13 may be tabulated as follows:

	IDELT=0	IDELT = 1	IDELT=2	IDELT=3
M=1	0.00000	0.12319	0.21202	0.26406
M=2	0.00000	1.17915	0.91358	0.60696
	0.00000	-1.11585	-0.74135	-0.36236
M=3	0.00000	3.57640#	0.00000	0.00000
	0.00000	-6.15536#	1.17917	0.91359
	0.00000	2.92811#	-1.11588	-0.74137
M=4	0.00000	3.57640#	0.00000	0.00000
	0.00000	-6.15536#	3.57640#	0.00000
	0.00000	2.92811#	-6.15535#	1.17919
	0.00000	0.00000	2.92811#	-1.11591

M=5	0.00000	3.57640#	0.00000	0.00001
	0.00000	-6.15536#	3.57640#	-0.00001
	0.00000	2.92811#	-6.15535#	3.57640#
	0.00000	-0.00001	2.92811#	-6.15535#
	0.00000	0.00001	0.00000	2.92811#

M=6	0.00000	3.57640#	0.00000	0.00001
	0.00000	-6.15536#	3.57640#	-0.00001
	0.00000	2.92811#	-6.15535#	3.57640#
	0.00000	-0.00002	2.92811#	-6.15535#
	0.00000	0.00002	-0.00001	2.92811#
	0.00000	-0.00001	0.00001	0.00000

In these results we note that the optimal coefficient vector for $\Delta = 0$ is always $\overline{\mathbf{B}} = 0$, as just discussed. From Eq. 14.47 we see that the channel transfer function can be written as

$$U(z) = \frac{z^{-1}}{e^{0.1}\csc(0.1\pi) - 2\,\mathrm{ctn}(0.1\pi)z^{-1} + e^{-0.1}\csc(0.1\pi)z^{-2}}$$

$$= \frac{z^{-1}}{3.57641 - 6.15537z^{-1} + 2.92812z^{-2}}$$

(14.49)

The coefficient vector, $\overline{\mathbf{B}}$, that exactly equalizes $\tilde{U}(z)$ must contain the three coefficients in the denominator of Eq. 14.49. The exact solutions are flagged where they appear in the tabulations just given. Note that the three coefficients in Eq. 14.49 appear in order, and at positions that are correct for each value of Δ. Notice also that there are slight roundoff errors in the tabulated coefficient vectors. These roundoff errors are acceptable in this case, and they allow the operation of SPSOLE, which is called by SPLESQ, beyond $M = 3$ without an "IERROR = 2" result, that is, a detected singularity in the **R**-matrix. The *theoretical* **R**-matrix in this example is of course singular for $M > 3$. Thus, in the preceding example, we have an equalizer which exactly cancels the effect of the unknown channel, $\tilde{U}(z)$ in Fig. 14.11, on an impulse input.

The next example in Fig. 14.14 is one in which we cannot exactly match the desired signal, that is, in which we cannot drive the TSE to zero. In this example, as in the previous example, the transfer functions $\tilde{U}(z)$ and $\tilde{B}(z)$ in Fig. 14.14 are

$$\tilde{U}(z) = \frac{e^{-0.1}\sin(0.1\pi)z^{-1}}{1 - 2e^{-0.1}\cos(0.1\pi)z^{-1} + e^{-0.2}z}$$

(14.50)

$$\tilde{B}(z) = \sum_{n=0}^{M-1} b_n z^{-n}$$

(14.51)

The input signal, f_k in Fig. 14.14, is in this case a random white signal with unit power. The input sequence is given by

ISEED = 12345

$$f_k = \sqrt{12}\,(\mathrm{SPUNIF\,(ISEED)} - 0.5); \qquad k = 0, 1, \ldots, K - 1$$

(14.52)

The random number function, SPUNIF, described in Appendix B, is used in this manner to produce K samples of f_k.

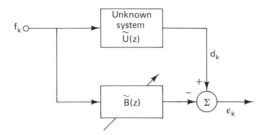

Figure 14.14 Least-squares modeling example.

As noted previously, the covariance elements, $r_{ff}(m, n)$ of the **R**-matrix computed within SPLESQ should approach scaled values of the correlation coefficients, $\phi_{ff}(m - n)$, as K increases. Since f_k is a white random sequence, the samples of f_k are independent and

$$\phi_{ff}(i) = 0; \quad i \neq 0$$
$$= 1; \quad i = 0$$
(14.53)

Therefore, the theoretical **R**-matrix is diagonal, that is, $\mathbf{R} = \mathbf{I}$ theoretically, and from Eq. 14.23, the optimal coefficients are given by

$$\overline{\mathbf{B}} = \mathbf{P}$$
(14.54)

The theoretical **P**-vector elements are the correlation coefficients $[\phi_{fd}(n)]$, as seen in Eq. 14.16. When we apply the linear filter equation in Table 13.2 to Fig. 14.14, since f_k is white with unit power, we have

$$\tilde{\Phi}_{fd}(z) = \tilde{U}(z)\tilde{\Phi}_{ff}(z)$$
$$= \tilde{U}(z)$$
(14.55)

Therefore, $\phi_{fd}(n)$, and thus the theoretical value of $\overline{b}_n$, is the impulse response of $\tilde{U}(z)$ which, from line 206 of Appendix A, is

$$\overline{b}_{n(\text{theoretical})} = e^{-0.1n} \sin(0.1n\pi)$$
(14.56)

Thus, with a broadband input signal, the least-squares coefficients $[b_n]$ will tend to match a finite portion of the impulse response of the unknown systems. One might suppose that a very large length (k) of the random input sequence would be required for the covariance solution to match the correlation solution, but in fact a short sequence suffices. As an example, the sequences of length 64 illustrated in Fig. 14.15 were used to derive a least-squares model of size $N = 50$. The calling statement for the least-squares design routine, SPLESQ, was

$$\text{CALL SPLESQ(F(-49),D(0),64,B,50,TSE,IE)}$$
(14.57)

The least-squares coefficients, $\overline{b}_0$ through $\overline{b}_{49}$, are plotted in Fig. 14.16 along with the continuous impulse response of $\tilde{U}(z)$, found by letting n vary continuously in Eq. 14.56; that is,

$$b(t) = e^{-0.1t} \sin(0.1\pi t)$$
(14.58)

The two plots in Fig. 14.16 are seen to agree exactly. The **R**-matrix generated during the execution of SPLESQ is nearly diagonal in this example, and SPSOLE, which is

Sec. 14.8   More Design Examples

357

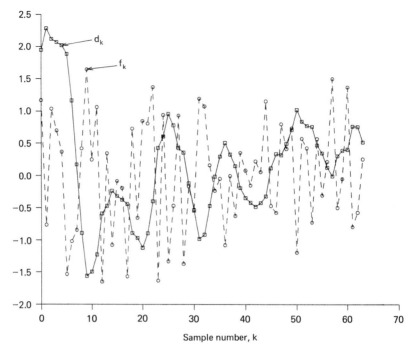

Figure 14.15 Sequences $[f_k]$ (connected with dashed lines) and $[d_k]$ (connected with solid lines) used to derive optimal coefficients $\bar{b}_0, \bar{b}_1, \ldots, \bar{b}_{49}$ in Fig. 14.14. Sequence length (K) is 64.

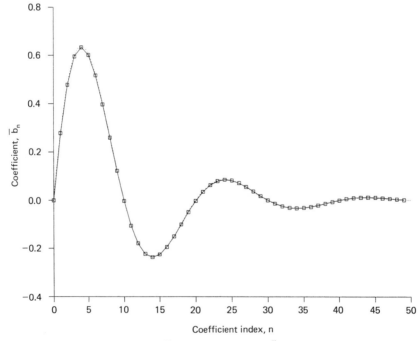

Figure 14.16 Optimal coefficients $[\bar{b}_n]$ in Fig. 14.14 with $\tilde{U}(z)$ as in Eq. 14.50, $N = 50$, and $K = 64$, along with the continuous impulse response in Eq. 14.58.

called by SPLESQ, has no difficulty in producing an accurate solution to the 50 linear equations contained in **RB** = **P**. The minimum total squared error, TSE in Eq. 14.57, was 0.0026 for the case plotted in Fig. 14.16. The minimum TSE in this example will in general decrease with increasing sequence length (k) as the covariance functions become more accurate correlation estimates, and will also decrease with increasing M as the length of the model, $\tilde{B}(z)$, approaches the infinite impulse response length of $\tilde{U}(z)$.

14.9 MULTIPLE INPUTS

Our discussion of least-squares system design to this point has applied mainly to the single-input, nonrecursive system in Fig. 14.6. Sometimes, as in *array processing* (Monzingo and Miller, 1980; Widrow and Stearns, 1985), we have a least-squares design problem in which there are multiple inputs combined and processed to produce a single output.

The general multiple-input case corresponding with Fig. 14.5 is illustrated in Fig. 14.17. Here we have a set of N sequences, $[f_{1k}]$ through $[f_{Nk}]$, each processed through a corresponding linear system. The outputs of the linear systems are summed to produce g_k, the single output to be compared with the desired signal, d_k.

The general result for the mean-squared-error performance surface for Fig. 14.17 follows from the derivation of the single-input result in Eq. 14.6. The revised versions of the power spectra in Eqs. 14.4 and 14.5 are now

$$\tilde{\Phi}_{gg}(z) = \sum_{m=1}^{N} \sum_{n=1}^{N} \tilde{H}_m(z)\tilde{H}_n(z^{-1})\tilde{\Phi}_{f_m f_n}(z) \qquad (14.59)$$

$$\tilde{\Phi}_{dg}(z) = \sum_{m=1}^{N} \tilde{H}_m(z)\tilde{\Phi}_{df_m}(z) \qquad (14.60)$$

These versions follow from Eqs. 14.4 and 14.5, and the linear nature of the configuration in Fig. 14.17. Equation 14.3 holds for Fig. 14.17 just as it did for Fig. 14.5.

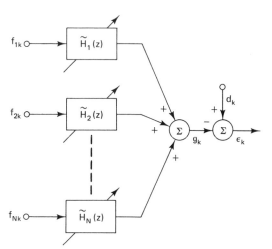

Figure 14.17 Multiple-input system; generalization of Fig. 14.5.

So if we substitute Eqs. 14.59 and 14.60 into Eq. 14.3, we obtain the multiple-input MSE formula:

$$\text{MSE} = \phi_{dd}(0) + \sum_{m=1}^{N} \sum_{n=1}^{N} \frac{1}{2\pi j} \oint [\tilde{H}_n(z^{-1})\tilde{\Phi}_{f_m f_n}(z) - \tilde{\Phi}_{d f_m}(z)]\tilde{H}_m(z)\frac{dz}{z} \qquad (14.61)$$

If we assume further that each linear system is a nonrecursive filter, we can obtain also a multiple-input version of the MSE in Eq. 14.10. We let

$$\tilde{H}_m(z) = \tilde{B}_m(z) = \sum_{n=0}^{M-1} b_{mn} z^{-n}; \qquad 1 \le m \le N \qquad (14.62)$$

Then, just as Eq. 14.10 followed from Eq. 14.6, we have

$$\text{MSE} = \phi_{dd}(0) + \sum_{m=1}^{N} \sum_{n=1}^{N} \sum_{i=0}^{M-1} \sum_{j=0}^{M-1} b_{mi} b_{mj} \phi_{f_m f_n}(i - j) - 2 \sum_{m=1}^{N} \sum_{i=0}^{M-1} b_{mi} \phi_{f_m d}(i) \qquad (14.63)$$

This result reduces to Eq. 14.10 with $N = 1$. The principal feature of Eq. 14.63 is that the MSE is still a *quadratic* function of the b's, because each b occurs only to the first and second degree.

Thus, the MSE performance function is quadratic whenever a linear combination of linear systems is being designed. In addition to array processing, this general result can be used in many cases in connection with nonlinear processing, distortion, and so on. For example, consider the system in Fig. 14.18. Here we have a nonlinear processor producing nonlinear functions, $[F_{mk}]$, of the input f_k. The F's could be any functions, for example,

$$F_{mk} = f_k^{m-1}; \qquad 1 \le m \le M \qquad (14.64)$$

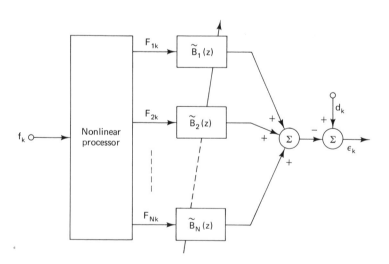

Figure 14.18 Nonlinear system in which the MSE is still a quadratic function of the coefficients, $[b_{mn}]$.

and so on. The **R**-matrix elements are correlations or covariances of the F's instead of f_k, but are still statistical functions of f_k. In any case, however, the performance surface is always a quadratic function of the filter coefficients. For an example, see Exercise 25.

14.10 EFFECTS OF INDEPENDENT BROADBAND NOISE

In many practical design cases, noise is added to the signals whose properties are used to design least-squares systems. The noise could be "plant noise" associated with the unknown system as in Figs. 14.2 and 14.3, or measurement noise added to the desired signal, d_k, or the input signal, f_k, in Fig. 14.5. In this section we derive some general effects of noise in least-squares design, assuming that the noise is independent, has a broad spectrum, and is additive, that is, adds to the signals. These assumptions are somewhat restrictive but nevertheless cover many cases of interest, and the approach used here is applicable with modifications if the noise is not independent or does not have a flat power spectrum.

The noisy least-squares design configuration is illustrated in Fig. 14.19. It is the same as Fig. 14.5 with noise added to the input signal, s_k, and the desired signal, d_k. Either of the two noise sequences, x_k and y_k, or both, may be present depending on the application of Fig. 14.19. If the noises are independent of the signals, and of each other, the cross-correlation and cross-spectral terms are zero, and we have

$$\tilde{\Phi}_{ff}(z) = \tilde{\Phi}_{ss}(z) + \tilde{\Phi}_{xx}(z)$$

$$\tilde{\Phi}_{vv}(z) = \tilde{\Phi}_{dd}(z) + \tilde{\Phi}_{yy}(z) \qquad (14.65)$$

$$\tilde{\Phi}_{vf}(z) = \tilde{\Phi}_{ds}(z)$$

When we apply the general MSE formula in Eq. 14.6 to Fig. 14.19, we have

$$\text{MSE} = \phi_{vv}(0) + \frac{1}{2\pi j} \oint [\tilde{H}(z^{-1})\tilde{\Phi}_{ff}(z) - \tilde{\Phi}_{vf}(z)]\tilde{H}(z) \frac{dz}{z} \qquad (14.66)$$

Using the relationship in Eq. 14.65, this MSE formula becomes

$$\text{MSE} = \phi_{dd}(0) + \phi_{yy}(0) + \frac{1}{2\pi j} \oint \left(\tilde{H}(z^{-1})[\tilde{\Phi}_{ss}(z) + \tilde{\Phi}_{xx}(z)] - \tilde{\Phi}_{ds}(z) \right) \tilde{H}(z) \frac{dz}{z} \qquad (14.67)$$

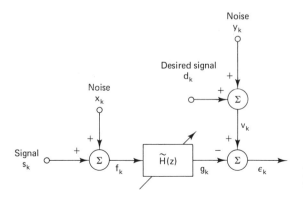

Figure 14.19 Least-squares design similar to Fig. 14.5 but with noisy signals.

This is the general expression for the mean-squared error when independent noise sequences, x_k and y_k, are added to the signal sequences, s_k and d_k.

If we assume that the input noise power spectrum, $\tilde{\Phi}_{xx}(z)$, is flat enough to be treated as a constant function of frequency, then the inverse transform, $\phi_{xx}(u)$ may be considered to be negligible everywhere except at $n = 0$. Under this condition, if we consider how Eq. 14.10 follows from Eq. 14.6 when $\tilde{H}(z)$ is the nonrecursive filter given by Eq. 14.8, we obtain from Eq. 14.67 the following:

$$\text{MSE} = \phi_{dd}(0) + \phi_{yy}(0) + \sum_{m=0}^{M-1} \sum_{n=0}^{M-1} b_m b_n \phi_{ss}(m - n)$$

$$+ \sum_{n=0}^{M-1} b_n^2 \phi_{xx}(0) - 2 \sum_{n=0}^{M-1} b_n \phi_{sd}(n) \tag{14.68}$$

That is, the only nonzero values of $\phi_{xx}(m - n)$ in the double summations are those for which $m = n$. In the vector notation defined in Eqs. 14.14 through 14.16 the elements of $\mathbf{P}$ are now $\phi_{sd}(n)$, that is,

$$\mathbf{P} = [\phi_{sd}(0) \quad \phi_{sd}(1) \quad \phi_{sd}(2) \quad \cdots \quad \phi_{sd}(M - 1)]^{\mathrm{T}} \tag{14.69}$$

With $\mathbf{R}$ and $\mathbf{B}$ as in Eqs. 14.14 and 14.15, the simplified expression of Eq. 14.68 becomes

$$\boxed{\text{MSE} = \phi_{dd}(0) + \phi_{yy}(0) + \mathbf{B}^{\mathrm{T}}(\mathbf{R} + \phi_{xx}(0)\mathbf{I})\mathbf{B} - 2\mathbf{P}^{\mathrm{T}}\mathbf{B}} \tag{14.70}$$

From this expression for the MSE we conclude the following:

- Adding independent white noise to the input signal is equivalent to adding a constant to the diagonal of the $\mathbf{R}$-matrix.
- Adding independent noise to the desired signal simply adds a constant to the MSE.

If the input correlation matrix, $\mathbf{R} + \phi_{xx}(0)\mathbf{I}$, is a diagonal matrix, then cross terms of the form $b_i b_j$ with $i \neq j$ do not appear in the MSE formula (e.g., see Eq. 14.68), and the error contours are circular. Furthermore, the $\phi_{xx}(0)$ term in Eq. 14.68 or in Eq. 14.70 increases the b_n^2 terms in the MSE formula, thereby causing the sides of the bowl-shaped error surface to become steeper. Thus we also conclude:

- Adding independent white noise to the input signal tends to steepen the sides of the performance function and to make its contours more circular.

If we set the derivative of Eq. 14.70 with respect to each b_n equal to zero, we obtain an expression equivalent to Eq. 14.17 for the optimal coefficients:

$$\overline{\mathbf{B}}(\mathbf{R} + \phi_{xx}(0)\mathbf{I}) = \mathbf{P}$$

$$\overline{\mathbf{B}} = (\mathbf{R} + \phi_{xx}(0)\mathbf{I})^{-1}\mathbf{P} \tag{14.71}$$

Thus, the optimal coefficients with input noise are modified by having the increased diagonal elements in the $\mathbf{R}$-matrix. We note that if the input is dominated by the noise sequence $[x_k]$, then the optimal coefficients are just scaled values of the cross-correlation terms; that is,

$$\overline{\mathbf{B}} = \frac{1}{\phi_{xx}(0)} \mathbf{P}; \qquad \phi_{xx}(0) >> \phi_{xx}(n) \tag{14.72}$$

Furthermore, if the input is entirely independent noise, then the cross-correlation terms are zero and

$$\overline{\mathbf{B}} = \mathbf{0} \tag{14.73}$$

We also observe in Eq. 14.71 that the solution for optimal coefficients is not affected at all by the noise $[y_k]$ added to the desired signal, which only affects the MSE and adds a constant to the minimum MSE which, similar to Eq. 14.20, is

$$(\mathrm{MSE})_{\min} = \phi_{dd}(0) + \phi_{yy}(0) - \mathbf{P}^{\mathrm{T}} \overline{\mathbf{B}} \tag{14.74}$$

Thus, in these results, we have expressed the effects of independent broadband noise on least-squares system design. Exercises 6 and 9 apply to this subject.

EXERCISES

1. In a least-squares interference-canceling design, suppose that the filter $\tilde{H}(z)$ has two weights, b_0 and b_1, that the desired signal power is σ_d^2, and that the correlation functions are

$$\mathbf{R} = \begin{bmatrix} r_0 & r_1 \\ r_1 & r_0 \end{bmatrix}, \qquad \mathbf{P} = \begin{bmatrix} p_0 \\ p_1 \end{bmatrix}$$

   (a) Express the MSE as a quadratic function of b_0 and b_1.
   (b) Express the optimal coefficients, $\overline{b}_0$ and $\overline{b}_1$, in terms of the r's and p's.
   (c) Express the minimum mean-squared error in terms of the signal statistics.

2. In Exercise 1, suppose that $E[f_k f_{k+1}] = r_1 = 0$.
   (a) What are the optimal coefficients?
   (b) What relationship between the signal statistics must hold in order for the minimum MSE to equal zero?

3. Consider a general sinusoid, $x_k = A \sin(\omega k + \alpha)$, sampled at K samples per cycle.
   (a) What is the average power in $[x_k]$?
   (b) Express the autocorrelation function of $[x_k]$ and show that it does not depend on the phase, α.
   (c) Show that the autocorrelation function of $[x_k]$ has a period equal to that of $[x_k]$.

4. A sequence $[x_k]$ with a period equal to eight samples is

$$[x_k] = [\dots, 1, 1, 1, 1, -1, -1, -1, -1, 1, 1, \dots]$$

   Plot $[x_k]$ and the autocorrelation function, $\phi_{xx}(k)$, for $-15 \le k \le 15$.

5. If the samples of two sequences $[x_k]$ and $[y_k]$ are *complex*, the correlation function $\phi_{xy}(n)$ is defined to be

$$\phi_{xy}(n) \overset{\Delta}{=} E[x_k^* y_{k+n}]$$

   That is, $\phi_{xy}(n)$ is the average product of the conjugate of x_k times y_{k+n}.
   (a) Show that $\phi_{xy}(n)$ agrees with the previous definition when $[x_k]$ is a real sequence.
   (b) Show that $\phi_{xy}^*(n) = \phi_{yx}(-n)$.
   (c) Express $\phi_{xx}(n)$ if x_k is a sinusoidal function, $x_k = A e^{j(\omega k + \alpha)}$. Show that $\phi_{xx}(n)$ does not depend on the phase of x_k and has a period equal to that of x_k.

6. A sequence $[x_k]$ has samples given by

$$x_k = A \cos(\omega k + \alpha) + r_k$$

where r_k is a uniform white random variate with variance σ_r^2. Express the autocorrelation function, $\phi_{xx}(n)$.

7. Suppose that x_k and y_k are sinusoidal signals, either real or complex, with frequencies ω_x and ω_y. What condition must hold in order for $\phi_{xy}(n)$ to be zero for all values of n?

8. Consider the two-step predictor shown here.
   (a) Let $s_k = \sqrt{2} \cos(2\pi k/15 + \pi/4)$. Determine the **R**-matrix and the **P**-vector.
   (b) Express the mean-squared error as a quadratic function of b_0 and b_1.
   (c) What are the optimal coefficients, $\bar{b}_0$ and $\bar{b}_1$?
   (d) What is the minimum of $E[\epsilon_k^2]$?

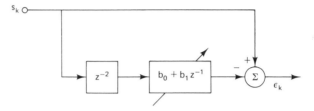

9. Do Exercise 8 with $s_k = \sqrt{2} \sin(2\pi k/15) + r_k$, where $[r_k]$ is a uniform white random sequence with samples in the range $[-1, 1]$. Plot the minimum MSE versus the power in $[r_k]$.

10. Consider the least-squares filter shown here, in which i is a positive integer.
   (a) What are the correlation functions in **R** and **P**?
   (b) If this filter is used in place of the filter in Exercise 8, what values of i will allow a solution for which $(\text{MSE})_{\min} = 0$?

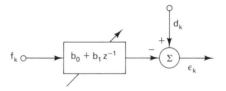

11. Give a simple expression for the performance function in Eq. 14.6 when the input signal, f_k, is a white random sequence.

12. A least-squares system has an adjustable "bias weight," which causes the constant c to be added to the filter output, as shown here. The bias weight is used in cases where the input sequence $[f_k]$ has a nonzero mean value. Suppose that $E[f_k] = \mu_f$ and $E[d_k] = 0$, and the filter weights, $[b_n]$, are optimized to $[\bar{b}_n]$. What is the optimal value of c in terms of μ_f and $[\bar{b}_n]$?

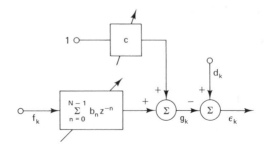

13. A noise-canceling system, used to cancel sinusoidal interference from a broadband signal, is shown here.

   (a) Let r_k be a uniform white random sequence with zero mean and $\sigma_r^2 = 2.0$. Find the signal power, σ_s^2.

   (b) Write an expression for the MSE, $E[\epsilon_k^2]$.

   (c) Find the optimal coefficients, $\bar{b}_0$ and $\bar{b}_1$.

   (d) Show that the minimum MSE equals σ_s^2.

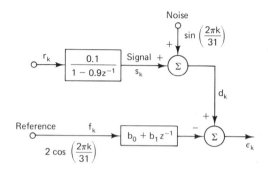

14. For the situation described in Exercise 8, draw contours for MSE = 1, 0.8, 0.6, 0.4, and 0.2 on the $b_0 b_1$ plane.

15. A least-squares filter is to be used to adjust the amplitude and phase of f_k to match the amplitude and phase of d_k as shown here.

   (a) Find $\bar{b}_0$ and $\bar{b}_1$.

   (b) Run the least-squares design routine, SPLESQ, with $M = 2$ and sequence lengths $K = 8$, 16, 32, and 64. For each value of K, print b_0, b_1, and TSEMIN. Compare the results with part (a) and explain.

16. Do Exercise 13 using SPLESQ. To generate $[r_k]$, use SPUNIF with seed = 123. Find $\bar{b}_0$, $\bar{b}_1$, and TSEMIN for $K = 10, 100, 1000$, and 10,000. Compare these results with the answers to Exercise 13(c) and (d), and explain.

17. Use subroutine SPSOLE to solve the following sets of linear equations.

   (a) $b_0 + 2b_1 + 3b_2 = 8$
   $4b_0 + 5b_1 + 6b_2 = 17$
   $9b_0 + 8b_1 + 6b_2 = 19$

   (b) $4b_0 + 3b_1 + 2b_2 + b_3 = 2$
   $3b_0 + 4b_1 + 3b_2 + 2b_3 = 0$
   $2b_0 + 3b_1 + 4b_2 + 3b_3 = 0$
   $b_0 + 2b_1 + 3b_2 + 4b_3 = -2$

   (c) Five equations of the form $\bar{\mathbf{B}} = \bar{\mathbf{R}}^{-1}\bar{\mathbf{P}}$, in which the correlation functions are

$$\phi_{ff}(n) = \cos\left(\frac{2\pi n}{15}\right) \quad \text{and} \quad \phi_{fd}(n) = 2\sin\left(\frac{2\pi n}{15}\right)$$

(d) $b_0 + b_1 = 1$
$b_0 + b_2 = 4$
$b_0 + b_3 = 0$
$b_0 + b_4 = 2$
$b_1 + b_2 = 3$

18. In this exercise we use the modeling configuration in Fig. 14.2 to design a lowpass FIR filter. The concept is illustrated here. The input signal, f_k, is composed of sine waves at frequencies evenly distributed between zero and half the sampling rate. The desired (real) gain characteristic, $\overline{H}(j\omega T)$, is applied to each of these components and the filtered version of f_k is then delayed to produce d_k.

(a) Write an expression for $\phi_{ff}(n)$.
(b) Compute the **R**-matrix for $N = 199$ and $M = 8$.
(c) Compute the **P**-vector for $N = 199$ and $M = 8$.
(d) Find $\overline{\mathbf{B}}$ for $N = 199$ and $M = 8$.
(e) Find $\phi_{dd}(0)$.
(f) For $N = 199$ and $M = 8$, compare the minimum MSE with $\phi_{dd}(0)$.

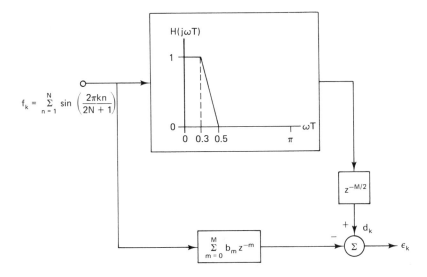

19. Do Exercise 18 by first generating the sequences $[f_k]$ and $[d_k]$ for $-M < k \leq 2N$ and then calling SPLESQ with $K = 2N + 1$ to design the optimal filter.

(a) Find $\overline{\mathbf{B}}$ for $N = 199$ and $M = 8$. Compare with the answer to Exercise 18(d).
(b) Using $N = 199$ and $M = 16$, plot the amplitude gain and the phase shift of the optimal filter.

20. A least-squares, linear-phase FIR filter, shown here, is to be designed.

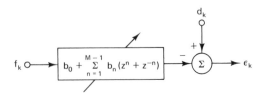

Least-Squares System Design   Chap. 14

(a) Express the error signal, ϵ_k.

(b) Express $E[\epsilon_k^2]$ in terms of correlation functions. Is the MSE a quadratic function of the b_n's?

(c) In the expression $\mathbf{R}\overline{\mathbf{B}} = \mathbf{P}$, what are the elements of $\mathbf{R}$ and $\mathbf{P}$ in this case?

21. Design the least-squares equalizer in Fig. 14.11 with $\tilde{U}(z)$ given by

$$\tilde{U}(z) = \frac{1}{1 - 1.82z^{-1} + 0.81z^{-2}}$$

Use the signal s_k shown in Fig. 14.12, and use $K = 50$. Tabulate $\overline{\mathbf{B}}$ as in the text before Eq. 14.49 for various values of M and Δ, and explain your results.

22. Do Exercise 21 with $M = 3$, $\Delta = 0$, $K = 100$, and

$$s_k = \sqrt{12}\,(r_k - 0.5); \qquad k = 0, 1, \ldots, 99$$

where r_k is a sample from SPUNIF beginning with ISEED = 345. Compare $\overline{\mathbf{B}}$ with the corresponding answer to Exercise 21.

23. The "equation error" method for IIR modeling is illustrated here. The transfer functions of the least-squares system is

$$\tilde{H}(z) = \frac{b_0 + b_1 z^{-1} + \cdots + b_{N-1}z^{-N+1}}{1 - a_1 z^{-1} - \cdots - a_{M-1}z^{-M+1}}$$

An error, ϵ_k' is formed by forward-modeling the numerator of $\tilde{H}(z)$ and inverse-modeling the denominator.

(a) Compare this scheme with the scheme in which the entire transfer function, $\tilde{H}(z)$, is used as in Fig. 14.2 with error ϵ_k. To do so, find $\tilde{E}'(z)$, the transform of ϵ_k', in terms of $\tilde{E}(z)$, the transform of ϵ_k. Are the two schemes equivalent? Are they equivalent if $E[\epsilon_k^{\prime 2}] = 0$?

(b) Discuss how one might adjust the a and b coefficients to minimize $E[\epsilon_k^{\prime 2}]$.

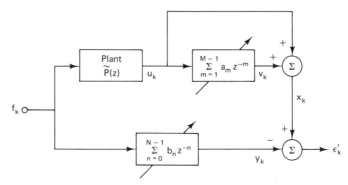

24. A simple direction-finding system is shown on the next page. A plane wave traveling at velocity c m/s arrives with direction angle θ at two receptors spaced x_0 meters apart, such that

$$s_{0k} = A\,\cos\!\left(\frac{2\pi k}{36}\right); \qquad s_{1k} = A\,\cos\!\left(\frac{2\pi(k - \Delta)}{36}\right)$$

where $\Delta = (x_0 \sin\theta)/c$ seconds. The second signal, s_{1k}, is sent through a "quadrature filter," in which there is a 90-degree (nine-sample) phase delay from b_0 to b_1. Show how to determine the direction angle, θ, from x_0, c, and the optimal coefficients, $\overline{b}_0$ and $\overline{b}_1$. Comment on the effects of adding mutually independent white noise at the two inputs.

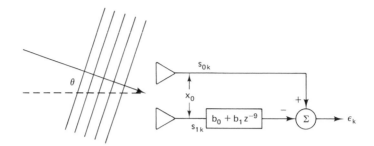

25. A type of multiple-input, nonlinear modeling system is shown here modeling a limiter whose output, d_k, is equal to f_k but limited to the range $(-1, 1)$.

(a) Show that the performance function, $E[\epsilon_k^2]$, is a quadratic function of the b's. What are the elements of **R** and **P** in this case?

(b) Experiment with the least-squares design process using the sequence

$$f_k = 5(\text{SPUNIF}(\text{ISEED}) - 0.5); \qquad k = 0, 1, \ldots, 999$$

Show the success of your results in terms of superimposed plots of f_k, d_k, and g_k for $k = 0, \ldots, 999$. In this exercise, use SPSOLE after determining the appropriate **R** and **P** elements.

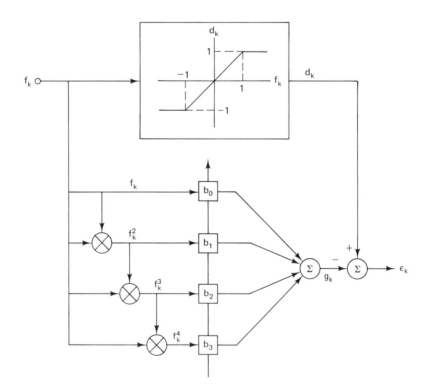

Least-Squares System Design   Chap. 14

26. In Exercise 13, suppose that random noise given by

$$x_k = 6(\text{SPUNIF(ISEED)} - 0.5)$$

is added to the reference input.
(a) Write the revised expression for the MSE.
(b) Find the optimal coefficients.
(c) Compare the minimum MSE with σ_s^2.

SOME ANSWERS

1. (a) $\text{MSE} = r_0(b_0^2 + b_1^2) + 2(r_1 b_0 b_1 - p_0 b_0 - p_1 b_1) + \sigma_d^2$

   (b) $\bar{b}_0 = \dfrac{r_0 p_0 - r_1 p_1}{r_0^2 - r_1^2}; \quad \bar{b}_1 = \dfrac{r_0 p_1 - r_1 p_0}{r_0^2 - r_1^2}$

2. (a) $\bar{b}_0 = p_0/r_0; \quad \bar{b}_1 = p_1/r_0$ (b) $r_0 \sigma_d^2 = p_0^2 + p_1^2$

3. (a) $\sigma_x^2 = A^2/2$ (b) $\phi_{xx}(n) = (A^2/2) \cos \omega n$ (c) period $= 2\pi/\omega$ samples

6. $\phi_{xx}(n) = (A^2/2) \cos \omega n + \sigma_r^2 \delta(n)$ **7.** Sampling such that $\pi/T > \omega_x$ and ω_y, and $\omega_x \neq \omega_y$

12. $c = -\mu_f \sum\limits_{n=0}^{N-1} \bar{b}_n$

13. (a) $\sigma_s^2 = 0.1053$ (b) $\text{MSE} = 2(b_0^2 + b_1^2) + 3.9181 b_0 b_1 - 0.4026 b_1 + 0.6053$
   c. $\bar{b}_0 = -2.4355, \, b_1 = 2.4863$

17. (a) $1, -1, 3$ (b) $1, -1, 1, -1$ c. $-2.6164, 0.8986, 2.4888, -0.2261$
   (d) $1, 0, 3, -1, 1$

REFERENCES

ATAL, B. S., and HANAUER, S. L., Speech Analysis and Synthesis by Linear Prediction of the Speech Wave. *J. Acoust. Soc. Amer.*, Vol. 50, No. 2, 1971, p. 637.

BLAHUT, R. E., *Fast Algorithms for Digital Signal Processing*, Chap. 11. Reading, Mass.: Addison-Wesley, 1985.

BORDLEY, T. E., Linear Predictive Coding of Marine Seismic Data. *IEEE Trans. Acoust. Speech Signal Process.*, Vol. ASSP-31, No. 4, August 1983, p. 828.

CLARK, G. A., and RODGERS, P. W., Adaptive Prediction Applied to Seismic Event Detection. *Proc. IEEE*, Vol. 69, No. 9, September 1981, p. 1166.

COWAN, C. F. N., and GRANT, P. M. (eds.), *Adaptive Filters*. Englewood Cliffs, N.J.: Prentice-Hall, 1985.

FRANKLIN, G. F., and POWELL, J. D., *Digital Control of Dynamic Systems*. Reading, Mass.: Addison-Wesley, 1980.

GERSHO, A., Adaptive Equalization of Highly Dispersive Channels for Data Transmission. *Bell System Tech. J.*, January 1969, p. 55.

GRIFFITHS, L. J., An Adaptive Lattice Structure for Noise-Cancelling Applications. *Proc. ICASSP-78*, Tulsa, Okla., April 1978, p. 87.

GRIFFITHS, L. J., SMOLKA, F. R., and TREMBLY, L. D., Adaptive Deconvolution: A New Technique for Processing Time-Varying Seismic Data. *Geophysics*, June 1977, p. 742.

HONIG, M. L., and MESSERSCHMITT, D. G., *Adaptive Filters*. Norwell, Mass.: Kluwer Academic Publishers, 1984.

ITAKURA, F., and SAITO, S., Digital Filtering Techniques for Speech Analysis and Synthesis. *Proc. 7th Int. Cong. Acoust.*, Budapest, 1971, p. 261.

JURY, E. I., A Note on the Reciprocal Zeros of a Real Polynomial with Respect to the Unit Circle. *IEEE Trans. Commun. Technol.,* Vol. CT-11, June 1964, p. 292.

KAILATH, T. (ed.), Special Issue on System Identification and Time Series Analysis. *IEEE Trans. Automatic Control,* Vol. AC-19, December 1974.

KELLY, L. G., *Handbook of Numerical Methods and Applications,* Chap. 8. Reading, Mass.: Addison-Wesley, 1967.

LANDAU, A. I., *Adaptive Control: The Model Reference Approach.* New York: Dekker, 1979.

LEVINSON, N., The Wiener RMS Error Criterion in Filter Design and Prediction. *J. Math. Phys.,* Vol. 25, 1946, p. 261.

LUCKY, R. W., Techniques for Adaptive Equalization of Digital Communication Systems. *Bell System Tech. J.,* February 1966, p. 255.

MAKHOUL, J., Spectral Analysis of Speech by Linear Prediction. *IEEE Trans. Audio Electro-acoust.,* Vol. AU-21, June 1973, p. 140.

MAKHOUL, J., Stable and Efficient Lattice Methods for Linear Prediction. *IEEE Trans. Acoust. Speech Signal Process.,* Vol. ASSP-25, No. 5, October 1977, p. 423.

MAKHOUL, J., and VISWANATHAN, R., Adaptive Lattice Methods for Linear Prediction. *Proc. ICASSP-78,* May 1978, p. 83.

MAKHOUL, J. L., and COSELL, L. K., Adaptive Lattice Analysis of Speech. *IEEE Trans. Acoust. Speech Signal Process.,* Vol. ASSP-29, No. 3, Pt. 3, June 1981, p. 654.

MONZINGO, R. A., and MILLER, T. W., *Introduction to Adaptive Arrays.* New York: Wiley, 1980.

RABINER, L. R., and SCHAFER, R. W., *Digital Processing of Speech Signals,* Chap. 8. Englewood Cliffs, N.J.: Prentice-Hall, 1978.

STEARNS, S. D., and DAVID, R. A., *Signal Processing Algorithms.* Englewood Cliffs, N.J.: Prentice-Hall, 1987.

STEARNS, S. D., and VORTMAN, L. J., Seismic Event Detection Using Adaptive Predictors. *Proc. ICASSP-81,* March 1981, p. 1058.

WIDROW, B., and STEARNS, S. D., *Adaptive Signal Processing.* Englewood Cliffs, N.J.: Prentice-Hall, 1985.

WIDROW, B., and WALACH, E., On the Statistical Efficiency of the LMS Algorithm with Nonstationary Inputs. *IEEE Trans. Inform. Theory,* Vol. IT-30, No. 2, Pt. 1, March 1984, p. 211.

WIDROW, B., ET AL., Adaptive Antenna Systems. *Proc. IEEE,* Vol. 55, December 1967, p. 2143.

WIDROW, B., ET AL., Adaptive Noise Cancelling: Principles and Applications. *Proc. IEEE,* Vol. 63, No. 12, December 1975, p. 1692.

WIDROW, B., McCOOL, J. M., LARIMORE, M. G., and JOHNSON, C. R., JR., Stationary and Nonstationary Learning Characteristics of the LMS Adaptive Filter. *Proc. IEEE,* Vol. 64, No. 8, August 1976, p. 1151.

WIDROW, B., TITCHENER, P. F., and GOOCH, R. P., Adaptive Design of Digital Filters. *Proc. ICASSP-81,* March 1981, p. 243.

WIENER, N., *Extrapolation, Interpolation and Smoothing of Stationary Time Series with Engineering Applications.* New York: Wiley, 1949.

ZEIDLER, J. R., SATORIUS, E. H., CHABRIES, D. M., and WEXLER, H. T., Adaptive Enhancement of Multiple Sinusoids in Uncorrelated Noise. *IEEE Trans. Acoust. Speech Signal Process.,* Vol. ASSP-26, No. 3, June 1978, p. 240.

ZOHAR, S., Toeplitz Matrix Inversion: The Algorithm of W. F. Trench. *J. Assoc. Comput. Mach.,* Vol. 16, No. 4, October 1969, p. 592.

15

Random Sequences and Spectral Estimation

15.1 INTRODUCTION

The analysis of random signals involves sequences of random data, that is, ordered sample sets of random functions in the case of digital processing. Often the objective in processing a random sequence is the measurement of the statistical properties of the sequence. As suggested by Chap. 13, estimation of the power spectrum is perhaps the central objective, because the power spectrum of $f(t)$ determines the variance σ_f^2, the distribution of σ_f^2 with frequency, all of the information in the autocorrelation function ϕ_{ff}, and so on.

The subject of spectral estimation is itself quite extensive, and there are texts devoted to this subject alone. (See the list of references.) Here the discussion is limited to the practical methods and procedures of spectral estimation with random sequences.

Another subject closely related to spectral estimation is the *generation of random sequences*. Random sequences are often required in digital simulations, for example. Physical variables such as electromagnetic noise, mechanical noise, Brownian motion, radar tracking errors, turbulence, and so on, are essentially unpredictable and must often be simulated as random functions with only their power spectra and amplitude pdf's known in advance. Such random sequences can be easily generated digitally by passing sampled white noise through an appropriate filter, as described below.

To cover these broad areas, that is, generating random sequences and estimating power spectra, this chapter proceeds as follows: First, there is a short discussion of sampled white noise viewed as a random sequence. Next the generation of other random sequences is viewed as a process of passing white noise through the appropriate filter, and an example is given in which a particular random sequence is generated. Finally methods of estimating the power spectrum are discussed, using the latter random sequence as an example and treating it as if its (known) power spectrum were unknown in advance.

15.2 WHITE NOISE

White noise is the name commonly given to a random function having a power spectrum that is flat for all practical purposes, or flat over a specified range of frequencies. The name by itself implies only a flat spectrum, and does not specify a distribution of amplitudes. The names *Gaussian* white noise, *uniform* white noise, and so on, are used to designate the form of the amplitude distribution.

In Fig. 15.1 for example, segments of two continuous white noise functions are shown. Both functions have the same flat power spectrum $\Phi(j\omega)$ and yet their amplitude distributions differ, with $f_1(t)$ being Gaussian and $f_2(t)$ being uniform. Although the total power and therefore the rms values of f_1 and f_2 are the same, they are concentrated at different amplitudes.

As given in Eq. 13.52, the total power in any noise function is $1/2\pi$ times the integral of $\Phi(j\omega)$. Thus if $\Phi(j\omega)$ were really the same (flat) at all frequencies, the value of $\Phi(j\omega)$ would have to be infinitesimal for the power to be finite. Hence it is helpful to think of $\Phi(j\omega)$ as in Fig. 15.2, that is, as a function that is flat out to some maximum frequency, ω_m. The total power, P in Fig. 15.2, is then $1/2\pi$ times the total area under $\Phi(j\omega)$. (Note that, in Fig. 15.1, we have $P = a^2/3$.)

If the continuous white noise function $f(t)$ is sampled at regular intervals of length T, and if only the sample values of $f(t)$ are of interest, then all of the power in

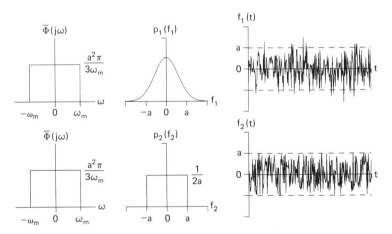

Figure 15.1 Gaussian and uniform white noise, $f_1(t)$ and $f_2(t)$, respectively.

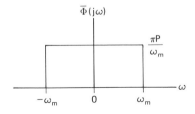

Figure 15.2 White noise spectrum; total power = P.

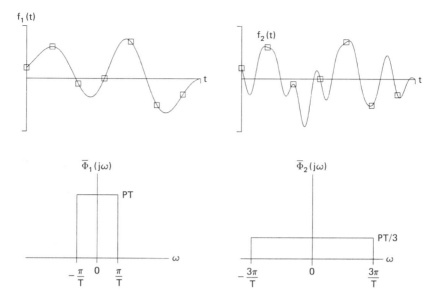

Figure 15.3 Continuous white noise functions with identical sample sets but differing power spectra.

$f(t)$ can be considered to be concentrated in the frequency interval $(-n\pi/T, n\pi/T)$, where n is any positive integer. In Fig. 15.3, for example, two white noise functions having identical sample sets are shown. All of the power in $f_1(t)$ is in the interval $(-\pi/T, \pi/T)$, while $f_2(t)$ has the power in interval $(-3\pi/T, 3\pi/T)$, the total power being the same in both cases. The important point is that (due to aliasing when n is greater than one) there is no way to distinguish the two functions using only the sample sets — power at frequencies above π/T is "folded in" to lower frequencies in the sampling process in a manner such that the spectrum remains flat between $-\pi/T$ and π/T.

Beginning with the premise that a sampled white noise function $f(t)$ must have a power spectrum $\Phi_{ff}(j\omega)$ like the spectrum $\Phi(j\omega)$ described above, one can specify some properties of the samples of $f(t)$. First the autocorrelation function $\phi_{ff}(\tau)$ is derived as follows:

$$
\begin{aligned}
\phi_{ff}(\tau) &= \frac{1}{2\pi} \int_{-\infty}^{\infty} \Phi_{ff}(j\omega) e^{j\omega\tau}\, d\omega \\
&= \frac{1}{2\pi} \int_{-n\pi/T}^{n\pi/T} \frac{PT}{n} e^{j\omega\tau}\, d\omega \\
&= P \frac{\sin(n\pi\tau/T)}{n\pi\tau/T}
\end{aligned}
\tag{15.1}
$$

Both $\phi_{ff}(\tau)$ and $\Phi_{ff}(j\omega)$ are illustrated in Fig. 15.4. Note that these are functions which, according to the discussion above, describe *any* regularly sampled white noise function.

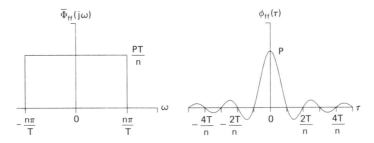

Figure 15.4 Properties of any sampled white noise function, $f(t)$. Usually, $n = 1$.

Given the properties illustrated in Fig. 15.4, three important deductions can be made concerning the *characteristics of any white noise function, $f(t)$*:

1. Since $\phi_{ff}(mT) = 0$ when m is any integer (not zero), *the samples of $f(t)$ must be statistically independent*. That is, $f[(m + 1)T]$ is independent of $f(mT)$, etc.

2. Since $\phi_{ff}(\tau)$ approaches zero as $|\tau|$ increases, *the sample distribution of $f(t)$ must have a mean value of zero*. One may therefore say that the mean of $f(t)$, μ_f, is zero. This property is affirmed in Eq. 13.44.

3. Since $\mu_f = 0$, the mean-squared value of $f(t)$ must be the variance, σ_f^2. Therefore, *the average power in $f(t)$ is $P = \sigma_f^2$.* (Of course, this is true whenever $\mu_f = 0$, whether or not the power spectrum is flat.)

Note also that changing n in Fig. 15.4 does not in any way affect these properties.

In digital signal processing the samples of $f(t)$ are generated using a computing routine such as SPUNIF or SPGAUS, both of which are described at the end of Chapter 13. These algorithms generate statistically independent samples of a uniform or Gaussian variate. Therefore, if the sequences produced by these routines have zero mean, they qualify as white noise sequences.

The output of SPGAUS has zero mean and unit variance, that is, unit power. So, to generate a white Gaussian sequence with power P, we would multiply each output by $\sqrt{P}$. Similarly, the output of SPUNIF has a mean of $\frac{1}{2}$ and a variance of $\frac{1}{12}$. Thus, to produce a white uniform sequence with power P, we would subtract $\frac{1}{2}$ from each sample and scale the result by $\sqrt{12P}$. In summary, with ISEED set initially, the computing algorithms for a white sequence $[f_n]$ are

---

White sequences with power P

$$\text{Uniform:} \quad f_n = \sqrt{12P}\,[\text{SPUNIF(ISEED)} - 0.5] \tag{15.2}$$

$$\text{Gaussian:} \quad f_n = \sqrt{P}\,[\text{SPGAUS(ISEED)}]$$

Random Sequences and Spectral Estimation   Chap. 15

Example 15.1

Show how to use SPUNIF to produce a uniform white sequence with time step $T = 0.1$ s and average power $P = 3$. Plot the theoretical probability density function and the theoretical power spectrum. In accordance with Eq. 15.2, we have

$$f_n = 6[\text{SPUNIF(ISEED)} - 0.5] \tag{15.3}$$

The resulting white noise samples are distributed from -3 to $+3$ as shown in Fig. 15.5. The power spectrum $\overline{\Phi}_{ff}(j\omega)$ is also shown in the figure. Note that $\overline{\Phi}_{ff}(j\omega)$ is equal to PT, so that the integral of Φ_{ff} is $2\pi P$, or 6π in this example.

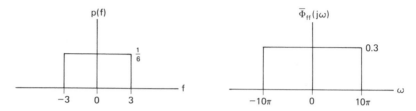

Figure 15.5 Distribution and spectrum for Example 15.1.

15.3 COLORED RANDOM SEQUENCES FROM FILTERED WHITE NOISE

When statistically independent white noise samples are filtered, the shape of the resulting power spectrum or "color" of the resulting sequence is determined by the transfer function of the filter, as illustrated in Fig. 15.6. The digital filtering algorithm in general produces samples that, unlike the white noise samples, are correlated, with a resulting nonwhite spectrum.

For the simulation of a random function with a predetermined power spectrum, it is convenient to make the input power density ($\overline{\Phi}_{ff}$ in Fig. 15.6) equal to one at frequencies below one-half the sampling rate. This is equivalent to setting the total input power, and therefore the variance of the input samples, as follows:

$$\sigma_f^2 = \frac{1}{2\pi} \int_{-\pi/T}^{\pi/T} \overline{\Phi}_{ff}(j\omega)\, d\omega = \frac{1}{T} \tag{15.4}$$

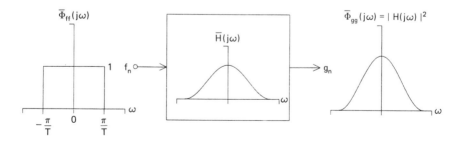

Figure 15.6 Colored noise from white noise via filtering.

where T is the sampling interval. Thus, if r_n is the nth independent uniform random sample between 0 and 1 coming from a computer library routine, then, as in Eq. 15.2,

$$f_n = \sqrt{\frac{12}{T}} \left(r_n - \frac{1}{2} \right) \tag{15.5}$$

would be the appropriate white noise sample in Fig. 15.6.

With $\overline{\Phi}_{ff} = 1$, the output power spectrum $\overline{\Phi}_{gg}$ is then

$$\overline{\Phi}_{gg}(j\omega) = \overline{\Phi}_{ff}(j\omega)|\overline{H}(j\omega)|^2 = |\overline{H}(j\omega)|^2 \tag{15.6}$$

that is, in this case the *power spectrum of the colored random function is equal to the squared magnitude of the filter spectrum.* The total power in the simulated function can be adjusted simply by multiplying $\overline{H}(j\omega)$ by a constant.

Example 15.2

Generate a random sequence with power density = 1 having all of its power concentrated at frequencies between 4 and 12 Hz. Use the sampling interval $T = 0.01$ s. This situation calls for the bandpass filter illustrated in Fig. 15.7. In a general-purpose computer, the process begins with the generation of independent random samples $[r_m]$ on the interval $(0, 1)$ in a library subroutine. Each sample r_m is then converted to f_m as in Eq. 15.5, and then f_m is fed through the bandpass filter to produce eventually the desired sample set, $[g_m]$.

Using the bandpass filter routine SPFIL2 in Appendix B with five sections in cascade (see Section 12.6), the details of the sequence generation process are illustrated in Fig. 15.8. The program causes a white uniform sequence to be generated and then filtered to produce the sequence $[g_m]$, which has the desired spectral property. Note that SPFIL2 is called with frequencies 0.04 and 0.12 Hz-s, and with NS = 10 two-pole sections.

The entire sequence $[g_m]$, a random sequence with all of its power concentrated between 4 and 12 Hz, is illustrated in Fig. 15.9 with straight lines connecting the sample points. Since the frequency units here are hertz and not radians per second, the total power is the integral of the righthand spectrum in Fig. 15.7, or 16, and so the rms value of g_m should be $\sqrt{16} = 4$. Note that this value, and thus the averaging power in g_m, is reduced slightly by the "startup" of g_m which results from energy storage in the digital filter at the beginning of the filtering process.

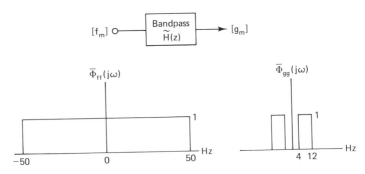

Figure 15.7 Random sequence generation in Example 15.2.

Random Sequences and Spectral Estimation   Chap. 15

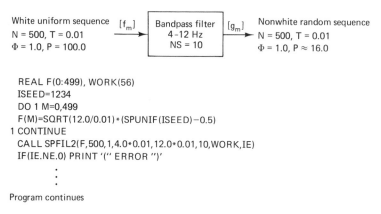

```
REAL F(0:499), WORK(56)
ISEED=1234
DO 1 M=0,499
F(M)=SQRT(12.0/0.01)*(SPUNIF(ISEED)−0.5)
1 CONTINUE
CALL SPFIL2(F,500,1,4.0*0.01,12.0*0.01,10,WORK,IE)
IF(IE.NE.0) PRINT '('' ERROR '')'
    .
    .
    .
Program continues
```

Figure 15.8 Generating the nonwhite random sequence in Example 15.2.

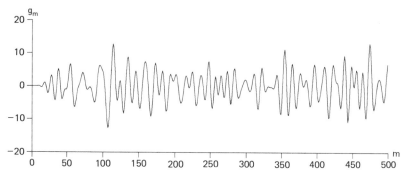

Figure 15.9 Random sequence with power distributed uniformly from 4 to 12 Hz; $N = 500$, $T = 0.01$.

15.4 POWER SPECTRAL ESTIMATES

Estimating the amplitude or power spectrum of a random function is an extensive subject within itself, involving some rather advanced points in probability and statistics. The list of references contains complete texts on this subject. In this section some practical methods for estimating the power spectrum using sampled data are discussed.

The classical theory of spectral estimation, both analog and digital, is given by Blackman and Tukey (1958). They discuss the question of how detailed and accurate the spectral estimate can be for a given amount of data, a question addressed again by Jenkins and Watts (1968a) and other authors as well. The Blackman–Tukey answer to this question is expressed in terms of R_x, the "x-percent probability" range, measured in dB and containing the estimated and actual spectral values with x-percent probability. (The "actual" spectral value can be viewed as the value obtained from a very long stationary record.) For example, if $R_{90} = 3$ dB, the spectral estimate for a given frequency band will, with probability 0.9, fall within a 3-dB interval around the actual spectral value for that same frequency band.

The range R_x is shown by Blackman and Tukey to depend on N/M, the ratio of the total number of samples to the number of frequency bands, that is, equal slices of the *positive* ω axis from 0 to π/T. Assuming that the set of N samples is treated as a single "piece" of data [see Blackman and Tukey (1958)], the approximate relation is

$$R_x \approx K(N/M - 0.833)^{-1/2};$$

x	80	90	96	98
K	11.2	14.1	17.7	20.5

(15.7)

so that $K = 11.2$ for the 80% range, and so on. The range R_x is plotted as a function of N/M in Fig. 15.10. As N/M becomes large, the range given by the figure or by Eq. 15.7 becomes approximately centered around the actual spectral value.

The curves in Fig. 15.10 give an answer to the question of why one cannot compute the power spectrum of a random sequence simply by taking the squared magnitude of its DFT. In this case M would equal $N/2$; therefore, N/M would be 2, and the confidence interval (i.e., the range R_x centered approximately around the estimated spectral value) about each computed DFT value would be very large. Clearly one would gain very little knowledge of the actual power spectrum without "smoothing" the computed DFT in some manner. It is this concept of smoothing the

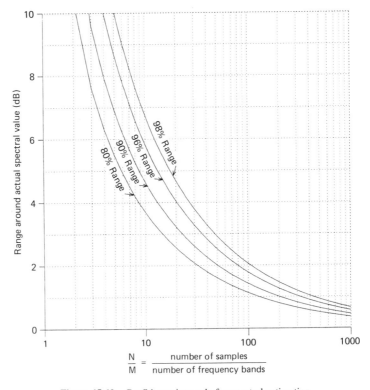

Figure 15.10 Confidence intervals for spectral estimation.

Random Sequences and Spectral Estimation   Chap. 15

spectral estimate plus a desire to limit the computing time for very long records that has led to the modern methods of spectral estimation.

Before proceeding to estimate the spectrum, in addition to choosing the number of frequency bands as described above, some precautions with the data itself are desirable [see Bingham et al. (1967)]

1. The estimate will be made as if the time series is *stationary;* therefore changes in the frequency content over the duration of the data record can give misleading results.

2. It is advisable to eliminate the mean value as well as any significant "trends" (straight-line changes) in the data before making the estimate. (If appropriate, the data can be highpassed to accomplish this.)

3. By multiplying the ends of the measured time series by a smoothing function, say a half-cosine bell extending into about one-tenth of the record, one can produce a smooth transition to zero at the ends of the sample set and reduce spurious ripples in the neighborhood of spectral peaks, as described in Chapter 8.

Regarding items 1 and 2, note that a time series generated as in Example 15.2, if long enough, will have close to zero mean and trend because components near zero frequency have been rejected by the bandpass filter.

15.5 DEMODULATION AND COMB-FILTER METHODS OF SPECTRAL ESTIMATION

One method of spectral estimation involves *complex demodulation* (Bingham, Godfrey, and Tukey (1967)). In this method the time series $f(t)$ is multiplied by a sinusoidal carrier $e^{-j\beta t}$. This is ordinary amplitude modulation and has the effect of shifting components around frequency β in $f(t)$ to around zero frequency in the demodulated time series. Then, by averaging the power output of a lowpassed version of $f(t)e^{-j\beta t}$, the power in $f(t)$ in the band of frequencies around β can be estimated.

The *comb filter* method illustrated in Fig. 15.11 is obviously quite similar to complex demodulation and is used now for purposes of illustration. The only difference is that here bandpassing replaces demodulation plus lowpassing. Each tooth of the comb is a bandpass filter whose output is squared and averaged to get the total power within the passband (at negative as well as positive frequencies). Each output is divided by the bandwidth $\Delta\nu$ so that the units of Φ are *squared magnitude per hertz.* Altogether there are M estimated spectral values, Φ_1 through Φ_M in Fig. 15.11. Thus the power spectrum is produced in the form of a histogram with M bars. Note that this is a "positive power spectrum," that is, as remarked above, each Φ_k represents combined power from negative and positive frequencies.

In Fig. 15.11 it is assumed that the spectral estimate is to be based on N samples of $f(t)$, so that each tooth output is squared, summed, and then divided by N to get the spectral estimate. Each sum should include all of the nonzero filter output. Thus there may be more than N terms in each sum to allow for any residual energy stored in the filter after the N samples of $f(t)$ have been inserted. (This is equivalent to defining $f_m = 0$ for $m > N$ and then letting m go from 0 to ∞ in each sum in Fig. 15.11.)

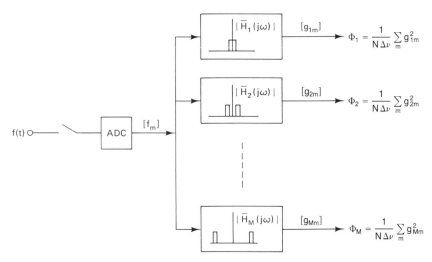

Figure 15.11 Comb filter for estimating the power spectrum, illustrating the effect of complex demodulation. Width of each passband $= \Delta \nu$ hertz. Summations go from $m = 1$ to N.

For computational efficiency, demodulation plus lowpassing is generally preferable to bandpassing. The important feature of either method, however, is that it is easy to vary N, the number of terms used to estimate Φ in Fig. 15.11. Thus one can adjust N to make either *local* estimates of a changing power spectrum or *global* estimates of a spectrum assumed to be stationary. Also, in contrast with the next method to be discussed, it is easy here to look only at certain portions of the spectrum, that is, to omit some of the passbands in Fig. 15.11, and also to use passbands of varying widths, thus providing varying degrees of spectral averaging.

Example 15.3

Measure the power spectrum of the sequence generated as in Fig. 15.7, using $T = 0.01$ s and $N = 5000$ samples of the random sequence. If the scheme in Fig. 15.11 is used with $\Delta \nu = 1$ Hz, then, since the Nyquist frequency is 50 Hz in this case, the total number of bands must be $M = 50$. Thus $N/M = 100$ and Fig. 15.10 gives the 90% range at about 1.42 dB. A typical result is shown in Fig. 15.12. As expected, the computation shows most of the power to be concentrated from 4 to 12 Hz, at around 2 squared magnitude units per hertz (i.e., 3 dB), all power being at positive frequencies as explained above. A minor point is that the *actual* spectrum is not quite rectangular in this case, the sequence having been generated by the digital filter shown in Fig. 15.8. The actual power spectrum is down to 1% of (or 20 dB below) its maximum at around 3.5 Hz and around 13.5 Hz, rather than 4 and 12 Hz exactly.

15.6 PERIODOGRAM METHODS OF SPECTRAL ESTIMATION

Other methods for estimating the power spectra of sampled signals are sometimes called periodogram methods. These methods are effectively the same as those above, but differ in procedure. (See Exercise 16 at the end of this chapter.) They are motivated by the definition of the power spectrum in Eq. 13.63. The approach is basically to estimate the autocorrelation function at a number of lags and then transform the resulting sequence. In this section we discuss several periodogram methods. They are compared using the standard measures of performance for statistical estimators.

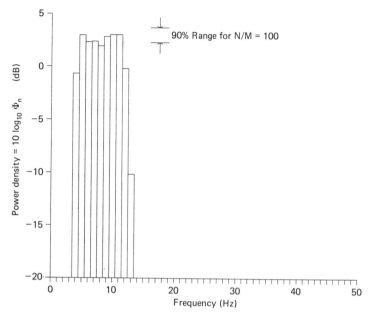

Figure 15.12 Computed power spectrum in Example 15.3; $N = 5000$. All power at positive frequencies.

The two quantities most often used in measuring the performance or accuracy of a statistical estimate are *bias* and *variance*. In the context of spectral estimators these quantities are defined

$$\text{bias} = B(\hat{\overline{\Phi}}(j\omega)) = \overline{\Phi}(j\omega) - E[\hat{\overline{\Phi}}(j\omega)] \tag{15.8}$$

and

$$\text{variance} = V(\hat{\overline{\Phi}}(j\omega)) = E[(\hat{\overline{\Phi}}(j\omega) - E[\hat{\overline{\Phi}}(j\omega)])^2] \tag{15.9}$$

where $\hat{\overline{\Phi}}(j\omega)$ denotes the estimate, $\overline{\Phi}(j\omega)$ the true power spectrum, and $E[\cdot]$ denotes the expected value. The variance portrays the same type of information about the estimate as the confidence intervals discussed in the preceding section. Estimates that have a large variance will also have a large confidence interval. It is desirable that both the bias and variance be as small as possible. An estimate is said to be *consistent* if, as the number of samples grows large, both the bias and the variance tend toward zero. Measures of bias and variance for the estimators discussed in this chapter are often signal-dependent and very difficult to obtain. In many cases the procedures required to obtain these measures are beyond the scope of this book. The results, however, are highly instructive, and will be presented here (mostly without derivation) to provide a basis for comparison. The derivation of these results can be found in the references.

The basis for all of the estimation methods discussed in this section is the periodogram. For an N-point data sequence $[x_m]$, the periodogram is defined to be

$$\overline{I}_N(j\omega) = \frac{1}{N} |\overline{X}_N(j\omega)|^2 \tag{15.10}$$

where

$$\overline{X}_N(j\omega) = \sum_{m=0}^{N-1} x_m e^{-jm\omega T} \tag{15.11}$$

It is easy to show that $\overline{I}_N(j\omega)$ is equivalent to the Fourier transform of a truncated autocorrelation estimate (see Exercise 15 at the end of this chapter). Specifically, if $\hat{\phi}_{xx}(l)$ is given by

$$
\begin{aligned}
\hat{\phi}_{xx}(l) &= \frac{1}{N} \sum_{m=0}^{N-l-1} x_m x_{m+l}; && 0 \le l < N \\
\hat{\phi}_{xx}(l) &= \hat{\phi}_{xx}(-l); && -N < l \le 0
\end{aligned}
\tag{15.12}
$$

then it can be shown that

$$\overline{I}_N(j\omega) = \sum_{l=-(N-1)}^{N-1} \hat{\phi}_{xx}(l) e^{-jl\omega T} \tag{15.13}$$

The periodogram itself can sometimes serve as a useful estimate of the power spectrum, although as mentioned in the preceding section, our confidence in this estimate is very poor. Let us derive expressions for the bias and variance of $\overline{I}_N(j\omega)$ starting first with the autocorrelation estimate. The bias of $\hat{\phi}_{xx}(l)$ is easily found using Eqs. 15.8 and 15.12,

$$
\begin{aligned}
B[\hat{\phi}_{xx}(l)] &= \phi_{xx}(l) - E\left[\frac{1}{N} \sum_{m=0}^{N-l-1} x_m x_{m+l} \right] \\
&= \phi_{xx}(l) - \frac{1}{N} \sum_{m=0}^{N-l-1} \phi_{xx}(l) \\
&= \phi_{xx}(l) \left[\frac{l}{N} \right]
\end{aligned}
\tag{15.14}
$$

Note that as $N \to \infty$ the bias approaches zero, so this estimate is *asymptotically unbiased*. An approximate expression for the variance of the correlation estimate is given by Jenkins and Watts (1974):

$$V[\hat{\phi}_{xx}(l)] \approx \frac{1}{N} \sum_{p=-\infty}^{\infty} [\phi_{xx}^2(p) + \phi_{xx}(p+1)\phi_{xx}(p-1)] \tag{15.15}$$

This expression holds when $N >> l$. Note that as $N \to \infty$, the variance approaches zero, so the correlation estimate is consistent.

For the periodogram the bias can be determined by taking the expected value of $\overline{I}_N(j\omega)$. Using the relationship in 15.13 and the result above, we find that

$$
\begin{aligned}
E[\overline{I}_N(j\omega)] &= E\left[\sum_{l=-(N-1)}^{N-1} \hat{\phi}_{xx}(l) e^{-jl\omega T} \right] \\
&= \sum_{l=-(N-1)}^{N-1} \left[\frac{N-|l|}{N} \right] \phi_{xx}(l) e^{-jl\omega T}
\end{aligned}
\tag{15.16}
$$

This is not equal to the true spectrum, $\overline{\Phi}(j\omega)$, because of the $(N - |l|)/N$ term and the finite duration of the summation. However, as $N \to \infty$, $E[\overline{I}_N(j\omega)]$ approaches $\overline{\Phi}(j\omega)$, so the estimate is asymptotically unbiased.

From Eq. 15.16 we see that $E[\overline{I}_N(j\omega)]$ can be interpreted as the Fourier transform of the true autocorrelation function multiplied by a window function of the form

$$w_l = \frac{N - |l|}{N}; \qquad |l| < N$$

This is the Bartlett window function described earlier in Chapter 8. Thus, the spectral estimate obtained using the periodogram can be viewed as the convolution of the true power spectrum with the spectral response of the Bartlett window, that is,

$$E[\overline{I}_N(j\omega)] = \frac{1}{2\pi} \int_{-\pi}^{\pi} \overline{\Phi}_{xx}(j\alpha)\overline{W}_{BT}(j(\omega - \alpha)) \, d\alpha \qquad (15.17)$$

where $\overline{W}_{BT}(j\omega)$ is given in Eq. 8.24. This convolution results in a smoothing of the true spectrum, and a rippling effect near the transitions known as Gibb's phenomenon, which was described in Chapter 8.

A general expression for the variance of the periodogram is difficult to obtain. However, if x_m is assumed to be a real, zero-mean, Gaussian process generated by passing white noise through a linear filter, Jenkins and Watts have shown

$$V[\overline{I}_N(j\omega)] = \overline{\Phi}_{xx}^2(j\omega)\left[1 + \left(\frac{\sin(N\omega T)}{N \sin \omega T}\right)^2\right] \qquad (15.18)$$

Note that this function does not decrease an N increases. This is illustrated in Fig. 15.13, where periodograms of a sinusoid plus white noise are shown for

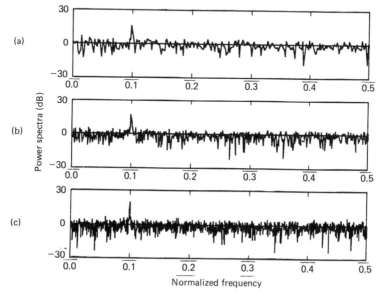

Figure 15.13 Periodograms of a sinusoid in white noise using (a) 512 data points, (b) 1024 data points, and (c) 2048 data points.

$N = 512$, 1024, and 2048. The variance of an estimate (which is a measure of the variation of the estimate around the true value of 0 dB) does not approach zero for larger values of N. Thus, the periodogram is not a consistent estimate.

A common method for reducing the variance is to average the estimate over several independent trials. The method of averaged periodograms which follows is often attributed to Bartlett. The N-length data sequence is partitioned into K blocks, each of length M (so $N = KM$). Members of the ith block are denoted

$$x^i_m = x_{m+(i-1)M}; \qquad 0 \leq m \leq M - 1, \quad 1 \leq i \leq K \qquad (15.19)$$

A periodogram is computed for each block,

$$\bar{I}^i_M(j\omega) = \frac{1}{M} \left| \sum_{m=0}^{M-1} x^i_m e^{-jm\omega T} \right|^2; \qquad 1 \leq i \leq K \qquad (15.20)$$

Then periodograms are averaged to produce a spectral estimate,

$$\hat{\Phi}_{xx}(j\omega) = \frac{1}{K} \sum_{i=1}^{K} \bar{I}^i_M(j\omega) \qquad (15.21)$$

If we assume that the periodograms $\bar{I}^i_M(j\omega)$ are independent, which is a reasonable assumption if $\phi_{xx}(l)$ is small for $l > M$, then the expected value of the estimate is given by

$$E[\hat{\Phi}_{xx}(j\omega)] = \frac{1}{K} \sum_{i=1}^{K} E[\bar{I}^i_M(j\omega)]$$

$$= E[\bar{I}_M(j\omega)] \qquad (15.22)$$

$$= \sum_{l=-(M-l)}^{M-l} \left[\frac{M - |l|}{M} \right] \phi_{xx}(l) e^{-jl\omega T}$$

which is similar to the expression in Eq. 15.16 except that $M < N$. The variance is approximately (Jenkins and Watts, 1974)

$$V[\hat{\Phi}_{xx}(j\omega)] = \frac{1}{K} V[\bar{I}_M(j\omega)]$$

$$= \frac{1}{K} \overline{\Phi}^2_{xx}(j\omega) \left[1 + \left(\frac{\sin(M\omega T)}{M \sin(\omega T)} \right)^2 \right] \qquad (15.23)$$

As $K \to \infty$ the variance approaches 0, so that by averaging periodograms we have obtained a consistent estimate. This is illustrated in Fig. 15.14. Here the periodogram of a sinusoid-plus-white-noise sequence of length $N = 4096$ is compared with an estimate obtained by averaging eight periodograms which are computed from eight distinct blocks of length $M = 512$. The reduction in variance is obvious.

Although averaging has decreased the variance, it has also caused an *increase* in bias as shown in Eq. 15.16, and a *decrease* in spectral resolution. The latter effect is associated with the increased width of the main lobe of the window spectrum, resulting from the reduction of the number of data points (from N to M) in the computation of the periodogram. It is reflected in the broadening of the sinusoidal spike in Fig. 15.14.

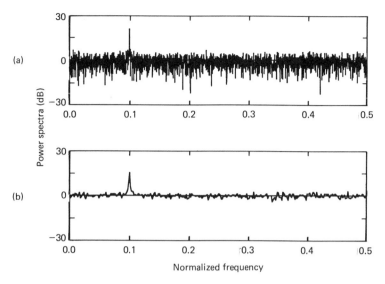

Figure 15.14 (a) Periodogram of a 4096 point data sequence, and (b) spectral estimate formed by averaging 8 periodograms of length 512.

We have shown that by averaging periodograms, the variance of the spectral estimate can be reduced at the expense of an increase in the bias and a decrease in spectral resolution. Now we will show that windowing can be used to achieve essentially the same effect. The basic idea is to compute a single periodogram using the entire N-length data sequence, then convolve the periodogram with a window function to obtain the estimated power spectrum:

$$\hat{\overline{\Phi}}_{xx}(j\omega) = \frac{1}{2\pi} \int_{-\pi}^{\pi} \overline{I}_N(j\alpha)\overline{W}_M(j(\omega - \alpha)\,d\alpha \tag{15.24}$$

where $\overline{W}_M(j\omega)$ represents the spectrum of a data window of length $2M - 1$ with $M < N$. Equivalently, we can compute the autocorrelation estimate in Eq. 15.12, multiply by the window sequence, and then transform, as follows:

$$\hat{\overline{\Phi}}_{xx}(j\omega) = \sum_{l=-(M-1)}^{M-1} \hat{\phi}_{xx}(l)w_l e^{-jl\omega T} \tag{15.25}$$

This is known as the Blackman–Tukey spectral estimator. With random signals, it is accurate when the window sequence of length $2M - 1$ is long enough to cover essentially all of the nonzero portions of the autocorrelation function.

To study the effect of windowing on the bias of the spectral estimate we examine the expected value of Eq. 15.24, which is

$$E[\hat{\overline{\Phi}}_{xx}(j\omega)] = \frac{1}{2\pi} \int_{-\pi}^{\pi} E[\overline{I}_N(j\alpha)]\overline{W}_M(j(\omega - \alpha))\,d\alpha$$

$$= \frac{1}{4\pi^2} \int_{-\pi}^{\pi} \int_{-\pi}^{\pi} \overline{\Phi}_{xx}(j\beta)\overline{W}_{BT}(j(\alpha - \beta))\overline{W}_M(j(\omega - \alpha))\,d\beta\,d\alpha \tag{15.26}$$

which is a double convolution involving the Bartlett window spectrum (which arises naturally due to the finite number of data points in the autocorrelation estimate), and the window spectrum of our choice. If M is small compared to N, then the main lobe of $\overline{W}_M(j\omega)$ will be wide compared to $\overline{W}_{BT}(j\omega)$, and the expression above can be approximated as

$$E[\hat{\overline{\Phi}}_{xx}(j\omega)] \approx \frac{1}{2\pi} \int_{-\pi}^{\pi} \overline{\Phi}_{xx}(j\alpha)\overline{W}_M(j(\omega - \alpha))\, d\alpha \qquad (15.27)$$

This result should be compared with Eq. 15.17 for the unwindowed periodogram. As with the previous method, we note an increase in the bias of the estimate due to the increase in the width of the main lobe of the window function.

An approximate expression for the variance of $\hat{\overline{\Phi}}_{xx}(j\omega)$ is given by Oppenheim and Schafer (1975). If we assume that the length $(2M - 1)$ of the window function is such that $\overline{W}_M(j\omega)$ is narrow with respect to variations in $\overline{\Phi}(j\omega)$, while at the same time wide compared to the function $(\sin[\omega TN/2]/\sin[\omega T/2])^2$, then the variance can be approximated by

$$V[\hat{\overline{\Phi}}_{xx}(j\omega)] \approx \frac{1}{2\pi N} \overline{\Phi}_{xx}^2(j\omega) \int_{-\pi}^{\pi} \overline{W}_M^2(j\alpha)\, d\alpha \qquad (15.28)$$

This expression should be compared with Eq. 15.18 for the periodogram. For large N, the expression in Eq. 15.18 is approximately $\overline{\Phi}_{xx}^2(j\omega)$. The ratio of Eq. 15.28 to this approximation is called the *variance ratio:*

$$\text{variance ratio} = \frac{1}{2\pi N} \int_{-\pi}^{\pi} \overline{W}_M^2(j\omega)\, d\omega \qquad (15.29)$$

To obtain a reduction in variance, the variance ratio must be less than 1. As seen in Table 15.1, with the proper selection of M, the ratio can be made to be less than 1 and a reduction in variance can be obtained. A reduction in variance is illustrated in Fig. 15.15, which shows the periodogram of a 4096-point sequence (sinusoid plus white noise as before), formed by transforming the autocorrelation estimate, compared with the spectral estimate obtained using the Blackman–Tukey method with a Hanning window of length 1025.

TABLE 15.1. Variance Ratio for Some Popular Window Functions

Window	Approximate Variance Ratio $\dfrac{1}{2\pi N} \displaystyle\int_{-\pi}^{\pi} \overline{W}_M^2(j\omega)\, d\omega$
Rectangular	$\dfrac{2M}{N}$
Bartlett	$\dfrac{2M}{3N}$
Hanning	$\dfrac{3M}{4N}$
Hamming	$\dfrac{4M}{5N}$

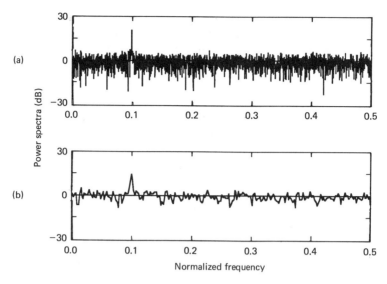

Figure 15.15 (a) Periodogram of a 4096-point data sequence, and (b) spectral estimate using Blackman–Tukey method with Hanning window of length 1025.

The method of Welch (1967) combines the concepts of averaging and windowing into a single unified approach. As in Bartlett's method, the data is partitioned into K blocks of length M. Then, however, the window function is applied directly to each data block. For each block, a modified periodogram is formed as follows:

$$\overline{P}^i_M(j\omega) = \frac{1}{MU} \left| \sum_{n=0}^{M-1} x^i_n w_n e^{-jn\omega T} \right|^2; \qquad i = 1, 2, \ldots, K \qquad (15.30)$$

where

$$U = \frac{1}{M} \sum_{n=0}^{M-1} w_n^2 \qquad (15.31)$$

The factor U is a scaling factor equal to the average power in the window function. Then, as before, the spectral estimate is formed by averaging the modified periodograms:

$$\hat{\overline{\Phi}}_{xx}(j\omega) = \frac{1}{K} \sum_{i=1}^{K} \overline{P}^i_M(j\omega) \qquad (15.32)$$

The expected value of this spectral estimate is

$$E[\hat{\overline{\Phi}}_{xx}(j\omega)] = \frac{1}{2\pi} \int_{-\pi}^{\pi} \Phi_{xx}(j\alpha)\overline{W}(j(\omega - \alpha)\,d\alpha \qquad (15.33)$$

which is identical to the result in Eq. 15.17 for the periodogram except that

$$\overline{W}(j\omega) = \frac{1}{MU} \left| \sum_{n=0}^{M-1} w_n e^{-jn\omega T} \right|^2 \qquad (15.34)$$

is the now square of the window spectrum, normalized by the factor MU in Eq. 15.31. The normalizing factor, MU, is required for the estimate to be asymptotically unbiased. Also, when the data sequence is partitioned into nonoverlapping blocks, Welch (1967) has shown that the variance of the spectral estimate is

$$V[\hat{\bar{\Phi}}_{xx}(j\omega)] \approx \frac{\overline{\Phi}^2_{xx}(j\omega)}{K} \qquad (15.35)$$

Thus, the estimate is consistent.

Welch shows that for a fixed number, N, of data points, it is possible to increase the reduction in variance by overlapping the blocks. In this case the expression in Eq. 15.35 becomes

$$V[\hat{\bar{\Phi}}_{xx}(j\omega)] \approx A \frac{\overline{\Phi}^2_{xx}(j\omega)}{K} \qquad (15.36)$$

where A is a positive constant slightly greater than 1 which depends on the amount of overlap. The variance in Eq. 15.36 is generally less than that in Eq. 15.35 because there are more blocks and hence the value of K is larger when overlapping is used. Thus, we have seen that Welch's method is a consistent estimate that provides some control over leakage through the choice of a window function.

An example comparing Welch's method with the averaged periodogram method and the Blackman–Tukey method is presented in Fig. 15.16. In all three cases a total of 4096 data points are used. Both Welch's method and the averaged periodogram method partition the data into blocks of size 512. In Welch's method the blocks are

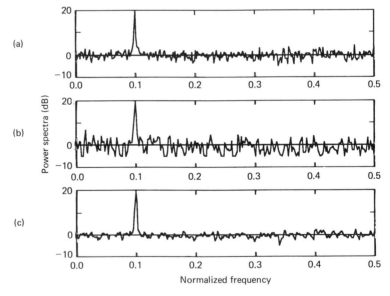

Figure 15.16 Spectral estimates of a 4096-point data sequence using (a) the averaged periodogram method with 8 blocks of 512 points, (b) the Blackman–Tukey method with a 1025-point Hanning window, and (c) Welch's method with blocks of size overlapped by 50% using a Hanning window.

overlapped by 50%. Both Welch's method and the Blackman–Tukey method use the Hanning window. Welch's method is seen in this example to be the best at reducing the variance and yet maintaining good spectral resolution.

15.7 PARAMETRIC METHODS OF SPECTRAL ESTIMATION

In this section we discuss *parameteric methods* of spectral estimation. These methods are distinctly different from those in the previous two sections in that they assume a specific model for the signal spectrum. That is, the signal spectrum is assumed to take on a specific functional form, the parameters of which are unknown. The spectral estimation problem is then one of estimating the unknown parameters of the model rather than one of estimating the spectrum itself. The spectrum is formed by substituting these parameters into the model.

The most popular models used in spectral estimation are finite parameter rational models of the form

$$\hat{\Phi}(j\omega) = \sigma^2 \left| \frac{1 + b_1 e^{-j\omega T} + b_2 e^{-j2\omega T} + \cdots + b_q e^{-jq\omega T}}{1 + a_1 e^{-j\omega T} + a_2 e^{-j2\omega T} + \cdots + a_p e^{-jp\omega T}} \right|^2 \qquad (15.37)$$

where σ^2 is a constant and the b_i and a_i coefficients are the $p + q$ unknown parameters to be estimated from the data. Equivalently, these models can be expressed

$$\hat{\Phi}(j\omega) = \sigma^2 |\tilde{H}(z)|^2_{z = e^{j\omega T}} \qquad (15.38)$$

where

$$\tilde{H}(z) = \frac{\tilde{B}(z)}{\tilde{A}(z)} \qquad (15.39)$$

and

$$\tilde{B}(z) = 1 + b_1 z^{-1} + b_2 z^{-2} + \cdots + b_q z^{-q}$$
$$\tilde{A}(z) = 1 + a_1 z^{-1} + a_2 z^{-2} + \cdots + a_p z^{-p} \qquad (15.40)$$

We can see that this model is representative of the spectrum of a discrete-time random process generated by passing white noise with variance σ^2 through a linear filter as shown in Fig. 15.17. This is precisely the method discussed previously in Section 15.3 for generating colored random processes. $\tilde{H}(z)$ is often called a "shaping" filter because it determines the spectral shape of the random process. The difference equation describing the input–output relation for this process is given by

$$x_n = w_n + b_1 w_{n-1} + b_2 w_{n-2} + \cdots + b_q w_{n-q} - a_1 x_{n-1}$$
$$- a_2 x_{n-2} - \cdots - a_p x_{n-p} \qquad (15.41)$$

where $[w_n]$ is a white noise process with variance σ^2.

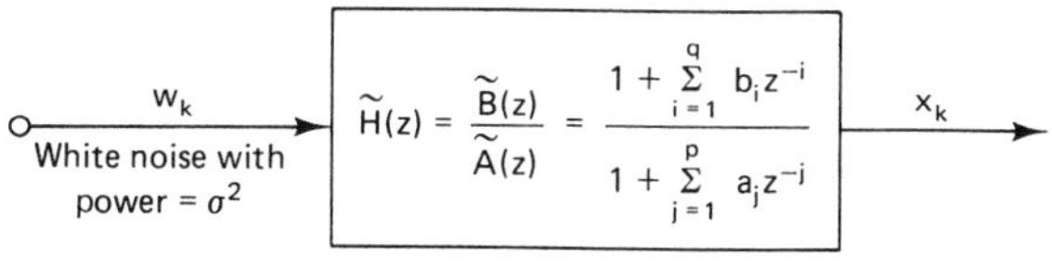

Figure 15.17 Finite-parameter signal model for stationary random process x_k.

Three different types of spectral models can be derived from Eq. 15.37, depending on the nature of the numerator and denominator coefficients. Models for which the a_i coefficients are zero are called moving average (MA) models. Those for which the b_i coefficients are zero are called autoregressive (AR) models. Models for which neither are zero (some of both are nonzero) are called autoregressive moving-average (ARMA) models. In practice, the choice of model should be based on prior knowledge of the manner in which the signal is generated. Unfortunately, we do not often have that knowledge. However we do often know the basic spectral shape of the signal, and this knowledge goes a long way toward helping us select a good signal model. To see this, let us look more closely at each of the three models.

MA models are best able to represent signals with spectral notches. They are called "all-zero" models because the shaping filter, $\tilde{H}(z) = \tilde{B}(z)$, is an all-zero FIR filter. They are designed or adjusted to approximate the shape of the true spectrum by properly placing the zeros of $\tilde{H}(z)$ in the complex plane. A notch in the signal spectrum can be formed by placing a pair of zeros near the unit circle at the appropriate angles, as illustrated in Fig. 15.18. Broadband signals can also be modeled with an MA filter, but the combination of broadband and notch spectral characteris-

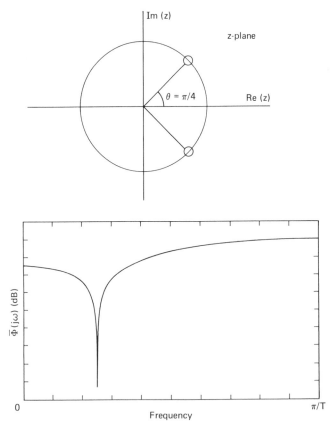

Figure 15.18 Representing a notch in the signal spectrum using a MA model with one zero pair on the unit circle.

Random Sequences and Spectral Estimation   Chap. 15

tics is more difficult to model in this fashion. Sharp spectral peaks are also difficult to form with an all-zero MA model. Several zeros are required to model a single spectral peak accurately, as illustrated in Fig. 15.19.

AR models are best able to represent narrowband signals. They are called "all-pole" models because $\tilde{H}(z) = 1/\tilde{A}(z)$ is an all-pole filter. In this case the shape of the true spectrum is approximated by properly placing the poles of $\tilde{H}(z)$ in the complex plane. These models can easily approximate a narrowband signal by placing a pole near the unit circle at the appropriate angle, as illustrated in Fig. 15.20. Broadband signals can also be modeled with an AR filter but, as with MA models, the combination of broadband and narrowband components is more difficult to model and generally requires more coefficients. Sharp notches are also difficult to form with an all-pole model. Several poles are required to accurately model a single notch, as shown in Fig. 15.21.

ARMA models (pole–zero models) are the most generally applicable models. They are capable of characterizing virtually all types of signals. However, the coefficients of the ARMA models are the most difficult to obtain.

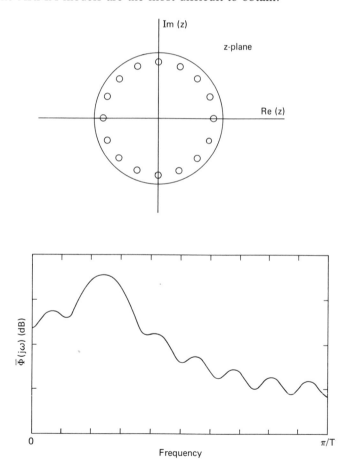

Figure 15.19 Using a MA model to represent a narrowband signal.

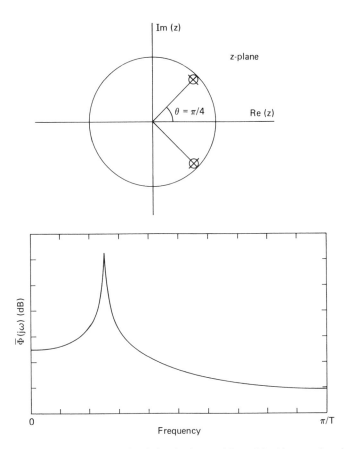

Figure 15.20 Representing a narrowband signal using an AR model with one pole pair near the unit circle.

In practice, one's choice of spectral model is often governed by the amount of computation required to form the estimate and the robustness of the method employed, rather than by the nature of the signal itself. The AR models are the easiest to obtain. The AR parameter estimates are determined by solving a set of simultaneous linear algebraic equations, as we shall see. The MA models are more difficult to obtain. Several methods are available for computing MA model parameters, one of which involves finding the roots of a qth order polynomial, a task which can be very difficult even for low-order polynomials. ARMA models are even more difficult to obtain. Again, a variety of methods is available for computing the model parameters. One approach uses nonlinear optimization algorithms, which are complicated and at times unstable, to search for the best set of parameters.

Since we are inclined to prefer simple and robust methods, the model that is used may not be representative of the true manner in which the signal was generated. Therefore it is important to note that any ARMA or MA model can be represented by an AR model of sufficient (possibly infinite) order, and similarly that any ARMA or AR process can be represented by an MA model of sufficient (possibly infinite) order. These properties assure us that a reasonable spectral model can be obtained

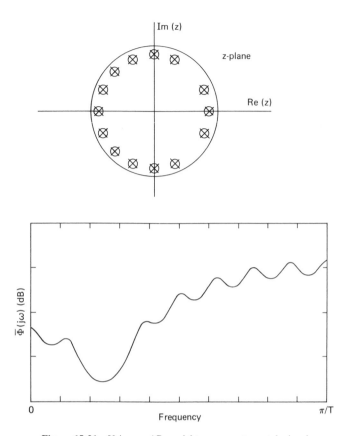

Figure 15.21 Using an AR model to represent a notch signal.

using either of the simpler methods, AR or MA, as long as we choose the parameter size large enough.

The key to solving for the parameters in the AR, MA, and ARMA models lies in the Yule–Walker equations of Yule (1927) and Walker (1931), which we now derive. If we set $a_0 = b_0 = 1$, Eq. 15.41 can be rewritten

$$\sum_{i=0}^{p} a_i x_{n-i} = \sum_{j=0}^{q} b_j w_{n-j} \tag{15.42}$$

Multiplying both sides by x_{n-l} and taking expected values yields

$$\sum_{i=0}^{p} a_i E[x_{n-l} x_{n-i}] = \sum_{j=0}^{q} b_j E[x_{n-l} w_{n-j}]$$

or

$$\sum_{i=0}^{p} a_i \phi_{xx}(l - i) = \sum_{j=0}^{q} b_j \phi_{xw}(l - j) \tag{15.43}$$

Sec. 15.7   Parametric Methods of Spectral Estimation   **393**

From Eq. 13.79 we know that the cross-correlation function $\phi_{xw}(k)$ can be written

$$\phi_{xw}(k) = \phi_{wx}(-k) = \sum_{m=-\infty}^{\infty} h_m \phi_{ww}(-k - m) \tag{15.44}$$

where h_m is the impulse response of the shaping filter $\tilde{H}(z)$. We assume that $\tilde{H}(z)$ is causal so that h_m is zero for $m < 0$. Since w_n is a white noise sequence, its autocorrelation function is given by

$$\phi_{ww}(i) = \sigma^2 \delta(i) \tag{15.45}$$

With this result, Eq. 15.44 becomes

$$\phi_{xw}(k) = \sigma^2 h_{-k} \tag{15.46}$$

Substituting Eq. 15.46 into Eq. 15.43 gives

$$\sum_{i=0}^{p} a_i \phi_{xx}(l - i) = \sigma^2 \sum_{j=0}^{q} b_j h_{j-l} \tag{15.47}$$

These equations over the variable l are known as the Yule–Walker equations. In spectral estimation the autocorrelation function $\phi_{xx}(l)$ is assumed to be known (actually, it is estimated from the data), and the unknowns are the a_i's, the b_i's, and the impulse response $[h_m]$ (which itself is a function of the a_i's and b_i's). As stated previously, solving for the ARMA coefficients (that is, both a_i's and b_i's) is difficult. The methods for ARMA solutions are beyond the scope of this text. Solving for the coefficients of an MA model, while not as difficult, also requires methods that we will not discuss here. Thorough treatments of these methods can be found in the references. For the AR model, however, a unique estimate can be generally found by solving a set of $p + 1$ linear equations in $p + 1$ unknowns. Thus, computationally, the AR method is by far the simplest of the three, and we will focus on this method in the remainder of this section.

The AR spectral model is formed by solving the Yule–Walker equations for the unknown parameters a_i with b_1 through b_q set to zero, and then substituting the solution into Eq. 15.37. Setting the b_i parameters to zero in Eq. 15.47 gives

$$\sum_{i=0}^{p} a_i \phi_{xx}(l - i) = \sigma^2 h_{-l} \tag{15.48}$$

We recall that h_m is causal so that $h_m = 0$ for $m < 0$. In addition we find that $h_0 = 1$ by using $a_0 = b_0 = 1$ in Eqs. 15.37 and 15.39. Thus, Eq. 15.48 can be written

$$\begin{aligned}
\sum_{i=0}^{p} a_i \phi_{xx}(l - i) &= \sigma^2; && l = 0 \\
&= 0; && l > 0
\end{aligned} \tag{15.49}$$

If we evaluate this equation for $l = 0, 1, \ldots, p$, we obtain $p + 1$ linear equations in $p + 1$ unknowns. (In practice, σ^2 is also an unknown.) Since the autocorrelation function $\phi_{xx}(l)$ is symmetric, these equations can be written in the following matrix form similar to Eq. 14.17:

$$
\begin{bmatrix}
\phi_{xx}(0) & \phi_{xx}(1) & \phi_{xx}(2) & \cdots & \phi_{xx}(p) \\
\phi_{xx}(1) & \phi_{xx}(0) & \phi_{xx}(1) & \cdots & \phi_{xx}(p-1) \\
\cdot & & \cdots & & \cdot \\
\cdot & & \cdots & & \cdot \\
\cdot & & \cdots & & \cdot \\
\phi_{xx}(p) & \phi_{xx}(p-1) & & \cdots & \phi_{xx}(0)
\end{bmatrix}
\cdot
\begin{bmatrix}
1 \\
a_1 \\
\cdot \\
\cdot \\
\cdot \\
a_p
\end{bmatrix}
=
\begin{bmatrix}
\sigma^2 \\
0 \\
\cdot \\
\cdot \\
\cdot \\
0
\end{bmatrix}
\qquad (15.50)
$$

Using matrix notation, we have

$$
\begin{bmatrix}
\phi_{xx}(0) & \mathbf{P}^T \\
\mathbf{P} & \mathbf{R}
\end{bmatrix}
\cdot
\begin{bmatrix}
1 \\
\mathbf{A}
\end{bmatrix}
=
\begin{bmatrix}
\sigma^2 \\
0
\end{bmatrix}
\qquad (15.51)
$$

where

$$
\begin{aligned}
\mathbf{A}^T &= [a_1 \quad a_2 \quad \cdots \quad a_p] \\
\mathbf{P}^T &= [\phi_{xx}(1) \quad \phi_{xx}(2) \quad \cdots \quad \phi_{xx}(p)]
\end{aligned}
\qquad (15.52)
$$

and

$$
\mathbf{R} =
\begin{bmatrix}
\phi_{xx}(0) & \phi_{xx}(1) & \phi_{xx}(2) & \cdots & \phi_{xx}(p-1) \\
\phi_{xx}(1) & \phi_{xx}(0) & \phi_{xx}(1) & \cdots & \phi_{xx}(p-2) \\
\cdot & & \cdots & & \cdot \\
\cdot & & \cdots & & \cdot \\
\cdot & & \cdots & & \cdot \\
\phi_{xx}(p-1) & \phi_{xx}(p-2) & & \cdots & \phi_{xx}(0)
\end{bmatrix}
\qquad (15.53)
$$

Carrying out the matrix multiplication in Eq. 15.51 above gives

$$
\phi_{xx}(0) + \mathbf{P}^T\mathbf{A} = \sigma^2 \qquad (15.54)
$$

$$
\mathbf{P} + \mathbf{R}\mathbf{A} = 0 \qquad (15.55)
$$

Solving Eq. 15.55 for the unknown coefficient vector $\mathbf{A}$ yields

$$
\mathbf{R}\mathbf{A} = -\mathbf{P} \quad \text{or} \quad \mathbf{A} = -\mathbf{R}^{-1}\mathbf{P} \qquad (15.56)
$$

The AR spectral estimate in Eq. 15.37 then takes the form

$$
\hat{\bar{\Phi}}(j\omega) = \frac{\phi_{xx}(0) + \mathbf{P}^T\mathbf{A}}{|1 + a_1 e^{-j\omega T} + a_2 e^{-j2\omega T} + \cdots + a_p e^{-jp\omega T}|^2} \qquad (15.57)
$$

In summary, the following steps are required to form the AR spectral estimate:

1. Use the data samples to compute an estimate of the autocorrelation function $\hat{\phi}_{xx}(l)$, for $0 \leq l \leq p$.
2. Use the autocorrelation estimates in Eq. 15.56 to solve for the AR coefficients. Methods for solving matrix equations of this type were discussed in Chapter 14.
3. Substitute the resulting AR coefficients into the expression in Eq. 15.57 to obtain the spectral estimate.

An example of an AR spectral estimate for a sinusoid in white noise is shown in Fig. 15.22. The estimate becomes more accurate as the number of parameters (AR coefficients) is increased, but even with a small number of parameters the nature of the signal spectrum is evident. When compared with the periodogram-based estimates in the previous section, the AR estimate is seen to be smoother, while the periodogram methods reveal more detail in the spectrum.

As noted previously, there is a strong similarity between steps 1 and 2 above and the method for solving the nonrecursive least-squares design problem discussed in Chapter 14. In fact, except for a difference in sign, the AR coefficients are identical to the least-squares solution for the weights of a one-step predictor (see Exercise 17). For this reason the AR method for computing the spectral estimate is also called the *linear prediction method*. This same spectral estimate can be derived using several different approaches, and it is therefore given several different names in the literature. The most common are the linear prediction method, the maximum entropy (ME) method, and the maximum likelihood (ML) method. Derivations that use the maximum entropy approach can be found in Burg (1975), Haykin (1979), Kay (1987), and Marple (1987), and those that use the maximum likelihood approach in Lim and Oppenheim (1988), Kay (1987), and Marple (1987).

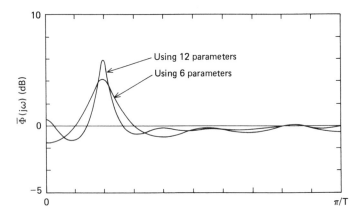

Figure 15.22 AR spectral estimates for a sinusoid in white noise.

It is difficult to obtain a general analytic expression for the statistical properties of the AR model. Kromer (1970), however, has studied its asymptotic properties and has shown that when p and N (the number of samples used to compute the correlation estimate) are sufficiently large and the true spectrum is reasonably smooth, the AR spectral estimate is asymptotically unbiased. Furthermore, the variance has been shown to be

$$V[\hat{\hat{\Phi}}(j\omega)] = \frac{2p}{N}\hat{\hat{\Phi}}(j\omega)^2 \qquad (15.58)$$

This expression is seen to approach zero as N increases. Thus, like most of the periodogram methods in the preceding section, the AR model produces a consistent estimate.

EXERCISES

1. Sketch the autocorrelation function corresponding to the power spectrum in Fig. 15.2. Label the significant values.

2. Explain why a white noise function must have a mean value of zero.

3. What is the total power in a noise function $f(t)$ if the amplitude distribution $p(f)$ is uniform on $(-1, +1)$?

4. What is the total power in $f(t)$ if $p(f)$ is given by
   (a) $p(f) = \begin{cases} 0; & |f| \geq K \\ (K - |f|)/K^2; & |f| \leq K \end{cases}$
   (b) $p(f) = \frac{1}{2}e^{-|f|}$

5. A computer program generates independent, equally likely random numbers between 0 and 100.
   (a) Give an algorithm that converts these numbers into white noise samples of a function with power $P = 10$.
   (b) Sketch the amplitude distribution and the power spectrum, given a sampling interval of $T = 1$ ms.

6. When the white noise sample variance is set equal to $1/T$:
   (a) What is the effect on the power density spectrum?
   (b) What is the result when the white noise is filtered?

7. Given the sample distributions in Exercise 4(a) and (b), state how the samples should be scaled so that the total white noise power is 1.

8. A computer generates independent, equally likely samples between 0 and 1. State a formula to convert these to white noise samples with unit power density. Assume a sampling interval of $T = 0.03$ s.

9. Show how to convert independent, normally distributed samples with $\mu = \sigma = 2$ into white noise samples with unit power density; sampling interval $= T$.

10. White noise samples $[f_m]$ having unit power density from $\omega = -\pi/T$ to $+\pi/T$ with $T = 0.5$ s are filtered as follows:

$$g_m = 10f_{m-1} + 0.8g_{m-1} - 0.16g_{m-2}$$

Sketch the power density spectrum $\overline{\Phi}_{gg}(j\omega)$.

11. Give the complete algorithmic design for a system with $T = 0.1$ s that produces a random sequence whose power spectrum is approximated by

$$\overline{\Phi}_{gg}(j\omega) = \frac{100}{\omega^2 + 1}$$

12. Suppose a sample array $[g_m]$ is stored in the Fortran array $G(M)$, $M = 1, \ldots, 5000$. Given $T = 10$ μs, write a Fortran program to estimate $\overline{\Phi}_{gg}(j\omega)$ in the bands 10–20 kHz and 20–30 kHz. Use subroutine SPFIL2 in Appendix B. Test using a white sequence.

13. In Exercise 12, what is the 96% confidence range on the spectral estimate?

14. As in Exercise 12, suppose again that $[g_m]$ is stored in $G(M)$, $M = 1, \ldots, 5000$, and $T = 10$ μs. Write and test a program to compute all values of Φ_n using the autocorrelation method, such that the 90% range is not more than one dB around the actual value.

15. Start with the definition of the periodogram in Eq. 15.10 and derive the result in Eq. 15.13. *Hint:* This problem not only requires a substitution of variables, but also a change in the range over which the summations are performed. This is most easily accomplished by rewriting the original expressions as infinite summations, making note of the fact that x_m is 0 outside the range $0 \leq m \leq N - 1$. Then, after the substitution of variables and rearrangement of terms, the summations can again be made finite to reflect only the nonzero terms.

16. In this exercise we show how a comb-filter spectral estimation is equivalent to the averaged periodogram method. Suppose that the comb filter method illustrated in Fig. 15.11 is implemented as shown below. The transfer functions $\tilde{G}_k(z)$ with $k = 1, \ldots, N$ are given by

$$\tilde{G}_k(z) = \frac{1}{1 - e^{j(2\pi k/N)}z^{-1}}$$

Note that this structure is reminiscent of the frequency sampling filter structure discussed in Section 12.6.

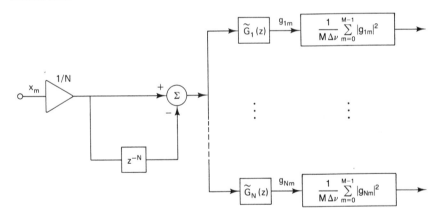

(a) Show that the transfer function from the input x_m to the kth filter output g_{km} is given by

$$\tilde{H}_k(z) = \frac{1}{N}\left(\frac{1 - z^{-N}}{1 - e^{j(2\pi k/N)}z^{-1}}\right)$$

and that this transfer function has a frequency response of the form

$$\overline{H}_k(j\omega) = A\left(\omega T - \frac{2\pi k}{N}\right)$$

in which

$$A(\omega T) = e^{-j[(N-1)/2]\omega T} \frac{\sin(N\omega T/2)}{N\,\sin(\omega T/2)}$$

(b) Sketch $|A(\omega T)|$ using $N = 11$ and verify that the magnitude response of $A(\omega T - 2\pi k/N)$ is that of a bandpass filter with unity gain at the center frequency $2\pi k/N$.

(c) The power density is formed by averaging the squared magnitude of the filter outputs and dividing by $\Delta\nu$ (the bandwidth of each filter) as shown in the figure. What is a reasonable value for $\Delta\nu$ in this implementation? Now, let the periodogram in Eq. 15.10 be evaluated at N discrete frequencies $\omega T = \pi k/N$, $k = 0, 1, \ldots, N - 1$ so that

$$\bar{I}_N(k) = \frac{1}{N}\,|\bar{X}_N(k)|^2$$

where

$$\bar{X}_N(k) = \sum_{m=0}^{N-1} x_m e^{-j(\pi mk/N)}$$

Consider Bartlett's method, which forms the spectral estimate by averaging periodograms. If we can show that $\bar{I}_N(k)$ is in some sense equivalent to $|g_{km}|^2/\Delta\nu$, then the method of averaged periodograms will be equivalent to the comb-filter method described above. The expression for $\bar{X}_N(k)$ can be viewed as a complex demodulation followed by a lowpass filter.

(d) Show that the spectrum of the demodulated signal

$$f_{km} = x_m e^{-j(\pi mk/N)}$$

is simply a frequency-shifted version of the spectrum of x_m, where frequencies previously centered around $\pi k/N$ in x_m are now centered around 0 (or dc) in f_{km}, that is,

$$\bar{F}_k(j\omega) = \bar{X}\left(j\left(\omega + \frac{\pi k}{NT}\right)\right)$$

(e) Show that the summation used to form $\bar{X}_N(k)$

$$\bar{X}_N(k) = \sum_{m=0}^{N-1} f_{km}$$

can be viewed as an Nth-order FIR filtering operation (with all filter coefficients equal to 1) with frequency response equal to $NA(\omega T)$.

(f) Use these results to argue that the periodogram is equivalent to $N|g_{km}|^2/\Delta\nu$, and that the two methods of spectral estimation are essentially the same.

17. Show that the least-squares solution (in Eq. 14.23) for the N weights of a one-step predictor [Fig. 14.1(b) with $\Delta = 1$] is identical (except in sign) to the solution for the coefficients of an Nth-order AR spectral model (Eq. 15.56). Further, show that the $(N + 1)$st parameter of the AR model, σ^2 in Eq. 15.54, is equal to the minimum mean-squared error of the linear predictor (Eq. 14.24). Suppose that the AR model provided by the linear predictor is sufficient to describe the spectral characteristics of the input signal. What can you say about the spectral characteristics of the error output of the predictor?

18. A discrete-time random process is generated by passing white noise through a discrete-time linear filter according to the expression

$$x_k = 0.8x_{k-1} + 0.2w_k$$

where w_k is a unit-variance white noise sequence.

(a) Derive an expression for the power spectral density of x_k, $\overline{\Phi_{xx}}(j\omega)$, and plot this function in dB versus frequency in Hz-s. Now generate 2048 samples of x_k using the expression above and SPUNIF for w_k. Compute and plot the power spectral estimate of x_k using each of the following methods:

(b) Periodogram

(c) Bartlett's method of averaging periodograms with a block size of 256

(d) The Blackman–Tukey method using a Hanning window (two-sided) of length 513

(e) Welch's method using a Hanning window, a block size of 256, and a 50% overlap of blocks

(f) The AR method with two coefficients

(g) The AR method with four coefficients

19. Repeat Exercise 18 for the case where the signal is generated according to:

(a) $x_k = -0.8x_{k-1} + 0.2w_k$

(b) $x_k = w_k - 1.27w_{k-1} - 0.81w_{k-2}$ (try using eight coefficients in the AR method)

20. Generate 512 samples of the sequence

$$x_k = \sin\left(\frac{k\pi}{4}\right) + w_k$$

where w_k is a unit-variance white noise sequence. Compute periodograms of length 128, 256, and 512. Plot your results (on both a regular and dB scale). How do they differ? Explain.

21. Repeat Exercise 20 with the sequence

$$x_k = \sin\left(\frac{k\pi}{8}\right) + w_k$$

How do these results differ from those in Exercise 20? Explain.

22. Generate 1024 samples of the sequence

$$x_k = A_1 \sin\left(\frac{k\pi}{5}\right) + A_2 \sin\left(\frac{3k\pi}{10}\right) + w_k$$

where w_k is a unit-variance white noise sequence and $A_1 = A_2 = 10$. Compute and plot the AR spectral estimate using:

(a) Four coefficients

(b) Six coefficients

(c) Eight coefficients

Based on these results what can you say about the ability of the AR spectral estimator to resolve multiple sinusoids?

23. Repeat Exercise 22 with $A_1 = 10$ and $A_2 = 5$.

24. Repeat Exercise 22 with the variance of the noise equal to 10.

SOME ANSWERS

1. $\phi(0) = P, \phi(\tau) = 0$ at $\tau = n\pi/\omega_m$ **2.** See Section 15.2   **3.** $\sigma_f^2 = \frac{1}{3}$
4.(b) $\sigma_f^2 = 2$ **5.(a)** $f_n = \sqrt{0.012}\,(r_n - 50)$ **8.** $f_n = 20(r_n - 1/2)$
9. $f_n = (r_n - 2)/(2\sqrt{T})$ **11.** *Hint:* See Fig. 15.6. Treat $H(j\omega)$ as a Butterworth filter.

REFERENCES

BINGHAM, C., GODFREY, M. D., and TUKEY, J. W., Modern Techniques of Power Spectrum Estimation. *IEEE Trans. Audio Electroacoust.* (Special Issue on Fast Fourier Transform and Its Applications to Digital Filtering and Spectral Analysis), Vol. AU-15, June 1967, p. 56.

BLACKMAN, R. B., and TUKEY, J. W., *The Measurement of Power Spectra.* New York: Dover, 1958.

BURG, J. P., *Maximum Entropy Spectral Analysis,* Ph.D. thesis. Department Geophysics, Stanford University, Stanford, Calif., May 1975.

CADZOW, J. A., Spectral Estimation: An Overdetermined Rational Model Equation Approach. *Proc. IEEE,* Vol. 70, September 1982, p. 907.

CHILDERS, D. G., and PAO, M.-T., Complex Demodulation for Transient Wavelet Detection and Extraction. *IEEE Trans. Audio Electroacoust.,* Vol. AU-20, No. 4, October 1972, p. 295.

DORIAN, L. V., *Digital Spectral Analysis,* Thesis, AD861229. Monterey, Calif.: Naval Post-Graduate School, September 1968.

EDWARD, J. A., and FITELSON, M. M., Notes on Maximum-Entropy Processing. *IEEE Trans. Inform. Theory,* Vol. IT-19, March 1973, p. 232.

HAYKIN, S. (ed.), *Topics in Applied Physics: Nonlinear Methods of Spectral Analysis.* New York: Springer-Verlag, 1979.

JENKINS, G. M., General Considerations in the Analysis of Spectra. *Technometrics,* Vol. 3, No. 2, May 1961, p. 133.

JENKINS, G. M., and WATTS, D. G., *Spectral Analysis.* San Francisco: Holden-Day, 1968a.

JENKINS, G. M., and WATTS, D. G., *Spectral Analysis and Its Applications.* San Francisco: Holden-Day, 1968b.

JENKINS, G. M., and WATTS, D. G., *Spectral Analysis of Time Series.* New York: Academic Press, 1974.

KAY, S., *Modern Spectral Estimation: Theory and Application.* Englewood Cliffs, N.J.: Prentice-Hall, 1987.

KROMER, R. E., *Asymptotic Properties of the Autoregressive Spectral Estimator,* Ph.D thesis. Department of Statistics, Stanford University, Stanford, Calif., 1970.

LANING, J. H., JR., and BATTIN, R. H., *Random Processes in Automatic Control.* New York: McGraw-Hill, 1956.

LIM, J. S., and OPPENHEIM, A. V. (eds.), *Advanced Topics in Signal Processing.* Englewood Cliffs, N.J.: Prentice-Hall, 1988.

MAKHOUL, J., Linear Prediction: A Tutorial Review. *Proc. IEEE,* Vol. 63, April 1975, p. 561.

MARPLE, S., *Digital Spectral Analysis with Applications.* Englewood Cliffs, N.J.: Prentice-Hall, 1987.

OPPENHEIM, A. V., and SCHAFER, R. W., *Digital Signal Processing.* Englewood Cliffs, N.J.: Prentice-Hall, 1975.

PARZEN, E., Mathematical Considerations in the Estimation of Spectra. *Technometrics,* Vol. 3, No. 2, May 1961, p. 167.

RICHARDS, P. I., Computing Reliable Power Spectra. *IEEE Spectrum,* January 1967, p. 83.

WALKER, G., On Periodicity in Series and Related Terms. *Proc. Roy. Soc.,* Vol. A131, 1931, p. 518.

WELCH, P. D., A Direct Digital Method of Power Spectrum Estimation. *IBM J. Res. Develop.*, Vol. 5, April 1961, p. 141.

WELCH, P. D., The Use of Fast Fourier Transform for the Estimation of Power Spectra: A Method Based on Time Averaging over Short, Modified Periodograms. *IEEE Trans. Audio Electroacoust.* (Special Issue on Fast Fourier Transform and Its Applications to Digital Filtering and Spectral Analysis), Vol. AU-15, June 1967, p. 70.

YULE, G. U., On a Method of Investigating Periodicities in Distributed Series, with Special Reference to Wolfer's Sunspot Numbers. *Philos. Trans.,* Vol. A226, 1927, p. 267.

Appendix A

Laplace and z-Transforms

The first part of this table contains operational transform pairs, such as the transforms of the derivative of $f(t)$. The second contains a list of explicit transforms. An important point, discussed in Chapter 7, is that $f(t)$ is always zero in this table for $t < 0$. Thus, if the shifting theorems are applied, $f(t)$ no longer "starts" at $t = 0$. For example, if $\tilde{F}(z) = Tz/(z - 1)^2$ is multiplied by z^{-n}, then the modified version of $f(t)$ is a ramp function that is zero until $t = nT$ and equal to $(t - nT)$ thereafter.

Line	Laplace Transform	$f(t)$	z-Transform
A	$F(s) = \int_0^\infty f(t)e^{-st}\,dt$	$f(t)$	$\bar{F}(z) = \sum_{m=0}^\infty f_m z^{-m}$
B	$AF(s)$	$Af(t)$	$A\bar{F}(z)$
C	$F(s) + G(s)$	$f(t) + g(t)$	$\bar{F}(z) + \bar{G}(z)$
D	$sF(s) - f(0+)$	$\dfrac{d}{dt}f(t)$	—
E	$\dfrac{F(s)}{s}$	$\int_0^t f(\tau)\,d\tau$	—
F	$-\dfrac{d}{ds}F(s)$	$tf(t)$	$-Tz\dfrac{d}{dz}[\bar{F}(z)]$
G	$F(s+a)$	$e^{-at}f(t); \quad a>0$	$\bar{F}(ze^{aT})$
H	$e^{-nsT}F(s)$	$f(t - nT); \quad n>0$	$z^{-n}\bar{F}(z)$
I	$aF(as)$	$f\left(\dfrac{t}{a}\right); \quad a>0$	$\bar{F}(z)$ with $\dfrac{T}{a} \longrightarrow T$

Line	$F(s)$	$f(t)$ for $t \geq 0$	$\bar{F}(z)$
100	1	$\delta(t)$; $f_m = \dfrac{1}{T}$ at $m = 0$	$\dfrac{1}{T}$
101	$\dfrac{1}{s}$	$u(t)$; $f_m = 1$ for $m \geq 0$	$\dfrac{z}{z-1}$
102	$\dfrac{1}{s^2}$	t	$\dfrac{Tz}{(z-1)^2}$
103	$\dfrac{1}{s^3}$	$\dfrac{1}{2!}t^2$	$\dfrac{T^2 z(z+1)}{2(z-1)^3}$
104	$\dfrac{1}{s^4}$	$\dfrac{1}{3!}t^3$	$\dfrac{T^3 z(z^2 + 4z + 1)}{6(z-1)^4}$
105	$\dfrac{1}{s^{k+1}}$	$\dfrac{1}{k!}t^k$	$\lim\limits_{a \to 0} \dfrac{(-1)^k}{k!}\dfrac{\partial^k}{\partial a^k}\left(\dfrac{z}{z - e^{-aT}}\right)$
150	$\dfrac{1}{s - (1/T)\ln a}$	$a^{t/T}$	$\dfrac{z}{z-a}$
151	$\dfrac{1}{s+a}$	e^{-at}	$\dfrac{z}{z - e^{-aT}}$
152	$\dfrac{1}{(s+a)^2}$	te^{-at}	$\dfrac{Tze^{-aT}}{(z - e^{-aT})^2}$
153	$\dfrac{1}{(s+a)^3}$	$\dfrac{t^2}{2}e^{-at}$	$\dfrac{T^2 e^{-aT} z}{2(z - e^{-aT})^2} + \dfrac{T^2 e^{-2aT} z}{(z - e^{-aT})^3}$
154	$\dfrac{1}{(s+a)^{k+1}}$	$\dfrac{t^k}{k!}e^{-at}$	$\dfrac{(-1)^k}{k!}\dfrac{\partial^k}{\partial a^k}\left(\dfrac{z}{z - e^{-aT}}\right)$
155	$\dfrac{a}{s(s+a)}$	$1 - e^{-at}$	$\dfrac{(1 - e^{-aT})z}{(z-1)(z - e^{-aT})}$
156	$\dfrac{a}{s^2(s+a)}$	$t - \dfrac{1 - e^{-at}}{a}$	$\dfrac{Tz}{(z-1)^2} - \dfrac{(1 - e^{-aT})z}{a(z-1)(z - e^{-aT})}$
157	$\dfrac{a}{s^3(s+a)}$	$\dfrac{1}{2!}\left(t^2 - \dfrac{2}{a}t + \dfrac{2}{a^2} - \dfrac{2}{a^2}e^{-at}\right)$	$\dfrac{T^2 z}{(z-1)^3} + \dfrac{(aT-2)Tz}{2a(z-1)^2} + \dfrac{z}{a^2(z-1)} - \dfrac{z}{a^2(z - e^{-aT})}$

Laplace and z-Transforms

Line	$F(s)$	$f(t)$ for $t \geq 0$	$\tilde{F}(z)$
158	$\dfrac{a}{s^{k+1}(s+a)}$	$\dfrac{1}{k!}\left[t^k - \dfrac{k}{a}t^{k-1} + \dfrac{k(k-1)}{a^2}t^{k-2} - \cdots \right.$ $\left. + (-1)^{k-1}\dfrac{k!}{a^k}t + (-1)^k\dfrac{k!}{a^k}\right] + (-1)^{k+1}\dfrac{e^{-at}}{a^k}$	$\dfrac{(-1)^{k+1}}{a^k}\dfrac{1}{1-e^{-aT}z^{-1}}$ $+ \dfrac{a}{k!}\lim_{x \to 0}\dfrac{\partial^k}{\partial x^k}\left[\dfrac{1}{(x+a)(1-e^{Tx}z^{-1})}\right]$
200	$\dfrac{b-a}{(s+a)(s+b)}$	$e^{-at} - e^{-bt}$	$\dfrac{z}{z-e^{-aT}} - \dfrac{z}{z-e^{-bT}}$
201	$\dfrac{(b-a)(s+c)}{(s+a)(s+b)}$	$(c-a)e^{-at} + (b-c)e^{-bt}$	$\dfrac{(c-a)z}{z-e^{-aT}} + \dfrac{(b-c)z}{z-e^{-bT}}$
202	$\dfrac{\beta}{s^2+\beta^2}$	$\sin \beta t$	$\dfrac{z \sin \beta T}{z^2 - 2z \cos \beta T + 1}$
203	$\dfrac{s}{s^2+\beta^2}$	$\cos \beta t$	$\dfrac{z(z - \cos \beta T)}{z^2 - 2z \cos \beta T + 1}$
204	$\dfrac{\beta}{s^2-\beta^2}$	$\sinh \beta t$	$\dfrac{z \sinh \beta T}{z^2 - 2z \cosh \beta T + 1}$
205	$\dfrac{s}{s^2-\beta^2}$	$\cosh \beta t$	$\dfrac{z(z - \cosh \beta T)}{z^2 - 2z \cosh \beta T + 1}$
206	$\dfrac{\beta}{(s+a)^2+\beta^2}$	$e^{-at} \sin \beta t$	$\dfrac{ze^{-aT} \sin \beta T}{z^2 - 2ze^{-aT} \cos \beta T + e^{-2aT}}$
207	$\dfrac{s+a}{(s+a)^2+\beta^2}$	$e^{-at} \cos \beta t$	$\dfrac{z^2 - ze^{-aT} \cos \beta T}{z^2 - 2ze^{-aT} \cos \beta T + e^{-2aT}}$
208	$\dfrac{\beta s}{(s+a)^2+\beta^2}$	$e^{-at}(\beta \cos \beta t - a \sin \beta t)$	$\dfrac{\beta z^2 - ze^{-aT}(\beta \cos \beta T + a \sin \beta T)}{z^2 - 2ze^{-aT} \cos \beta T + e^{-2aT}}$
300	$\dfrac{ab}{s(s+a)(s+b)}$	$1 + \dfrac{b}{a-b}e^{-at} - \dfrac{a}{a-b}e^{-bt}$	$\dfrac{z}{z-1} + \dfrac{bz}{(a-b)(z-e^{-aT})}$ $- \dfrac{az}{(a-b)(z-e^{-bT})}$

Line	$F(s)$	$f(t)$ for $t \geq 0$	$\bar{F}(z)$
301	$\dfrac{ab(s+c)}{s(s+a)(s+b)}$	$c + \dfrac{b(c-a)}{a-b}e^{-at} + \dfrac{a(b-c)}{a-b}e^{-bt}$	$\dfrac{cz}{z-1} + \dfrac{b(c-a)z}{(a-b)(z-e^{-aT})} + \dfrac{a(b-c)z}{(a-b)(z-e^{-bT})}$
302	$\dfrac{1}{(s+a)(s+b)(s+c)}$	$\dfrac{e^{-at}}{(b-a)(c-a)} + \dfrac{e^{-bt}}{(a-b)(c-b)} + \dfrac{e^{-ct}}{(a-c)(b-c)}$	$\dfrac{z}{(b-a)(c-a)(z-e^{-aT})} + \dfrac{z}{(a-b)(c-b)(z-e^{-bT})} + \dfrac{z}{(a-c)(b-c)(z-e^{-cT})}$
303	$\dfrac{s+d}{(s+a)(s+b)(s+c)}$	$\dfrac{(d-a)}{(b-a)(c-a)}e^{-at} + \dfrac{(d-b)}{(a-b)(c-b)}e^{-bt} + \dfrac{(d-c)}{(a-c)(b-c)}e^{-ct}$	$\dfrac{(d-a)z}{(b-a)(c-a)(z-e^{-aT})} + \dfrac{(d-b)z}{(a-b)(c-b)(z-e^{-bT})} + \dfrac{(d-c)z}{(a-c)(b-c)(z-e^{-cT})}$
304	$\dfrac{a^2}{s(s+a)^2}$	$1 - (1+at)e^{-at}$	$\dfrac{z}{z-1} - \dfrac{z}{z-e^{-aT}} - \dfrac{aTe^{-aT}z}{(z-e^{-aT})^2}$
305	$\dfrac{a^2(s+b)}{s(s+a)^2}$	$b - be^{-at} + a(a-b)te^{-at}$	$\dfrac{bz}{z-1} - \dfrac{bz}{z-e^{-aT}} + \dfrac{a(a-b)Te^{-aT}z}{(z-e^{-aT})^2}$
306	$\dfrac{(a-b)^2}{(s+b)(s+a)^2}$	$e^{-bt} - e^{-at} + (b-a)te^{-at}$	$\dfrac{z}{z-e^{-bT}} - \dfrac{z}{z-e^{-aT}} + \dfrac{(a-b)Te^{-aT}z}{(z-e^{-aT})^2}$
307	$\dfrac{(a-b)^2(s+c)}{(s+b)(s+a)^2}$	$(c-b)e^{-bt} + (b-c)e^{-at} - (a-b)(c-a)te^{-at}$	$\dfrac{(c-b)z}{z-e^{-bT}} + \dfrac{(b-c)z}{z-e^{-aT}} - \dfrac{(a-b)(c-a)Te^{-aT}z}{(z-e^{-aT})^2}$

Line	$F(s)$	$f(t)$ for $t \geq 0$	$\bar{F}(z)$
308	$\dfrac{\beta^2}{s(s^2 - \beta^2)}$	$\cosh \beta t - 1$	$\dfrac{z(z - \cosh \beta T)}{z^2 - 2z \cosh \beta T + 1} - \dfrac{z}{z - 1}$
309	$\dfrac{\beta^2}{s(s^2 + \beta^2)}$	$1 - \cos \beta t$	$\dfrac{z}{z - 1} - \dfrac{z(z - \cos \beta T)}{z^2 - 2z \cos \beta T + 1}$
310	$\dfrac{\beta^2(s + a)}{s(s^2 + \beta^2)}$	$a - a \sec \theta \cos(\beta t + \theta)$ where $\theta = \tan^{-1}\left(\dfrac{\beta}{a}\right)$	$\dfrac{az}{z - 1} - \dfrac{az^2 - az \sec \theta \cos(\beta T - \theta)}{z^2 - 2z \cos \beta T + 1}$
311	$\dfrac{a^2 + \beta^2}{s[(s + a)^2 + \beta^2]}$	$1 - e^{-at} \sec \theta \cos(\beta t + \theta)$ where $\theta = \tan^{-1}\left(-\dfrac{a}{\beta}\right)$	$\dfrac{z}{z - 1} - \dfrac{z^2 - ze^{-aT} \sec \theta \cos(\beta T - \theta)}{z^2 - 2ze^{-aT} \cos \beta T + e^{-2aT}}$
312	$\dfrac{(a^2 + \beta^2)(s + b)}{s[(s + a)^2 + \beta^2]}$	$b - be^{-at} \sec \theta \cos(\beta t + \theta)$ where $\theta = \tan^{-1}\left(\dfrac{a^2 + \beta^2 - ab}{b\beta}\right)$	$\dfrac{bz}{z - 1} - \dfrac{b[z^2 - ze^{-aT} \sec \theta \cos(\beta T - \theta)]}{z^2 - 2ze^{-aT} \cos \beta T + e^{-2aT}}$
313	$\dfrac{(a - b)^2 + \beta^2}{(s + b)[(s + a)^2 + \beta^2]}$	$e^{-bt} - e^{-at} \sec \theta \cos(\beta t + \theta)$ where $\theta = \tan^{-1}\left(\dfrac{b - a}{\beta}\right)$	$\dfrac{z}{z - e^{-bT}} - \dfrac{z^2 - ze^{-aT} \sec \theta \cos(\beta T - \theta)}{z^2 - 2ze^{-aT} \cos \beta T + e^{-2aT}}$
314	$\dfrac{[(a - b)^2 + \beta^2](s + \alpha)}{(s + b)[(s + a)^2 + \beta^2]}$	$(\alpha - b)e^{-bt} - (\alpha - b)e^{-at} \sec \theta \cos(\beta t + \theta)$ where $\theta = \tan^{-1}\left(\dfrac{(\alpha - a)(b - a) + \beta^2}{(\alpha - b)\beta}\right)$	$\dfrac{(\alpha - b)z}{z - e^{-bT}} - \dfrac{(\alpha - b)[z^2 - ze^{-aT} \sec \theta \cos(\beta T - \theta)]}{z^2 - 2ze^{-aT} \cos \beta T + e^{-2aT}}$
315	$\dfrac{[(a - b)^2 + \beta^2](s^2 + \alpha s + \gamma)}{(s + b)[(s + a)^2 + \beta^2]}$	$(b^2 - b\alpha + \gamma)e^{-bt} + k^2 e^{-at} \sec \theta \cos(\beta t + \theta)$ where $k^2 = a^2 + \beta^2 - 2ab + b\alpha - \gamma$ $\theta = \tan^{-1}\left(\dfrac{ak^2 - (a^2 + \beta^2)(\alpha - b) + \gamma(2a - b)}{\beta k^2}\right)$	$\dfrac{(b^2 - b\alpha + \gamma)z}{z - e^{-bT}} + \dfrac{k^2[z^2 - ze^{-aT} \sec \theta \cos(\beta T - \theta)]}{z^2 - 2ze^{-aT} \cos \beta T + e^{-2aT}}$

Line	$F(s)$	$f(t)$ for $t \geq 0$	$\bar{F}(z)$
400	$\dfrac{a^3}{s^2(s+a)^2}$	$at - 2 + (at + 2)e^{-at}$	$\dfrac{(aT+2)z - 2z^2}{(z-1)^2} + \dfrac{aTe^{-aT}z}{(z - e^{-aT})^2} + \dfrac{2z}{z - e^{-aT}}$
401	$\dfrac{a^2 b^2}{s^2(s+a)(s+b)}$	$abt - (a+b) - \dfrac{b^2}{a-b}e^{-at} + \dfrac{a^2}{a-b}e^{-bt}$	$\dfrac{abTz}{(z-1)^2} - \dfrac{(a+b)z}{z-1} + \dfrac{a^2 z}{(a-b)(z - e^{-bT})} - \dfrac{b^2 z}{(a-b)(z - e^{-aT})}$
402	$\dfrac{a^2 b^2(s+c)}{s^2(s+a)(s+b)}$	$abct + [ab - c(a+b)] - \dfrac{b^2(c-a)}{a-b}e^{-at} - \dfrac{a^2(b-c)}{a-b}e^{-bt}$	$\dfrac{abcTz}{(z-1)^2} + \dfrac{ab - c(a+b)z}{z-1} - \dfrac{a^2(b-c)z}{(a-b)(z - e^{-bT})} - \dfrac{b^2(c-a)z}{(a-b)(z - e^{-aT})}$
403	$\dfrac{abc}{s(s+a)(s+b)(s+c)}$	$1 - \dfrac{bc}{(b-a)(c-a)}e^{-at} - \dfrac{ca}{(c-b)(a-b)}e^{-bt} - \dfrac{ab}{(a-c)(b-c)}e^{-ct}$	$\dfrac{z}{z-1} - \dfrac{bcz}{(b-a)(c-a)(z - e^{-aT})} - \dfrac{caz}{(c-b)(a-b)(z - e^{-bT})} - \dfrac{abz}{(a-c)(b-c)(z - e^{-cT})}$
404	$\dfrac{abc(s+d)}{s(s+a)(s+b)(s+c)}$	$d - \dfrac{bc(d-a)}{(b-a)(c-a)}e^{-at} - \dfrac{ca(d-b)}{(c-b)(a-b)}e^{-bt} - \dfrac{ab(d-c)}{(a-c)(b-c)}e^{-ct}$	$\dfrac{dz}{z-1} - \dfrac{bc(d-a)z}{(b-a)(c-a)(z - e^{-aT})} - \dfrac{ca(d-b)z}{(c-b)(a-b)(z - e^{-bT})} - \dfrac{ab(d-c)z}{(a-c)(b-c)(z - e^{-cT})}$
405	$\dfrac{a^2 b}{s(s+b)(s+a)^2}$	$1 - \dfrac{a^2}{(a-b)^2}e^{-bt} + \dfrac{ab + b(a-b)}{(a-b)^2}e^{-at} + \dfrac{ab}{a-b}te^{-at}$	$\dfrac{z}{z-1} - \dfrac{a^2 z}{(a-b)^2(z - e^{-bT})} + \dfrac{[ab + b(a-b)]z}{(a-b)^2(z - e^{-aT})} + \dfrac{abTe^{-aT}z}{(a-b)(z - e^{-aT})^2}$
406	$\dfrac{a^2 b(s+c)}{s(s+b)(s+a)^2}$	$c + \dfrac{a^2(b-c)}{(a-b)^2}e^{-bt} + \dfrac{ab(c-a) + bc(a-b)}{(a-b)^2}e^{-at} + \dfrac{ab(c-a)}{a-b}te^{-at}$	$\dfrac{cz}{z-1} + \dfrac{a^2(b-c)z}{(a-b)^2(z - e^{-bT})} + \dfrac{[ab(c-a) + bc(a-b)]z}{(a-b)^2(z - e^{-aT})} + \dfrac{ab(c-a)Te^{-aT}z}{(a-b)(z - e^{-aT})^2}$

Line	$F(s)$	$f(t)$ for $t \geq 0$	$\bar{F}(z)$
407	$\dfrac{(a^2 + \beta^2)^2}{s^2[(s+a)^2 + \beta^2]}$	$(a^2 + \beta^2)t - 2a + 2ae^{-at}\sec\theta\cos(\beta t + \theta)$ where $\theta = \tan^{-1}\left(\dfrac{\beta^2 - a^2}{2a\beta}\right)$	$\dfrac{[(a^2 + \beta^2)T + 2a]z - 2az^2}{(z-1)^2}$ $+ \dfrac{2a[z^2 - ze^{-aT}\sec\theta\cos(\beta T - \theta)]}{z^2 - 2ze^{-aT}\cos\beta T + e^{-2aT}}$
408	$\dfrac{(a^2 + \beta^2)^2(s+b)}{s^2[(s+a)^2 + \beta^2]}$	$b(a^2 + \beta^2)t + k^2 - k^2 e^{-at}\sec\theta\cos(\beta t + \theta)$ where $k^2 = a^2 + \beta^2 - 2ab$ $\theta = \tan^{-1}\left(-\dfrac{ak^2 + b(a^2 + \beta^2)}{\beta k^2}\right)$	$\dfrac{[bT(a^2 + \beta^2) - k^2]z + k^2 z^2}{(z-1)^2}$ $- \dfrac{k^2[z^2 - ze^{-aT}\sec\theta\cos(\beta T - \theta)]}{z^2 - 2ze^{-aT}\cos\beta T + e^{-2aT}}$
500	$\dfrac{(abc)^2}{s^2(s+a)(s+b)(s+c)}$	$abct - (bc + ca + ab) + \dfrac{b^2 c^2}{(b-a)(c-a)}e^{-at}$ $+ \dfrac{c^2 a^2}{(c-b)(a-b)}e^{-bt} + \dfrac{a^2 b^2}{(a-c)(b-c)}e^{-ct}$	$\dfrac{abcTz}{(z-1)^2} - \dfrac{(bc+ca+ab)z}{z-1} + \dfrac{b^2 c^2 z}{(b-a)(c-a)(z - e^{-aT})}$ $+ \dfrac{c^2 a^2 z}{(c-b)(a-b)(z - e^{-bT})} + \dfrac{a^2 b^2 z}{(a-c)(b-c)(z - e^{-cT})}$
501	$\dfrac{(abc)^2(s+d)}{s^2(s+a)(s+b)(s+c)}$	$abcdt + [abc - (bc + ca + ab)d] + \dfrac{b^2 c^2(d-a)}{(b-a)(c-a)}e^{-at}$ $+ \dfrac{c^2 a^2(d-b)}{(c-b)(a-b)}e^{-bt} + \dfrac{a^2 b^2(d-c)}{(a-c)(b-c)}e^{-ct}$	$\dfrac{abcdTz}{(z-1)^2} + \dfrac{[abc - (bc + ca + ab)d]z}{z-1}$ $+ \dfrac{b^2 c^2(d-a)z}{(b-a)(c-a)(z - e^{-aT})} + \dfrac{c^2 a^2(d-b)z}{(c-b)(a-b)(z - e^{-bT})}$ $+ \dfrac{a^2 b^2(d-c)z}{(a-c)(b-c)(z - e^{-cT})}$
502	$\dfrac{(a^2 b)^2}{s^2(s+b)(s+a)^2}$	$a^2 bt - [ab + a(a+b)] + \dfrac{a^4}{(a-b)^2}e^{-bt}$ $- \dfrac{ab^2(3a - 2b)}{(a-b)^2}e^{-at} - \dfrac{a^2 b^2}{a-b}te^{-at}$	$\dfrac{a^2 bTz}{(z-1)^2} - \dfrac{[ab + a(a+b)]z}{z-1} + \dfrac{a^4 z}{(a-b)^2(z - e^{-bT})}$ $- \dfrac{ab^2(3a - 2b)z}{(a-b)^2(z - e^{-aT})} - \dfrac{a^2 b^2 Te^{-aT}z}{(a-b)(z - e^{-aT})^2}$

Appendix B

Computing Algorithms in Fortran-77

This appendix contains a complete listing of the Fortran-77 source code for all of the computing routines described in the text. The code is also included on the 362-kB floppy disk accompanying the text, which was written from MS-DOS version 3.0. The floppy disk file directory includes

```
DSAB.FOR
TDSAB.FOR
```

The first file, DSAB.FOR, is the Fortran-77 source library of routines listed in this appendix. The second file, TDSAB.FOR, is the Fortran-77 source code for a program that tests the routines in DSAB. The tests are by no means exhaustive, but they do indicate via PRINT statements that the routines in DSAB are operating correctly under given conditions.

```
c-
c-                          DSAB.FOR
c-              DIGITAL SIGNAL ANALYSIS, Appendix B.
c-              Source code for all routines in Fortran-77.
c-
c-
c-
c-******************************************************************************
c-*                      CHAPTER 6 ROUTINES                                    *
c-******************************************************************************
c-
      SUBROUTINE SPFFT(X,N)
c-Latest date: 12/04/87
c-FFT using time decomposition with input bit reversal.
c-X(0:N+1)=REAL array holding REAL input sequence X(0),---,X(N-1), plus
c-        2 extra elements X(N),X(N+1), which need not be initialized.
c-Number of samples (N) must be a power of 2 greater than 2.
c-After execution, 1st FFT component is X(0)=real, X(1)=imaginary,
c-                 2nd FFT component is X(2)=real, X(3)=imaginary,
c-                                    etc.,
c-          N/2 (last) FFT component is X(N)=real, X(N+1)=imaginary.
c-
      COMPLEX X(0:N/2),T,U
      PI=4.*ATAN(1.)
      MR=0
      DO 2 M=1,N/2-1
       L=N/2
    1  L=L/2
        IF(MR+L.GE.N/2) GO TO 1
        MR=MOD(MR,L)+L
        IF(MR.LE.M) GO TO 2
        T=X(M)
        X(M)=X(MR)
        X(MR)=T
    2 CONTINUE
      L=1
    3 IF(L.LT.N/2) THEN
        DO 5 M=0,L-1
          DO 4 I=M,N/2-1,2*L
           T=X(I+L)*EXP(CMPLX(0.,-M*PI/FLOAT(L)))
           X(I+L)=X(I)-T
           X(I)=X(I)+T
    4     CONTINUE
    5   CONTINUE
        L=2*L
        GO TO 3
      ENDIF
    6 X(N/2)=X(0)
      DO 7 M=0,N/4
       U=CMPLX(SIN(M*2.*PI/N),COS(M*2.*PI/N))
       T=((1.+U)*X(M)+(1.-U)*CONJG(X(N/2-M)))/2.
       X(M)=((1.-U)*X(M)+(1.+U)*CONJG(X(N/2-M)))/2.
       X(N/2-M)=CONJG(T)
    7 CONTINUE
      RETURN
      END
c-
      SUBROUTINE SPIFFT(X,N)
c-Latest date: 08/24/88
c-Inverse FFT. Transforms SPFFT output back into N times original input.
c-X(0:N+1) is a REAL array holding the complex DFT components with
c-         indices from 0 through N/2 in order as described in SPFFT.
c-As in SPFFT, N must be a power of 2 greater than 2.
c-
```

```
      COMPLEX X(0:N/2),T,U
      PI=4.*ATAN(1.)
      DO 1 M=0,N/4
       U=CMPLX(SIN(M*2.*PI/N),-COS(M*2.*PI/N))
       T=(1.+U)*X(M)+(1.-U)*CONJG(X(N/2-M))
       X(M)=(1.-U)*X(M)+(1.+U)*CONJG(X(N/2-M))
       X(N/2-M)=CONJG(T)
    1 CONTINUE
      MR=0
      DO 3 M=1,N/2-1
       L=N/2
    2  L=L/2
        IF(MR+L.GE.N/2) GO TO 2
       MR=MOD(MR,L)+L
       IF(MR.LE.M) GO TO 3
       T=X(M)
       X(M)=X(MR)
       X(MR)=T
    3 CONTINUE
      L=1
    4 IF(L.LT.N/2) THEN
        DO 6 M=0,L-1
         DO 5 I=M,N/2-1,2*L
          T=X(I+L)*EXP(CMPLX(0.,M*PI/FLOAT(L)))
          X(I+L)=X(I)-T
          X(I)=X(I)+T
    5    CONTINUE
    6    CONTINUE
        L=2*L
        GO TO 4
      ENDIF
      RETURN
      END
c-******************************************************************
c-*                    CHAPTER 8 ROUTINES                          *
c-******************************************************************
c-
      FUNCTION SPWIND(ITYPE,N,K)
c-Latest date: 08/24/88
c-This function returns the Kth sample of a data window ranging from
c-  K=-N through K=N.
c-ITYPE=1: Rectangular   2: Bartlett   3: Hanning
c-      4: Hamming        5: Blackman   6: Kaiser
c-   (Note:  Windows are defined in Table 8.1 of Digital Signal Anal.
c-           Kaiser window has "beta" fixed internally at 5.44.)
c-N=Half-size of window as in Table 8.1.  Total window size = 2N+1.
c-K=Sample number within window, from -N thru N.
c-   (If K is outside this range or if N<1, SPWIND is set to 0.)
c-
c-Example (Rectangular):
c-                            |
c-                            |
c-                    _____|_____  1
c-                    |       |       |
c-                    |       |       |
c-                    |       |       |
c-                    |       |       |
c-            _____|       |       |_____
c-                   -N       0   K   N
c-
      BETA=5.44
      PI=4.*ATAN(1.)
      SPWIND=0.
      IF(ITYPE.LT.1.OR.ITYPE.GT.6) RETURN
      IF(N.LT.1.OR.ABS(K).GT.N) RETURN
      GO TO (1,2,3,4,5,6), ITYPE
```

Computing Algorithms in Fortran-77

413

```
      1 SPWIND=1.0
        RETURN
      2 SPWIND=1.0-ABS(K)/FLOAT(N)
        RETURN
      3 SPWIND=0.5*(1.0+COS(PI*K/FLOAT(N)))
        RETURN
      4 SPWIND=0.54+0.46*COS(PI*K/FLOAT(N))
        RETURN
      5 SPWIND=0.42+0.5*COS(PI*K/FLOAT(N))+0.08*COS(2.*PI*K/FLOAT(N))
        RETURN
      6 SPWIND=BESSEL(BETA*(1.-(K/FLOAT(N))**2)**.5)/BESSEL(BETA)
        RETURN
        END
c-Bessel function for Kaiser window in SPWIND.
      FUNCTION BESSEL(X)
      BESSEL=1.
      TERM=1.
      DO 1 I=2,50,2
        TERM=TERM*(X/I)**2
        IF(TERM.LT.1.E-8*BESSEL) GO TO 2
        BESSEL=BESSEL+TERM
      1 CONTINUE
      2 RETURN
        END
c-************************************************************************
c-*                        CHAPTER 9 ROUTINES                           *
c-************************************************************************
c-
      SUBROUTINE SPLTOD(KAPPA,NU,N,B,A)
c-Latest date: 06/16/88
c-Converts a digital filter from symmetric lattice to direct form.
c-KAPPA(0:N-1)  =REAL Lattice coefficients in DSA, Fig. 9.13.
c-NU(0:N)       = "      "         "        "   "   "   "   .
c-N             =Number of lattice stages =order of filter.
c-B(0:N),A(1:N) =Coefficients in direct form, defined by
c-
c-            B(0)+B(1)*Z**(-1)+...........+B(N)*Z**(-N)
c-   H(Z) = ---------------------------------------------
c-            1+A(1)*Z**(-1)+.........+A(N)*Z**(-N)
c-
c-(Note carefully that array A is dimensioned A(1:N), not A(0:N).
c-
      REAL KAPPA(0:N-1),NU(0:N),B(0:N),A(1:N)
      B(0)=NU(0)
      DO 1 NS=1,N
        B(NS)=0.
      1 CONTINUE
      DO 4 NS=1,N
        IF(NS.GT.2) THEN
          DO 2 K=1,(NS-1)/2
            TEMP=A(K)
            A(K)=TEMP+KAPPA(NS-1)*A(NS-K)
            A(NS-K)=A(NS-K)+KAPPA(NS-1)*TEMP
      2     CONTINUE
        ENDIF
        IF(MOD(NS,2).EQ.0) A(NS/2)=(1.+KAPPA(NS-1))*A(NS/2)
        A(NS)=KAPPA(NS-1)
        DO 3 K=0,NS-1
          B(K)=B(K)+NU(NS)*A(NS-K)
      3   CONTINUE
        B(NS)=B(NS)+NU(NS)
      4 CONTINUE
      RETURN
      END
```

```
c-
      SUBROUTINE SPDTOL(B,A,N,KAPPA,NU,IERROR)
c-Latest date: 06/16/88
c-Converts a digital filter from direct to symmetric lattice form.
c-B(0:N),A(1:N) =Numerator coefficients in direct form, defined by
c-
c-          B(0)+B(1)*z**(-1)+..........+B(N)*z**(-N)
c-   H(z) = -------------------------------------------
c-          1+A(1)*z**(-1)+..........+A(N)*z**(-N)
c-
c-N                =Number of lattice stages =order of filter.
c-Note the dimensions:  B(0:N), A(1:N).  NOT A(0:N).
c-KAPPA(0:N-1)  =REAL Lattice coefficients in DSA, Fig. 9.13.
c-NU(0:N)       = "      "        "      "   "   "   "   " .
c-Lattice coeff. are returned in REAL arrays KAPPA(0:N-1) and NU(0:N).
c-IERROR=0     Conversion with no errors detected.
c-      1       Unstable H(z) due to absolute kappa >1.
c-
      REAL B(0:N),A(1:N),KAPPA(0:N-1),NU(0:N)
      IERROR=1
      DO 1 NS=0,N
       NU(NS)=B(NS)
       IF(NS.LT.N) KAPPA(NS)=A(NS+1)
    1 CONTINUE
      DO 6 NS=N,1,-1
       DO 2 K=0,NS-1
        NU(K)=NU(K)-NU(NS)*KAPPA(NS-K-1)
    2  CONTINUE
       IF(ABS(KAPPA(NS-1)).GE.1.0) RETURN
       DIV=1.-KAPPA(NS-1)**2
       IF(NS-2) 6,5,3
    3  DO 4 K=0,(NS-3)/2
        TEMP=KAPPA(K)
        KAPPA(K)=(TEMP-KAPPA(NS-1)*KAPPA(NS-K-2))/DIV
        KAPPA(NS-2-K)=(KAPPA(NS-2-K)-KAPPA(NS-1)*TEMP)/DIV
    4  CONTINUE
    5  IF(MOD(NS,2).EQ.0) KAPPA(NS/2-1)=KAPPA(NS/2-1)/(1.+KAPPA(NS-1))
    6 CONTINUE
      IERROR=0
      RETURN
      END
c-
      SUBROUTINE SPNLTD(KAPPA,N,B)
c-Latest date: 06/16/88
c-Converts a nonrecursive digital filter from lattice to direct form.
c-KAPPA(0:N-1)  =REAL Lattice coefficients in DSA, Fig. 9.14.
c-The lattice stage equations for n=1,---,N are
c-
c-           |    | |                           | | |      |
c-           |X(n)| |1            KAPPA(n-1)z**(-1)| |X(n-1)|
c-           |    |=|                             |*|      |
c-           |Y(n)| |KAPPA(n-1)        z**(-1)     | |Y(n-1)|
c-           |    | |                           | | |      |
c-
c-N                =Number of lattice stages =order of filter.
c-B(1:N)           =Direct-form coefficients, defined by
c-
c-   H(Z) = 1.0+B(1)*Z**(-1)+..........+B(N)*Z**(-N)
c-
c-(Note carefully that array B is dimensioned B(1:N), not B(0:N).
c-
      REAL KAPPA(0:N-1),B(1:N)
      DO 2 NS=1,N
       IF(NS.GT.2) THEN
```

```fortran
      DO 1 K=1,(NS-1)/2
        TEMP=B(K)
        B(K)=TEMP+KAPPA(NS-1)*B(NS-K)
        B(NS-K)=B(NS-K)+KAPPA(NS-1)*TEMP
    1   CONTINUE
      ENDIF
      IF(MOD(NS,2).EQ.0) B(NS/2)=(1.+KAPPA(NS-1))*B(NS/2)
      B(NS)=KAPPA(NS-1)
    2 CONTINUE
      RETURN
      END
```

```fortran
c-
      SUBROUTINE SPNDTL(B,N,KAPPA,IERROR)
c-Latest date: 06/16/88
c-Converts a nonrecursive digital filter from direct to lattice form.
c-B(1:N)        =Direct-form coefficients, defined in
c-
c-    H(Z) = 1.0+B(1)*z**(-1)+..........+B(N)*z**(-N)
c-
c-N             =Number of lattice stages =order of filter.
c-              (Note the dimension: B(1:N), NOT B(0:N).)
c-KAPPA(0:N-1)  =REAL Lattice coefficients in DSA, Fig. 9.14.
c-The lattice stage equations for n=1,---,N are
c-
c-
c-                |     |  |                                  |  |       |
c-                |X(n) |  |1          KAPPA(n-1)z**(-1)|  |X(n-1) |
c-                |     |=|                                   |*|       |
c-                |Y(n) |  |KAPPA(n-1)      z**(-1)  |  |Y(n-1) |
c-                |     |  |                                  |  |       |
c-
c-Lattice coeff. are returned in REAL array KAPPA(0:N-1).
c-IERROR     = 0: Conversion complete with no absolute kappas >1.0.
c-           = 1:      "      incomplete due to abs. kappa near 1.0.
c-
      REAL B(1:N),KAPPA(0:N-1)
      IERROR=1
      DO 1 NS=0,N-1
       KAPPA(NS)=B(NS+1)
    1 CONTINUE
      DO 5 NS=N,1,-1
       IF(ABS(1.0-ABS(KAPPA(NS-1))).LT.1.E-6) RETURN
       DIV=1.-KAPPA(NS-1)**2
       IF(NS-2) 5,4,2
    2  DO 3 K=0,(NS-3)/2
        TEMP=KAPPA(K)
        KAPPA(K)=(TEMP-KAPPA(NS-1)*KAPPA(NS-K-2))/DIV
        KAPPA(NS-2-K)=(KAPPA(NS-2-K)-KAPPA(NS-1)*TEMP)/DIV
    3  CONTINUE
    4  IF(MOD(NS,2).EQ.0) KAPPA(NS/2-1)=KAPPA(NS/2-1)/(1.+KAPPA(NS-1))
    5 CONTINUE
      IERROR=0
      RETURN
      END
```

```fortran
c-
      SUBROUTINE SPALTD1(KAPPA,LAMDA,GAIN,N,B,A)
c-Latest date: 08/24/88
c-Converts a digital filter from asymmetric lattice to direct form.
c-KAPPA(0:N-1)  =REAL lattice coefficients in DSA, Table 9.2 (left).
c-LAMDA(0:N-1)  = "        "       "       "        "  "  "      "   .
c-GAIN          =coefficient in series with lattice.
c-
c-The lattice stage equations for n=1,---,N are
c-
```

```
c-                     |     | |                                   | | |     |
c-                     |X(n)| |1              LAMDA(n-1)z**(-1)|  |X(n-1)|
c-                     |    |=|                                  |*|      |
c-                     |Y(n)| |KAPPA(n-1)              z**(-1)|  |Y(n-1)|
c-                     |     | |                                   | | |     |
c-
c-N                =Number of lattice stages =order of filter.
c-B(0:N),A(1:N) =Coefficients in direct form, defined by
c-
c-             B(0)+B(1)*Z**(-1)+..........+B(N)*Z**(-N)
c-   H(Z) = ----------------------------------------
c-             1+A(1)*Z**(-1)+..........+A(N)*Z**(-N)
c-
c-(Note carefully that array A is dimensioned A(1:N), not A(0:N).
c-
      REAL KAPPA(0:N-1),LAMDA(0:N-1),B(0:N),A(1:N)
      B(1)=1
      B(0)=KAPPA(0)
      A(1)=LAMDA(0)
      IF(N.GT.1) THEN
        DO 2 NS=2,N
         B(NS)=1
         A(NS)=LAMDA(NS-1)
         DO 1 K=1,NS-1
          B(NS-K)=KAPPA(NS-1)*A(NS-K)+B(NS-K-1)
          A(NS-K)=A(NS-K)+LAMDA(NS-1)*B(NS-K-1)
    1    CONTINUE
         B(0)=KAPPA(NS-1)
    2   CONTINUE
      ENDIF
      DO 3 NS=0,N
       B(NS)=B(NS)*GAIN
    3 CONTINUE
      RETURN
      END
c-
      SUBROUTINE SPDTAL1(B,A,N,KAPPA,LAMDA,GAIN,IERROR)
c-Latest date: 06/10/88
c-Converts a digital filter from direct to asymmetric lattice form.
c-B(0:N),A(1:N) =Numerator coefficients in direct form, defined by
c-
c-             B(0)+B(1)*z**(-1)+...........+B(N)*z**(-N)
c-   H(z) = -----------------------------------------
c-             1+A(1)*z**(-1)+..........+A(N)*z**(-N)
c-
c-N                =Number of lattice stages =order of filter.
c-Note the dimensions:  B(0:N), A(1:N).  NOT A(0:N).
c-KAPPA(0:N-1)  =REAL lattice coefficients in DSA, Table 9.2 (left).
c-LAMDA(0:N)    = "      "      "      "    "    "    "   "   .
c-GAIN             =gain coefficient in series with computed lattice.
c-
c-The lattice stage equations for n=1,---,N are
c-
c-                     |     | |                                   | | |     |
c-                     |X(n)| |1              LAMDA(n-1)z**(-1)|  |X(n-1)|
c-                     |    |=|                                  |*|      |
c-                     |Y(n)| |KAPPA(n-1)              z**(-1)|  |Y(n-1)|
c-                     |     | |                                   | | |     |
c-
cThe lattice and gain coefficients are returned by this routine.
c-IERROR=0      Conversion with no errors detected.
c-      1       Lattice form does not exist -- D close to 0 in
c-                                           Table 9.2.
```

Computing Algorithms in Fortran-77   **417**

```
        REAL B(0:N),A(1:N),KAPPA(0:N-1),LAMDA(0:N-1)
        IERROR=1
        B0=B(0)
        GAIN=B(N)
        IF(ABS(GAIN).LT.1.E-10) RETURN
        DO 1 NS=0,N-1
         KAPPA(NS)=B(NS+1)/GAIN
         LAMDA(NS)=A(NS+1)
   1  CONTINUE
        DO 3 NS=N,1,-1
         TK=B0
         TL=LAMDA(NS-1)
         D=1.-TK*TL
         IF(ABS(D).LT.1.E-10) RETURN
         B0=(KAPPA(0)-TK*LAMDA(0))/D
         DO 2 K=0,NS-2
          LAMDA(K)=(LAMDA(K)-TL*KAPPA(K))/D
          KAPPA(K)=(KAPPA(K+1)-TK*LAMDA(K+1))/D
   2     CONTINUE
         LAMDA(NS-1)=TL
         KAPPA(NS-1)=TK
   3  CONTINUE
        IERROR=0
        RETURN
        END
c-
        SUBROUTINE SPALTD2(KAPPA,LAMDA,GAMMA,N,B,A)
c-Latest date: 06/09/88
c-Converts a digital filter from asymmetric lattice to direct form.
c-KAPPA(0:N-1)  =REAL lattice coefficients in DSA, Table 9.2 (right).
c-LAMDA(0:N-1)  = "      "      "      "    "   "   "   "     " .
c-GAMMA         = "      "      coefficient at right end of lattice.
c-
c-The lattice stage equations for n=1,---,N are
c-
c-              |    |  |                                  |  |        |
c-              |X(n)|  |1                       z**(-1)|  |X(n-1)|
c-              |    |=|                                  |*|        |
c-              |Y(n)|  |KAPPA(n-1)   LAMDA(n-1)z**(-1)|  |Y(n-1)|
c-              |    |  |                                  |  |        |
c-
c-N               =Number of lattice stages =order of filter.
c-
c-B(0:N),A(1:N) =Coefficients in direct form, defined by
c-
c-         B(0)+B(1)*Z**(-1)+..........+B(N)*Z**(-N)
c-   H(Z) = ----------------------------------------
c-         1+A(1)*Z**(-1)+..........+A(N)*Z**(-N)
c-
c-(Note carefully that array A is dimensioned A(1:N), not A(0:N).
c-
        REAL KAPPA(0:N-1),LAMDA(0:N-1),B(0:N),A(1:N)
        B(1)=LAMDA(0)*GAMMA
        B(0)=KAPPA(0)
        A(1)=GAMMA
        IF(N.LT.2) RETURN
        DO 2 NS=2,N
         B(NS)=LAMDA(NS-1)*B(NS-1)
         A(NS)=B(NS-1)
         DO 1 K=1,NS-1
          B(NS-K)=KAPPA(NS-1)*A(NS-K)+LAMDA(NS-1)*B(NS-K-1)
          A(NS-K)=A(NS-K)+B(NS-K-1)
   1     CONTINUE
         B(0)=KAPPA(NS-1)
```

```
      2 CONTINUE
        RETURN
        END
c-
        SUBROUTINE SPDTAL2(B,A,N,KAPPA,LAMDA,GAMMA,IERROR)
c-Latest date: 08/05/88
c-Converts a digital filter from direct to asymmetric lattice form.
c-B(0:N),A(1:N) =Numerator coefficients in direct form, defined by
c-
c-           B(0)+B(1)*z**(-1)+..........+B(N)*z**(-N)
c-   H(z) = -----------------------------------------
c-           1+A(1)*z**(-1)+.........+A(N)*z**(-N)
c-
c-N                =Number of lattice stages =order of filter.
c-Note the dimensions:  B(0:N), A(1:N). NOT A(0:N).
c-KAPPA(0:N-1)   =REAL lattice coefficients in DSA, Table 9.2 (right).
c-LAMDA(0:N-1)   = "      "        "      "   "   "   "   "   "   .
c-GAMMA          = "      "        coefficient at right end of lattice.
c-The lattice stage equations for n=1,---,N are
c-
c-               |    |  |                              |  |        |
c-               |X(n)|  |1                    z**(-1)| |X(n-1)|
c-               |    |=|                              |*|        |
c-               |Y(n)|  |KAPPA(n-1)  LAMDA(n-1)z**(-1)| |Y(n-1)|
c-               |    |  |                              |  |        |
c-
cThe lattice coefficients are returned by this routine.
c-IERROR=0        Conversion with no errors detected.
c-      1         Lattice form does not exist -- D or p(n,n) close to 0
c-                                                    in Table 9.2.
        REAL B(0:N),A(1:N),KAPPA(0:N-1),LAMDA(0:N-1)
        IERROR=1
        B0=B(0)
        DO 1 NS=0,N-1
         KAPPA(NS)=B(NS+1)
         LAMDA(NS)=A(NS+1)
      1 CONTINUE
        DO 3 NS=N,1,-1
         TK=B0
         IF(ABS(LAMDA(NS-1)).LE.1.E-10*ABS(KAPPA(NS-1))) RETURN
         TL=KAPPA(NS-1)/LAMDA(NS-1)
         D=TL-TK
         IF(ABS(D).LT.1.E-10) RETURN
         B0=(KAPPA(0)-TK*LAMDA(0))/D
         DO 2 K=0,NS-2
          LAMDA(K)=(TL*LAMDA(K)-KAPPA(K))/D
          KAPPA(K)=(KAPPA(K+1)-TK*LAMDA(K+1))/D
      2  CONTINUE
         LAMDA(NS-1)=TL
         KAPPA(NS-1)=TK
      3 CONTINUE
        GAMMA=B0
        IERROR=0
        RETURN
        END
c-
        SUBROUTINE SPCFIL(F,N,B,A,LI,NS,WORK,IERROR)
c-Latest date: 05/11/88
c-Cascade filtering of the sequence F(0:N-1).
c-Initial conditions are set to zero.  Output replaces input.
c-F(0:N-1)      =N-sample data sequence, replaced with filtered version.
c-B(0:LI,NS)    =Numerator coefficients of H(z).
c-A(1:LI,NS)    =Denominator    "        "  " .
c-LI & NS       =Last coef. index & number of filter sections in cascade.
```

```
c-The transfer function of the nth filter section is
c-
c-            B(0,n)+B(1,n)*z**(-1)+ ... +B(LI,n)*z**(-LI)
c-   Hn(z) = -------------------------------------------------
c-            1+A(1,n)*z**(-1)+ ... +A(LI,n)*z**(-LI)
c-
c-WORK(2*LI+2) =Work array of size >=(2*LI+2). No need to initialize.
c-IERROR       =0: no errors.
c-             1: NS<1 or LI<1.
c-             2: filter output is over 1.E10 times max. input.
c-
      REAL F(0:N-1),B(0:LI,NS),A(1:LI,NS),WORK(0:LI,0:1)
c-Check for error 1 and initialize.
      IERROR=1
      IF(NS.LT.1.OR.LI.LT.1) RETURN
      IERROR=2
      FMAX=1.E10*ABS(F(0))
      DO 1 I=1,N-1
       FMAX=MAX(FMAX,1.E10*ABS(F(I)))
    1 CONTINUE
c-Do the filtering. Outer loop is section; inner loop is sample nmbr.
      DO 5 J=1,NS
       DO 2 I=1,LI
        WORK(I,0)=0.
        WORK(I,1)=0.
    2  CONTINUE
       DO 4 K=0,N-1
        WORK(0,0)=F(K)
        WORK(0,1)=B(0,J)*F(K)
        DO 3 I=LI,1,-1
         WORK(0,1)=WORK(0,1)+B(I,J)*WORK(I,0)-A(I,J)*WORK(I,1)
         WORK(I,0)=WORK(I-1,0)
         WORK(I,1)=WORK(I-1,1)
    3   CONTINUE
        F(K)=WORK(0,1)
        IF(ABS(F(K)).GT.FMAX) RETURN
    4  CONTINUE
    5 CONTINUE
      IERROR=0
      RETURN
      END
c-
      SUBROUTINE SPNFIL(F,N,B,LI,WORK,IERROR)
c-Latest date: 05/11/88
c-Nonrecursive (FIR) filtering of the sequence F(0:N-1).
c-Initial conditions are set to zero. Output replaces input.
c-F(0:N-1)     =N-sample data sequence, replaced with filtered version.
c-B(0:LI)      =Filter coefficients of H(z).
c-LI           =Last coef. index, at least 1.
c-The transfer function of the nonrecursive filter is
c-
c-   H(z) = B(0)+B(1)*z**(-1)+ ... +B(LI)*z**(-LI)
c-
c-WORK(LI+1)   =Work array of size >= LI+1. No need to initialize.
c-IERROR       =0: no errors.
c-             1: LI<1.
c-
      REAL F(0:N-1),B(0:LI),WORK(0:LI)
c-Check for error 1 and initialize. Set past values to 0.
      IERROR=1
      IF(LI.LT.1) RETURN
      DO 1 I=1,LI
       WORK(I)=0.
```

```fortran
    1 CONTINUE
      DO 3 K=0,N-1
       WORK(0)=F(K)
       F(K)=B(0)*F(K)
       DO 2 I=LI,1,-1
        F(K)=F(K)+B(I)*WORK(I)
        WORK(I)=WORK(I-1)
    2  CONTINUE
    3 CONTINUE
      IERROR=0
      RETURN
      END
c-
      SUBROUTINE SPLFIL(F,N,KAPPA,NU,NS,WORK,IERROR)
c-Latest date: 07/18/88
c-Lattice filtering of the sequence F(0:N-1) using lattice in Fig. 9.13.
c-Initial conditions are set to zero.  Output replaces input.
c-F(0:N-1)        =N-sample sequence, replaced with filtered version.
c-KAPPA(0:NS-1)   =REAL lattice elements in Fig. 9.13 of Dig. Sig. Anal.
c-NU(0:NS)        =REAL   "     "      "    "    "    "    "    "    "   .
c-NS              =Number of lattice stages.
c-WORK(0:NS)      =Work array of size >=NS+1.  No need to initialize.
c-IERROR          =0: no errors.
c-                 1: NS<1.
c-                 2: filter output is over 1.E10 times max. input.
c-
      REAL F(0:N-1),KAPPA(0:NS-1),NU(0:NS),WORK(0:NS)
c-Check for error 1 and initialize.  Set past values to 0.
      IERROR=1
      IF(NS.LT.1) RETURN
      IERROR=2
      FMAX=1.E10*ABS(F(0))
      DO 1 I=1,N-1
       FMAX=MAX(FMAX,1.E10*ABS(F(I)))
    1 CONTINUE
      DO 2 I=0,NS
       WORK(I)=0.
    2 CONTINUE
c-Do the filtering.  Outer loop is sample nmbr.; inner loop is section.
      DO 4 K=0,N-1
       SUM=0.
       X=F(K)
       DO 3 I=NS-1,0,-1
        X=X-KAPPA(I)*WORK(I)
        Y=WORK(I)+KAPPA(I)*X
        SUM=SUM+NU(I+1)*Y
        WORK(I+1)=Y
    3  CONTINUE
       WORK(0)=X
       F(K)=SUM+NU(0)*X
       IF(ABS(F(K)).GT.FMAX) RETURN
    4 CONTINUE
      IERROR=0
      RETURN
      END
c-
      SUBROUTINE SPNLFIL(F,N,KAPPA,NS,WORK,IERROR)
c-Latest date: 08/24/88
c-Nonrecursive lattice filtering of the sequence F(0:N-1).
c-Initial conditions are set to zero.  Output replaces input.
c-F(0:N-1)        =N-sample sequence, replaced with filtered version.
c-KAPPA(0:NS-1)   =REAL lattice elements in Fig. 9.14 of Dig. Sig. Anal.
c-The lattice stage equations for n=1,---,NS are
```

```
c-
c-                |    | |                                      | |         |
c-                |X(n)|  |1                KAPPA(n-1)z**(-1)|  |X(n-1)|
c-                |   |=|                                     |*|         |
c-                |Y(n)|  |KAPPA(n-1)          z**(-1)|       |Y(n-1)|
c-                |    | |                                      | |         |
c-
c-NS              =Number of lattice stages.
c-WORK(NS)        =Work array of size >=NS.  No need to initialize.
c-IERROR          =0: no errors.
c-                 1: NS<1.
c-
        REAL F(0:N-1),KAPPA(0:NS-1),WORK(0:NS-1)
c-Check for error 1 and initialize.  Set past values to 0.
        IERROR=1
        IF(NS.LT.1) RETURN
        DO 2 I=0,NS-1
         WORK(I)=0.
      2 CONTINUE
c-Do the filtering.  Outer loop is sample nmbr.; inner loop is section.
        DO 4 K=0,N-1
         X=F(K)
         Y=X
         DO 3 I=0,NS-1
          Y1=Y
          Y=KAPPA(I)*X+WORK(I)
          X=X+KAPPA(I)*WORK(I)
          WORK(I)=Y1
      3  CONTINUE
         F(K)=X
      4 CONTINUE
        IERROR=0
        RETURN
        END
c-****************************************************************************
c-*                        CHAPTER 12 ROUTINES                              *
c-****************************************************************************
c-
        SUBROUTINE SPLHBW(ITYPE,FC,NS,B,A,IERROR)
c-Latest date: 03/04/88
c-Low- and highpass Butterworth digital filter design routine.
c-ITYPE=1(lowpass) or 2(higpass)
c-FC   =Cutoff frequency in Hz-s.  (Sampling frequency=1.0.)
c-NS   =Number of 2-pole filter sections.  The nth section has
c-
c-             B(0,n) + B(1,n)*z**(-1) + B(2,n)*z**(-2)
c-     Hn(z)=-------------------------------------------
c-             1.0 + A(1,n)*z**(-1) + A(2,n)*z**(-2)
c-
c-B(0:2,NS) and A(2,NS) are where the routine stores the coefficients.
c-IERROR=0: no errors            2: FC not between 0.0 and 0.5
c-       1: ITYPE not valid      3: NS not greater that 0
c-
        REAL B(0:2,NS),A(2,NS)
        COMPLEX SN,ZN
        PI=4.*ATAN(1.)
        IERROR=1
        IF(ITYPE.NE.1.AND.ITYPE.NE.2) RETURN
        IERROR=2
        IF(FC.LE.0..OR.FC.GE..5) RETURN
        IERROR=3
        IF(NS.LE.0) RETURN
        WCP=TAN(PI*FC)
        DO 1 N=1,NS
         SN=WCP*EXP(CMPLX(0.,PI*(2.*N+2.*NS-1.)/(4.*NS)))
         ZN=(1.+SN)/(1.-SN)
```

```
              B(0,N)=((2-ITYPE)*WCP*WCP+(ITYPE-1)*1.)/(1.-2.*REAL(SN)+WCP*WCP)
              B(1,N)=(3-2*ITYPE)*2.*B(0,N)
              B(2,N)=B(0,N)
              A(1,N)=-2.*REAL(ZN)
              A(2,N)=REAL(ZN)**2+AIMAG(ZN)**2
        1 CONTINUE
              IERROR=0
              RETURN
              END
c-
              SUBROUTINE SPBBBW(ITYPE,F1,F2,NS,B,A,IERROR)
c-Latest date: 03/09/88
c-Bandpass and bandstop Butterworth digital filter design routine.
c-ITYPE=1(bandpass) or 2(bandstop)
c-F1,F2 =3-dB cutoff frequencies in Hz-s; F2>F1.  (Sampling freq.=1.0.)
c-NS    =EVEN number of 2-pole filter sections.  The nth section has
c-
c-               B(0,n) + B(1,n)*z**(-1) + B(2,n)*z**(-2)
c-      Hn(z)=-------------------------------------------
c-               1.0 + A(1,n)*z**(-1) + A(2,n)*z**(-2)
c-
c-B(0:2,NS) and A(2,NS) are where the routine stores the coefficients.
c-IERROR=0: no errors          2: 0.0<F1<F2<0.5 not true
c-      1: ITYPE not valid      3: NS not even or NS not >0
c-
              REAL B(0:2,NS),A(2,NS)
              COMPLEX SL,SB,ZB
              PI=4.*ATAN(1.)
              IERROR=1
              IF(ITYPE.NE.1.AND.ITYPE.NE.2) RETURN
              IERROR=2
              IF(F1.LE.0..OR.F2.LE.F1.OR.F2.GE..5) RETURN
              IERROR=3
              IF(NS.LE.0.OR.MOD(NS,2).NE.0) RETURN
              W1P=TAN(PI*F1)
              W2P=TAN(PI*F2)
              DO 2 NL=1,NS/2
               SL=(W2P-W1P)*EXP(CMPLX(0.,PI*(2.*NL+NS-1.)/(2.*NS)))
                DO 1 M=0,1
                 SB=(SL+(1-2*M)*SQRT(SL*SL-4.*W1P*W2P))/2.
                 ZB=(1.+SB)/(1.-SB)
                 FACTOR=1.-2.*REAL(SB)+ABS(SB)**2
                 IF(ITYPE.EQ.1) THEN
                  B(0,2*NL-M)=(W2P-W1P)/FACTOR
                  B(1,2*NL-M)=0.
                  B(2,2*NL-M)=-B(0,2*NL-M)
                 ELSEIF(ITYPE.EQ.2) THEN
                  B(0,2*NL-M)=(1.+W1P*W2P)/FACTOR
                  B(1,2*NL-M)=-2.*(1.-W1P*W2P)/FACTOR
                  B(2,2*NL-M)=B(0,2*NL-M)
                 ENDIF
                 A(1,2*NL-M)=-2.*REAL(ZB)
                 A(2,2*NL-M)=ABS(ZB)**2
        1    CONTINUE
        2 CONTINUE
              IERROR=0
              RETURN
              END
c-
              SUBROUTINE SPFIL1(X,N,ITYPE,FC,NS,WORK,IERROR)
c-Latest date: 03/25/88
c-Filter 1.  Applies lowpass or highpass Butterworth filter to data X.
c-X(0:N-1) =data array.  Filtering is in place.  Initial conditions=0.
c-N        =length of data sequence.
c-ITYPE    =1(lowpass) or 2(highpass).
```

Computing Algorithms in Fortran-77   **423**

```
c-FC        =cutoff (3-dB) frequency in Hz-s.  (Sampling freq. = 1.0.)
c-NS        =number of 2-pole sections; rolloff = 12 dB/octave/section.
c-WORK      =work array, dimensioned WORK(5*NS+6) or larger.
c-IERROR    =0: no errors.
c-             1: ITYPE not 1 or 2.          3: NS not greater than 0.
c-             2: FC not between 0.0 and 0.5.   3+I: SPCFIL error I.
c-
      REAL X(0:N-1),WORK(5*NS+6)
c-Initialize filter coeff. and set initial signal values to zero.
      CALL SPLHBW(ITYPE,FC,NS,WORK(1),WORK(3*NS+1),IERROR)
      IF(IERROR.NE.0) RETURN
c-Do the filtering using the SPCFIL routine.
      CALL SPCFIL(X,N,WORK(1),WORK(3*NS+1),2,NS,WORK(5*NS+1),IERROR)
      IERROR=IERROR+3
      IF(IERROR.NE.3) RETURN
      IERROR=0
      RETURN
      END
c-
      SUBROUTINE SPFIL2(X,N,ITYPE,F1,F2,NS,WORK,IERROR)
c-Latest date:  03/08/88
c-Filter 2.  Applies bandpass or bandstop Butterworth filter to data X.
c-X(0:N-1)   =data array.  Filtering is in place.  Initial conditions=0.
c-N          =length of data sequence.
c-ITYPE      =1(bandpass) or 2(bandstop).
c-F1,F2      =corner (3-dB) frequencies in Hz-s.  (Sampling freq. = 1.0.)
c-NS         =even number of 2-pole sections; rolloff = 6 dB/octave/sect.
c-WORK       =work array, dimensioned WORK(5*NS+6) or larger.
c-IERROR     =0: no errors.
c-             1: ITYPE not 1 or 2.   3: NS not even & greater than 0.
c-             2: F1,F2 not valid.   3+I: SPCFIL error I.
c-
      REAL X(0:N-1),WORK(5*NS+6)
c-Initialize filter coeff. and set initial signal values to zero.
      CALL SPBBBW(ITYPE,F1,F2,NS,WORK(1),WORK(3*NS+1),IERROR)
      IF(IERROR.NE.0) RETURN
c-Do the filtering using the SPCFIL routine.
      CALL SPCFIL(X,N,WORK(1),WORK(3*NS+1),2,NS,WORK(5*NS+1),IERROR)
      IERROR=IERROR+3
      IF(IERROR.NE.3) RETURN
      IERROR=0
      RETURN
      END
c-****************************************************************************
c-*                    CHAPTER 13 ROUTINES                                  *
c-****************************************************************************
c-
      FUNCTION SPUNIF(ISEED)
c-Latest date: 04/14/88
c-Uniform random number within the interval (0.0,1.0).
c-Initialize by setting ISEED to any integer, then leave ISEED alone.
c-This routine has a cycle length equal to 16,777,216.
c-
      DOUBLE PRECISION DP
      DP=53 84 2 21.D0*ABS(ISEED)+1
      DP=DP-INT(DP/16 777 216.D0)*16 777 216.D0
      ISEED=DP
      SPUNIF=(ISEED+1)/16 777 217.0
      RETURN
      END
c-
      FUNCTION SPGAUS(ISEED)
c-Latest date: 04/13/88
c-Approximately Gaussian random number with mean=0 and variance=1.
c-Initialize by setting ISEED, then leave ISEED alone in your program.
```

```
c-Note:  A sequence will be correlated after 1,398,101 samples.
c-
      SPGAUS=-6.
      DO 1 I=1,12
       SPGAUS=SPGAUS+SPUNIF(ISEED)
    1 CONTINUE
      RETURN
      END
c-
      FUNCTION SPEXV(X,N,I)
c-Latest date: 08/25/88
c-Expected or mean value of sequence [X(0)**I, X(1)**I, ---, X(N-1)**I].
c-Inputs are REAL array X(0:N-1), array size N, and exponent I.
c-If N<=0, SPEXV is set to X(0)**I.
c-
      REAL X(0:N-1)
      SPEXV=X(0)**I
      IF(N.LE.0) RETURN
      DO 1 J=1,N-1
       SPEXV=SPEXV+X(J)**I
    1 CONTINUE
      SPEXV=SPEXV/N
      RETURN
      END
c-**********************************************************************
c-*                    CHAPTER 14 ROUTINES                             *
c-**********************************************************************
c-
      SUBROUTINE SPSOLE(DD,LR,LC,M,IERROR)
c-Latest date: 12/09/88
c-Solution of M linear equations via Gauss-Jordan elimination.
c-Written for linear least-squares design with R-matrix and P-vector,
c-  but may be used with any set of linear equations provided the
c-  diagonal elements are large and coeff. matrix is well-conditioned.
c-DD=augmented matrix [R,P], DOUBLE PRECISION DD(0:LR,0:LC).  An element
c-  DD(I,J) is the double precision coefficient on row I, column J.
c-M=No. of equations.  Thus DD(0,M) thru DD(M-1,M) is the P-vector in
c-  column M of DD, which is replaced by the solution (optimal weights).
c-IERROR=0: No errors.
c-      1: LR<M-1 OR LC<M, that is, DD is not dimensioned large enough.
c-      2: No solution due to small diagonal element(s) or near-singula1
c-         coefficient matrix.
c-
      DOUBLE PRECISION DD(0:LR,0:LC),DDMAX
      IERROR=1
      IF(LR.LT.M-1.OR.LC.LT.M) RETURN
      IERROR=2
      DDMAX=ABS(DD(0,0))
      DO 1 I=1,M-1
       DDMAX=MAX(DDMAX,ABS(DD(I,I)))
    1 CONTINUE
      DO 5 I=0,M-1
       IF(ABS(DD(I,I)).LE.1.D-10*DDMAX) RETURN
       DO 2 J=I,M-1
        DD(I,J+1)=DD(I,J+1)/DD(I,I)
    2  CONTINUE
       DO 4 K=0,M-1
        IF(K.EQ.I) GO TO 4
        DO 3 J=I,M-1
         DD(K,J+1)=DD(K,J+1)-DD(K,I)*DD(I,J+1)
    3   CONTINUE
    4  CONTINUE
    5 CONTINUE
      IERROR=0
      RETURN
      END
```

```
c-
      SUBROUTINE SPLESQ(F,D,K,B,M,TSEMIN,IERROR)
c-Latest date: 09/07/89
c-
c-                           ^                    D(0:K-1)
c-                         __|_____                + |
c-                        |        |               - |
c-   F(-M+1:K-1)----->| B(0:M-1) |--------->0------|
c-                        |_____|                      |
c-                           ^                            |
c-                           |------------------------|
c-
c-This routine finds the M least-squares coefficients, B(0:M-1),
c- based on the segments F(-M+1:K-1) and D(0:K-1).  F and D could refer
c- to points in the same array, as F(k)=D(k-1) in a one-step predictor.
c-Note carefully:  The "F" in the calling sequence points to F(-M+1).
c-                  The "D" "  "      "        "        "   "   D(0).
c-R-matrix covariance elements R(I,J)=[Sum(F(k-I)*F(k-J)); k=0,K-1].
c-P-vector       "          "         P(I)  =[Sum(D(k)*F(k-I));  k=0,K-1].
c-The maximum filter size (M) is MMAX.  The SPSOLE routine is called.
c-TSEMIN is set to the minimum TSE, or to 0 if negative due to roundoff.
c-IERROR=0:   No errors.
c-        1-2: Error in SPSOLE execution.
c-        3:   M out of range [1,MMAX].
c-
      PARAMETER (MMAX=50)
      REAL F(-M+1:K-1),D(0:K-1),B(0:M-1),P(0:MMAX-1)
      DOUBLE PRECISION DD(0:MMAX-1,0:MMAX)
      IERROR=3
      IF(M.LE.0.OR.M.GT.MMAX) RETURN
      DO 4 I=0,M-1
       DO 2 J=0,I
        DD(I,J)=0.
        DO 1 KK=0,K-1
         DD(I,J)=DD(I,J)+F(KK-I)*F(KK-J)
    1   CONTINUE
        DD(J,I)=DD(I,J)
    2  CONTINUE
       DD(I,M)=0.
       DO 3 KK=0,K-1
        DD(I,M)=DD(I,M)+D(KK)*F(KK-I)
    3  CONTINUE
       P(I)=DD(I,M)
    4 CONTINUE
      CALL SPSOLE(DD,MMAX-1,MMAX,M,IERROR)
      IF(IERROR.NE.0) RETURN
      DO 5 I=0,M-1
       B(I)=DD(I,M)
    5 CONTINUE
      SUM=0.
      DO 6 KK=0,K-1
       SUM=SUM+D(KK)**2
    6 CONTINUE
      PTBOPT=0.
      DO 7 I=0,M-1
       PTBOPT=PTBOPT+P(I)*B(I)
    7 CONTINUE
      TSEMIN=MAX(0.,SUM-PTBOPT)
      RETURN
      PROGRAM TDSAB
c-Latest date: 07/18/88
C-
C-******************************************************************************
C-                           TDSAB.FOR                                        *
C-               DIGITAL SIGNAL ANALYSIS SUBPROGRAM TESTS                      *
C-                     Source code in Fortran-77                              *
C-******************************************************************************
```

```
c-
c-This program tests the subroutines and functions in DIGITAL SIGNAL
c- ANALYSIS, Appendix B.  In each test the message "passed" or "failed"
c- is printed after the name of the routine by the SPOUT subroutine at
c- the end of this program.
c-Each test checks for correct performance for a specific input, not for
c- all inputs, so the test is a check rather than a guarantee that the
c- routine is working correctly.
c-A perfomance factor, Q, is computed as a function of the subroutine
c- output and compared with its correct integer value in the "CALL
c- SPOUT" statement.  For example, the correct integer value of Q in the
c- test of the SPFFT suroutine is 1490.
c-If a test fails, check the listings of both the subroutine and this
c- program against those in Appendix B of DIGITAL SIGNAL ANALYSIS.  If
c- the listings agree, make sure your compiler meets the Fortran-77
c- standard.  Next, try printing the real value of Q for this test.
c- There may be a roundoff error.  If all else fails to yeild a reason
c- for the failure, call one of the authors at a reasonable hour.
c-
      REAL X1(0:9),X2(0:99),B1(0:2,4),A1(2,4),B2(0:2,4),A2(2,4)
      REAL BD(0:3),AD(3),KAPPA(0:2),LAMDA(0:1),NU(0:3),WORK(6)
      DOUBLE PRECISION DD(0:3,0:4)
      DATA DD/9.,4.,-1.,-4.,1.,8.,-2.,-5.,2.,5.,7.,1.,3.,6.,-3.,
     +        6.,29.,59.,4.,13./
C-
C-**********************************************************************
C-                          CHAPTER 6.                                *
C-**********************************************************************
C-
C-TEST OF SPFFT.  PERFORM A TRANSFORM AND TEST A CHECKSUM (Q).
      DO 610 K=0,7
       X1(K)=MOD((K+1)**3,20)-10
  610 CONTINUE
      CALL SPFFT(X1,8)
      Q=0.
      DO 611 K=0,9
       Q=Q-(K+1)**2*X1(K)
  611 CONTINUE
      CALL SPOUT('SPFFT  ',1490-INT(Q))
C-
C-TEST OF ISPFFT.  PERFORM AN INVERSE TRANSFORM AND TEST A CHECKSUM (Q).
      DO 620 K=0,9
       X1(K)=MOD((K+1)**3,20)-10
  620 CONTINUE
      CALL SPIFFT(X1,8)
      Q=0.
      DO 621 K=0,7
       Q=Q+(K+1)**3*X1(K)
  621 CONTINUE
      CALL SPOUT('SPIFFT ',2726-INT(Q))
C-
C-**********************************************************************
C-                          CHAPTER 8.                                *
C-**********************************************************************
C-
C-TEST OF SPWIND.  COMPUTE SOME WINDOW VALUES.
      Q=0.
      DO 1 ITYPE=1,6
       Q=Q+6**ITYPE*SPWIND(ITYPE,128,99)
    1 CONTINUE
      CALL SPOUT('SPWIND ',8851-INT(Q))
C-
C-**********************************************************************
C-                          CHAPTER 9.                                *
C-**********************************************************************
C-
```

Computing Algorithms in Fortran-77

```
C-TEST OF SPLTOD.  CONVERT FROM DIRECT TO LATTICE AND TEST CHECKSUM.
      DO 910 I=0,3
      NU(I)=I+1
      IF(I.GT.0) KAPPA(I)=(-.5)**I
  910 CONTINUE
      CALL SPLTOD(KAPPA,NU,3,BD,AD,IE)
      IF(IE.NE.0) PRINT '('' SPLTOD ERROR'',I3)', IE
      Q=BD(0)
      DO 911 I=1,3
      Q=Q+10**I*AD(I)+9**(I+1)*BD(I)
  911 CONTINUE
      CALL SPOUT('SPLTOD ',28265-INT(Q))
C-
C-TEST OF SPDTOL.  CONVERT FROM DIRECT TO LATTICE AND TEST CHECKSUM.
      DO 920 I=0,3
      BD(I)=.5-.5*I+3.*(I/2)+2.*(I/3)
      IF(I.GE.1) AD(I)=-.125-.375*(I/2)+.75*(I/3)
  920 CONTINUE
      CALL SPDTOL(BD,AD,3,KAPPA,NU,IE)
      IF(IE.NE.0) PRINT '('' SPDTOL ERROR'',I3)', IE
      Q=NU(3)
      DO 921 I=0,2
      Q=Q+10**I*KAPPA(I)+9**(I+2)*NU(I)
  921 CONTINUE
      CALL SPOUT('SPDTOL ',21246-INT(Q))
C-
C-TEST OF SPNLTD.  CONVERT FROM NONRECURSIVE LATTICE TO DIRECT AND TEST.
      KAPPA(0)=+2.
      KAPPA(1)=-.5
      KAPPA(2)=+3.
      CALL SPNLTD(KAPPA,3,BD(1))
      Q=-BD(1)+10.*BD(2)+100.*BD(3)
      CALL SPOUT('SPNLTD ',325-INT(Q))
C-
C-TEST OF SPNDTL.  CONVERT FROM NONRECURSIVE DIRECT TO LATTICE AND TEST.
      BD(1)=-0.5
      BD(2)=+2.5
      BD(3)=+3.0
      CALL SPNDTL(BD(1),3,KAPPA,IE)
      IF(IE.NE.0) PRINT '('' SPNDTL ERROR'',I3)', IE
      Q=KAPPA(0)-KAPPA(1)+10.*KAPPA(2)
      CALL SPOUT('SPNDTL ',32-INT(Q))
C-
C-TEST OF SPALTD1.  CONVERT FROM ASYMMETRIC LATTICE 1 TO DIRECT AND TEST.
      KAPPA(0)=7.5
      KAPPA(1)=1.0
      LAMDA(0)=-6.0
      LAMDA(1)=0.8
      GAIN=2.0
      CALL SPALTD1(KAPPA,LAMDA,GAIN,2,BD,AD)
      Q=BD(0)+10.*BD(1)+100.*BD(2)+1000.*AD(1)+10000.*AD(2)
      CALL SPOUT('SPALTD1',8232-INT(Q))
C-
C-TEST OF SPDTAL1.  CONVERT FROM DIRECT TO ASYMMETRIC LATTICE 1 AND TEST.
      BD(0)=1.0
      BD(1)=3.0
      BD(2)=2.0
      AD(1)=0.0
      AD(2)=0.8
      CALL SPDTAL1(BD,AD,2,KAPPA,LAMDA,GAIN,IE)
      IF(IE.NE.0) PRINT '('' SPDTAL1 ERROR'',I3)', IE
      Q=KAPPA(0)+10.*KAPPA(1)-100.*LAMDA(0)+10000.*LAMDA(1)+10000.*GAIN
      CALL SPOUT('SPDTAL1',28617-INT(Q))
C-
```

```
C-TEST OF SPALTD2.   CONVERT FROM ASYMMETRIC LATTICE 2 TO DIRECT AND TEST.
      KAPPA(0)=2.0
      KAPPA(1)=1.0
      LAMDA(0)=-0.4
      LAMDA(1)=2.5
      GAMMA=-2.0
      CALL SPALTD2(KAPPA,LAMDA,GAMMA,2,BD,AD)
      Q=BD(0)+10.*BD(1)+100.*BD(2)+1000.*AD(1)+10000.*AD(2)
      CALL SPOUT('SPALTD2',8231-INT(Q))
C-
C-TEST OF SPDTAL2.   CONVERT FROM DIRECT TO ASYMMETRIC LATTICE 2 AND TEST.
      BD(0)=1.0
      BD(1)=3.0
      BD(2)=2.0
      AD(1)=0.0
      AD(2)=0.8
      CALL SPDTAL2(BD,AD,2,KAPPA,LAMDA,GAMMA,IE)
      IF(IE.NE.0) PRINT '('' SPDTAL2 ERROR'',I3)', IE
      Q=KAPPA(0)+10.*KAPPA(1)-100.*LAMDA(0)+10000.*LAMDA(1)+10000.*GAMMA
      CALL SPOUT('SPDTAL2',5052-INT(Q))
C-
C-TEST OF SPCFIL.   RUN A DATA SEQUENCE THRU 2 SECTIONS AND TEST CHECKSUM.
      DO 930 K=0,5
 930  X1(K)=K+2.
      DO 931 I=0,3
      B1(I,1)=I+1
      B1(I,2)=4-I
      IF(I.GE.1) A1(I,1)=.5**I
      IF(I.GE.1) A1(I,2)=(-.5)**I
 931  CONTINUE
      CALL SPCFIL(X1,6,B1,A1,3,2,WORK,IE)
      IF(IE.NE.0) PRINT '('' SPCFIL ERROR'',I3)', IE
      Q=0.
      DO 932 K=0,5
 932  Q=Q+X1(K)
      CALL SPOUT('SPCFIL ',422-INT(Q))
C-
C-TEST OF SPNFIL.   RUN A SEQUENCE THRU AN FIR FILTER AND TEST CHECKSUM.
      DO 940 K=0,5
 940  X1(K)=K+2.
      DO 941 I=0,3
      BD(I)=I+1
 941  CONTINUE
      CALL SPNFIL(X1,6,BD,3,WORK,IE)
      IF(IE.NE.0) PRINT '('' SPNFIL ERROR'',I3)', IE
      Q=0.
      DO 942 K=0,5
 942  Q=Q+X1(K)
      CALL SPOUT('SPNFIL ',145-INT(Q))
C-
C-TEST OF SPLFIL.   RUN A SEQUENCE THRU A 3-SECTION LATTICE AND TEST Q.
      DO 950 K=0,5
 950  X1(K)=K+1.
      DO 951 I=0,3
      IF(I.LE.2) KAPPA(I)=.5**I
      NU(I)=I+1.
 951  CONTINUE
      CALL SPLFIL(X1,6,KAPPA,NU,3,WORK,IE)
      IF(IE.NE.0) PRINT '('' SPLFIL ERROR'',I3)', IE
      Q=0.
      DO 952 K=0,5
 952  Q=Q+X1(K)
      CALL SPOUT('SPLFIL ',147-INT(Q))
C-
```

```
C-TEST OF SPNLFIL.  RUN SEQUENCE THRU NONRECURSIVE LATTICE AND TEST Q.
      DO 960 K=0,5
  960  X1(K)=3.-K
      DO 961 I=0,2
       KAPPA(I)=.5**(I+1)
  961 CONTINUE
      CALL SPNLFIL(X1,6,KAPPA,3,WORK,IE)
      IF(IE.NE.0) PRINT '('' SPNLFIL ERROR'',I3)', IE
      Q=0.
      DO 962 K=0,5
  962  Q=Q+X1(K)
      CALL SPOUT('SPNLFIL',9-INT(Q))
C-
C-*********************************************************************
C-                           CHAPTER 12.                            *
C-*********************************************************************
C-
C-TEST OF SPLHBW.  DESIGN LOW- AND HIGHPASS FILTERS AND TEST CHECKSUM.
      CALL SPLHBW(1,.4,3,B1,A1,IE)
      IF(IE.NE.0) PRINT '('' SPLHBW ERROR'',I3)', IE
      CALL SPLHBW(2,.4,3,B2,A2,IE)
      IF(IE.NE.0) PRINT '('' SPLHBW ERROR'',I3)', IE
      Q=0.
      DO 1211 I=1,3
       DO 1210 J=0,2
        Q=Q+B1(J,I)+B2(J,I)*7**I*9**J
 1210  CONTINUE
        Q=Q+A1(1,I)+A1(2,I)+A2(1,I)+A2(2,I)*5**I
 1211 CONTINUE
      CALL SPOUT('SPLHBW ',1650-INT(Q))
C-
C-TEST OF SPBBBW.  DESIGN BANDPASS AND BANDSTOP FILTERS AND TEST CHECKSUM.
      CALL SPBBBW(1,.3,.4,4,B1,A1,IE)
      IF(IE.NE.0) PRINT '('' SPBBBW ERROR'',I3)', IE
      CALL SPBBBW(2,.3,.4,4,B2,A2,IE)
      IF(IE.NE.0) PRINT '('' SPBBBW ERROR'',I3)', IE
      Q=0.
      DO 1221 I=1,4
       DO 1220 J=0,2
        Q=Q+B1(J,I)+B2(J,I)*7**I*9**J
 1220  CONTINUE
        Q=Q+A1(1,I)+A1(2,I)+A2(1,I)+A2(2,I)*5**I
 1221 CONTINUE
      CALL SPOUT('SPBBBW ',24672-INT(Q/10.+.4))
C-
C-TEST OF SPFIL1.  LOWPASS A DATA SEQUENCE AND TEST CHECKSUM.
      DO 1230 K=0,9
 1230  X1(K)=10000*SIN(1.5*K)
      CALL SPFIL1(X1,10,1,.4,2,X2,IE)
      IF(IE.NE.0) PRINT '('' SPFIL1 ERROR'',I3)', IE
      Q=0.
      DO 1231 K=0,9
 1231  Q=Q+X1(K)
      CALL SPOUT('SPFIL1 ',559-INT(Q))
C-
C-TEST OF SPFIL2.  BANDSTOP A DATA SEQUENCE AND TEST CHECKSUM.
      DO 1240 K=0,9
 1240  X1(K)=10000*SIN(1.5*K)
      CALL SPFIL2(X1,10,2,.3,.4,2,X2,IE)
      IF(IE.NE.0) PRINT '('' SPFIL2 ERROR'',I3)', IE
      Q=0.
      DO 1241 K=0,9
 1241  Q=Q+X1(K)
      CALL SPOUT('SPFIL2 ',1076-INT(Q))
C-
```

```
C-***********************************************************************
C-                              CHAPTER 13.                             *
C-***********************************************************************
C-
C-TEST OF SPUNIF.   TEST A PSEUDORANDOM SEQUENCE OF LENGTH 5.
      ISEED=1235
      Q=100.*SPUNIF(ISEED)
      DO 1310 I=2,5
 1310 Q=Q+I*100.*SPUNIF(ISEED)
      CALL SPOUT('SPUNIF ',798-INT(Q))
C-
C-TEST OF SPGAUS.   TEST A PSEUDORANDOM SEQUENCE OF LENGTH 5.
      ISEED=1235
      Q=100.*SPGAUS(ISEED)
      DO 1320 I=2,5
 1320 Q=Q+I*100.*SPGAUS(ISEED)
      CALL SPOUT('SPGAUS ',608-INT(Q))
C-
C-***********************************************************************
C-                              CHAPTER 14.                             *
C-***********************************************************************
C-
C-TEST OF SPSOLE.   TEST FOR SOLUTION = 1,2,3,4.
      CALL SPSOLE(DD,3,4,4,IE)
      IF(IE.NE.0) PRINT '('' SPSOLE ERROR'',I3)', IE
      Q=1000*DD(0,4)+100*DD(1,4)+10*DD(2,4)+DD(3,4)
      CALL SPOUT('SPSOLE ',1234-INT(Q))
C-
C-TEST OF SPLESQ.   TEST FOR COEF. = 1,2,3,4 USING SEQUENCES WITH K=6.
      DO 1410 K=0,8
       IF(K.LT.3) X1(K)=0.
       IF(K.GE.3) X1(K)=K-1
       IF(K.GE.3) X2(K)=10.*(K-3)-4.*(5/K)+1.*(4/K)+5.*(3/K)
 1410 CONTINUE
      CALL SPLESQ(X1(0),X2(3),6,BD,4,TSE,IE)
      IF(IE.NE.0) PRINT '('' SPLESQ ERROR'',I3)', IE
      Q=BD(3)+10.*BD(2)+100.*BD(1)+1000.*BD(0)+1000.*TSE
      CALL SPOUT('SPLESQ ',1234-INT(Q))
      STOP
      END
C-
      SUBROUTINE SPOUT(NAME,I)
      CHARACTER NAME*7
      IF(I.EQ.0) PRINT '(1X,A7,'' PASSED.'')', NAME
      IF(I.NE.0) PRINT '(1X,A7,'' FAILED. !!!'')', NAME
      RETURN
      END
```

Computing Algorithms in Fortran-77

Index

Rectangular window, 153–155, 157
Recursive systems, 147, 179
    algorithm for, 180
    diagrams, 186
    impulse response, 183
    lattice form, 194
    transfer function, 181
Rejection frequency, 184, 261
Residues, 134–135
Residue theorem, 134
Reversal of time series, 211
Rolloff rate, 228, 287
Root-locus plot, 228
Roundoff errors, 53, 300

S

s-plane plots, 226–227
Sampled-data systems, 49
Sample sequence, 4, 20, 44–45, 60
Sample space, 307
Sample width, finite, 50, 54, 85
Sampling interval, 4, 20, 45
Sampling theorem, 45, 48
Sampling window, 54
Seismic waveform, 95
Shifting theorem, 62, 128, 182
Sidelobes, reduction of, 155, 379, 385
Signal analysis, 1
Signal flow diagrams, 106
Signal space, 17
Simulation, 232
    bilinear, 242
    classes of, 234
    of closed-loop control system, 248
    comparisons, 244
    error bounds, 240–241
    error measures, 234, 245
    error models, 233, 236, 238, 240
    input-invariant, 235
    of nonlinear elements, 250
    of nonlinear systems, 248
    of a product of transfer functions, 249
    ramp-invariant, 238, 244
    role in systems analysis, 232, 257
    step-invariant, 235, 250 (*see also* Modeling)

Smoothing, 149. *See also* Interpolation
Spectral estimation, 377
    AR, ARMA, MA methods, 389
    autocorrelation method, 385
    Bartlett method, 384
    Blackman-Tukey method, 385
    comb filter method, 379
    confidence intervals, 378
    linear prediction method, 396
    maximum entropy, maximum likelihood methods, 396
    parametric methods, 389
    periodogram methods, 380, 387
    Welch method, 387 (*see also* Power spectrum)
Spectrum, 29, 71
    examples of, 94 (*see also* Amplitude spectrum, Power spectrum, Fourier transform, DFT, FFT)
Speech waveforms, 96
SPALTD1 and SPALTD2 (asymmetric lattice to direct), 416, 418
SPBBBW and SPLHBW (Butterworth design), 285, 422–423
SPCFIL (Butterworth filter), 213, 419
SPDTAL1 and SPDTAL2 (digital to asymmetric lattice), 201, 417, 419
SPDTOL (direct to lattice), 197, 415
SPEXV (expected value, moment), 332, 425
SPFFT, SPIFFT (fast Fourier transform routines), 119, 412
SPFIL1 and SPFIL2 (Butterworth filters), 288, 423–424
SPGAUS (Gaussian variate), 331, 424
SPLESQ (least-squares design), 352, 426
SPLFIL (lattice filter), 213, 421
SPLTOD (lattice to direct), 195, 414
SPNDTL (nonrecursive direct to lattice), 200, 416
SPNFIL (nonrecursive filter), 213, 420
SPNLFIL (nonrecursive lattice filter), 213, 421
SPNLTD (nonrecursive lattice to direct), 200 415
SPSOLE (solution to linear equations), 15, 352, 425
    examples of, 25–26, 353
SPUNIF (uniform variate), 331, 424
SPWIND (window functions), 156, 413